Renewable Energy

Martin Kaltschmitt · Wolfgang Streicher
Andreas Wiese

Renewable Energy

Technology, Economics
and Environment

With 270 Figures and 66 Tables

 Springer

Editors

Prof. Dr.-Ing. Martin Kaltschmitt
Institute of Environmental Technology
and Energy Economics
Hamburg University of Technology
Germany
Institute for Energy and Environment (IE) gGmbH
Leipzig, Germany

Ao. Univ.-Prof. Dipl.-Ing. Dr. techn. Wolfgang Streicher
Institute of Thermal Engineering
Graz University of Technology
Austria

Dr.-Ing. Andreas Wiese
Lahmeyer International GmbH
Bad Vilbel, Germany

Library of Congress Control Number: 2007923414

ISBN 978-3-540-70947-3 Springer Berlin Heidelberg New York

Springer is a part of Springer Science+Business Media
springer.com
© Springer-Verlag Berlin Heidelberg 2007

Typesetting: by the editors
Production: Integra Software Services Pvt. Ltd., India
Cover design: wmxDesign GmbH, Heidelberg

Printed on acid-free paper   SPIN: 10942611   42/3100/Integra   5 4 3 2 1 0

Preface

The utilisation of renewable energies is not at all new; in the history of mankind renewable energies have for a long time been the primary possibility of generating energy. This only changed with industrial revolution when lignite and hard coal became increasingly more important. Later on, also crude oil gained importance. Offering the advantages of easy transportation and processing also as a raw material, crude oil has become one of the prime energy carriers applied today. Moreover, natural gas used for space heating and power provision as well as a transportation fuel has become increasingly important, as it is abundantly available and only requires low investments in terms of energy conversion facilities. As fossil energy carriers were increasingly used for energy generation, at least by the industrialised countries, the application of renewable energies decreased in absolute and relative terms; besides a few exceptions, renewable energies are of secondary importance with regard to overall energy generation.

Yet, the utilisation of fossil energy carriers involves a series of undesirable side effects which are less and less tolerated by industrialised societies increasingly sensitised to possible environmental and climate effects at the beginning of the 21st century. This is why the search for environmental, climate-friendly and social acceptable, alternatives suitable to cover the energy demand has become increasingly important. Also with regard to the considerable price increase for fossil fuel energy on the global energy markets in the last few years, not only in Europe, high hopes and expectations are placed on the multiple possibilities of utilising renewable sources of energy.

Against this background, this book aims at presenting the physical and technical principles of the main possibilities of utilising renewable energies. In this context, firstly the main characteristics of the available renewable energy streams are outlined. Subsequently, the technologies of heat provision from passive and active solar systems, ambient air, shallow geothermal energy as well as energy from deep geothermal sources are presented. Also the processes of electricity generation from solar radiation (photovoltaic and solar thermal power plant technologies), wind energy, hydropower and geothermal energy are addressed. Furthermore, the possibilities of harnessing ocean energies are briefly discussed. Only the possibilities of the energetic exploitation of biomass are not explained in detail; in this regard, please refer to /1-4/.

For the main possibilities of renewable energies utilisation, in addition, parameters and data are provided which allow for an economic and environmental assessment of the discussed options. The assessment thus enables a better judgment of the possibilities and limits of the various options of utilising renewable sources of energy.

The present English version is a corrected and enlarged update of the 4th German edition published in early 2005. In contrast to the German edition, all comments and information have been adapted to include the framework conditions outside of Central Europe. Additionally the presentations on the possibilities of solar thermal power generation have been significantly enhanced.

The elaboration of the present book would not have been possible without the assistance of a number of the most varied persons and institutions. First of all we would like to thank very much Lahmeyer International GmbH for sponsoring the translation services; without this support this English edition would not have been possible. We would like to extend our great gratitude to Dipl.-Ing. Ilka Sedlacek, Dr. Olaf Goebel, Dipl.-Öko. Rosa Mari Tarragó, MSc Richard Lawless and Dr.-Ing. Eckhard Lüpfert for their valuable text contributions and their helpful support. We also would like to thank very much Zandia Viebahn for the English translation. Additionally we would like to express our sincerest thanks to Barbara Eckhardt, Petra Bezdiak and Alexandra Mohr who assisted in the layout of this book. However, there are many more persons, not forgetting the publisher, whom we need to thank for their cooperative and fruitful cooperation and assistance. Finally, and most importantly, we owe the most to the highly committed and collaborative authors.

Hamburg/Leipzig, Graz, Frankfurt, January, 2007

Martin Kaltschmitt, Wolfgang Streicher and Andreas Wiese

List of Authors

Dipl.-Ing. Stephanie Frick
    Institute for Energy and Environment (IE) gGmbH, Leipzig, Germany
Dr. Ernst Huenges
    GeoForschungsZentrum (GFZ), Potsdam, Germany
Prof. Dr.-Ing. Klaus Jorde
    Center for Ecohydraulics Research, University of Idaho, Boise (ID), USA
Dr. Reinhard Jung
    GGA-Institute, Hannover, Germany
Dr.-Ing. Frank Kabus
    Geothermie Neubrandenburg GmbH, Neubrandenburg, Germany
Prof. Dr.-Ing. Martin Kaltschmitt
    Institute of Environmental Technology and Energy Economics, Hamburg University of Technology, Germany
    Institute for Energy and Environment (IE) gGmbH, Leipzig, Germany
Prof. Dr. Klaus Kehl
    University of Applied Science Oldenburg/Ostfriesland/Wilhelmshaven, Emden, Germany
Dipl.-Ing. Dörte Laing
    German Aerospace Centre; Institute of Technical Thermodynamics, Stuttgart, Germany
Dr. Iris Lewandowski
    Copernicus Institute for Sustainable Development and Innovation, Department of Science, Technology and Society, Utrecht University, The Netherlands (presently working at: Shell Global Solutions International BV, Amsterdam, The Netherlands)
Dipl.-Ing. Winfried Ortmanns
    SunTechnics GmbH, Hamburg, Germany
Dr. habil. Uwe Rau
    Institute of Physical Electronics, University Stuttgart, Germany
Dr. Burkhard Sanner
    UBeG GbR, Wetzlar, Germany
Prof. Dr. Dirk Uwe Sauer
    Electrochemical Energy Conversion and Storage Systems Group, Institute for Power Electronics and Electrical Drives, RWTH Aachen University, Germany

Dipl.-Ing. Sven Schneider
    Institute for Energy and Environment (IE) gGmbH, Leipzig, Germany
Dipl.-Ing. Gerd Schröder
    Institute for Energy and Environment (IE) gGmbH, Leipzig, Germany
Dr.-Ing. Peter Seibt
    Geothermie Neubrandenburg GmbH, Neubrandenburg, Germany
Dr.-Ing. Martin Skiba
    REpower Systems AG, Hamburg, Germany
Ao. Univ.-Prof. Dipl.-Ing. Dr. techn. Wolfgang Streicher
    Institute of Thermal Engineering, Graz University of Technology, Austria
Dr.-Ing. Gerhard Weinrebe
    Schlaich Bergermann und Partner, Structural Consulting Engineers, Stutt-
    gart, Germany
Dr.-Ing. Andreas Wiese
    Lahmeyer International GmbH, Bad Vilbel, Germany

Summary of Contents

Table of Contents

List of Symbols

a	Index
A	Scaling factor
A	Width of the focal line
A_{abs}	Absorber surface
A_{ap}	Solar aperture surface
A_n	Surfaces n of a building
A_G	Albedo
A_{WEC}	Minimum space required around a wind energy converter
b	Profile thickness
B	Specific angle
c_d	Drag coefficient
$c_{d,operation}$	Drag coefficient at a certain point of operation
c_l	Lift coefficient
$c_{l,0}$	Lift coefficient of a vaulted profile shape for an inflow angle of $0°$
$c_{l,operation}$	Lift coefficient at a certain point of operation
c_p	Coefficient of pressure
c_p	Power coefficient
c_p	Specific thermal (heat) capacity
$c_{p,ideal}$	Ideal power coefficient
$c_{p,max}$	Maximum power coefficient
$c_{p,th}$	Theoretical power coefficient
$c_{p,Air}$	Specific thermal (heat) capacity of ambient air
C	Capacitor
C	Concentration ratio
C_1	Constant 1 for the calculation of the utilisable heat of an absorber
C_2	Constant 2 for the calculation of the utilisable heat of an absorber
C_{flux}	Concentration ratio defined by the radiation flux density ratio
C_{geom}	Geometrically determined concentration ratio
$C_{ideal,2D}$	Maximum concentration ration for a single-axis concentrator
$C_{ideal,3D}$	Maximum concentration ration for a two-axis concentrator
CET	Central European Time
COP	Coefficient of Performance
d_{Rot}	Rotor diameter of a wind energy converter

d_S	Diameter of the sun
d_T	Diameter of the chimney tube (tower)
D	Diode
e_0	Elementary charge
E	(Electron) Energy
E	Equation of time
E	Evaporation
E_g	Energy gap
E_C	Energy level of the conduction band
E_{Ocean}	Evaporation from the ocean
E_S	Evaporation from a defined surface (S)
E_{SC}	Solar constant
E_V	Energy level of the valence band
E_{Wa}	Energy of the water
E_{WEC}	Energy yield of a wind energy converter
E_{Wi}	Energy of the wind
f	Arch of a profile
f	Frequency of the grid
f/l	Relative curvature
F	Flow
F	Force
$F_{a,S}$	Flow above ground from a defined surface (S)
$F_{b,S}$	Flow below ground from a defined surface (S)
F_f	Shading factor by side overhangs
F_h	Shading factor by the horizon
F_o	Shading factor by overhangs
F_s	Solar fractional saving
F_C	Reduction factor of the g-value due to flexible shading
$F_{Centrifugal}$	Centrifugal force
$F_{Coriolis}$	Coriolis force
F_D	Drag force
F_D	Reduction factor of the g-value due to dust on the glass pane
$F_{D,a}$	Axial component of the drag force
$F_{D,t}$	Tangential component of the drag force
F_F	Reduction factor of the g-value due to the window frame
$F_{Gradient}$	Gradient force
F_L	Lifting force
$F_{L,a}$	Axial component of the lifting force
$F_{L,t}$	Tangential component of the lifting force
F_R	Overall force on the rotor blade
F_S	Reduction factor of the g-value due to fixed shading
F_T	Tangential force

$F_{Wi,slow}$	Force given by the wind energy converter slowing down the wind flow
$F_{Wi,WEC}$	Overall wind force affecting the wind energy converter
FF	Fill factor
g	Acceleration of gravity
g	Energy transmittance factor (g-value)
$g_{diffuse}$	Diffuse energy transmittance factor (diffuse g-value)
G_{abs}	Radiation flux density of the absorber
G_{ap}	Radiation flux density at the aperture level
G_b	Direct (beam) radiation
$G_{b,n}$	Direct (beam) normal radiation
$G_{b,t,a}$	Direct (beam) radiation on the tilted, aligned surface
G_d	Diffuse radiation
$G_{d,t,a}$	Diffuse radiation on the tilted, aligned surface
G_g	Global radiation
$\dot{G}_g$	Global radiation incident
$\dot{G}_{g,abs}$	Global radiation incident on the absorber surface
$G_{g,t,a}$	Global radiation on the tilted, aligned surface
$G_{r,t,a}$	Reflected radiation on the tilted, aligned surface
$\dot{G}_{g,rel}$	Global radiation on an absorber area of one square metre
$GMST$	Greenwich Mean Summer Time
GMT	Greenwich Mean Time
h	Altitude
h	Enthalpy
h	Head
h_1	Geodetic level at reference point 1
h_2	Geodetic level at reference point 2
h_3	Geodetic level at reference point 3
h_4	Geodetic level at reference point 4
h_5	Geodetic level at reference point 5
h_{abs}	Thermal loss coefficient of a absorber
h_i	Wind occurrence probability within a certain wind speed interval i
h_{ref}	Reference altitude
h_{util}	Utilisable head
h_H	Height of a hill above the surroundings
h_{HW}	Geodetic level of the headwater
h_T	Height of the chimney tube (tower)
h_{TW}	Geodetic level of the tailwater
H	Heat production
i	Discount rate
i	Reference point within a stream-tube
i	Wind speed interval

I	Current
I_0	Saturation current
I_{10}	Ten hour discharge current
I_a	Annual share of the total investment
I_{total}	Total investment
I_C	Current through the capacitor (C)
I_D	Current through the diode (D)
I_{MPP}	Current within the maximum power point (MPP)
I_{Ph}	Photocurrent
I_{SC}	Short-circuit current
k	Boltzmann constant
k	Shape parameter
k_A	Distance factor
$k_{A,x}$	Distance factor with regard to the main wind direction
$k_{A,y}$	Distance factor crosswise to the main wind direction
l	Profile length
$l_{1,2}$	Half value length of a hill
L	Technical life time
L_{ES}	Distance between the sun and the earth
LT	Local time
m	Index
m	Mass
$\dot{m}$	Mass flow
$\dot{m}_{Air}$	Mass flow of air
$\dot{m}_{in}$	Water circulation in the pool (water-)mass stream
$\dot{m}_{out}$	Water circulation from the pool (water-)mass stream
m_{Wi}	Air mass
$\dot{m}_{Wi}$	Air mass flow
$\dot{m}_{Wi,free}$	Air mass flow without any energy extraction
$\dot{m}_{Wi,i}$	Air mass flow at reference point i
M	Drive torque
M_S	Radiant flux density of the sun
n	Day of the year
n	Index
n	Number of rotor revolutions
n_G	Number of rotor revolutions of the generator
p	Pressure
p_0	Initial pressure
p_1	Pressure at site 1 respectively reference point 1

p_2	Pressure at site 2 respectively reference point 2
p_3	Pressure at reference point 3
p_4	Pressure at reference point 4
p_5	Pressure at reference point 5
p_{bottom}	Pressure below a cross-section profile
p_d	Dynamic pressure component
p_i	Pressure at reference point i
p_s	Static pressure component
p_{top}	Pressure above a cross-section profile
$p_{Wi,0}$	Wind pressure within the atmosphere
$p_{Wi,1}$	Wind pressure at reference point 1
$p_{Wi,2}$	Wind pressure at reference point 2
$p_{Wi,i}$	Wind pressure at reference point i
P	Power
P	Precipitation
P_{el}	Electric power
$P_{el,i}$	Electric power corresponding to a defined wind speed interval i
P_{Abs}	Power of the absorber or collector of a solar updraft tower power plant
P_{Drive}	Compressor drive power
P_{Flow}	Power at the air flow within a chimney tube (tower)
P_{Rot}	Power at the rotor shaft
$P_{Rot,th}$	Theoretical power at the rotor shaft
P_S	Precipitation on a defined surface (S)
$P_{Turbine}$	Power at the turbine shaft
P_{Wa}	Water power
$P_{Wa,act}$	Actual useable water power
$P_{Wa,th}$	Theoretical water power
P_{Wi}	Wind power
$P_{Wi,1}$	Wind power at reference point 1
$P_{Wi,2}$	Wind power at reference point 2
$P_{Wi,i}$	Wind power at reference point i
$P_{Wi,ext}$	Wind power extracted by the wind energy converter
P_{WEC}	Power of a wind energy converter
P_{Ocean}	Precipitation on the ocean
q_i	Secondary heat flow
q_{in}	Heat flow added to a structural element
$\dot{q}$	Heat flow density
$\dot{q}_{conductive}$	Conductive proportion of the heat flow density
$\dot{q}_{convective}$	Convective proportion of the heat flow density
$\dot{q}_{rad}$	Heat flow density from radiogenic heat
$\dot{q}_t$	Terrestrial heat flow density
$\dot{q}_{Wa}$	Volume-related flow rate of water
Q_i	Internal heat

Q_H	Energy requirement for heating
Q_L	Heating losses of a building
Q_S	Solar heat generated within a certain time period within a specific building
$\dot{Q}$	Heat flow
$\dot{Q}_{abs}$	Heat released into the pool by the absorbers
$\dot{Q}_{aux}$	Conventional energy carrier saving
$\dot{Q}_{conv}$	Convection losses
$\dot{Q}_{conv,abs}$	Convection losses of the absorber
$\dot{Q}_{conv,cov}$	Convection losses of the absorber cover
$\dot{Q}_{cond}$	Thermal conductivity losses
$\dot{Q}_{cond,abs}$	Thermal conductivity losses of the absorber
$\dot{Q}_{conv,frame}$	Convection losses of the absorber frame
$\dot{Q}_{demand}$	Heat demand
$\dot{Q}_{evap}$	Evaporation losses at the water surface
$\dot{Q}_{human}$	Heat gain by the pool users
$\dot{Q}_l$	Thermal losses of a building
$\dot{Q}_{rad}$	Long-wave radiation losses
$\dot{Q}_{rad,abs}$	Long-wave radiation losses of the absorber
$\dot{Q}_{rad,cov}$	Long-wave radiation losses of the absorber cover
$\dot{Q}_{refl,abs}$	Reflection losses of the absorber
$\dot{Q}_{refl,cov}$	Reflection losses of the absorber cover
$\dot{Q}_{trans}$	Transmission losses
$\dot{Q}_{trans,abs}$	Thermal conductivity losses of the absorber
$\dot{Q}_{useful}$	Utilisable thermal flow
$\dot{Q}_{Cond.}$	Heat flow delivered by the condenser
$\dot{Q}_{Evap.}$	Heat flow to the evaporator
$\dot{Q}_g$	Heat gain created by radiation impinging on the pool
$\dot{Q}_T$	Transmission losses of a building
$\dot{Q}_V$	Ventilation and infiltration losses of a building
r	Radius
R	Rotor radius
R_P	Parallel or shunt resistance
R_S	Retention from a defined surface (S)
R_S	Series resistance
Rot	Rotor plane
s	Entropy
s	Slip
s	Speed-up-ratio
s_{max}	Maximum speed-up-ratio
S	Surface, wind passage area
S_1	Cross-section through a stream-tube at reference point 1

S_2	Cross-section through a stream-tube at reference point 2
S_{abs}	Absorber area
S_i	Cross-section through a stream-tube at reference point i
S_{Rot}	Rotor surface
S_W	Correction factor for window orientation
SPF	Seasonal Performance Factor
t	Time period
T	Absolute temperature
T_{abs}	Absolute temperature of the absorber
T_e	Absolute external (ambient) temperature
TST	True solar time
U	Thermal transmittance coefficient (U-value)
U	Voltage
U^*_{coll}	Temperature-independent heat transfer coefficient of the collector
U_{coll}	Temperature-dependent heat transfer coefficient of the collector
U_{eq}	Equivalent thermal transmittance coefficient (equivalent U-value)
U_n	Thermal transmittance coefficient referring to the surface n of a building
U_D	Diffusion voltage
U_G	Thermal transmittance coefficient referring to the glazing (U_G-value)
U_{MPP}	Voltage within the maximum power point (MPP)
U_{OC}	Open-circuit voltage
U_W	Thermal transmittance coefficient referring to the window (U_W-value)
v	Velocity
v_1	Velocity at reference point 1
v_2	Velocity at reference point 2
v_c	Velocity of light
v_u	Blade tip speed
v_{Air}	Velocity of the air inside a chimney tube (tower)
$v_{Air,max}$	Maximum velocity of the air inside a chimney tube (tower)
v_I	Inflow wind velocity
v_R	Wind velocity in rotor direction
v_S	Velocity of a wind-blown surface (S)
v_{Wa}	Water velocity
$v_{Wa,1}$	Water velocity at reference point 1
$v_{Wa,2}$	Water velocity at reference point 2
$v_{Wa,3}$	Water velocity at reference point 3
$v_{Wa,4}$	Water velocity at reference point 4
$v_{Wa,5}$	Water velocity at reference point 5
$v_{Wa,i}$	Water velocity at reference point i
v_{Wi}	Wind velocity

$v_{Wi,1}$	Wind velocity at reference point 1
$v_{Wi,2}$	Wind velocity at reference point 2
$v_{Wi,a}$	Wind velocity at a defined site (a)
$v_{Wi,h}$	Wind velocity at a defined altitude (h)
$v_{Wi,i}$	Wind velocity at reference point i
$v_{Wi,North}$	Wind velocity, northern hemisphere
$v_{Wi,ref}$	Wind velocity at a reference altitude (h_{ref})
$v_{Wi,Rot}$	Wind velocity within the rotor level
$v_{Wi,South}$	Wind velocity, southern hemisphere
$v_{Wi,x}$	Wind velocity at a defined site (x)
V	Volume
V_0	Initial Volume
$V_{Reservoir}$	Within a reservoir stored water volume
x	Index
y	Index
z	Layer thickness
z	Number of rotor blades
α	Absorption coefficient
α	Inclination of the surface with respect to the horizontal level
α	Inflow angle
α	Slope or tilt angle
α	Solar altitude
α_{abs}	Absorption coefficient of the absorber
α_{ideal}	Ideal absorption coefficient
$\alpha_{operation}$	Inflow angle at a certain point of operation
α_{real}	Real absorption coefficient
α_{th}	Thermal penetration coefficient
α_{Hell}	Hellmann exponent, roughness exponent
α_S	Absorption coefficient in the spectrum of visible solar radiation
α_I	Absorption coefficient in the infrared radiation spectrum
β	Angle between the cord and the arc of a circle
β	Azimuth angle
β	Work rate of a heat pump
β_a	Annual work rate of a heat pump
γ	Angle between the inflow velocity and the rotor direction
γ	Collector azimuth angle
γ	Ratio of heat gain to heat loss
δ	Solar declination
δ	Angle of the profile
ε	Efficiency rate of a heat pump

ε	Emission coefficient
ε	Ratio of the lift to the drag coefficient (L/D ratio)
ε_{abs}	Emission coefficient of the absorber
ε_{min}	Minimum ratio of the lift to the drag coefficient (L/D ratio)
ε_I	Emission coefficient in the infrared radiation spectrum
ε_S	Emission coefficient in the spectrum of visible solar radiation
ζ	Heat rate of a heat pump
ζ_a	Annual heat rate of a heat pump
η	Efficiency
η	Utilisation factor
$\eta_{mech.-elec.}$	Efficiency taking mechanical and electrical losses into consideration
η_{Coll}	Collector efficiency
η_{PP}	Power plant efficiency
η_{Rot}	Efficiency factor of the rotor (i.e. twisting and friction losses)
η_{Tower}	Efficiency factor of the chimney tube
$\eta_{Turbine}$	Efficiency factor of the turbine
θ	Temperature
θ_1	Lower temperature of a thermodynamic cycle
θ_2	Higher temperature of a thermodynamic cycle
θ_a	Acceptance semi-angle
θ_{abs}	Temperature of the absorber
$\theta_{ambient}$	Ambient temperature
θ_{ce}	Temperature of the ceiling
θ_e	External (ambient) temperature
θ_{fl}	Temperature of the floor
θ_i	Internal (room, living space) temperature
$\theta_{i,n}$	Internal temperature of the respective surface (n) of a building
θ_{in}	Inlet temperature
θ_{out}	Outlet temperature
θ_{ss}	Temperature of the sun space
θ_{super}	Temperature of a overheated medium
θ_{Air}	Air temperature
λ	Heat (thermal) conductivity
λ	Longitude
λ	Tip speed ratio
λ_O	Reference meridian
ξ	Loss coefficient
ξ_{IS}	Loss coefficient for the intake structure
ξ_{PS}	Loss coefficient of the penstock
π	Relation of circle circumference to circle diameter (Ludolf number)
ρ	Reflection coefficient
ρ_{abs}	Reflection coefficient of the absorber
ρ_{ideal}	Ideal reflection coefficient
ρ_{real}	Real reflection coefficient

$\rho_{Air,amb}$	Density of the ambient air
$\rho_{Air,Tower}$	Air density inside a chimney tube (tower)
ρ_I	Reflection coefficient in the infrared radiation spectrum
ρ_S	Reflection coefficient in the spectrum of visible solar radiation
ρ_{SM}	Density of the storage medium
ρ_{Wa}	Water density
$\rho_{Wa,1}$	Water density at reference point 1
$\rho_{Wa,2}$	Water density at reference point 2
ρ_{Wi}	Density of the moving air (wind)
$\rho_{Wi,1}$	Density of the moving air (wind) at reference point 1
$\rho_{Wi,2}$	Density of the moving air (wind) at reference point 2
σ	Stefan-Boltzmann-constant
τ	Transmission coefficient
τ_{cov}	Transmission coefficient of the absorber cover
τ_e	Transmission coefficient of a transparent structural element
τ_e^*	Transmission coefficient of a transparent structural element referring to the vertical radiation incidence
τ_G	Transmission coefficient of the global radiation
τ_{GA}	Transmission coefficient of adsorption within gases
τ_I	Transmission coefficient in the infrared radiation spectrum
τ_{MD}	Transmission coefficient of the Mie diffusion
τ_{PA}	Transmission coefficient of absorption within particles
τ_{RD}	Transmission coefficient of the Rayleigh diffusion
τ_S	Transmission coefficient in the spectrum of visible solar radiation
φ	Latitude
ψ	Incident angle
ψ_z	Zenith angle
ω_h	Hourly angle
Δ	Difference

1 Introduction and Structure

The aim of this book is to outline and discuss the main fields of renewable energy applications, and thereby create a solid basis for their evaluation. For this purpose, both the physical foundations and the technical bases are presented. Additionally, key figures which allow for a classification of these options according to the demands of the energy system are elaborated. To ensure a simple, comprehensible and transparent presentation of the different options of using renewable energy for the provision of heat and power, the individual chapters describing the various options have been similarly structured whenever possible and sensible.

In this context, the global energy system is presented and thus also the framework into which energy provision based on renewable energy sources has to be incorporated. Subsequently, the basic structure of the chapters of this book is explained more in detail, and technical terms used frequently throughout this book, are defined. Also the underlying methodological approaches with regard to the key figures, which are characteristic for the individual renewable energy application, are presented. To conclude, most important technologies for exploiting fossil energy carriers for the provision of heat and electricity are also briefly described and characterised; they set a base standard for evaluating energy provision options based on renewable energy sources.

1.1 Energy system

Our current living standard could not be maintained without energy. The provision of energy or – more precisely – of the related energy services (e.g. heated living spaces, information, and mobility) involves a huge variety of environmental impacts which are increasingly less tolerated by the society of the 21st century. This is why the "energy problem" in conjunction with the underlying "environmental problem" continues to be a major topic in energy engineering, as well as in the energy and environmental policies of Europe and, to some extent, also worldwide. From the current viewpoint, this attitude is not expected to change within the near future; the worldwide controversy about the potential risks of the anthropogenic greenhouse effect is only one example. On the contrary, in view of the increasing knowledge and recognition of the effects associated with energy utilisation in the broadest sense of the term, increased complexity has to be expected.

Against this background, the dimensions of global energy system are illustrated and discussed as follows. However, first some energy terms are defined.

1.1.1 Energy terms

According to Max Planck, energy is defined as the ability of a system to cause external action. In this respect the following forms of energy are distinguished: mechanical energy (i.e. potential or kinetic energy), thermal, electric and chemical energy, nuclear energy and solar energy. In practical energy appliances, the ability to perform work becomes visible by force, heat and light. The ability to perform work from chemical energy, as well as nuclear and solar energy is only given if these forms of energy are transformed into mechanical and/or thermal energy.

The term energy carrier – thus a carrier of the above defined energy – is a substance that could be used to produce useful energy, either directly or by one or several conversion processes. According to the degree of conversion, energy carriers are classified as primary or secondary energy carriers and as final energy carriers. The respective energy content of these energy carriers consists of primary energy, secondary energy and final energy. Definitions of the individual terms are as follows (Fig. 1.1) /1-1/, /1-2/.

- Primary energy carriers are substances which have not yet undergone any technical conversion, whereby the term primary energy refers to the energy content of the primary energy carriers and the "primary" energy flows. From primary energy (e.g. wind power, solar insulation) or primary energy carriers (e.g. hard coal, lignite, crude oil, and biomass) secondary energy or secondary energy carrier can either be produced directly or by one or several conversion steps.
- Secondary energy carriers are energy carriers that are produced from primary or other secondary energy carriers, either directly or by one or several technical conversion processes (e.g. gasoline, heating oil, rape oil, electrical energy), whereby the term secondary energy refers to the energy content of the secondary energy carrier and the corresponding energy flow. This processing of primary energy is subject to conversion and distribution losses. Secondary energy carriers and secondary energies are available to be converted into other secondary or final energy carriers or energies by the consumers.
- Final energy carrier and final energy respectively are energy streams directly consumed by the final user (e.g. light fuel oil inside the oil tank of the house owner, wood chips in front of the combustion oven, district heating at the building substation). They result from secondary and possibly from primary energy carriers, or energies, minus conversion and distribution losses, self-consumption of the conversion system and non-energetic consumption. They are available for the conversion into useful energy.
- Useful energy refers to the energy available to the consumer after the last conversion step to satisfy the respective requirements or energy demands (e.g.

space heating, food preparation, information, transportation). It is produced from final energy carrier or final energy, reduced by losses of this last conversion (e.g. losses due to heat dissipation by a light bulb to generate light, losses of wood chip fired stove to provide heat).

The entire energy quantity available to humans is referred to as energy basis. It is composed by the energy of the (predominantly exhaustible) energy resources and the (largely renewable) energy sources.

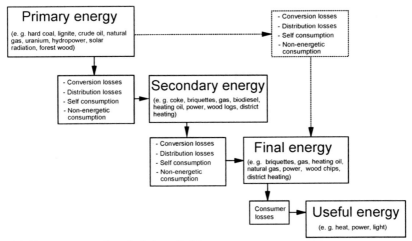

Fig. 1.1 Energy conversion chain (see /1-1/)

In terms of energy resources, generally fossil and recent resources are distinguished.

− Fossil energy resources are stocks of energy that have formed during ancient geologic ages by biologic and/or geologic processes. They are further subdivided into fossil biogenous energy resources (i.e. stocks of energy carrier of biological origin) and fossil mineral energy resources (i.e. stocks of energy carrier of mineral origin or non-biological origin). The former include among others hard coal, natural gas and crude oil deposits, whereas the latter comprise for instance the energy contents of uranium deposits and resources to be used for nuclear fusion processes.

− Recent resources are energy resources that are currently generated, for instance, by biological processes. They include, among others, the energy contents of biomass and the potential energy of a natural reservoir.

Energy sources, by contrast, provide energy streams over a long period of time; they are thus regarded as almost "inexhaustible" in terms of human (time) dimensions. But these energy flows are released by natural and technically uncontrollable processes from exhaustible fossil energy resources (like fusion process within the sun). Even if such processes take place within very long time periods − and are thus unlimited in human (time) dimensions – they are nevertheless exhaustible (e.g. burn-out of the sun sometime in the future).

The available energies or energy carriers can be further subdivided into fossil biogenous, fossil mineral and renewable energies or fossil biogenous, fossil mineral and renewable energy carriers.

– Fossil biogenous energy carriers primarily include the energy carriers coal (lignite and hard coal) as well as liquid or gaseous hydrocarbons (such as crude oil and natural gas). A further differentiation can be made between fossil biogenous primary energy carriers (e.g. lignite) and fossil biogenous secondary energy carriers (e.g. gasoline, Diesel fuel).

– Fossil mineral energy carriers comprise all substances that provide energy derived from nuclear fission or fusion (such as uranium, thorium, hydrogen).

– The term renewable energy refers to primary energies that are regarded as inexhaustible in terms of human (time) dimensions. They are continuously generated by the energy sources solar energy, geothermal energy and tidal energy. The energy produced within the sun is responsible for a multitude of other renewable energies (such as wind and hydropower) as well as renewable energy carriers (such as solid or liquid biofuels). The energy content of the waste can only be referred to as renewable if it is of non-fossil origin (e.g. organic domestic waste, waste from the food processing industry). Properly speaking, only naturally available primary energies or primary energy carriers are renewable but not the resulting secondary or final energies or the related energy carriers. For instance, the current generated from renewable energies by means of a technical conversion process itself is not renewable, since it is only available as long as the respective technical conversion plant is operated. However, in everyday speech secondary and final energy carriers derived from renewable energy are often also referred to as renewable.

1.1.2 Energy consumption

In 2005, the worldwide consumption of fossil primary energy carriers and of hydropower amounted to approximately 441 EJ /1-3/. Roughly 28 % of this overall energy consumption accounts for Europe and Eurasia, approximately 27 % for North America, about 5 % for Central and South America, roughly 5% for the Middle East, approximately 3 % for Africa and about 32% for Asia and the Pacific region (primarily Australia and New Zealand). North America, Europe and Eurasia as well as Asia and the Pacific region consume about 90 % of the currently used primary energy derived from fossil energy carriers and hydropower.

Fig. 1.2 illustrates the evolution of primary energy consumption of fossil energy carriers and hydropower according to regions over the past 40 years. According to the figure, the worldwide primary energy consumption has increased by more than the factor of 2.5 over this period of time. All illustrated regions show a considerable increase. It is also perceivable that the increase is by far not linear, but has been noticeably influenced by the two oil price crises in 1973 and 1979/80. Also at the beginning of the 1990's, the increase of the worldwide energy consumption slowed down significantly. This is partly attributable to the downturn of the global economy and the restructuring of the former Eastern block including the former USSR. At the same time a significant increase of fossil primary energy consumption could be perceived for Asia. Only towards the middle of 1990's, the worldwide primary energy consumption started to increase again more quickly. Towards the end of the 1990s the increase of the primary energy consumption slowed down again to increase noticeably at the beginning of the first decade of the 21st century.

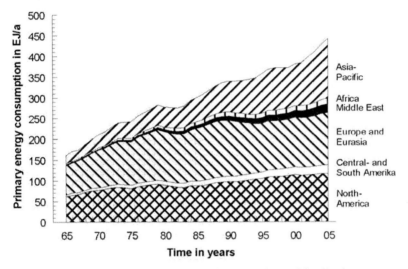

Fig. 1.2 Evolution of the worldwide consumption by regions of fossil primary energy carriers and hydropower (data according to /1-3/)

In 2005, the overall energy consumption of fossil energy carriers and hydropower was covered by 36 % by crude oil, by 24 % by natural gas, by 28 % by coal and always by 6 % electrical energy generated by nuclear and hydropower respectively. On a regional level these fractions are strongly dependent on local and national characteristics due to varying national energy politics or available primary energy resources differing from region to region (Fig. 1.3). For instance, in Asia the major share of the given demand for fossil primary energy carriers is covered by coal (this applies in particular to the People's Republic of China), whereas this energy carrier is of almost no importance in regions such as the Middle East. Due to the abundance of crude oil and natural gas mainly liquid and gaseous fossil hydrocarbons are used. In line with this observation, the high use of natural gas in Russia is attributable to its abundant natural gas resources.

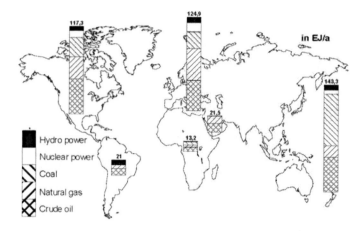

Fig. 1.3 Worldwide consumption of fossil primary energy carriers and hydropower according to regions and energy carriers in the year 2005 (data according to /1-3/)

Within the past 40 years, the composition of energy carriers applied worldwide has changed tremendously (Fig. 1.4). This applies in particular to natural gas. While this energy carrier only had a share of roughly 17 % in the overall consumption of fossil energy carriers and hydropower in 1965, it contributed with about 24 % to cover the overall primary energy demand in 2005. In 1965, nuclear energy had still no importance on a global scale; in the year 2005; however, it covered roughly 6 % of the global primary energy demand and still has a strong tendency to increase. Although coal consumption significantly increased from 62 EJ (1965) to 123 EJ (2005) its consumption diminished from 40 % in the year 1965 to scarcely 28 % in 2005, when compared to the overall consumption of fossil energy carriers and hydropower. Within the same period, the consumption of crude oil increased from about 65 EJ (1965) to approximately 161 EJ (2005). The crude oil consumption has thus more than doubled within scarcely four decades; however, its share in the overall primary energy consumption has more or less remained the same.

The above indications only include power generated by hydropower and nuclear energy as well as energy carriers traded on the commercial world energy markets. All other kind of renewable and un-conventional energies, such as firewood and other kinds of biomass (e.g. straw, dried manure) or wind have thus not been considered. Currently, there are only very rough estimations available regarding the amount and regional distribution of the energetic use of for, instance, biomass and wind. For biomass as the most important non-commercial renewable energy carrier, for instance, these estimations cover a very wide range and reach from 20 to scarcely 60 EJ/a. According to these estimations, biomass contributes 5 to 15 % of the world-wide primary energy utilisation of fossil energy carriers and hydropower to satisfy the given energy demand.

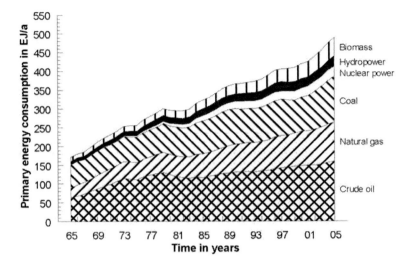

Fig. 1.4. Worldwide consumption of fossil primary energy carriers, hydropower and biomass according to energy carriers (for data also refer among others to /1-3/)

1.2 Applications of renewable energies

Provision of final or useful energy using renewable energies is based on energy flows originated by the movement and gravitation of planets (i.e. tidal energy), heat stored and released by the earth (i.e. geothermal energy) and in particular energy radiated by the sun (i.e. solar radiation) (Chapter 2.1). There is thus a great variety of renewable energies in terms of energy density, variations of the available forms of energy and the related secondary or final energy carriers and final energy to be provided. Each technical option for utilizing the described renewable energy flows or carriers must be adapted to the corresponding characteristics of the available renewable energy; there is thus a broad range of current and also future technical processes and methods to exploit the renewable energy options most successfully.

In the following, the different renewable energy sources are classified. Subsequently it will be discussed which energy utilisation options will be analysed more in detail throughout this book.

1.2.1 Renewable energies

The three sources of renewable energies give rise to a multitude of very different energy flows and carriers due to various energy conversion processes occurring in nature. In this respect, for instance, wind energy and hydropower, as well as, ocean current energy (as energy flows) and solid or liquid biofuels (as energy carrier; i.e. stored solar energy) all represent more or less conversions of solar energy (Fig. 1.5).

The energy flows available on earth that directly or indirectly result from these renewable energy sources vary tremendously, for instance, in terms of energy density or with regard to spatial and time variations. Since the following explanations are limited to the most important utilisation methods of renewable energy supply, only the energy flows and the corresponding conversion methods and technologies will be discussed in detail. They mainly include
– solar radiation,
– wind energy,
– hydropower,
– photosynthetically fixed energy and
– geothermal energy.
The above options of using renewable energies are characterised by very different properties. For this purpose, within the explanations throughout this book, first the corresponding physical and chemical principles of the formation of the respective renewable energy flow are discussed in depth. Following an excursus on the given possibilities of measuring the magnitudes relating to the respective energy flow on site (e.g. measuring the wind speed with an anemometer or the solar radiation with a radiation meter) spatial and time variations of the renewable energy supply are addressed in detail.

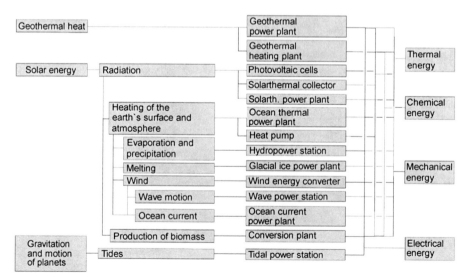

Fig. 1.5 Options of using renewable energies for the provision of useful energy (Solarth. Solarthermal)   (see /1-1/)

1.2.2 Investigated possibilities

Appropriate techniques permit the exploitation and conversion of different renewable energy flows or energy carriers into secondary or final energy, energy carriers or useful energy, respectively. Currently, there are tremendous variations in terms of utilisation methods, status of technology and given perspectives. Moreover, not all options are possible for every site and every set of boundary conditions. Therefore only those opportunities that are most promising from the current viewpoint will be investigated in more detail in the following. They include:
– solar heat provision by passive systems (i.e. architectural measures to use solar energy),
– solar thermal heat provision by active systems (i.e. solar thermal collector systems),
– solar thermal electrical power provision (i.e. solar tower plants, solar farm plants, Dish/Stirling and Dish/Brayton systems, solar chimney plants),
– photovoltaic conversion of solar radiation into electrical energy (i.e. photovoltaic systems),
– power generation by wind energy (i.e. wind turbines),
– power generation by hydropower to provide electrical energy (i.e. hydropower plants),
– utilisation of ambient air and shallow geothermal energy for heat provision (i.e. utilisation of low thermal heat by means of heat pumps),
– utilisation of deep geothermal energy resources for heat and/or power provision (i.e. utilisation of the energy stored in deep porous-fractured reservoirs by means of open and closed systems) and
– utilisation of photosynthetically fixed energy to provide heat, power and transportation fuels (i.e. energy provision on the basis of biomass).
With the exception of biomass utilisation all of the possibilities of utilising renewable energies mentioned above will be outlined and discussed in detail in the following. In addition, within the scope of an excursus, selected options of harnessing ocean energy (such as tidal energy, ocean current energy) are addressed. Furthermore, the possibilities of utilising biomass, and thus photo-synthetically fixed energy, are outlined briefly as an overview in the annex; please refer to /1-4/ for a more detailed and comprehensive description of utilising organic material for energetic purposes.

1.3 Structure and procedure

Due to the great variety of possibilities to use renewable energy sources with the aim to fulfil the demand for end or useful energy, it is very difficult to present the different possibilities in a similar manner. It is thus highly important to explain the different utilisation methods in a flexible manner. Therefore the basic procedure of addressing the various options is outlined in the following. Furthermore, the main terms on which the respective explanations are based are defined.

1.3.1 Principles

The possibilities and boundaries to convert renewable energies into end or useful energy largely depend on the respective physical and technical conditions. They will thus be explained and discussed in detail for each utilisation method. Whenever possible, among others, also the theoretically or technically maximum achievable efficiencies and technical availabilities will be indicated. These technical key figures are defined as follows.

– Efficiency. The efficiency is defined as the ratio of useful power output (e.g. electricity, heat) to the power input (e.g. solar radiation, geothermal energy). It depends on the respective operating conditions of the conversion plant, as well as a series of other factors, which vary over time (e.g. in case of a heating boiler the efficiency varies with ambient temperature, among other factors).
– Utilisation ratio. The utilisation ratio is defined as the ratio of the total output of useful energy to the total energy input within a certain period of time (e.g. one year). The observed time periods may include part load periods and breaks as well as start-up and shutdown times. Utilisation ratios are thus usually smaller compared to the efficiency of conversion plants indicated for the design point at full load.
– Technical availability. The technical availability describes that portion of the time period under observation, within which a plant has actually been available for its intended purpose and thus considers time periods during which the plant has been unavailable due to malfunctions.

1.3.2 Technical description

The renewable energy supply, whose physical principles and supply characteristics are described first (Chapter 2), can be transformed by appropriate conversion plants into secondary or end energy carriers, or directly into useful energy. The methods and procedures appropriate under the given conditions are described within the following chapters; explanations are based on state-of-the-art technology and current conditions.

First, the system components referring to the respective utilisation method are discussed and subsequently explained within the context of the overall energy conversion system. This includes the characteristic curve, the energy flow as well as the respective losses given within the entire provision or conversion chain. In addition, further aspects related to the respective conversion technology are discussed in depth.

1.3.3 Economic and environmental analysis

The different options to convert renewable energy sources into end or useful energy are assessed economically as well as environmentally by means of selected reference plants. In the following, the pertinent terms, definitions as well as the approach are discussed on which the economic and environmental assessment is based.

Definition of reference plants. Based on the current market spectrum, appropriate reference plants are defined according to the present state of technology. These reference technologies are described and discussed within each chapter. In this respect, heat and power provision must be distinguished. For the options of heat provision, supply tasks are also defined, because no nation-wide heat distribution grids exist and heat provision must always be considered in the context of secured consumer supply. In addition, the respective renewable energy supply to be tapped by the reference plants is defined. These typical plants for the current situation will later on serve as a basis for the actual economic and environmental analyses.

Heat provision. As supply tasks for the heat provision, three different single family houses (SFH) with a different heat demand, one multiple family house (MFH), as well as three district heating networks (DH) of different sizes are analysed. According to Table 1.1 these supply tasks are characterised by heat demand for domestic hot water and space heating (SFH and MFH) or the corresponding total heat demand (DH). The analysed single-family-houses represent the heat demand of a low-energy house (SFH-I), a building realised with state-of-the-art heat insulation (SFH-II), as well as a building with heat insulation typical for Central Europe (SFH-III). The multiple family house is a building of 15 flats built in compliance with state-of-the-art insulation standards.

Table 1.1 Supply tasks for heat provision

Demand case		Small scale systems				Large scale systems		
		SFH-I[a]	SFH-II[b]	SFH-III[c]	MFH	DH-I	DH-II	DH-III
Domestic hot water demand	in GJ/a	10.7	10.7	10.7	64.1	8,000	26,000	52,000
Space heating demand[d]	in GJ/a	22	45	108	432			
Building / Total heat load[e]	in kW	5	8	18	60	1,000	3,600	7,200

[a] corresponds to low-energy housing construction; [b] corresponds to state-of-the-art heat insulation; [c] corresponds to average heat insulation in Central Europe; [d] excluding transmission losses of boiler and domestic hot water storages or distribution losses (district heating network and house substations); [e] in case of district heating networks of all connected consumers.

The system boundary for the economic as well as environmental investigations is defined by the respective feed-in points into the house distribution network for domestic hot water (e.g. storage tank exit) or space heating (e.g. boiler exit). However, heat distribution losses within the respective buildings as well as the power consumption of the circulating pumps for the heating system and the domestic hot water system have not been considered. These system elements have been assumed to be the same for all observed technologies based on fossil and on renewable energies. These conditions are defined to allow for a direct comparison of the obtained results.

The investigated district heating networks are three systems that have been designed to deliver heat for space heating of dwelling houses (i.e. buildings that are characterised by consumer behaviour comparable to household customers). The energetic key figures of the different district heating systems are illustrated in Table 1.2. However, the district heating networks are characterised by transfer losses. Due to these losses, as well as the losses at the building substation and at the domestic heat water storage of the supplied buildings, the heat to be fed into the district heating network by the heating station exceeds the overall head demand of all connected consumers. With an average utilisation ratio of the district heating network of 85 %, and an average utilisation ratio of 95 % for the building substations i.e. domestic hot water provision, the heat to be fed into the heating network by the heating station amounts to 9,900 (DH-I), 32,200 (DH-II) and 64,400 GJ/a (DH-III).

Table 1.2 Key figures of the investigated district heating networks

Demand case		DH-I	DH-II	DH-III
Heat demand[a]	in GJ/a	8,000	26,000	52,000
Heat at heating station[b]	in GJ/a	9,900	32,200	64,400
Utilisation ratio of the network[c]	in %	0.85	0.85	0.85
Utilisation ratio of the substations[d]	in %	0.95	0.95	0.95

[a] all connected consumers; [b] including the losses of the heat distribution network and building substations; [c] average value of the entire year; [d] average utilisation ratio of all connected consumers (hot water 80 %, space heating 98 %).

Electricity provision. For power generation systems no supply tasks have been defined. The system boundary is the feed-in point into the power grid. For this reason, potential requirements for net reinforcements and modifications within the conventional power plant park have not been considered. Capacity effects have not been investigated either.

Economic analysis. Key figures of any energy generation opportunity are the costs. They will thus be analysed in detail for every option discussed in depth. For this purpose, first the investments for the most important system components of the applied conversion technology as well as the overall investment volume will be addressed; these figures will be discussed in the context of the current conditions given in Europe.

The specific energy provision costs of the different investigated energy provision options are calculated on the basis of the monetary value of the year 2005; hence, inflation-adjusted costs are presented. For this purpose, an inflation-adjusted discount rate i of 0.045 (i.e. 4.5 %) is assumed for the economic calculations throughout this book. In general, the indicated costs refer to the overall economy; i.e. plants are depreciated over the technical lifetime L of the respective plant or respective plant component that may vary according to the applied technology or system. However, calculations do not consider taxes (e.g. value added tax), subsidies (e.g. granted within the scope of market launches, credit from public bodies which reduce interest rates) and extraordinary depreciation possibilities. Annual annuities are always based on the initial overall investments. Based on the total investment I_{total}, the annual share I_a (i.e. the annuity) throughout the entire technical lifetime is calculated according to Equation (1.1) (also refer to /1-5/).

$$I_a = I_{total} \frac{i\,(1+i\,)^L}{(1+i\,)^L - 1} \qquad (1.1)$$

On the basis of the yearly annuity (i.e. the share of the total investment cost mature each year throughout the overall technical lifetime), the overall annual costs can be calculated by considering the additional respective variable costs (e.g. maintenance costs, operation costs, fuel costs (if applicable)). From these overall annual costs the specific energy provision costs (i.e. electricity production costs in €/kWh, heat provision costs in €/GJ) can be calculated considering the mean annual energy provision at plant exit (e.g. electrical energy of a wind turbine fed into the grid, the caloric energy of a heat pump fed into a heat supply system within a dwelling house required to utilise shallow geothermal heat).

The approach of using constant monetary values leads to lower costs (due to inflation adaptation) compared to the more common calculation based on nominal values; nevertheless, order and relation of costs of different alternatives remain unchanged. However, such calculations based on real costs offer the advantage of yielding results in a known monetary value – in this case of the year 2005.

The energy production costs indicated in the following may differ tremendously from the results of other studies and analyses due to cost calculation procedures or different baseline assumptions for the costs calculation or the consideration of possible external effects. The indicated costs should thus be regarded as an average magnitude affecting the entire overall economy. Considerable deviations from the indicated figures, both elevated and depressed values, may be observed in concrete cases under different site-specific circumstances.

As far as options primarily designed for power generation based on renewable energies are concerned, specific electricity production costs are determined and indicated. Otherwise, the heat production costs at the conversion plant are calculated and discussed.

Environmental analysis. Within energy politics and energy industry, discussions on environmental effects caused by the use of a certain energy source or energy carrier are of major importance. This is why for every option of using renewable energy sources for the provision of useful energy; also selected environmental effects will be addressed. This assessment will be performed for environmental effects related to manufacturing, ordinary operation, malfunctions and the end of operation.

1.4 Conventional energy provision systems

The addressed renewable sources of energy and the corresponding technologies can be applied for the provision of heat and/or power, by substituting conversion technologies based of fossil fuel energy. It is thus sensible to compare the assessed options of using renewable sources of energy with the corresponding fossil fuel-based options which they are to substitute.

For this purpose, standard conventional energy provision technologies based on fossil fuel energy, used for comparison, are defined and discussed in the following. In addition, cost figures are provided allowing for a comparison with the cost parameters assessed for the option of using renewable energy sources.

1.4.1 Boundary conditions

Energy prices are an important parameter to describe the availability of fossil energy carriers. The average expenditures direct to the consumer, including consumption taxes (e.g. mineral oil tax), but excluding value added tax, are indicated in Table 1.3. In the year 2006, for hard coal or natural gas used in power plants, the plant operators had to pay approximately 2.2 and 6.0 €/GJ respectively. Compared to this, the price level for final consumers is comparatively higher. Even though the prices varied in a wide range between 2004 and 2006, constant prices are assumed for the lifetime of the plants.

Table 1.3 Energy prices for fossil energy carriers (for households including value added tax (VAT); average prices have been determined over recent years)

Light heating oil, household customers	12 – 18 €/GJ
Natural gas, household customers	15 – 19 €/GJ
Natural gas, power plants	5 – 7 €/GJ
Hard coal, power plants	1.7 – 2.7 €/GJ

1.4.2 Power generation technologies

In the following, the main system elements of conventional power generation systems are outlined. In addition, two reference systems are defined, for which the costs are calculated on the basis of investment, operation, maintenance, and fuel costs. These calculations allow for a comparison of power generation by conventional plants with the corresponding renewable energy option. Additionally, selected environmental effects are discussed.

Economic analysis. Thermal power stations convert part of the energy contained in fuels (e.g. hard coal, lignite, natural gas, crude oil) into electrical energy. For this purpose, often power plants based on steam cycles and/or gas turbines are used. In the following, the main system characteristics of such technologies are briefly described. However, engines used for power generation by stand-alone systems (e.g. mountain lodges), for standby power supply, and partly for covering the peak demand, have not been addressed.

- Steam power plant. The main components of steam power plants based on coal, natural gas or crude oil include furnace, steam generator, turbine, generator, water cycle, flue-gas cleaning (depending on the applied fuel dust filter, flue-gas desulphurisation and denitrification) as well as control and electro technical devices. Coal-fired power plants additionally require fuel treatment (such as coal grinding). To date, for the common modern hard coal and lignite fired power plants, fuel is combusted in a pulverised way. For power plants with a capacity below 500 MW, also fluidised-bed systems are in use. Oil and gas-fired boilers are generally equipped with a conventional burner firing. The downstream steam generator transfers the energy released during the oxidation of the fuel to the water cycle, thereby generating steam which is subsequently relieved by a multi-stage steam turbine. The thermal energy which has been converted into mechanical energy is then transferred to a generator which transforms it into electrical energy. To close the cycle, the steam that exits the turbine is condensed by a cooling system and re-transferred to the steam generator by a feed-water pump. Currently, steam turbine power plants are characterised by net efficiencies of up to 45 % and above.
- Gas turbine power plant. Gas turbine power plants mainly consist of turbo-compressor, combustion chamber, turbine and generator. First, the compressor pressurises ambient air sucked in from outside, which is subsequently transferred to the combustion chamber. Within this combustor the pressurised air chemically reacts with the fuel under release of heat. Downstream, within the turbine, the flue gas is relieved to ambient pressure and thus released at relatively high temperatures into the atmosphere; a generator coupled to the turbine shaft transforms the provided mechanical energy into electrical energy. With 38 %, the net efficiencies of gas turbine power plants are slightly below those

of steam power plants. Therefore, this power plant type is of decreasing importance.

– Gas and steam power plant. A gas turbine power plant can be ideally combined with a steam turbine power plant. Within such gas and steam power plants the hot exhaust gases released from the gas turbine are transferred to a heat recovery boiler which generates superheated steam for a steam process. The so-called gas and steam plants allow achieving efficiencies of above 58 %.

The provision of electrical energy by such conventional power plants is characterised by corresponding costs. They are briefly discussed in the following. However, first one hard coal-fired steam power plant and one natural gas-fired gas and steam power plant designed according to state-of-the-art technology (i.e. new plant construction) are defined.

– For the hard coal-fired steam power plant with an electrical capacity of 600 MW and a mean annual efficiency of 45 % (Table 1.4) pulverised coal-firing is assumed. This power plant thus represents plants to be built under European conditions in the years to come. Furthermore a typical application for medium load power generation of roughly 5,000 full load hours per year has been assumed.

Table 1.4 Technical and economic parameters of the investigated power generation systems based on fossil fuel energy

		Hard coal-fired power plant	Natural gas-fired power plant
Fuel		Hard coal	Natural gas
Power plant type		Pulverised coal firing	GaS[a]
Nominal electric capacity	in MW	600	600
Technical life time	in a	30	25
Annual mean system efficiency	in % (net)	45	58
Full load hours	in h/a	5,000	5,000
Fuel consumption	in TJ/a	24,000	18,600

[a] Gas and steam power plant.

– As a further characteristic option to generate power from fossil energy carriers a gas-fired gas and steam turbine power plant also of a block size of 600 MW and an annual mean efficiency of roughly 58 % is assessed (Table 1.4). Also for this case medium load power generation (approximately 5,000 full load hours) is assumed.

For the comparison of the economic figures of these power generation technologies with generation technologies based on renewable sources of energies (e.g. wind power plants), for conventional technologies mean and not maximum full load hours are assumed; from a technical viewpoint, for the latter they are considerably higher and could reach 8,000 hours per year and more (a maximum of 8,760 h/a is theoretically possible). Power generation by renewable energies, by contrast, always depends on the availability of the renewable energy source (e.g.

available wind or water supply). The achievable full load hours are thus dependent on the energy availability of the respective site. Although such plants could theoretically also be operated according to the given power demand within the grid, they are for economic reasons optimised to such an extent so that they maximise power generation. However, conventional power plants operated within a conventional-renewable power plant system usually supply as much electrical energy to cover the given demand according to their respective task within the overall power plant system (provision of base, medium or peak load). Therefore, the full load hours of conventional power plants are defined here on the basis of the demand unlike systems based on renewables, where full load hours are determined by the availability of primary energy. Additionally, it has to be considered that the energy provided by conversion plants based on renewable sources of energy, usually substitutes medium load.

To estimate the costs of power generation by fossil fuel energy, in the following, variable and fixed costs as well as the specific power generation costs of the reference plants outlined in Table 1.4 are discussed (Table 1.5).

Investments and operation costs. When compared to natural gas-fired gas and steam turbine power plants, hard coal-fired power plants are characterised by considerably higher investment and operation costs due to the higher expenditures for e.g. coal preparation and flue-gas treatment (Table 1.5). The operation costs are due to, for instance, personnel, maintenance, flue-gas cleaning, and disposal of combustion residues (i.e. ashes) and insurance and, in particular, fuels. Due to the specific lower fuel costs the assessed hard coal-fired power plant has lower expenditures for fuels than the gas and steam turbine power plant.

Table 1.5 Costs of power generation from hard coal and natural gas (see Table 1.4)

		Hard coal-fired power plant	Natural gas-fired power plant
Total investments	in €/kW	1,100	500
Annual costs			
Annual investments	in Mio. €/a	40.5	20.2
Operation costs	in Mio. €/a	22.3	8.8
Fuel costs	in Mio. €/a	52.0	111.7
Total	in Mio. €/a	114.8	140.8
Electricity generation costs	in €/kWh	0.038	0.047

Electricity generation costs. According to the discussed boundary conditions, power production costs at the power plant (Table 1.5) are calculated on the basis of the indicated assumption (Table 1.4) and the assumed economic boundary conditions (i.e. interest rate of 4.5 % assumed throughout the physical life of 30 or 25 years respectively).

According to this assumption, the coal-fired power plant shows power generation costs of approximately 0.038 €/kWh. Power generation by natural gas is comparatively more expensive. Power production costs amount to 0.047 €/kWh (Table 1.5). For power generation by natural gas, the applied fuel accounts for the
major share of the annual costs, whereas for hard coal-fired power plants costs are
distributed relatively evenly among investments, fuels and further operation.

Environmental analysis. Besides the airborne emissions of harmful substances,
like SO_2 and NO_x as well as greenhouse gas and dust emissions, additional pollutants (such as heavy metals) are released during ordinary operation of the assessed
power plants. In addition, the provision of fossil fuels is associated with a series
of further environmental effects. In the following, some of these environmental
effects are discussed exemplarily.

– For a long time, coal-fired power plants have been a considerable source of anthropogenic dust and SO_2 emissions in Europe. Only after more strict legal
  emission regulations had been introduced, requiring the installation of extensive flue-gas cleaning systems, these emissions have been considerably reduced.
– Above ground mining of lignite greatly affects the scenery due to its space-
  consuming measures and the transfer of large material quantities. However,
  these effects can be partly compensated by re-cultivation measures once the
  available lignite is extracted from the ground; in some cases, the establishment
  of lakes can even enhance the recreational value of a particular site. Hard coal
  underground mining may cause downfalls of the created hollow spaces and
  thus lower the earth's surface. Due to this fact, the ground water layer may be
  disturbed, buildings located on the surface may show fissures, small streams
  may be diverted, and thus the use of the affected areas may be significantly
  limited.
– The residues remaining after coal combustion (i.e. ashes or dusts) may contain
  heavy metals and radioactive elements. Depending on the coal composition,
  especially the particulate matter, partly released with the flue gas into the atmosphere, may be charged with these harmful substances. Such contaminated
  dusts must therefore be safely removed from the flue gas and securely evacuated.
– During the exploitation of natural gas reservoirs the ground (onshore) or the
  ocean (offshore) may be polluted, for instance, due to chemical auxiliary and
  operating materials (e.g. drilling fluid) released during drilling of wells and/or
  production of gas.

1.4.3 Heat provision technologies

In the following, conversion technologies currently applied for heat generation from fossil fuel energy are outlined and discussed. First, the techniques and systems are presented according to the current state of technology. Subsequently, these techniques are assessed according to economic and environmental parameters.

Economic analysis. The main system elements for plants providing heat using crude oil or natural gas are, beside the heating boiler with the respective burner, the fuel storage and supply as well as the domestic hot water provision.

Natural gas-fired heating boilers are usually supplied with the fuel from the natural gas grid. Besides this option, there is also the possibility of operating heating gas-fired boilers with producer gas or liquefied gas (e.g. propane). Oil-fired systems are supplied with fuels by steel or plastic tanks, located underground or above ground, which are in turn supplied by trucks.

Inside the boiler the liquid or gaseous fuel is oxidised by released heat. A heat exchanger transfers this heat to an appropriate heat transfer or distributing medium (in most cases water), which transfers the thermal energy to the consumer.

For space heating and domestic hot water supply, today mainly low-temperature and condensing boilers are applied. Within these systems burners with and without a fan are used.

- Gas burners equipped with and without fan. Gas burners equipped with a fan add combustion air to the gaseous fuel before the combustion takes place. Gas burners without a fan, so-called atmospheric gas burners, run by self-priming (i.e. combustion air is transferred to the combustion chamber by the thermal lift). The chimney must thus generate enough flue to overcome all resistances of the heating system. The gas/air mixture is burnt inside the corresponding nozzles. Subsequently, heat released due to oxidation is separated from the exhaust gas and can subsequently be utilised.
- Oil burners equipped with fan. Oil burners are supposed to atomise or vaporise the liquid fuel (i.e. the heating oil) as fine as possible, to mingle it intensively with combustion air added by a fan, and to burn the mixture by producing as low emissions as possible. For heating purposes mainly pressurised atomising burners are used. Common oil burners ignite the hydrocarbon molecules inside the flame and burn them with a yellowish flame (so-called yellow flame burners). Compared to this within so-called blue burner the oil drops are gasified prior to the actual combustion inside the burner pipe by recirculation of the hot fuel gas. This technology is characterised by combustion-related advantages.
- Low-temperature boiler. Depending on the ambient temperature, low-temperature boilers are operated at variable out-flow temperatures between 75 and 40 °C or lower. Particularly the flue gas and stand-by losses of boilers equipped with a domestic hot water heating can thereby be considerably re-

duced during the non-heating summer period. Annual efficiencies between 91 and 93 % (in relation to the heating value of the fuel) can thus be achieved.
- Condensing boiler. Condensing boilers allow for the most extensive exploitation of the energy contained in the fuel. By extensive cooling of hot flue gases with the in-flow of the heating system, the perceptible heat of the exhaust gases as well as the latent heat (evaporation heat) of steam contained in the flue gas can almost entirely be utilised. Yet, this heat can only be used if the return (in-flow) temperature of the heating system is below the dew point temperature of the flue gas released by the boiler; only in this case part of the steam contained in the flue gas can be condensed by releasing energy (heat). Such condensing boilers are available for oil and gas. Natural gas-fired condensing boilers show annual efficiencies of up to 104 %, referred to the heating value of the gas.

Domestic hot water generation is mainly performed by means of storage domestic hot water heaters, located on top, below, or beside the heating boiler. The water is either heated by a heat exchanger located inside the storage facility (i.e. directly heated storage tank) or by an external heat exchanger (i.e. indirectly heated storage tank). In addition, electrically heated domestic hot water storages are in use.

Heat provision for domestic hot water generation and/or space heating by oil-fired or natural gas-fired boilers is characterised by corresponding costs, which are briefly discussed in the following. However, first the assessed reference plants are defined.

For the heat provision according to the defined supply tasks (Table 1.1) – depending on the required thermal capacity – the application of natural gas-fired boilers with condensing technology (5 kW for SFH-I; 8 kW for SFH-II; 18 kW for SFH-III; 60 kW for MFH), of atmospheric low temperature natural gas-fired boilers (9 kW for SFH-II), and of oil-fired low temperature boilers (20 kW for SFH-III; 67 kW for MFH) have been assumed (Table 1.6). Domestic hot water generation is provided via a storage system which is charged by an external heat exchanger in case of the multi-family house (MFH) and by internal heat exchangers for the single-family houses (SFH-I, SFH-II, SFH-III) respectively.

The applied amount of fossil fuel energy is determined by the amount of heat provided to the domestic hot water storage and to the supply point of the heat distribution system within the supplied building, and by the overall system efficiency of the heat generation system. Therefore the losses of the domestic hot water storage as well as the lower boiler efficiency for domestic hot water generation during the non-heating summer months have been considered. Especially for buildings with specific low heat demand (e.g. SFH-I) the annual mean system efficiency may be significantly lower than the efficiency of the boiler.

The defined district heating networks (Table 1.2) will not be analysed within the scope of this book for the use of fossil fuel energy. For economic as well as for environmental reasons, to date preferably decentralised solutions have been implemented.

Table 1.6 Parameters of the assessed heat generation systems based on fossil fuel energy

Demand case		SFH-I	SFH-II		SFH-III			MFH
Hot water demand	in GJ/a	10.7	10.7	10.7	10.7	10.7	64.1	64.1
Heat demand	in GJ/a	22	45	45	108	108	432	432
Boiler capacity	in kW	5	8	9	18	20	60	67
Fuel		NG[h]	NG[h]	NG[h]	NG[h]	FO[a]	NG[h]	FO[a]
Technology		CB[b]	CB[b]	LT[c]	CB[b]	LT-BB[d]	CB[b]	LT-BB[d]
Technical life time[e]	in a	15	15	15	15	15	15	15
Boiler efficiency	in %	104	104	93	104	93	104	93
System efficiency[f]	in %	95	98	88	101	91	100	90
Fuel input[g]	in GJ/a	34.5	56.6	63.2	117.2	131	495.5	553.5
Hot water storage	in l	160	160	160	160	160	800	800

[a] domestic fuel oil; [b] condensing boiler; [c] atmospheric low temperature gas-fired boiler; [d] low temperature oil-fired boiler with blue burner; [e] burner, boiler, and hot water storage; [f] in addition to the boiler efficiency the system efficiency also considers the losses of domestic hot water generation system; [g] including losses; [h] natural gas.

To estimate the costs of heat generation based on fossil fuel energy, in the following investment and operation costs as well as the specific heat provision costs are outlined for the reference systems defined in Table 1.6.

Investments and operation costs. To determine the investment costs the expenses for the boiler, the burner, the domestic hot water storage, civil engineering structures (e.g. furnace room design, chimney, oil tank, connection to the natural gas grid) as well as assembly and installation costs (Table 1.7) have been considered for the systems indicated in Table 1.6.

Table 1.7 Heat provision costs

Demand case		SFH-I	SFH-II	SFH-II	SFH-III	SFH-III	MFH	MFH
Boiler capacity in kW		5	8	8	18	20	60	67
Technology		CB[a]	CB[a]	LT[b]	CB[a]	LT-BB[c]	CB[a]	LT-BB[c]
Investments								
Boiler, burner etc.	in €	2,800	2,800	2,600	3,100	3,000	7,000	7,000
Tank, chimney etc.	in €	2,600	2,600	2,600	2,600	3,200	5,880	6,200
Assembly, installation	in €	830	830	830	830	950	1,180	1,200
Total	in €	6,230	6,230	6,030	6,530	7,150	14,060	14,400
Operation costs	in €/a	225	233	174	283	349	422	422
Fuel costs	in €/a	575	943	1,053	1,953	1,965	8,258	8,303
Heat provision costs	in €/GJ	40.0	31.0	28.3	24.3	22.7	20.2	18.2
	in €/kWh	0.144	0.112	0.102	0.087	0.082	0.073	0.065

[a] gas condensing boiler; [b] atmospheric low temperature natural gas-fired boiler; [c] low temperature oil-fired boiler with blue flame burner.

The operation costs of the assessed heat provision plants include, among other costs, expenses for maintenance and repair, as well as for electric power required for plant operation (including, among others, burner, fan, and self-ignition system). In addition, fuel costs have to be taken into consideration. They are shown separately from the other operation costs within Table 1.7.

Heat generation costs. The specific heat production costs of these plants are calculated on the basis on the same economic boundary conditions compared to the electricity generation costs (interest rate of 4.5 %, amortization period corresponding to the plant's technical life time) and the technical assumptions outlined in Table 1.7. They are summarised as follow.

For instance, the assessed oil-fired heating system is characterised by heat production costs at the plant of approximately 18 €/GJ (MFH) and 23 €/GJ (SFH-III) respectively. The assessed gas-fired heating system, by contrast, has slightly higher heat provision costs, amounting to scarcely 20 €/GJ (MFH) and 24 €/GJ (SFH-III; Table 1.7) respectively. In both cases, the costs are caused to an equal share by the expenses for fuels, and for erection and operation of the heat provision system. The assessment also reveals that heat production costs largely depend on the installed capacity. Therefore, they increase significantly with decreasing thermal capacity (i.e. SFH-II respectively SFH-I). For instance, for SFH-I (i.e. low-energy detached family house) they amount to about 40 €/GJ.

Environmental analysis. Besides the consumption of fossil fuel energy and emissions of harmful substances, pollutants are released into soil, air, and water during the operation of oil-fired or natural gas-fired heating installations. These pollutants show very different environmental effects. One example of such environmental effects is unburned hydrocarbons contributing to the creation of near-surface ozone when exposed to UV irradiation (i.e. summer smog).

In addition, also the provision of fossil fuels is associated with a series of effects which may damage the environment.
- When drilling wells or extracting crude oil and natural gas, chemical auxiliary material, for oil-well drilling and oil production, crude oil itself may enter the surrounding soil (onshore) or the sea (offshore); this could have significant effects on the environment.
- During the transportation of crude oil or crude oil products by sea, tanker accidents have recurrently had disastrous effects on the aquatic flora and fauna.
- When processing crude oil within refineries, a series of non-usable products are created that need to be disposed of as hazardous waste. Also, volatile hydrocarbons may be released into air, soil, and water when processing crude oil.
- Also, potential hazards during the transport of fuel oil from the refinery to the consumers represent a major source of danger in particular for soils and waters. If, for instance, oil tanks stored in the basement of dwelling houses are over flown in case of flooding, fuel oil may leak from the tank and enter surface waters. The environmental hazards due to this oil leakage are often more serious than the environmental effects caused by the flood itself.

2 Basics of Renewable Energy Supply

2.1 Energy balance of the earth

The energy flows of the earth are fed from various sources described below. Solar energy has a share of more than 99.9 % of all the energy converted on earth. The solar radiation incident on the earth is weakened within the atmosphere and partially converted into other energy forms (e.g. wind, hydro power). Therefore the structure and the main attributes of the earth's atmosphere will also be described in more detail, followed by balancing the global energy flows.

2.1.1 Renewable energy sources

Solar energy. The sun is the central body of our planetary system; it is the star closest to the earth. Its schematic structure with its main parameters is shown in Fig. 2.1. Accordingly, the nucleus has temperatures of approximately 15 Mio. K. Energy is released by nuclear fusion where hydrogen is melted to helium. The resulting mass loss is converted into energy E. According to Einstein, it can be calculated multiplying mass m and the square of the speed of light v_c (Equation (2.1)). Approximately 650 Mio. t/s of hydrogen are converted into approximately 646 Mio. t/s of helium. The difference of approximately 4 Mio. t/s is converted into energy.

$$E = m v_c^2 \tag{2.1}$$

The energy released within the nucleus of the sun is initially transported by radiation to approximately 0.7 times the solar radius. Further transport to the surface of the sun takes place through convection. Afterwards, the energy is released into space. This energy stream released by the sun is differentiated as radiation of matter on the one hand and electromagnetic radiation on the other hand (see /2-1/).
- The radiation of matter consists of protons and electrons released by the sun at a speed of approximately 500 km/s. However, only a few of these electrically charged particles reach the earth's surface, as most of them are deflected by the terrestrial magnetic field. This is of particular importance for life on earth, as this harsh matter radiation would not allow organic life in its current form.
- Electromagnetic radiation mainly released by the photosphere (Fig. 2.1) covers the entire frequency from short-wave to long-wave radiation. This type of solar

radiation is approximately equivalent to that of a black body. The radiant flux density of the sun M_S can be derived from the temperature within the photosphere (approximately 5,785 K), the degree of emission, and the Stefan-Boltzmann-constant; it is approximately 63.5 10^6 W/m².

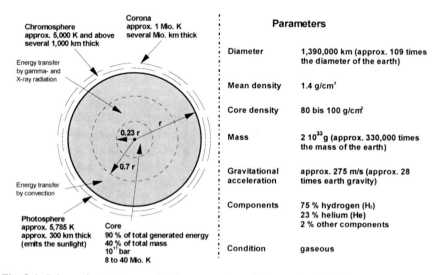

Fig. 2.1 Schematic structure and main parameters of the sun (see 2-25/)

The radiant flux density of the sun decreases – if losses are not considered – with the square of the distance travelled. Thus the radiant flux density at the outer rim of the earth atmosphere E_{SC} can be calculated according to Equation (2.2).

$$E_{SC} = \frac{M_S \pi d_S^2}{\pi (2 L_{ES})^2}$$
(2.2)

If the diameter of the sun d_S is assumed up to the photosphere (approximately 1.39 10^9 m) and a mean distance between the sun and the earth (L_{ES}) of approximately 1.5 10^{11} m is taken into consideration, a radiant flux density of approximately 1,370 W/m² can be calculated at the top rim of the earth atmosphere (see /2-3/). This mean value is called the solar constant. Over several years it varies less than 0.1 % due to a fluctuation in solar activity.

The solar radiation incident on the atmospheric rim throughout the course of the year is nevertheless characterised by seasonal variations. They are caused by the elliptical orbit, where the earth moves around the sun during the course of one year (Fig. 2.2). This changes the distance between the two celestial bodies. And this distance variation leads to a fluctuation in the radiation incident on the atmospheric rim; this results in the course of the solar constants shown in Fig. 2.3.

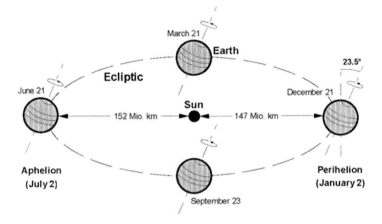

Fig. 2.2 Elliptical orbit of the earth around the sun (see /2-2/)

Thus, the solar constant reaches its maximum in January at almost 1,420 W/m², due to reaching the shortest distance between the sun and the earth (Perihelion) on January, 2nd. The opposite takes place on June, 2nd, when it reaches its minimum with approximately 1,330 W/m² (Aphelion).

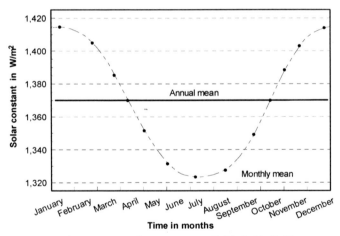

Fig. 2.3 Solar constant in the course of one year (see /2-2/, /2-3/, /2-5/)

In spite of the higher radiation intensity at the outer atmospheric rim, on average there are significantly lower temperatures on the Northern hemisphere during the winter than during the summer. The reason for this is that the rotation axis of the earth forms an angle of 66.5° with the orbital plane (Fig. 2.2). Thus during the winter the Southern hemisphere is facing the sun more than the Northern hemisphere. This leads to a higher solar altitude and longer periods of sunshine.

On the Northern hemisphere, however, the solar radiation incidents at a generally flatter angle during this season with comparably short days. Areas close to the North Pole are sometimes not facing the sun throughout the entire day. During winter solstice, all places between 66.5° N and the pole have "eternal polar night". Correspondingly, on the Southern hemisphere the sun never disappears below the horizon south of 66.5° S ("midnight sun").

As the earth continues its orbit around the sun, its relative position changes. For the Northern hemisphere the sun starts to rise higher and higher, whereas the midday altitudes get increasingly lower on the Southern hemisphere. On March, 21st, solar radiation incidents on both poles. The Northern hemisphere is now more sun-facing, i.e. the mean solar position above the horizon gets increasingly higher. This continues until summer solstice (June, 21st), when the midnight sun then lights up the North Pole areas, and the Antarctic region sinks into "eternal night".

Due to these interrelations, and thus, primarily due to the angle of the earth axis towards the ecliptic, the solar radiation in different regions of this earth is subject to significant seasonal fluctuations.

Geothermal energy. The energy flowing from the interior of the earth to its surface is fed by three different sources. On the one hand this is the energy stored in the interior of the earth resulting from the gravitational energy generated during the formation of the earth. The primordial heat that had even existed before that time is added as a second source. Thirdly, the process of decay of radioactive isotopes in the earth (in particular in the earth's crust) releases heat. Due to the generally low heat conductivity of rocks, this heat resulting from these three sources is to a large extent still stored in the earth.

The formation of the earth took place approximately 4.5 Billion years ago. It was a step-by-step accumulation of matter (rocks, gases, dust) within an existing fog. This process started off at low temperatures, which changed due to the increasing mechanical force of the matter amassing. During this aggregation of matter, gravitational energy was probably converted almost entirely into heat. Towards the end of this accumulation of mass, after approximately 200 Mio. years, the top level of the earth had melted. Due to this melting process, a large amount of the released heat was emitted into space again. In spite of all uncertainties about the accumulation of mass and the energy emission during this phase, the energy that remained in the earth during this phase was between 15 and 35 10^{30} J /2-4/. The smaller value reflects a cold to warm primordial earth, the higher value a warm to hot primordial earth.

The earth contains radioactive elements (i.e. uranium (U^{238}, U^{235}), thorium (Th^{232}), potassium (K^{40})). Due to radioactive decay processes, they release energy over a period of millions of years. The mass fraction of uranium or thorium in granite is, for example, approximately 20 ppm and in basalt 2.7 ppm. With the appropriate half-life, a released energy of approximately 5.55 MeV for a decay event and approximately 6 (thorium) or 8 (uranium) decay events until a stable condition is reached, an amount of heat of around 1 J/(g a) is generated. This results, for example, in a radiogenic heat production efficiency of approximately 2.5 $\mu W/m^3$ in granite rock and of approximately 0.5 $\mu W/m^3$ in basalt rock.

The decay of such natural, long-living isotopes in the earth permanently generates heat. The involved isotopes in the near-surface layers of the earth are mainly enriched in the continental earth crust. Due to such radioactive decay processes, the earth has received around $7 \cdot 10^{30}$ J of radiogenic heat since its formation. The potential radiogenic heat of still existing radioactive isotopes is approximately $12 \cdot 10^{30}$ J /2-4/. These figures are rather vague as little is known about the distribution of radioactive isotopes in the interior of the earth.

The currently available heat resulting from the earth's formation, respectively the primordial heat, as well as the heat already released to date and the heat attributable to the further decay of radioactive isotopes, all result in a total heat of the earth between 12 and $24 \cdot 10^{30}$ J; in the exterior earth crust up to a depth of 10,000 m this amounts to approximately 10^{26} J. This energy potential is equal to the solar radiation incident on the earth in the course of many millions of years.

Energy from planetary gravitation and planetary motion. The earth and the moon rotate around a mutual centre of gravity. Due to the disproportional character of the overall mass between these two celestial bodies it is located within the body of the earth. When earth and moon rotate around this gravitational centre, all points on these celestial bodies move in circles with the same radius around the gravitational centre. Within the centre of the earth, the gravitational force of the moon equals the centripetal force required for the rotational motion of the earth. On the side facing the moon, the gravitational force is stronger, therefore all matter on this side of the earth attempts to move towards the moon. In contrast, on the side facing away from the moon, the mass gravitational force of the moon is smaller than the centripetal force required for the movement of all matter in this orbit. Therefore, all matter on earth attempts to move away from the moon. This effect can, for example, be observed with the tides of movable water masses on the surface of the earth.

The earth's body is stretched to a certain extent under the influence of these forces. The response time of this deformation that change its direction by 360 degrees within 24 hours is too long to allow the earth's body to stretch fully. Therefore a complete formation of the theoretical distortion does not take place. The water, however, follows this deformation with a small delay due to the inner friction of the water masses, the friction with the ground of the sea, the clash with the continental rims, and from entering straits and bays. These delaying forces thus lead to a phase shift between the highest moon position and the high tide, and thus also to a reduction in the speed of the rotation of the earth.

The energy source causing the tides is mainly a result of the combined planetary motions and the mass gravitational effect that the celestial bodies, earth and moon have on each other.

2.1.2 Atmosphere

The atmosphere of the earth is defined as the gaseous atmosphere held by the gravitational force of the earth. It is divided into different "layers" (Fig. 2.4). Only the lower layers are of particular interest for the use of renewable energies on the surface of the earth. For the use of wind power, the atmosphere up to an altitude of, at most, several 100 m is significant.

The lower section of the atmosphere is called the troposphere. It is the atmospheric layer that influences the weather and where formation of clouds and precipitation mainly takes place. In a timely and spatial average, it is characterised by a temperature decrease with increasing altitude. The extent of this temperature change depends on location and time. The temperature gradient can fluctuate within relatively large boundaries around a mean value of 0.65 K/100 m. Under certain meteorological circumstances, vertically sharply defined layers occur where the temperature does not decrease with increasing altitude, but increases instead. Such inversions occur particularly often at altitudes between 1,000 and 2,000 m at the planetary boundary layer (Chapter 2.3.1) as well as immediately above the surface of the earth ("soil inversions") and at the tropopause.

The boundary of the troposphere is the tropopause connected to the stratosphere. Within the stratosphere, a temperature maximum is reached at an altitude of 40 to 50 km. The next atmospheric layer is the mesosphere; it reaches the next temperature extreme, a minimum at an altitude of approximately 80 km. Above it, demarcated by the mesopause, is located the thermosphere.

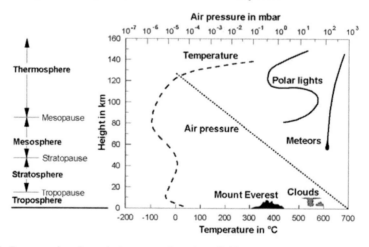

Fig. 2.4 Cross-section through the atmosphere (see /2-2/)

Up to an altitude of around 100 km, the atmosphere consists of a mixture of different gases (Table 2.1). The proportion of its ingredients is as defined as a chemical compound. This gaseous mix is – especially in the troposphere – mixed with water vapour and interspersed with aerosol suspensions. However, this is subject to strong fluctuations depending on time and location.

The basic mass, dry, pure air, consists of a mix of gases that cannot enter into the liquid or solid phase under atmospheric conditions (i.e. permanent gases). Their condensation or solidification temperatures are far below the temperatures occurring within the atmosphere. Apart from the main components nitrogen (N_2) and oxygen (O_2), there are also small traces of argon (Ar) and carbon dioxide (CO_2). Additionally, there are traces of further noble gases such as neon (Ne), helium (He), krypton (Kr) and xenon (Xe) as well as small amounts of ozone (O_3) and hydrogen (H_2) in the atmosphere. Particularly the proportions of the two last gases vary depending on time and location (see also /2-2/, /2-3/, /2-5/, /2-5/).

Table 2.1 Composition of the air (see /2-2/)

Components independent from time and space	
nitrogen (N_2)	78.08 vol. %; 75.53 mass %
oxygen (O_2)	20.95 vol. %; 23.14 mass %
argon (Ar)	0.93 vol. %; 1.28 mass %
further noble gases (He, Ne, Kr, Xe)	traces
Components dependent on time and space	
steam (H_2O)	depending on meteorological conditions up to 4 %
carbon dioxide (CO_2)	0.03 vol. %; 0.05 mass %; tendency currently increasing
admixtures	
gases ozone (O_3)	from the high atmosphere
radon (Rn)	from radioactive soil respiration
sulphur dioxide (SO_2)	from e.g. volcanoes, post-volcanic activities
carbon monoxide (CO)	oxidizes into carbon dioxide (CO_2) in the short term
methane (CH_4)	from e.g. animal digestion, anaerobic fermentation
VOC	from plants
Aerosols gaseous aerosols	from gaseous reactions (sulphates, nitrates etc.)
dust	for example plain, desert, or volcano dust
plant ash	from forest and steppe fires
sea water salt	transferred into the air with breaking wave crests
biomass	for example micro-organisms, pollen

mass % mass percentage; vol. % volume percentage.

2.1.3 Balance of energy flows

Energy from the three primary renewable energy sources sun, geothermal heat, and planetary gravitation and motion, occurs on the earth in very different forms (e.g. heat, fossil energy carriers or biomass) or causes very different effects (e.g. waves, evaporation and precipitation). Fig. 2.5 shows a flow diagram allocating these forms and effects to the corresponding energy sources. Additionally, the non-renewable energies or energy carriers are also shown to give a complete impression of the situation. However, only the main routes and coherences are shown in this diagram, as it is not always possible to make a definite allocation. Thus wind energy, for example, is a result of air movements within the atmos-

phere caused by solar radiation and influenced by the rotation of the earth. The
heat on the crust of the earth that is accessible to human beings consists of solar
energy and geothermal heat.

According to Fig. 2.5, besides the renewable energy flows from sun, geothermal heat and planetary gravitation and motion, additional primary non-renewable
energy sources are the atomic nucleus. They can be used to generate heat either
through nuclear fusion or nuclear fission. The solar energy flow causes a large
number of additional types of energy manifestation and effects. Over the past
millions of years, among others the fossil biogenous energy carriers such as coal,
crude oil and natural gas have been generated through solar radiation. Together
with the energy from the atomic nucleus (i.e. the fossil mineral energy carriers)
they are the non-renewable energies or energy carrier available to human beings.
All other forms are renewable energies or energy carriers. One part of the energy
currently incident on the earth's surface from the sun is transformed within the
atmosphere and causes – among other effects – evaporation and precipitation,
wind and waves. The global radiation incident on the earth's surface heats up the
surface of the sea and of the land. This warming is responsible – among others –
for the oceanic currents and the plant growth. Together with these manifestations,
also geothermal heat and tidal energy caused by planetary gravitation and motion
are counted among the renewable energies.

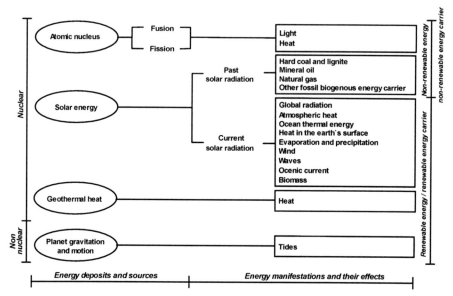

Fig. 2.5 Energy sources, types and effects (see /2-7/)

As the earth is almost in an energetic balance, the added energy has to be balanced by a corresponding withdrawal. This energy balance of the earth is shown
in Fig. 2.6. The very largest part of the energy converted on the earth every year

thus originates from the sun (over 99.9 %). Planetary gravitation and motion as well as geothermal energy additionally only account for approximately 0.022 % of the energy balance. The global use of primary energy from fossil biogenous and fossil mineral energy reserves and resources adds a further 0.006 % or approximately 413 EJ (2005) per year (see /2-38/).

Sun radiation incident on the earth every year is around $5.6 \cdot 10^{24}$ J. Around 31 % of this radiation is directly reflected back into space at the surface of the atmospheric rim. The remaining 69 % enters the atmosphere. A larger part reaches the surface of the earth, whereas a smaller part is absorbed within the atmosphere. A small part (on average approximately 4.2 %) of the radiation that reaches the surface of the earth is immediately reflected back into the atmosphere. The majority of the radiation that reaches the surface of the earth is available for evaporation, convection and radiation. It is to this end turned into long-wave heat radiation and as so, radiated back into space. A small part of the radiation reaching the earth's surface is converted into organic matter through photosynthesis.

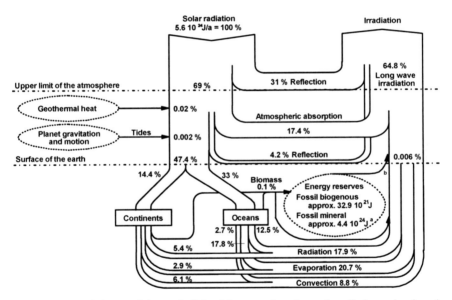

Fig. 2.6 Energy balance of the earth ([a] for this example only nuclear fission using breeder technology (1.5 TJ/kg uranium), additionally – not presented here – fusion would be possible; [b] global use of primary energy – i.e. fossil biogenous and fossil mineral energy carriers – of approximately 413 EJ per year in 2005 /2-38/; according to different sources)

Thus, there is almost a balanced situation between the energy input and output on the surface of the earth. But an insignificantly higher amount of energy is added than withdrawn because part of the energy is stored as biomass. If this organic sub-stance is not decomposed organically, burnt or converted in any other way by human beings it can be converted into fossil biogenous energy carriers within geological periods. This mainly concerns the plankton growing within the sea, which partly sinks down to the sea bottom. On the other hand, more energy can be released using fossil biogenous and fossil mineral energy carriers in the short term than is added to the earth by the described renewable energy flows.

2.2 Solar radiation

Part of the energy incident on the earth from the sun can be directly received as radiation on the surface on the earth and be converted into different utilisable forms of energy. Therefore, the main principles of solar radiation and its main characteristics will be discussed below.

2.2.1 Principles

Optical windows. The atmosphere is to a large extent impermeable for solar radiation. Only within the optical spectral range (0.3 to 5.0 µm; window I) and within the low-frequency range (10^{-2} to 10^{2} m, window II) radiation can pass the atmosphere (so-called optical windows of the atmosphere; Fig. 2.7). Due to energetic reasons only window I is relevant for the technical use of solar energy. The most important part of the optical window I covers the range of visible light between 0.38 and 0.78 µm.

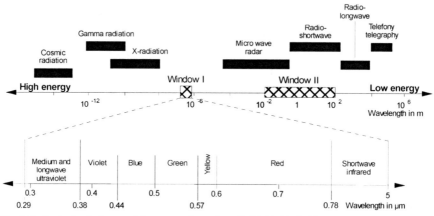

Fig. 2.7 Optical windows of the atmosphere (see /2-2/, /2-3/, /2-4/)

Weakening of radiation. Within the atmosphere radiation is weakened; this process is called extinction. Various mechanisms are involved.
– Diffusion. Diffusion is diversion of radiation from its original radiation angle without energy transfer and thus without a loss of energy. Such diffusion takes

place i.e. in air molecules, water drops, ice crystals, and aerosol particles. Rayleigh and Mie diffusion are differentiated. Rayleigh diffusion is diffusion at particles with a radius significantly smaller than the wavelength of the incident light (e.g. air molecules). The Mie diffusion takes place at particles with a radius within the wavelength of the incident light and larger (e.g. aerosol particles). The larger the particles at which the sunlight is diffused, the more they diffuse into a forward direction. Mie diffusion then turns into diffraction.

- Absorption. Absorption is the conversion of solar radiation into other energy forms. In general, solar radiation is converted into heat during this process. Such absorption can take place in aerosol, cloud and precipitation particles. Additionally, a selective absorption is possible; here selected spectral and wave-length ranges of solar radiation are absorbed by some gases existing within the atmosphere. This is especially the case for ozone (O_3) and water vapour (H_2O). Ozone, for example, almost completely absorbs the spectral range between 0.22 and 0.31 μm. Carbon dioxide (CO_2), in comparison, only minimally absorbs solar radiation.

This weakening is described by the so-called transmission factor τ_G (Equation (2.3)); it covers all weakening effects affecting solar global radiation incident on the outer atmospheric layer passing through the atmosphere. G_g is the global radiation and E_{SC} the solar constant.

$$G_g = E_{SC} \cdot \tau_G \qquad (2.3)$$

The transmission factor consists of the Rayleigh diffusion τ_{RD}, the Mie diffusion τ_{MD} and the absorption within gases τ_{GA} as well as the absorption within particles τ_{PA} (Equation (2.4)).

$$\tau_G = \tau_{RD}\, \tau_{MD}\, \tau_{GA}\, \tau_{PA} \qquad (2.4)$$

Spectral range. Due to the weakening of the radiation within the atmosphere of the earth, the energy distribution spectrum of sunlight is changed. Fig. 2.8 shows the spectrum of the solar radiation before and after passing through the earth's atmosphere.

Due to the described radiation weakening processes within the atmosphere of the earth, the energy distribution of solar radiation reaching the earth shows the following characteristics.

- The energy maximum is within the visible spectral range between 0.5 and 0.6 μm (green to yellow light).
- With a decreasing wave length (i.e. in the ultraviolet spectrum) the radiated power decreases rapidly.
- With an increasing spectral range (i.e. in the infrared spectrum) the radiation decreases more slowly.

- Some specific wave lengths show deep cuts in the energy distribution curve ("dark ranges"). They are caused by selective absorption of the sunlight by selected elements within the atmosphere.

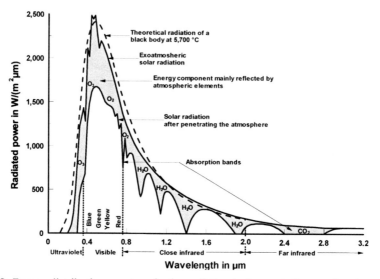

Fig. 2.8 Energy distribution spectra of solar radiation before and after passing through the atmosphere of the earth (see /2-3/)

Direct, diffuse and global radiation. The diffusion mechanisms within the atmosphere cause diffuse and direct radiation to incident on the surface of the earth. Direct radiation is the radiation incident on a particular spot, having travelled a straight path from the sun. In contrast, diffuse radiation is the radiation emerged by diffusion in the atmosphere and thus indirectly reaching a particular spot on the earth's surface. The sum of direct (beam) radiation G_b and diffuse radiation G_d, always related to the horizontal receiving surface, is called global radiation G_g (Equation (2.5)). The diffuse radiation G_d consists of the radiation diffused in the atmosphere, the atmospheric counter-radiation, and the radiation reflected by the neighbourhood.

$$G_g = G_b + G_d \qquad (2.5)$$

For calculations of the overall solar radiation on a particular receiving surface (e.g. surface of a solar collector), direct and diffuse radiation have to be differentiated as they incident at different mean angles on the receiving surface. The atmospheric counter-radiation and the radiation reflected by the neighbourhood on the receiving surface generally have little impact. In wintertime or in mountainous areas, reflected radiation can, however, contribute to a larger extent to global radiation, for example caused by a snow blanket.

The proportion of diffuse and direct radiation within the overall global radiation incident on a particular spot is subject to daily and seasonal fluctuations. Fig. 2.9 therefore shows, as an example, the annual course of direct, diffuse and global radiation at one particular site in South Germany. According to that, the annual mean share of diffuse radiation in Central European regions considerably exceeds the amount of direct radiation. During the winter months, global radiation almost exclusively consists of diffuse radiation. In summer, the share of direct radiation increases significantly, but is on average always smaller than the share of diffuse radiation. This can be completely different in other places throughout the world. In the desert, the share of direct radiation is in most cases very high. On the other hand in regions with a very high precipitation and/or lots of fog, the contribution of diffuse radiation to the overall global radiation might be well above 80 %. Also, this might vary dramatically according to different times of the day and/or different seasons of the year.

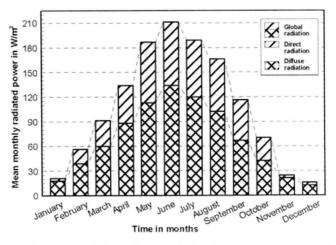

Fig. 2.9 Annual course of diffuse, direct, and global radiation, exemplary for a site in South Germany (see /2-3/)

Direct radiation on tilted, aligned surfaces. The direct radiation incident on a tilted surface is determined by its incident angle ψ (Fig. 2.10). This angle in turn is dependent on the alignment and the location of the receiving surface and the position of the sun (Equation (2.6)).

Here α is the slope or tilt angle of the surface (horizontal 0), β the surface azimuth angle (i.e. diversion from the South alignment, south 0, west positive), φ the latitude (north positive), δ the solar declination and ω_h the hourly angle of the sun; this angle is at $0°$ when the sun is at its highest position and is negative in the morning and positive in the afternoon.

$$\cos\psi = (\cos\alpha\sin\varphi - \cos\varphi\cos\beta\sin\alpha)\sin\delta +$$
$$(\sin\varphi\cos\beta\sin\alpha + \cos\alpha\cos\varphi)\cos\delta\cos\omega_h + \tag{2.6}$$
$$\sin\beta\sin\alpha\cos\delta\sin\omega_h$$

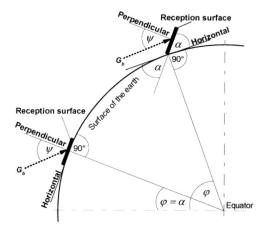

Fig. 2.10 Geometric interrelationships of radiation incident on tilted surfaces (see /2-8/)

ω_h can be calculated using True Solar Time (*TST*), resulting from the local time *LT* (in h) according to international conventions (winter time) and the equation of time *E*, which takes into account the perturbations in the earth's rate of rotation which affect the time the sun crosses the observer's meridian, see Equation (2.7) (according to /2-8/).

$$\omega_h = (LT\cdot 60 + 4\cdot(\lambda_0 - \lambda) + E)/60\cdot 15 - 180 \quad \text{with}$$

$$E = 229.2\cdot(0.000075 + 0.001868\cdot\cos B - 0.032077\cdot\sin B -$$
$$-0.014615\cdot\cos(2\,B) - 0.04089\cdot\sin(2\,B)) \tag{2.7}$$

$$B = (n-1)\frac{360}{365}$$

n is the observed day of the year (1 ... 365), λ_0 the reference meridian (-15° at Greenwich Mean Time (GMT), -30° at Central European Time (CET)) and λ the longitude of the site. The declination of the sun δ, describing the angular distance of the sun at its highest point from the equator of the sky, is calculated according to Equation (2.8). It assumes values between -23.45° on December, 22nd, and +23.45° on June, 22nd.

$$\delta = -23,45 \cos \frac{2\pi}{365,25}(n+10)$$ (2.8)

If the receiving surface is horizontal to the solar radiation, the zenith angle ψ_z can be calculated according to Equation (2.9).

$$\cos \psi_z = \sin \varphi \sin \delta + \cos \varphi \cos \delta \cos \omega_h$$ (2.9)

Conversion of the direct solar radiation on a tilted, aligned surface $G_{b,t,a}$ (i.e. oriented towards a particular direction) can be calculated from the direct radiation G_b on the horizontal surface, using the angle of radiation incidence ψ, the inclination of the surface with respect to the horizontal level α, the solar azimuth angle β, and the collector azimuth angle (alignment of the surface normal according to the direction) γ, using Equation (2.10).

$$G_{b,t,a} = G_b \left(\sin \psi \cos \alpha - \sin \alpha \cos \psi \sin(\beta - \gamma) \right)$$ (2.10)

Diffuse radiation on tilted, aligned surfaces. The conversion of the diffuse proportion of the solar radiation on the tilted and aligned surface $G_{d,t,a}$ depends on a number of influencing factors and cannot be described entirely analytically. If it is assumed, as a simplification, that diffuse radiation is evenly distributed within space, the same proportion incidents from all directions on a particular point of the surface of the earth (isotropic model). Under these simplified boundary conditions, the diffuse radiation incident on tilted, aligned surfaces is calculated with the diffuse radiation incident on the horizontal surface G_d and the angle of inclination of the receiving surface against the horizontal α according to Equation (2.11).

$$G_{d,t,a} = \tfrac{1}{2} G_d (1 + \cos \alpha)$$ (2.11)

Assuming an isotropic distribution of radiation only describes the given circumstances to a limited extent. If the atmosphere is solely filled with diffuse radiation due to a heavy and homogenous cloud cover, the area around the position of the sun is nevertheless generally brighter than the rest of the sky. This aspect is considered in Equation (2.12), that assumes an even distribution of isotropic radiation in space, superimposed by a so-called circumsolar share (see /2-9/).

$$G_{d,t,a} = G_d \left(\frac{1}{2} \left(1 - \frac{G_{b,t,a}}{E_{sc}} \right)(1 + \cos \alpha) + \left(\frac{G_{b,t,a}}{E_{sc}} \frac{\cos \psi}{\cos(90° - \alpha)} \right) \right)$$ (2.12)

Reflection radiation on tilted, aligned surfaces. A certain proportion of the global radiation incident in the neighbourhood of a defined surface is reflected onto the receiving tilted, aligned surface $G_{r,t,a}$. This reflected radiation can be calculated using the albedo A_G (i.e. the ratio of reflected to the incident global radiation), the global radiation incident on the horizontal receiving surface G_g, and the inclination angle towards the horizontal α according to Equation (2.13).

$$G_{r,t,a} = A_G\, G_g\, \sin^2(\alpha/2)$$
(2.13)

The albedo depends on site-specific conditions. Values can, for example, range between 0.7 and 0.9 with snow, between 0.25 and 0.35 with sand, and between 0.1 and 0.2 in forest and farmland areas.

Global radiation on tilted, aligned surfaces. The global radiation incident on an tilted and aligned surface, for example the surface of a photovoltaic module, consists of the incident direct ($G_{b,t,a}$, Equation (2.10)) and diffuse radiation ($G_{d,t,a}$, Equation (2.11) and (2.12)) as well as the radiation reflected by the neighbourhood on this receiving surface ($G_{r,t,a}$, Equation (2.13)). The total global radiation incident $G_{g,t,a}$ on an tilted and aligned surface is thus calculated according to Equation (2.14).

$$G_{g,t,a} = G_{b,t,a} + G_{d,t,a} + G_{r,t,a}$$
(2.14)

2.2.2 Supply characteristics

Measuring radiation. In order to be able to measure short and long-wave radiation fluxes through the atmosphere, a number of different measuring instruments are available. Relative and absolute instruments are generally differentiated /2-3/, /2-10/.

If radiation energy has to be measured in absolute terms, the incident solar energy has to be converted into a measurable parameter first. Most of these radiation measurement instruments thus absorb the radiation energy on a blackened surface which converts it into heat. Due to this process the temperature of the surface increases. Correspondingly, an amount of heat per time unit is released by heat transfer in the instrument or into the air by thermal radiation. Based on this balanced conditions are established. In this case, the resulting temperature increase is a measure for the radiation energy. Types of such absolute instruments are for example, the Moll-Gorcynsky pyranometer, and the compensation-pyrheliometer named after Angström. A relative instrument can be used to calibrate such an absolute instrument. One example of a relative instrument is the Michelson-Marten Actinometer.

In order to measure direct solar radiation (i.e. the proportion of direct radiation related to the global radiation) the pyrheliometer is used. One of two equal, blackened thin Manganin surfaces is exposed to direct solar radiation which, in turn, heats it up. The other surface, without incident solar radiation, is heated with electrical energy to the same temperature as the exposed surface. The production of heat is proportional to the square of the electric current. Thus the electric current is equivalent to the absorbed radiation energy. Pyrheliometers are normally adjusted to the incident radiation, and are built in a way that enables only direct radiation to incident on the receiving surface, for example by positioning the plane surface within a tube.

In order to measure global radiation, pyranometers are used (e.g. the Moll-Gorcynsky pyranometer). This instrument has a radiation thermopile as receiving surface. Its counter junctions are thermally connected with the casing. The temperature difference caused by heating of the receiving surface due to the solar radiation generates pressure, which serves as a measure for global radiation. In order to avoid atmospheric influences on the measuring process, the receiving surface is protected with spherical calotte made of different materials, in accordance with the spectral range to be measured. In order to measure for example short-wave radiation fluxes, hemispheres made of silica glass are used. To measure long and short-wave radiation fluxes, spherical caps made of polyethylene, and for long-wave radiation fluxes silicon hemispheres are used. Pyranometers are mainly aligned horizontally. If the direct radiation proportion of the global radiation is cut out, e.g. by shielding from direct solar radiation by a circular disk or a fixed shade ring (shadow band), these instruments can also be used to measure diffuse radiation.

In order to calculate the radiation balance, one pyranometer is required for the upper and the lower hemisphere. Depending on the type of cover, the balance can be drawn for different spectral ranges.

Often, only the duration of sunshine is measured. It is mostly captured by a sunshine recorder named after Campbell-Stokes, employing a glass sphere to focus the sun's rays to an intense spot or a focal point, which will burn a mark on a curved card mounted concentrically with the sphere.

Distribution of radiation. Worldwide, global radiation is measured at numerous sites. If these measured radiation values available as hourly, daily or monthly mean values are added up over one year and the long-term mean values are calculated, the average expected radiation for this particular site is obtained. Fig. 2.11 shows the global solar radiation distribution on earth.

On a first glance the graphic reveals that the highest global radiation occurs north and south of the equator. Especially within the deserts and on mountains located here, the yearly global radiation reaches its maximum. North and south of these zones the global radiation irradiated throughout the year decreases.

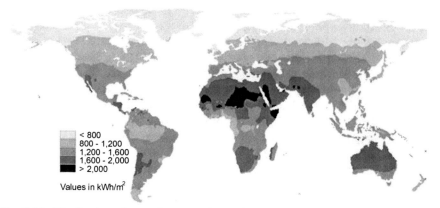

Fig. 2.11 Distribution of annual mean values of total global radiation world wide (data according to /2-11/)

Maps as shown in Fig. 2.11 can only give a rough estimate of the global radiation to be expected within a larger region. Locally the solar radiation can thus vary within the boundaries indicated in Fig. 2.11. Therefore Fig. 2.12 shows exemplarily the distribution of the mean solar global radiation over many years within Germany and Austria.

It becomes apparent that South Germany is characterised by the highest supply of solar radiation. In North Germany – with the exception of the North and Baltic Sea islands – the radiation totals are sometimes significantly lower. The reason for the higher radiation supply in South Germany is the more southern location, and the proximity to the equator. Additionally, the area is less cloudy on average. Both taken together lead to an increased global radiation and a longer average sunshine duration. The mean global radiation total over many years varies between around 290 and 470 kJ/(cm^2 a) or approximately 800 and 1,300 kWh/m^2 due to regional differences.

The situation is also basically similar for Austria. But as this country is located further south the average global radiation is slightly higher. This is also true for the maximum solar radiation which is measured on some mountain tops. This is attributable to the fact that the atmosphere on high mountains is thinner compared to lowlands, and thus the solar radiation is less weakened.

Time variations. The solar radiation supply in one location is subject to significant time fluctuations. Some of these fluctuations are of deterministic, some of them of stochastic nature.

Fig. 2.13 shows the timely differences of the global solar radiation supply exemplarily with radiation data measured in one location in North Germany. The annual course of the mean daily radiated power is thus characterised by a lower radiation supply during the winter months and a higher supply during the summer.

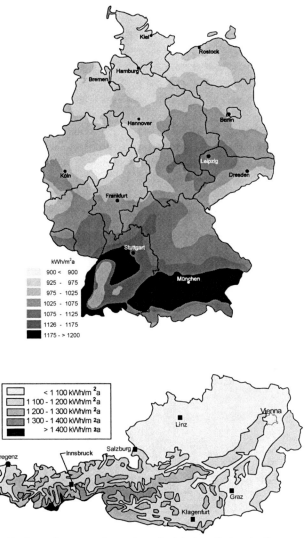

Fig. 2.12 Distribution of mean values of total global radiation in Germany (above) and Austria (below) (see /2-12/ (above) and /2-13/ (below))

The two courses of the daily radiated power are also shown in Fig. 2.13, taking January, 30th and October, 30th as examples. These parts of the graphic clarify how the radiation supply is distributed over the course of one day. The temporal distribution of the radiant flux density for the day in January, for example, was defined by an overcast sky throughout the entire day. The almost exclusively diffuse radiation is characterised by a very low intensity. The 30th of October, however, was almost cloudless; only the fall of the global radiation around mid-day indicates that clouds are passing over.

Furthermore, the course of the average radiated power per minute at mid-day confirms that the analysed day in January was characterised by an evenly overcast sky with a small amount of solar radiation that only varied a little. In contrast, on the exemplarily investigated day in October, the solar radiation incident was generally higher and thus subject to larger variations due to changes in the level of cloudiness.

Similar curves will also be found at other sites on earth. But significant variations may occur. For example the order of magnitude of the measured global solar radiation might vary depending among other things on the latitude of the measuring spot. The daily cycle is also influenced by the latitude and by the time of the year. And the radiated power is very much controlled by the micro-meteorological conditions (i.e. blue sky or cloudy weather).

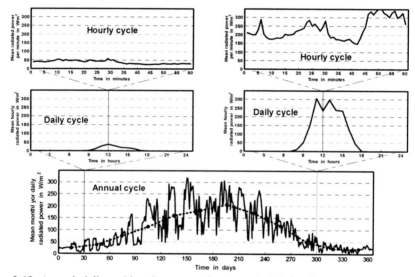

Fig. 2.13 Annual, daily and hourly curves of measured global radiated power based on an exemplary site in North Germany (see /2-7/)

The fact that for example the annual cycle of the solar radiation can vary significantly is also shown in Fig. 2.14. This graphic displays the global radiation at Stockholm/Sweden, Kuala Lumpur/Malaysia, Victoria/Argentina, Béchar/Algeria and Manaus/Brazil. The relatively far north located city of Stockholm shows a well developed annual cycle. The same characteristic also shows the city of Victoria but mirror-inverted. This is due the fact that Victoria is located in the southern hemisphere and Stockholm on the northern side of the earth. Compared to these cities, Kuala Lumpur and Manaus show a much less pronounced annual cycle because both cities are close to the equator. But the level of the measured global radiation is different because of unequal micro-meteorological condition.

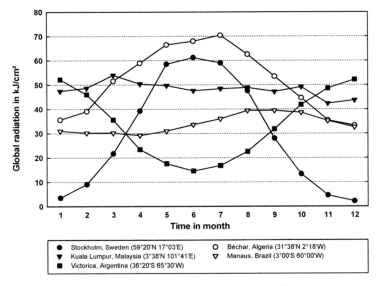

Fig. 2.14 Annual curves of measured global radiation at different sites throughout the world (data according to different sources)

The solar radiation supply is also characterised by clear variations over several years. This becomes apparent in Fig. 2.15 showing the annual totals for global radiation, exemplarily for four sites in Germany, in the course of the last decades. Additionally, the mean global radiation, the corresponding standard deviation, and the minimum and the maximum global radiation total, are shown. From the figure, the following conclusions can be drawn.

– Of the represented sites, the southernmost station, Hohenpeißenberg, shows the largest mean global radiation total over several years, whereas in Hamburg, the northernmost site, the lowest mean annual global radiation was recorded. The high radiation total on the Hohenpeißenberg is not only due to its southernmost location compared to all other stations, it is also due to the exposed position of the test station (mountain station).

– Only the comparison of the standard deviations of the annual global radiation totals at the four sites, with regard to the respective mean value of global radiation, is meaningful. This shows that they are relatively similar at the four stations, with a maximal standard deviation of +/-10 %. Within Germany, the relative standard deviation of the annual global radiation totals is almost independent of the annual totals of global radiation. In principle, this is not only true for the standard deviations, but also for minimum and maximum values.

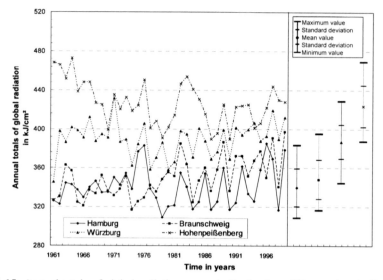

Fig. 2.15 Annual totals of global radiation exemplarily for four different sites in Germany between 1961 and 1998 (data see /2-12/)

Fig. 2.16 shows the course of the monthly mean totals of the global radiation over thirty years, using the example of the stations shown in Fig. 2.15. With reference to Fig. 2.15 the mean values, the standard deviation, as well as, the minimum and the maximum values are also shown here. Together with the seasonal dependence of the solar radiation supply, the summer months are mostly characterised by larger fluctuations in solar radiation supply than the winter months.

The solar radiation is additionally characterised by a distinctive daily course. Fig. 2.17 therefore shows the mean diurnal course of the mean hourly radiated power for various months, using the example of two sites in Germany for the average over 10 years.

The figure shows the known typical course of the day, with an increase in the solar radiation supply during the morning hours, a maximum at mid-day and a decrease in the hours of the afternoon and the evening. The radiation maxim, the daily radiation time, and thus the areas surrounded by the curves showing the incident daily radiation, are at the peak during the summer months. During the winter, they are at a correspondingly low level. This characteristic might change depending among others on the latitude of the site. Analyses of the global radiation of locations in the southern hemisphere (e.g. Australia, Chile) reveal that global radiation reaches its peak during the winter months and becomes significantly lower during summer.

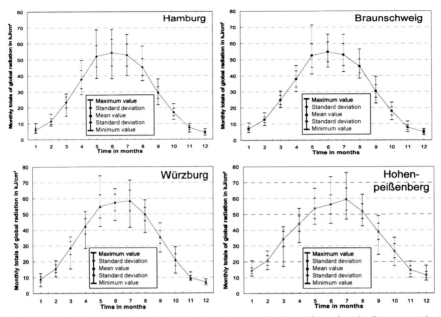

Fig. 2.16 Monthly mean global radiation totals exemplarily at four sites in Germany (data captured between 1961 and 1998, data see /2-12/)

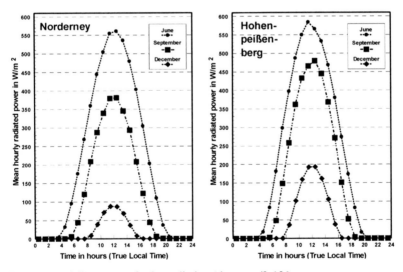

Fig. 2.17 Mean daily course of solar radiation (data see /2-12/)

The typical annual course, as well as, the different daily courses in spring, summer, autumn and winter are, to a large extent, a result of the oblique inclination of the earth's rotation axis towards the sun (approximately 23.5° deviation from the vertical, Fig. 2.2). This causes the mean solar position above the horizon on the northern hemisphere to be significantly lower during the winter than during the summer. Additionally, north of the equator the time spans when the sun shines throughout the course of one day, are shorter than in the summer. This becomes clear in Fig. 2.18; the left side of the graphic shows the mean hourly solar position above the horizon at a monthly average in Central Europe. For the winter months, the short period of time when the sun is above the horizon, as well as, the very low solar position above the horizon compared to the summer become apparent. This changes during the summer, as the northern hemisphere is facing the sun. The reverse is true for the southern hemisphere.

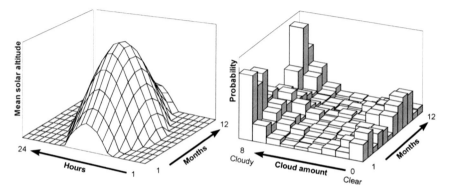

Fig. 2.18 Monthly mean solar position (on the left) and monthly mean degree of cloudiness (on the right) exemplarily for one site in South Germany (for data see /2-12/)

In a first approximation, the weakening of solar radiation within the atmosphere of the earth is proportional to the radiation course through the atmosphere. During the summer a larger part of the radiation incident at the top atmospheric rim thus reaches the earth surface at most places of the globe. This is due to the higher solar position, and thus, the on-average shorter radiation path. Additionally, the radiation absorption and the reflection in the atmosphere depend on the water content of the atmosphere, and thus, on how overcast the sky is. Due to predominant meteorological conditions in many parts of the earth, they are subject to significant seasonal fluctuations. Fig. 2.18 (on the right) therefore shows the likelihood of the occurrence of certain cloud amounts on a monthly average, exemplarily for a South-German site. The cloud amounts which are a meteorological measure for the mean hourly cloudiness, fluctuate between 0 (no clouds) and 8 (completely overcast). When comparing the likelihood of occurrence shown in Fig. 2.18 (on the right) it becomes apparent that the atmosphere at the investigated spot is on average decidedly cloudier during the winter months than during the summer. This statement can generally be applied to other sites and years throughout the wide parts of the northern and southern hemisphere. Altogether, shorter periods of sunshine, as well as, small radiation incident angles, an over-proportional cloudiness, and thus a weakening of radiation in the atmosphere characterise the winter in comparison to the summer.

These contexts show that the radiation is composed of a deterministic and a stochastic element. The former is the radiation component that is always present (i.e. the proportion of diffuse radiation incident during the course of the entire day when the sky is completely overcast). The latter describes the proportion between the deterministic component and the maximum possible radiation (i.e. the maximum possible radiation at a completely clear sky during the entire course of the day) that is likely to occur to a certain extent. Both components vary depending on the season and the time of day. Fig. 2.19 shows schematically the course of the minimum and the maximum possible mean hourly radiation intensity (i.e. the radiation intensity at a completely clear or completely overcast sky) for the days of winter and summer solstice at a site in Central Europe. Furthermore, the graphic contain an exemplary possible course of solar radiation intensity. According to that, the spectrum within which the solar radiation can fluctuate during daytime hours is very high. On the other hand, the substantial influence of the cloud amount on the solar efficiency becomes apparent.

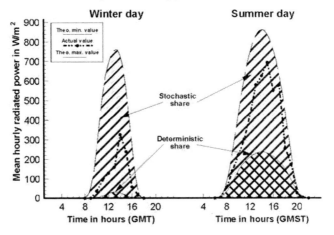

Fig. 2.19 Deterministic and stochastic part of solar radiation (Theo. theoretical, min. minimum, max. maximum) (see /2-7/)

Thus the hourly radiated power is deterministic and predictable within certain boundaries. However, within these boundaries which can be very far apart from each other depending on the time of the day or the year, the radiation supply is mainly stochastic. The stochastic character of solar radiation is significantly influenced by the current macro and micro-meteorological conditions. These variations are therefore interdependent at different points in time that lie closely together. Thus the cloudiness at a certain point in time has a significant influence on how overcast the atmosphere is going to be in the following hour. This influence decreases with an increasing time distance. This is also true for the space dependency. Cloudiness at different, geographically close sites is coupled – dependent on the local conditions – via the large and small-scale interdependencies within the atmosphere.

In order to be able to evaluate a concrete site, e.g. for the installation of a solar panel, the shading of direct solar radiation by mountains, buildings and trees has to be considered. So-called solar-position plots (Fig. 2.20) can be used to measure the shading. For every 21st day of each month, for a particular latitude, the solar position (i.e. the angle between the solar radiation incident and the horizontal) above the solar azimuth (i.e. the deviations of the solar position from the Southern direction) is shown on these diagrams. Additionally, the corresponding time for the respective solar position is included.

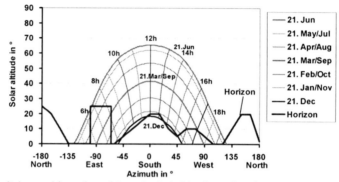

Fig. 2.20 Solar-position plot with an included horizon for sites at 48° Northern latitude (see /2-14/)

The outlines of surrounding elevations can now be entered into such a solar-position plot. Afterwards, the relevant daily and seasonal shading effects can be read from the diagram. Such a graphic can, for example, help identify the ideal positioning of a house which is supposed to generate high passive solar yields for an optimum use of solar radiation, keeping shading as low as possible during the times when solar energy should be used.

The alignment of a solar panel is also decisive for its energy output. Fig. 2.21 therefore shows the monthly overall global radiation (i.e. the total of direct and diffuse radiation) on surfaces with different alignments for a location in Central Europe. According to that, during the heating period in the winter months, the highest radiation amount of all vertical surfaces, incidents on the vertical surface aligned towards the South. Outside the heating period in the summer months, the radiation amount on the southern vertical surface is lower than on vertical surfaces aligned towards East or West. During the heating period, only diffuse radiation incidents on vertical surfaces aligned to the North. Skylights inclined 45° to the South have a very high radiation incident during the summer. During the winter, however, radiation is similar to the vertical Southern wall. Therefore sunspaces with glazing slanting towards the South, for example, often show the problem of overheating during the summer. The top curve shows the theoretical maximum of a surface 2-axis-tracked surface. In winter only little less solar radiation is incident on the vertical Southern surface.

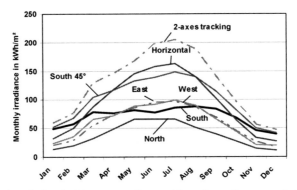

Fig. 2.21 Global radiation incident on surfaces with various alignments in Central Europe (Climate Graz/Austria) (see /2-14/)

2.3 Wind energy

In addition to the global water cycle, solar radiation also maintains the movement of the air masses within the atmosphere of the earth. Of the total solar radiation incident on the outer layer of the atmosphere, approximately 2.5 % or $1.4 \cdot 10^{20}$ J/a are utilised for the atmospheric movement. This leads to a theoretical overall wind power of approximately $4.3 \cdot 10^{15}$ W. The energy contained in the moving air masses, which for example can be converted into mechanical and electrical energy by wind mills, is a secondary form of solar energy. The aim of the following discourse is to show the main basic principles of the supply in wind energy and to discuss its supply characteristics.

2.3.1 Principles

Mechanisms. Wind is generated as equalising currents, essentially as a result of varying temperature levels on the surface of the earth, by which differences in air pressure have been created. The air masses then flow from higher pressure areas to lower pressure areas.

The so-called gradient force caused by the pressure gradient between such a high and a low-pressure zone impacts on an air particle. Additionally, the Coriolis force impacts on each particle within a rotating reference system. This Coriolis force is always vertical to the direction of movement and to the rotational axis.

If there is a large pressure difference at great altitudes, an air particle exposed to this pressure difference starts moving from a point of higher air pressure to a point of lower air pressure. It thus wants to migrate from an isobar with the pressure p_1 towards an isobar with the pressure p_2 (Fig. 2.22). Through the movement from p_1 towards the lower pressure level p_2, the gradient force accelerates the particle at constantly increasing speed. At the same time, the influence of the Coriolis force increases. This force is defined as the result of the product of particle mass, angular velocity of the rotating system, and the particle velocity relative to the rotating reference system. As it impacts always vertically to the direction of the movement (Fig. 2.22), it constantly causes a change of the direction of the velocity vector. The resulting change of the directional movement lasts as long as the magnitude of the Coriolis force is equal to the gradient force. The particle is then not any longer subject to a resultant force; it is in equilibrium. Its speed and the Coriolis force hence remain unchanged, it is moving in parallel to the isobars. The consequent wind category, where air moves alongside the isobars, is called the geostrophic wind (Fig. 2.23, left).

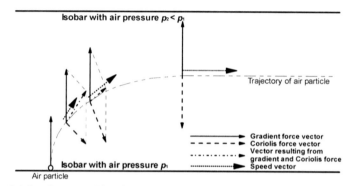

Fig. 2.22 Origin of geostrophic wind (Northern hemisphere; see /2-6/)

The larger the pressure gradient, the closer the isobars are together and the higher is the gradient force. Therefore the air particles are accelerated more, and the speed of the particle moving from isobar with the pressure p_1 to the isobar with the pressure p_2 increases. The amount of the Coriolis force, however, increases proportionally to the speed of the particle the force impacts on.

Therefore for isobars located in parallel the equilibrium of forces between Coriolis force and gradient force, and thus the straight movement of the particle along the isobars always occur, independently of the difference in pressure or the gradient force. The speed of the geostrophic wind itself depends solely on the size of the pressure differences.

In areas with a low or a high-pressure core the isobars are curved. In addition to the two forces already mentioned, a third force, the centrifugal force, also acts on the air particle. It points radially outwards (Fig. 2.23, centre and right). The resulting wind is called the gradient wind. In the northern hemisphere, it blows anti-clockwise, and clockwise in the southern hemisphere, around low pressure areas. For a high pressure area, the situation is reversed. As the centrifugal force reinforces the gradient force in a high pressure zone and weakens it in a low pressure zone, the gradient wind velocity in the high pressure area is greater than in the low pressure area /2-3/, /2-5/.

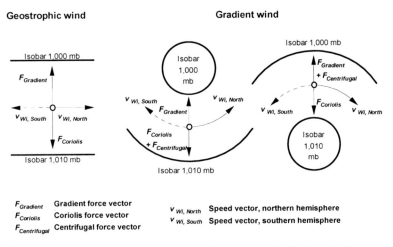

Fig. 2.23 Geostrophic and gradient wind (according to /2-2/, /2-3/, /2-5/, /2-6/)

Global air circulation systems. The described mechanisms of an air movement within the atmospheric layers are the precondition for the global air circulation system existing on earth (Fig. 2.24).

On a global scale, the surface of the earth heats up most where the sun is in the zenith (i.e. in the area around the equator). This causes a low pressure zone close to the equator, where air streams in from the North and the South. Without the influences caused by the continents this equatorial convergence zone would stretch around the equator like a belt, shifting parallel to the seasonally changing of the position of the sun between the tropic of Cancer and Capricorn with a certain time delay.

Because of the actual given influence of the oceans and the continents, this convergence zone is almost always North of the equator; however, it moves slightly with changing seasons. If the earth did not rotate, the air would flow close to the ground from the polar areas towards the equator. Here, it would be lifted up into the convergence zone and flow towards the poles again in higher atmospheric layers. By sinking down in high-pressure zones above the poles, the circulation process would finish.

Such simple flow conditions cannot occur on a rotating planet. Therefore, on a first glance only an "ideally" rotating planet is examined, without looking at the influence of sea and land. Thus a plant is taken into consideration where the temperature is only determined by the latitude. The air then flows closely to the equator towards the tropical convergence zone. It is however diverted by the Coriolis force. This leads to an air current, blowing practically at the same force throughout the entire year, from North-East and South-East (North-East and South-East trade winds). The trade winds flow from the so-called subtropical high pressure cells, which are located in the vicinity of the 30th latitude on each hemisphere. This subtropical high pressure belt is characterised by weak winds and clear weather. On the polar side follows a zone where the Western winds of the medium latitudes are predominant. For example Central Europe is located within this zone of influence. Wind direction and wind velocity change significantly depending on the wandering cyclones and anticyclones in this area. This West wind area is limited by a low pressure trough towards each of the poles. In the polar areas the wind conditions show strong fluctuations. On average, weak high pressure areas prevail in the deeper strata.

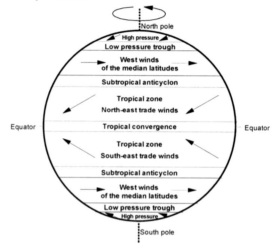

Fig. 2.24 Planetary air flow at high altitudes (see /2-3/)

Due to these complex interdependencies, influenced additionally to a large extent by sea and land, as well as seasonal and other effects, a global air circulation system is created. It is responsible for the global exchange of air. For energetic utilisation, these air movements are only of minor importance, as it is currently almost impossible to use the energy of the moved atmosphere at great altitudes, where these air circulation systems are active.

Local air circulation systems. The forces responsible for wind generation are active everywhere within the atmosphere. But with increasing closeness to the surface of the earth, they are also increasingly influenced by local effects. Therefore, the so-called free atmosphere at great altitudes, where the described global air circulation systems are active, and the planetary boundary layer, located close to the surface of the earth, are differentiated.

Geostrophic wind and gradient wind only occur if the pressure gradient and the Coriolis force are predominant. This is only the case within the free atmosphere.

Global air circulation systems can therefore only occur there. Below this free atmosphere is the planetary boundary layer that ends at the surface of the earth.

Even within this boundary layer, and thus close to the ground, air currents are generated that are called local winds. Thermal up and down currents, land and sea winds, mountain and valley winds belong to this category. Such air movements are mostly generated in the same way. Ascending air masses are found on areas heating up quickly due to the incident solar radiation, i.e. areas having a low heat capacity (e.g. land) and descending air masses, however, are above the neighbouring areas with a larger heat capacity (e.g. sea). During the day the wind blows from the latter zones to the former zones (e.g. sea wind) and during the night this process is reversed (e.g. land wind) /2-15/.

These local air circulation systems with a different force occur practically everywhere on earth. Due to their closeness to the surface of the earth they can be utilised for wind power generation. Examples of the use of such land and sea winds are some of the large wind parks in California (USA).

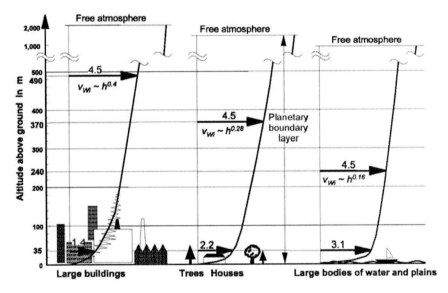

Fig. 2.25 Dependence of wind velocity v_{Wi} on altitude h (see /2-7/)

Due to friction with the (rough) surface of the earth, the geostrophic wind, or the air circulation due to local effects, is reduced within the planetary boundary layer (therefore often also called friction layer) almost to a point of standstill at direct proximity to the ground. The resulting vertical mean wind velocity profile for selected surface conditions is described in Fig. 2.25. The vertical course of the wind, and thus the altitude of the planetary boundary layer depend on the weather conditions, the roughness of the soil and the character of the topography. The thickness of the boundary layer varies between approximately 500 and 2,000 m above ground.

The roughness of the soil is determined by vegetation and land development. Above surfaces with a low level of roughness (e.g. water surfaces) the wind velocity increases very quickly with increasing altitude in the lower 10 % of the planetary boundary layer. Under these conditions the thickness of the planetary boundary layer shows low values. In contrast, above areas with a high level of roughness (e.g. settlements) the wind velocity of the free atmosphere is reached at higher altitudes; the vertical increase of the wind velocity above ground is slower in this case. Thus the roughness of the ground is a measure for the rate of increase of the wind velocity in the vertical direction above ground. It is generally described by the term roughness length. Table 2.2 shows some typical values.

Table 2.2 Exemplary roughness lengths of different surfaces /2-16/

Type of ground cover	Roughness length in cm
Smooth surface	0.002
Snowy surface	0.01 – 0.1
Sandy surface	0.1 – 1.0
Grassland (depending on vegetation)	0.1 –10
Corn fields	5 – 50
Forests and cities	50 – 300

Besides the roughness length, the thermal stratification also has an impact on the vertical change in wind velocity in the planetary boundary layer. If, for example, the vertical temperature decrease is in the range of 0.98 K/100 m, this is called a (dry-)adiabatic temperature gradient. The atmosphere is then layered neutrally. In this case the thermal stratification of the atmosphere has no influence on the vertical wind profile. If, however, the vertical temperature gradient is smaller than the adiabatic gradient, stratification is stable. The wind velocity increases more quickly with increasing altitude above ground under these conditions. In the case of unstable stratification (i.e. a larger vertical temperature gradient compared to the adiabatic case) the wind velocity increase is smaller with increasing altitudes.

The stability of thermal stratification varies with the approaching air mass, but also, within such an air mass, during the course of a day. Above the sea, however, no diurnal course of layer stability worth mentioning can be observed, as the high specific heat capacity of water in connection with the turbulent heat transfer within the water lead to very little change of the water surface temperature during the course of a day. In the course of a year, on the other hand, due to the delayed change in the water surface temperature, in spring a tendency towards stable, and in late autumn a tendency towards unstable conditions can be observed. In comparison, above land areas, the diurnal course of the stratification stability is much more pronounced at times of strong incident solar radiation.

However, at higher wind speeds a neutral stratification can generally be assumed, as deviations from the neutral condition are less pronounced due to the turbulent mixing of atmospheric layers. In the lower velocity segment including the stratification stability into the observation of the vertical wind profile can be useful. For wind mills, this is the segment from the start speed until reaching nominal power.

For a quantitative description of the vertical wind profile of the planetary boundary layer, various approaches have been developed in the past. However, many descriptions of the vertical wind profile are unsuitable for general use due to parameters being too difficult to determine. For engineering application purposes a semi-empirical formula is commonly used.

The Hellmann approach /2-17/ (the so-called Hellmann altitude formula) is a relatively simple approximation; it is defined according to Equation (2.15). $v_{Wi,h}$ is the mean wind velocity at an altitude h and $v_{Wi,ref}$ is the wind speed at a reference altitude h_{ref} (mostly 10 m). α_{Hell} is the altitude wind exponent (Hellmann-Exponent, the roughness exponent) and a function of the roughness length as well as the thermal stability in the planetary boundary layer.

$$v_{Wi,h} = v_{Wi,ref} \left(\frac{h}{h_{ref}} \right)^{\alpha_{Hell}}$$
(2.15)

Table 2.3 shows approximations of α_{Hell} for different surfaces near the coast and for different stratification within the planetary boundary layer.

The exact estimate of the size of the exponent is nevertheless difficult. For long-term observations of the mean wind velocity value to be expected at a certain altitude of the planetary boundary layer, the exponent α_{Hell} primarily is to be seen as a function of the roughness length, as other influences reach equilibrium throughout the course of a year.

Table 2.3 Approximation values for the Hellmann-Exponent dependent on location in coastal regions and stratification stability /2-18/

Stability	Open water surface	Flat, open coast	Cities, villages
Unstable	0.06	0.11	0.27
Neutral	0.10	0.16	0.34
Stable	0.27	0.40	0.60

In spite of the blurring with regard to Equation (2.15), the approximation is still used in practice, as it delivers useful results for conditions that are not too extreme and altitudes that are not too high /2-16/, /2-17/, /2-18/.

Influence of topography. The flow processes within the planetary boundary layer are additionally influenced by the orography, as due to the low level of compressibility of the air, the flow field above the orography is changed. Due to the impact of the surface of the earth, vertical movements of the streaming air masses are generated on both sides of an obstacle. Additionally, the horizontal flow is accelerated on the upwind side and slowed down at the downwind side. Horizontal flow deviations of the air current are also caused /2-20/.

Nevertheless, a closed analytical description of the air flow conditions above all types of area elevations could only be realised with significant difficulties, as the concrete shape of the respective obstacle can hardly be captured exactly in practice. Furthermore, the wind profile is additionally influenced by the initial flow direction, the stratification stability and the roughness of the ground, among others.

Therefore the effect of the change in velocity (Fig. 2.26) caused by orography e.g. above the crest of escarpments, hills or ridges, is often named in relative terms and defined as the Speed-Up-Ratio Δs or in short, the Speed-Up (Equation (2.16)). v_{Wi} is the mean wind velocity and Δh the corresponding altitude above ground. The index x defines the cross-section through the elevation and the index a a point on the upwind side of the hill, where the current is not influenced by it.

$$\Delta s = \frac{v_{Wi,x}(\Delta h) - v_{Wi,a}(\Delta h)}{v_{Wi,a}(\Delta h)} \tag{2.16}$$

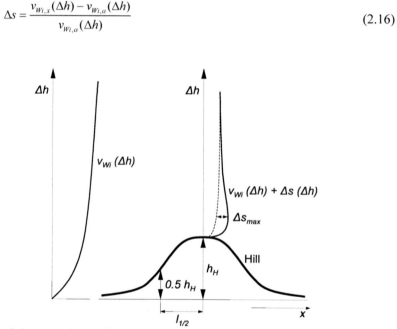

Fig. 2.26 Coherences for a hill overflowed by air (for an explanation of the symbols see text)

For flat two-dimensional chains of hills the approximation $\Delta s = 2\, h_H / l_{1/2}$ resp. $\Delta s = 1.6\, h_H / l_{1/2}$ can be used. h_H is the height of the hill above the surroundings and $l_{1/2}$ the so-called half-value length and thus the horizontal distance between the peak and the half-value height (i.e. half the peak height of the hill). For example, a maximum speed increase of around 60 % or more can result for typical values ($h_H = 100$ m; $l_{1/2} = 250$ m), which then has a significant influence on, for example, the energy yield of a wind mill.

The altitude of the maximal increase in wind speed above the peak for most hills is between 2.5 and 5.0 m /2-10/, /2-14/.

Wind power. Due to the described context, the air masses within the atmosphere are permanently moving. The kinetic energy E_{Wi} of these moving air masses depends on the air mass m_{Wi} and the square value of the wind velocity v_{Wi} (Equation (2.17)).

$$E_{Wi} = \frac{1}{2} m_{Wi} v_{Wi}^2 \qquad (2.17)$$

The air mass flow entering through a particular surface m_{Wi} is defined by the surface S it flows through, the density of the air ρ_{Wi} and the wind velocity v_{Wi} in line with Equation (2.18).

$$m_{Wi} = S \rho_{Wi} \frac{dx}{dt} = S \rho_{Wi} v_{Wi} \qquad (2.18)$$

With the Equations (2.17) and (2.18) the power contained in wind P_{Wi} can thus be calculated (Equation (2.19)). According to that, the wind power is proportional to the cube of the wind velocity; furthermore it depends on the density of the air ρ_{Wi} and the cross-sectional area S of the wind flow in question.

$$P_{Wi} = \frac{1}{2} m_{Wi} v_{Wi}^2 = \frac{1}{2} S \rho_{Wi} v_{Wi}^3 \qquad (2.19)$$

2.3.2 Supply characteristics

Measuring wind direction and wind speed. The direction of wind is measured using wind vanes aligning themselves, under wind pressure, to the respective direction of the wind. The result can be transferred mechanically or electronically to a registration instrument.

Instruments that measure the wind speed (anemometers) are distinguished as either measuring the instantaneous or the mean value. Instruments that measure the instantaneous value of the wind are

- Plate-anemometers, where the wind pressure directs a pendulous plate aligned vertically to the wind direction;
- Dynamic air speed indicators, where either the pitot pressure (i.e. the pressure at the front stagnation point of a body flowing against (Pitot-tube)) or the dynamic pressure (i.e. the difference between the Pitot-pressure and the static surrounding pressure (Prandtl's Pitot tube)) are measured;
- Thermal anemometers, where the temperature of e.g. heating wires changes as a result of the air masses flowing past. This change can easily be measured.

Instruments to measure the mean value are
- Cup anemometers, anemometers which either measure the mean wind velocity over a few seconds (approximately 10 to 30 s) or the wind path, i.e. the product of the mean wind velocity and the time. Cup anemometers are currently the most widely used measuring instruments for 10-minute mean values as well as the 2-second-gust);
- Impeller anemometers, offering principally the same features as the cup anemometers.

Wind distribution. The measuring instruments outlined above are in use worldwide. The measured wind speeds can be analysed and the annual mean value can be calculated. If the yearly mean wind velocity is averaged over various years, areas of similar wind speeds can be identified. Fig. 2.27 shows these values on a worldwide scale referring to 10 m above ground.

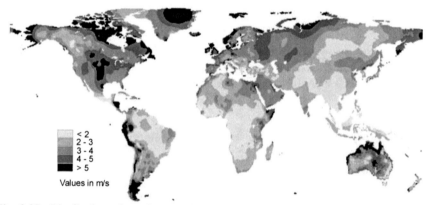

Fig. 2.27 Distribution of yearly annual mean values of the wind speed for 10 m above ground worldwide (data according to /2-11/)

According to that high average wind velocities are given especially in coastal regions; this is for example the case at the north coast of Australia, the Atlantic coast in Patagonia/Argentina, and the west coast of Canada. But also some mountain areas like the west side of the Andes in e.g. Peru or Bolivia as well as some plains (like the Midwest of the USA) are characterised by considerable wind speeds. When interpreting such a wind speed distribution on a worldwide scale, we need to consider that the data bases for some areas are rather poor.

If, for instance, the average mean wind velocity at 50 m above ground measured at different sites throughout Germany is applied to the entire area of Germany we arrive at the conditions outlined in Fig. 2.28. According to this, the North Sea is characterised by an annual mean wind velocity of over 8 m/s far before the coastline. On the East and West Frisian Islands and the mud flats, the long-term mean air-flow speed varies between 7 and 8 m/s and above. At the coast and in the inland, the speed is generally lower. For example mean wind velocities between 6 and 7 m/s are measured at the North Sea coast and the adjacent inland and speed between 5 and 6 m/s are observed for the Baltic coast. Inland, wind velocities of these dimensions only occur at higher altitudes (or the highest elevations of the low mountain ranges and on top of some well-positioned individual mountains). In the remaining area of Germany there are average wind velocities between 4 and 5 m/s in the Northern part, and sometimes even below 3 m/s in the Southern part. In protected river valleys in South Germany, the annual mean velocities can sometimes even be below 2 m/s.

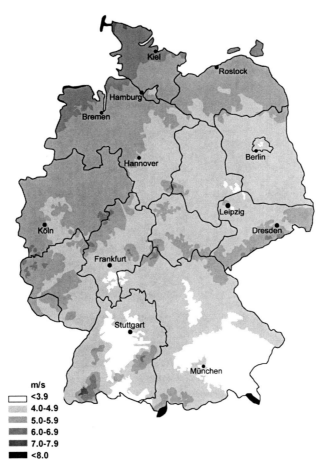

Fig. 2.28 Zones of similar wind velocities at 50 m above ground for the example of Germany (see /2-21/)

The distribution of wind velocity shown in Fig. 2.27 and 2.28 was identified by using comparatively few measurement points. For a small-scale analysis, as it is required for the utilisation of wind power and the technical and economic evaluation of concrete sites, these interdependencies are likely to shift due to local influencing parameters; this is particularly true for low mountain ranges with a very complex topography.

Therefore Fig. 2.29 additionally shows the distribution of the mean annual wind velocity for a small terrain with a complex topography at 50 m above ground. The inserted contour lines clearly delimit the hills and valleys in the area presented in Fig. 2.29. In the given example, the mean annual wind velocity in the valleys has been below 4 m/s due to shading effects. As a contrast, there is free flow against the hilltops; thus in these areas higher wind velocities of above 6 m/s can occur at times. Additionally, the moved air masses are accelerated when overflowing the hill formations. This aspect and the nature of the flow against the hills lead to these comparatively high wind velocities on mountain tops. In contrast, the mean wind velocities for plains are between 4 and 5 m/s.

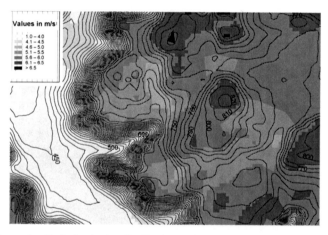

Fig. 2.29 Example of wind velocity distribution at 50 m above ground in a complex area

Due to their inherent indistinctness, such maps showing the regional distribution of the long-term mean annual wind velocity only serve to initially identify areas, and potential wind sites, with a high wind energy supply. For a concrete site evaluation they cannot replace the measurement of the local wind velocities, as these depend to a large extent on the surface roughness of the site, potential obstacles in the local environment, the relief of the area, and the altitude above ground. These parameters are very dependent on local site conditions. However, the wind velocity distribution maps shown exemplarily in Fig. 2.29, enable the identification of areas and possible sites for which more detailed examinations and measurements of the wind conditions make sense.

Time variations. Using the example of a site in North Germany, Fig. 2.30 shows the annual course with monthly and daily mean wind velocities. Additionally two daily courses with mean hourly wind speeds (30th and 300th day of the year) and two hourly course curves on the basis of mean wind speed valued per minute (for the 12th hour respectively) are shown. Accordingly, the wind velocity at this site is characterised by a weak annual course. The mean hourly wind speed values and the velocities per minute, in contrast, are only characterised to a very limited extent by a typical course.

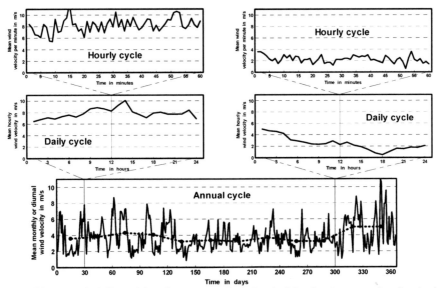

Fig. 2.30 Annual, daily and hourly courses of wind velocities for the example of a site in North Germany (see /2-7/)

The diagram also shows the large variations of the mean daily, hourly and minute wind velocities during the given period of time. If, for example, the variations of the mean minute wind speed were analysed with regard to the mean hourly value, variations between +/- 30 and 40 % would be obtained.

Similar curves will also be found for other sites throughout the globe, but very high variations may occur. For example, the order of magnitude of the measured wind speed might vary according to the local given situation at the measuring spot and the investigated height above ground, among other factors. The wind speed is also very much controlled by micro-meteorological conditions (i.e. stable or unstable weather conditions).

The fact that, for instance, the annual cycle of the wind speed is subject to significant fluctuations is also shown in Fig. 2.31. This graphic displays the monthly average wind speeds at Helgoland Island/Germany, Kuala Lumpur/Malaysia, Wellington/New Zealand, Lanzarote/Canary Islands/Spain and Buenos Aires/Argentina. According to these data all annual cycles of the wind speed displayed show more or less pronounced seasonal differences. The wind velocities on the island of Lanzarote, for example, reach their maximum during summer time and fall off in winter.

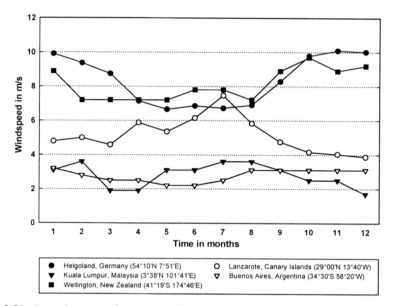

Fig. 2.31 Annual curves of measured wind speeds at different sites throughout the world (data according to various sources)

The mean wind velocities show very large variations over different years. Fig. 2.32 shows the mean annual wind velocities for the example of four sites. On the Feldberg, for example, annual mean wind velocities fluctuated between 6.1 and 8.5 m/s within the period under consideration, showing an average value of 7.2 m/s; variations thus amount to approximately one fifth. Similar fluctuations also occur at other sites.

Additionally, the annual course may vary tremendously over the years due to significantly different meteorological conditions. This becomes apparent when looking at the monthly mean wind velocities outlined in Fig. 2.33 for the same locations represented in Fig. 2.32. Furthermore, the maximum and the minimum values and the standard deviations are shown.

All weather stations shown in Fig. 2.33 are characterised by a particular annual cycle. At these sites during the summer months the wind velocities are below the annual average and are almost independent of local conditions. However, during the course of the winter, above average air flow speeds prevail when looking at the long-term average lifespan. But the respective monthly mean wind velocities can fluctuate significantly. And this observation is almost independent of a specific site.

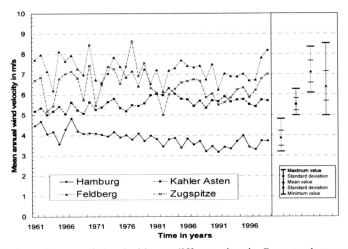

Fig. 2.32 Annual mean wind velocities at different sites in Germany between 1961 and 1998 (data see /2-12/)

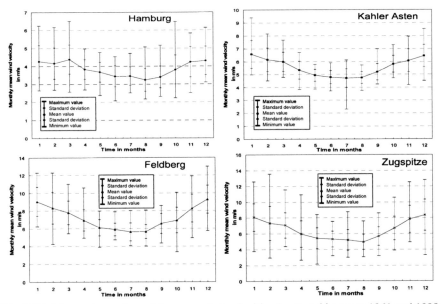

Fig. 2.33 Mean values of monthly mean wind velocities (captured between 1961 and 1998; data see /2-12/)

Often, a site is characterised by a typical mean diurnal course over the year. This daily cycle can sometimes be clearly observed on a particular day. However, most of the time it can only be observed to a very limited extent, or not at all. With the exception of very few sites worldwide, the mean diurnal course is in line with the so-called low-land or ground type (e.g. Norderney/Germany, Fig. 2.34).

During the night hours, until 6 o'clock in the morning, the wind velocity is at its minimum. Thereafter the velocity of the air movement increases slowly. At 9 o'clock in the morning the diurnal mean value is normally reached. In the early afternoon, between 2 and 4 o'clock, the wind speed reaches its maximum and then decreases again. Between 7 and 8 o'clock in the afternoon it again reaches its diurnal mean value and at midnight or afterwards its diurnal minimum.

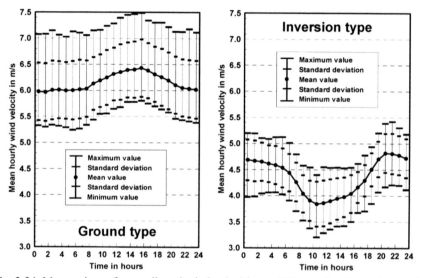

Fig. 2.34 Mean values of mean diurnal wind velocities at different sites (ground type: data from Norderney/Germany; inversion type: data from Hohenpeißenberg/Germany; data according to /2-12/)

On some days, the diurnal course of the wind velocity can deviate significantly from the mean diurnal course given in the yearly average due to changing local weather conditions. However, in the course of one year, stationary weather conditions generally prevail. The diurnal course exemplarily described for the site of Norderney/Germany is equal to the diurnal course at a thermal stratification during robust weather conditions, with a strong mixing of the air layer close to the ground during the day, due to the incident solar radiation and a stable stratification during the night /2-15/.

At exposed mountain top sites (e.g. Hohenpeißenberg/Germany, Fig. 2.34) and less topographically structured areas, between 50 to 100 m above ground, the diurnal course of the wind velocity is reversed compared to the ground type. It is referenced as the inversion type. The velocity maximum is reached during the night, the minimum at mid-day or in the afternoon. The reversal of the mean diurnal course of wind velocity can be explained with the varying thermal stratification between day and night. During the day, with an unstable thermal stratification due to the insulated solar energy and the reflected thermal radiation, the planetary boundary layer expands. This causes the wind to slow down. If layering is stable at night, the top air flow is decoupled from the layer close to the ground and thus often reaches very high wind speeds /2-15/.

In the segment between 50 and 100 m above ground, the mean diurnal course of the wind velocity is the so-called "transition type". At this altitude, a double wave with two velocity maxima occurs at mid-day and at mid-night. Two minima can be observed in the morning and in the evening, but the amplitudes are relatively small in that case /2-15/.

Frequency distribution. In spite of the diurnal and the annual course observed on average, measured wind velocities can vary significantly at different times, in different locations, and at different heights above ground. This is why it is very difficult to compare measured time series. Therefore time series of measured wind velocity, with varying time dissipation, are characterised by their distribution functions. On this basis they can easily and reliably be compared among each other. For that purpose, the measured wind speed values are classified to different wind speed classes. For each class the probability of occurrence of the measured wind speed values, allocated to this class with regard to the overall amount of measured wind velocity values, is calculated. This frequency distribution always shows a typical course.

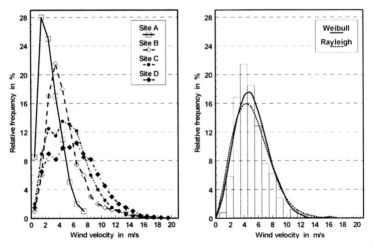

Fig. 2.35 Frequency distribution of wind velocity time series for different sites (left) and a corresponding mathematical approximation for site B (right) (see /2-7/)

Fig. 2.35, left, shows distributions of hourly mean wind velocity values measured at different sites. According to these findings, in the region of the annual mean wind velocity the respective highest occurrence probability is given. If the mean wind speed is relatively low, these probabilities show comparatively high values. They are, however, only limited to a small range of velocities. With an increasing average wind velocity, the absolute height of the maximum occurrence probability decreases; at the same time the frequency distribution is significantly more balanced.

Mathematical approximations to such probability distributions can be made with different functions that can be described by means of only a few parameters. For the distribution of wind velocities either the Weibull or the Rayleigh Distribution can be used (Fig. 2.35, on the right). Today, mainly the Weibull Distribution is used being the more generally defined distribution function. The corresponding density function is defined according to Equation (2.20). k is the so-called shape parameter and A the scaling factor (Table 2.4). v_{Wi} describes the wind velocity.

This results, for example, in the calculated shape and scaling factors compiled for various sites in Germany and shown in Table 2.4. According to this table, the shape factor is generally characterised by smaller values and a decreasing mean annual velocity.

Table 2.4 Shape and scaling factors of the annual mean wind velocity for the example of various sites in Germany /2-22/

Site	Annual mean wind speed in m/s	Scaling factor in m/s	Shape factor
Helgoland	7.2	8.0	2.09
List	7.1	8.0	2.15
Bremen	4.3	4.9	1.85
Brunswick	3.8	4.3	1.83
Saarbrücken	3.4	3.9	1.82
Stuttgart	2.5	2.8	1.24

$$f(v_{W_i}) = \frac{k}{A}\left(\frac{v_{W_i}}{A}\right)^{(k-1)} e^{-\left(\frac{v_{W_i}}{A}\right)^k} \tag{2.20}$$

2.4 Run-of-river and reservoir water supply

Of the total solar energy incident on earth approximately 21 % or 1.2×10^6 EJ/a are used for maintaining the global water cycle of evaporation and precipitation. But only scarcely 0.02 % or 200 EJ/a out of this amount of energy are finally available as kinetic and potential energy stored in the rivers and lakes of the earth /2-23/.

2.4.1 Principles

Water reserves of the earth. The water reserves of the earth available in solid (ice), liquid (water) and gaseous (water vapour) condition have a total volume of scarcely $1.4 \cdot 10^9 \text{ km}^3$ /2-23/.

Table 2.5 shows the global distribution of the different occurrence modifications of water in terms of volume. Accordingly, with 0.001 % water vapour within the atmosphere only accounts for very little of the entire water reserve of the earth. with 2.15 % the proportion of ice is also comparably small. Thus, the overwhelming majority of the global water reserves is liquid water with around 97.8 %, mainly concentrated in the oceans.

Table 2.5 Water reserves on the earth /2-23/

		Volume in 10^3 km^3	Volume proportion in %
Water vapour in the atmosphere (volatile)		ca. 13	ca. 0.001
Water (liquid) in	Rivers and streams	ca. 1	ca. 0.00001
	Fresh-water lakes	ca. 125	ca. 0.009
	Groundwater	ca. 8,300	ca. 0.61
	Oceans	ca. 1,322,000	ca. 97.2
Ice (solid) in polar ice and glaciers		ca. 29,200	ca. 2.15
Total		ca. 1,360,000	100.0

Water cycle. The described water reserve on the earth is continuously cycled by the incident solar energy. This global water cycle is mainly fed by evaporation of water from the oceans and, among other factors, the plants and continental waters (Fig. 2.36). This evaporated water is transferred within the atmosphere as water vapour by global and local wind circulation, and afterwards precipitates as e.g. in the form of rain, snow, soft hail, or dew. Above oceans a little less water rains down than is evaporated. This leads to correspondingly higher precipitation levels on land and results in a net import of water from the oceans to the continents. The resulting precipitation feeds snow fields, glaciers, streams, rivers, lakes and the groundwater.

If the global total precipitation is related to the surface of the earth, a mean annual precipitation of around 972 mm (1 mm equals 1 l/m^2) is obtained. But depending on the local conditions these values may vary significantly. For example, it never rains in certain desert areas for many years (e.g. Sahara Desert, Kalahari Desert). At exposed sites however, for example at mountain ascents, up to 5,000 mm and more rainfall can be observed in the course of one year. On average, on the continents the mean annual precipitation is approximately 745 mm. Of the water volume precipitating for example under Central European conditions approximately 62 % evaporate directly or indirectly; the remaining 38 % flow off as surface or groundwater. But these proportions are subject to change under other conditions.

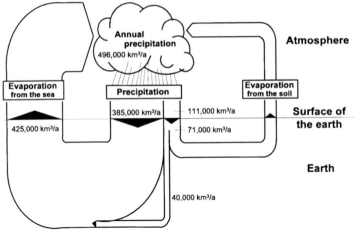

Fig. 2.36 Water cycle of the earth (see /2-23/)

The potential energy that the water reaches in the cloud and rain-forming layers of the atmosphere is only utilisable to a very limited extent. For the part raining down over the oceans, this energy is mainly converted into non-utilisable low temperature heat. Of the proportion raining down on the continents, around 64 % drain into the ground; the resulting potential energy can thus not be entirely used either, as portions of the drained water only surface again at sources located further downhill at a lower elevation.

Therefore, ultimately only the water flowing off the surface is available for potential utilisation, corresponding to approximately 36 % of the total rainfall on continents. Hence, only the potential energy resulting from the respective differences in altitude between the respective precipitation site and the sea level is theoretically useful. Without any technical utilisation measures, the energy stored in waterways or lakes in the form of potential energy would be converted into thermal energy at a temperature level close to ambient temperature by erosion in the riverbed and whirling.

Precipitation. Within the atmosphere of the earth different atmospheric layers contain different steam volumes. This steam is converted into a visible form, if the air temperature sinks below the dew point, and the water molecules bind to condensation nuclei (i.e. fine, floating aerosol particles), forming small water droplets. If the temperatures are above the freezing point, precipitation occurs, if the cloud droplets attach among each other (coalescence) and can no longer be carried by the air current /2-24/.

Liquid precipitation is called rain. It is distinguished between drizzle with a drop radius between 0.05 and 0.25 mm and rain with a drop radius between 0.25 and 2.5 mm. Liquid precipitation can also fall from clouds containing ice crystals (e.g. thunderstorm clouds). On its way from the atmosphere to the surface of the earth, such precipitation is generally transferred from a solid to a liquid phase. If, however, the air temperature in the layers below the cloud is also below the freezing point, solid precipitation can occur. Types of such solid precipitation are, among others, snow, snow grass, frazil ice, soft hail, and hail. Snow can either occur as "ordinary snowfall" or as snow showers, and consists of snow stars or other types of ice crystals which fall down either separately or merged together into flakes /2-24/.

Dew or rime, both forms of precipitation, are caused by condensation or deposition of gaseous water, if the temperature of a surface very close to the ground is below the dew or rime point. Additionally, fog or clouds can form water deposits or, at temperatures below the freezing point, form various types of frost deposits /2-24/.

From precipitation to flow. The water precipitating on an area of land within a defined period of time is stored in the ground, evaporates again or flows off in streams and rivers. With the water resources equation (Equation (2.21)), the total water available in the course of a particular period and for a particular area can be described (Fig. 2.37). P_S is the precipitation on the surface S of a particular area, $F_{a,S}$ is the flow above ground and $F_{b,S}$ the flow below ground from this area during the investigated period of time Δt. E_S is the evaporation from the surface of this area and R_S the retention (i.e. the proportion of the precipitation stored in the ground as groundwater) during the same period of time.

$$P_S = (F_{a,S} + F_{b,S}) + E_S \pm R_S \tag{2.21}$$

Above ground, precipitation could be snow, ice and/or water (i.e. rain), however below ground exclusively water occurs. If the time span Δt is large enough, retention can be neglected as it is sufficiently balanced.

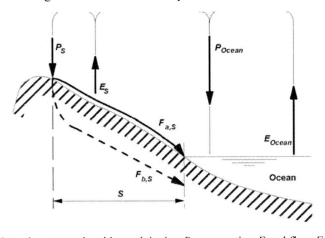

Fig. 2.37 Natural water cycle with precipitation P, evaporation E and flow F below (b) or above (a) ground in an area with the surface S (see /2-23/)

Commonly, precipitation is expressed in terms of the precipitation level (i.e. as the ratio of precipitation volume and surface (P_s/S) in mm). This also applies to the runoff level describing the share of the overall precipitation level that is effectively drained and does not evaporate, or flows off the observed area with the groundwater stream. The so-called flow coefficient, defined as the ratio of flow and precipitation, describes which proportion of the precipitation is finally drained. Together with a number of other parameters, the flow coefficient is particularly dependent on the precipitation and the condition of the analysed area (i.e. vegetation, permeability, topography). However, the flow coefficient generally increases with an increasing precipitation level.

On the basis of these interdependencies, the flow characteristics in a certain area can be described. With the knowledge of precipitation, evaporation and retention, the flow regime can also be explained, at least qualitatively. The flow regime is thus defined as the timely behaviour and the discharge volume of a creek or river throughout the course of one year within a certain area. In order to obtain this value, the catchment area needs to be unambiguously assigned to the corresponding outflowing stream or river. The "inflow" streaming towards a defined point along the course of a stream or a river is thereby equal to the "outflow" of the respective catchment area. It shows time variations that generally fluctuate significantly, in dimension and course, within one year and from year to year.

The flow of a potential catchment area is, among other factors, dependent on the size of the area and the precipitation. It has to be taken into consideration that the watershed (above ground) defining the catchment area can be very different from the watershed below ground in the case of sloping, impermeable layers. This can significantly enlarge or reduce the flow of a certain catchment area.

Precipitation and runoff of a particular catchment area are only indirectly linked, as only parts of the falling rain only flow off immediately. In times of high precipitation the runoff is delayed due to the formation of reserves, whereas in times of low precipitation there is more flow due to these reserves being used. At temperatures below the freezing point, additional delays of the runoff occur due to the water stored in form of snow and ice. Additionally, parts of the precipitation are completely lost through i.e. immediate evaporation, indirect evaporation caused by plant growth, or through increased evaporation due to irrigation measures. Thus, no direct conclusion can be drawn from the amounts of rain to the runoff.

Snowfall is also important for the runoff activity of a particular area, as the water stored in snow only reaches the runoff with a certain delay. The snow blanket is influenced, among other factors, by air temperature, global radiation, wind and topography. Sudden melting of snow – as well as heavy rain exceeding the storage capacity of an area – can lead to flooding. Extreme floods therefore often occur when melting snow and heavy rainfalls occur at the same time (like it is often the case in early spring in some parts of Central Europe).

Power and work capacity of water. Due to gravitation water flows within a stream or a river from a higher geodesic site to a lower geodetic site. At both sites the water is characterised by a particular potential and kinetic energy which is different from each other. In order to identify this energy difference of the outflowing water, in an approximation, a stationary and friction-free flow with incompressibility can be assumed. With these preconditions the hydrodynamic Bernoulli pressure equation can be applied, and written according to Equation (2.22).

$$p + \rho_{Wa}\, g\, h + \tfrac{1}{2}\rho_{Wa}\, v_{Wa}^2 = const.$$

(2.22)

p is the hydrostatic pressure, ρ_{Wa} the water density, g the acceleration of gravity, h the head and v_{Wa} the velocity of the water flow. Equation (2.22) can be converted in a way that the first term expresses the pressure level, the second term the site level and the third term the velocity level (Equation (2.23)).

$$\frac{p}{\rho_{Wa}\, g} + h + \frac{1}{2}\frac{v_{Wa}^2}{g} = const.$$

(2.23)

This Equation (2.23) for example, allows determining the utilisable head h_{util} of a particular section of a stream or river. It is calculated according to Equation (2.24) from the pressure differences, the geodesic difference in height and the different flow velocities of the water. When applying this formula we need to bear in mind that this is an idealised form of analysis that does not consider any actual losses. Under actual conditions therefore, from the utilisable head, also the head losses resulting from friction of the individual water molecules among each other and the surrounding matter, have to be subtracted (see Chapter 7).

$$h_{util} = \frac{p_1 - p_2}{\rho_{Wa}\, g} + \left(h_1 - h_2\right) + \frac{v_{Wa,1}^2 - v_{Wa,2}^2}{2g}$$

(2.24)

As the terms defined by the differences in pressure and in velocity are usually relatively small, the geodesic head between the two water surfaces in a water course (e.g. stream, river) can generally be used as the utilisable head in a first rough estimation. The other elements of Equation (2.24) mainly occur within the hydraulic system of a hydro power station.

Starting from this assumption, the power P_{Wa} resulting from the respective water supply can be calculated using Equation (2.25). q_{Wa} is the volume-related flow rate. According to this formula the product of flow and utilisable head basically determine the power of the water. Large heads can generally be achieved in mountainous areas, whereas in lowland areas mainly the flow assumes high values.

$$P_{Wa} = \rho_{Wa}\, g\, q_{Wa}\, h_{util}$$ (2.25)

By integrating Equation (2.25) over time, the corresponding work capacity of water power is obtained. It is thus calculated on the basis of the water density, the gravitational acceleration, the flow rate related to a defined time period and the utilisable head.

A certain water volume $V_{Reservoir}$ can be held back in times of high precipitation. Also for that type of storage, the water resources equation (Equation (2.21)), including inflow, outflow, and occurring leakage and evaporation losses, applies. Additionally the discussed conditions apply to the power and work that can be provided by the reservoir. The power depends on the time period required for emptying the reservoir. The stored energy E_{Wa} is defined by the size of the reservoir and thus by the stored water volume and the utilisable head. Equation (2.26) applies.

$$E_{Wa} = \rho_{Wa}\, g\, h_{util}\, V_{Reservoir}$$ (2.26)

2.4.2 Supply characteristics

Measuring water-technical parameters. The utilisation options for hydro power can only be reliably estimated on the basis of measured values. Therefore, measuring precipitation, runoff, or flow is of great importance. The individual techniques will be briefly explained in the following (see /2-23/).

Measuring precipitation. Currently, common precipitation measuring instruments represent tiny catchment areas for which the water balance is drawn. These are relatively simple collecting basins designed as a rain gauge (pluviometer) or as a recording rain gauge (pluviograph).

The simplest standard model of a pluviometer measures the precipitation caught within a certain time period. This is done by a cylindrical gauging vessel, where every increase of the level equals a certain precipitation height.

Pluviographs continuously record the retention. This can for example be done by a float gauge inside the collecting basin or by permanent weighing of this basin.

Of the solid precipitation deposit (e.g. snow precipitation), a vertical column with a specific volume is withdrawn from the fallen snow blanket using a so-called snow core cylinder. This amount of snow is melted and subsequently expressed as precipitation level.

Runoff measurement. In order to measure the flow of running waters, currently primarily three fundamentally different methods are applied.

- Measuring the stream velocity. The flow rate is defined as the integral of the stream velocity across the cross-section of the stream. If this cross-section is known, a horizontal arranged propeller driven by the current measures the flow speed. The measured values allow calculating the through-flow. Also magnetic induction gauges or acoustic Doppler can be used for the same purpose. If this is done at different flow levels, the location-specific water level-flow ratio can be derived as well.
- Measuring the water level. If the relation between the water level and the flow (e.g. from measurements, hydraulic calculations or sample experiments) is available, measuring the water level with slat, floating or other level gauges is sufficient. The respective flow can then be derived from the flow curve. Because of its simplicity, this method of measuring the water level is most suitable for the automatic recording of level data and flows over many years.
- Measuring tracer concentration. For this method, salts or dyes are added to the river upstream. Downstream, the respective concentration of the added substances is measured. Assuming that the concentration of the salts or the dye remains almost constant throughout the entire cross-section of the stream or river and that the flow is stationary, the flow can be calculated on the basis of a salt or dye balance.

Flow measurement. Measuring the through-flow in pipelines is mainly done with a built-in gauge. The method is normally much simpler compared to measuring the flow of running waters. The following methods are used.
- The pressure difference before and behind a cross-section change within a pressure pipeline can be measured relatively easily thanks to technical facilities. It correlates directly with a change of the flow speed. If the diameter of the pipe is known, the through-flow can be calculated (Venturi pipe).
- The flow resistance of a moving body can also be measured relatively easily. If a body is included in a vertical pipeline transporting e.g. water, a balance between the static weight and the sum of static and dynamic buoyancy will set up. Thus if the weight of the body is known, the through-flow can be identified. As this measurement method leads to hydraulic losses in the pipeline, it is hardly applied nowadays.
- When moving an electrical conductor vertically to the power lines of a magnetic field, voltage is induced in the conductor that can be measured. This change in voltage is proportional to the movement of the electrical conductor. Ions in the water function as such conductors. If the pipe diameter and the magnetic inductively are known, the through-flow is calculated on the basis of the measured voltage.
- The flow through a pipeline can also be measured, with or against the flow direction, using ultrasound. This is due to the varying sound speeds of water in relation to the pipeline.

– The through-flow of a pipeline can moreover be measured with Hydrometric vanes firmly installed into a pipeline, driving a meter via a mechanical connection (e.g. water meters).

Distribution and variations of precipitation. The instruments for measuring the precipitation (such as the pluviometer) have been in use worldwide since centuries. The measured values serve to calculate the annual precipitation. If the yearly precipitation is averaged over many years, areas of similar precipitation can be identified. Fig. 2.38 shows these values on a worldwide scale.

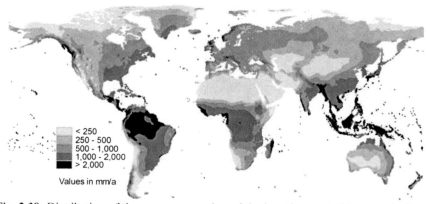

Fig. 2.38 Distribution of the average annual precipitation (time period between 1961 and 1990; some of the data points within the ocean are probably from weather buoys; data according to /2-11/)

According to this map, areas with a high precipitation are primarily located close to the equator. On the other side, the deserts existing on earth are notorious for being regions with very low rainfall.

Such maps of the average annual precipitation can only provide a first rough estimation about the given coherences on a worldwide scale. To analyse the given dependencies in more detail Fig. 2.39 shows the distribution of precipitation exemplarily for Germany. Here, the annual rainfall is mainly influenced by the orography and the increasing continentality of the climate towards the East. Precipitation varies here between approximately 500 mm in the Northern part of the Rhine Valley and over 2,500 mm at the rim of the Alps. There are also very rainy areas in the low mountain ranges.

The mean values of precipitation levels over many years shown in Fig. 2.39 can be subject to large differences throughout one year. This reveals Fig. 2.40 showing the annual precipitation levels at four different sites (Hamburg, Hanover, Stuttgart, Munich; all sites are located in Germany) between 1961 and 1995. The annual precipitation levels in this period over 35 years fluctuated within +/- 25 % in Munich for example and within +/- 50 % in Hanover by the respective mean value.

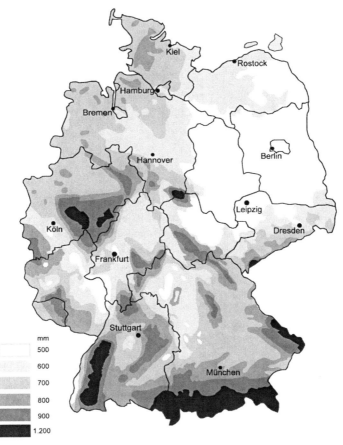

Fig. 2.39 Mean levels of precipitation in Germany (see /2-12/, /2-24/)

The precipitation levels not only vary between different years, considerable fluctuations can often also be observed within one year. Therefore Fig. 2.41 shows the monthly mean precipitation levels for Helgoland/Germany, Abidjan/Ivory Coast, Belém/Brazil, Béchar/Algeria, and Kuala Lumpur/ Malaysia. Firstly, it becomes obvious that the annual cycles are on a very different level of precipitation. This is due to the location of e.g. Béchar within the desert and Belém within the tropics. Secondly, the plot shows clearly different annual cycles. They are influenced by the seasons; e.g. Abidjan is characterised by a rainy season during May/June/July which is responsible for the significant peak within the annual cycle. The same also applies to the other sites but is much less pronounced there.

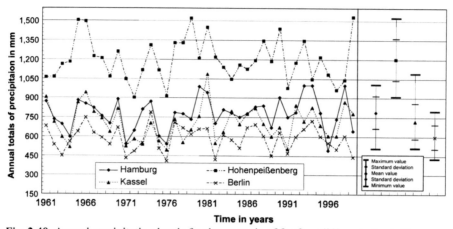

Fig. 2.40 Annual precipitation levels for the example of for four different sites in Germany between 1961 and 1999 (see /2-12/)

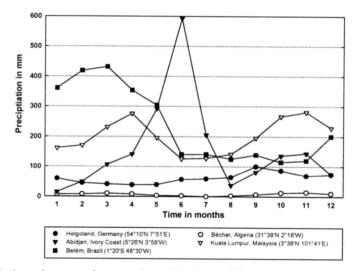

Fig. 2.41 Annual curves of measured precipitation at different sites throughout the world (data according to various sources)

But within a specific year these annual cycles may vary considerably due to specific micro-meteorological conditions. Therefore Fig. 2.42 shows the corresponding standard deviations, for minimum and maximum values at the sites see Fig. 2.40. According to these representations, independent of the location, the monthly mean precipitation levels can be visibly different. This can be observed when looking at the standard deviations, the minimum and the maximum values.

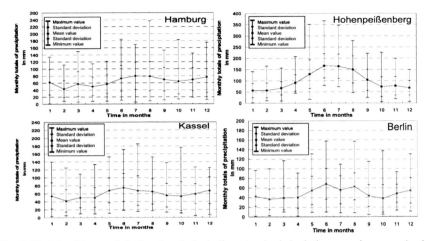

Fig. 2.42 Monthly precipitation levels including standard deviation, maximum and minimum values for the example of four sites in Germany (captured between 1961 and 1999; data see /2-12/)

Due to the very irregular precipitation during the day, stating mean daily curves would not make sense. In an approximation, a stochastic occurrence of precipitation can be assumed for Central Europe. This might be different in tropical areas, for instance.

River systems, runoff and runoff characteristic. The runoff from a specific area is determined more or less pronouncedly by the precipitation outlined above. This flowing off water is drained in streams and rivers. Here, the runoff is concentrated and thus bodies of flowing water are possible sites for using hydro power for energy provision.

Fig. 2.43 shows the main rivers systems with their respective catchment areas for the example of Austria. The map clearly indicates that Austria is well drained by numerous streams and rivers interconnected with each other. Due to the relative high precipitation level, the density of flowing water bodies is relatively high. Also because of the country's mountainous character, the possibilities to use the energy stored within these rivers and lakes are quite promising. This is one of the reasons why Austria is a country with a very high share of electricity generation related to the worldwide electricity generation.

Some of the streams and rivers shown in Fig. 2.43 are partly filled with water that has fallen as precipitation in areas outside of Austria. This share in external water is not part of the energy supply resulting from precipitation over Austria. When developing an overall energy balance such effects have to be taken into consideration.

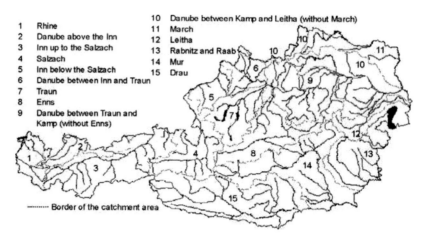

1   Rhine
2   Danube above the Inn
3   Inn up to the Salzach
4   Salzach
5   Inn below the Salzach
6   Danube between Inn and Traun
7   Traun
8   Enns
9   Danube between Traun and
    Kamp (without Enns)

10 Danube between Kamp and Leitha (without March)
11 March
12 Leitha
13 Rabnitz and Raab
14 Mur
15 Drau

-------- Border of the catchment area

Fig. 2.43 River systems in Austria and their catchment areas (according to /2-13/)

At these streams and rivers, the daily mean flows can sometimes vary tremendously. Fig. 2.44 shows the flows of different water levels of the Neckar river located in the South-West of Germany as an example. Particularly the spring is characterised by a high level of average through-flow due to the snowmelt in the Black Forest, which can also vary significantly especially when the snowmelt is coupled with heavy rainfalls. The last part of the year, however, is characterised by low through-flows. The flow peaks are caused by storms with large amounts of precipitation. These times are marked by a fast increase and an ensuing fast decrease in the through-flows.

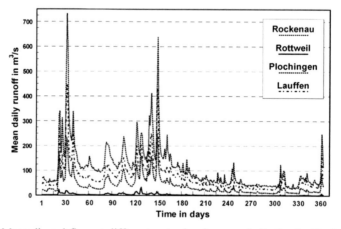

Fig. 2.44 Mean diurnal flow at different water levels exemplarily for the Neckar river for one specific year (see /2-7/)

Looking at the daily or monthly mean runoff values over several years, the run-off is characterised by a seasonal curve depending on the analysed river and the particular stretch of the river. Depending on the river and the meteorological conditions such annual cycles might be visible within one year or on the long-term average only. As an example for such characteristic annual cycles Fig. 2.45 shows mean monthly runoffs at selected gauges of Austrian rivers. According to this the Danube as well as the Gurgler Ache shows significant differences in the runoff between different months. The maximum runoff is given here during summer because of the snow melting on the higher altitudes of the Alps. This is not the case for the Mattig; the peak runoff here occurs during spring because the catchment area of this stream is located in lower mountains where the snow melts earlier in the year compared to the Gurgler Ache.

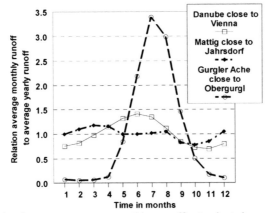

Fig. 2.45 Examples for average mean monthly runoffs at selected gauges in Austria (according to /2-13/)

But the long term average of the annual cycle is subject to significant changes between different years. Therefore Fig. 2.46 shows the mean values of the monthly mean through-flows over many years and the corresponding average and maximum fluctuation range using the example of a water level at Rockenau for the Neckar river and at Maxau for the Upper Rhine. According to this plot the variations that might occur between different years on a monthly average are tremendous. In some months the maximum measured runoff is 10-times higher compared to the average for the Neckar river and some 4-times for the Rhine. Such flood events have often the unwanted effect of flooding villages and cities as well as agricultural land.

Reservoirs. In line with the topographic conditions, the possibilities of storing water vary substantially with regard to different regions.

In the high mountains, conditions are in most cases quite favourable. Therefore, for example in the Alps, there are already quite many reservoirs. The

Forggensee, for example, serves as a head reservoir for the run-of-river power stations at the Lech. One of the oldest high-pressure power stations in Germany for example operates between the Lake Kochel and Lake Walchen. Most of the reservoirs, have several functions and energy generation is not always their main function.

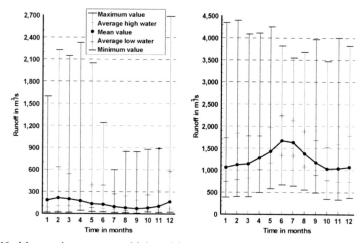

Fig. 2.46 Mean values, average high and low water as well as minimum and maximum fluctuations of the monthly mean runoff at the river levels Rockenau/Neckar (left) and Maxau/Rhine (right) (data see /2-25/)

In the low mountain ranges there might exist reservoir lakes which sometimes mainly generate energy, but sometimes have the main purpose of ensuring the supply of drinking and domestic water. This is mainly the case in those parts where – without a reservoir – supply shortages would mainly occur during the summer due to the low water supply. Often both aspects are interconnected.

Typical examples of water reservoirs for energy generation are the annual reservoirs in the Alps. They are usually empty by the end of winter and are refilled by melting snow in the catchment area and possibly by water transfers from other valleys during the summer months. During this time, they only release little water. At the end of the summer, the reservoirs should be entirely refilled. During the winter months, when energy is high in demand, and there is practically no inflow into the reservoir, the stored water is used. These reservoirs also increase the flow of the downstream rivers.

2.5 Photosynthetically fixed energy

Biomass in the broader sense is all phytomass and zoomass. An estimated total of $1.84 \; 10^{12}$ t of dry mass exists on the continents. The major part of phytomass or plant mass is generated by autotrophic organisms that can generate the energy they need to survive by exploiting solar energy through the process of photosynthesis. Heterotrophic organisms, though, of which zoo mass primarily consists of, need to consume organic substances to produce energy.

Biomass can be divided into primary and secondary products. The former are produced by direct use of solar energy through photosynthesis. In terms of energy supply, these are farm and forestry products from energy crop cultivation (i.e. fast-growing trees, energy grasses) or plant by-products, residues, and waste from farming and forestry including the corresponding downstream industry and private households (i.e. straw, residual and demolition wood, organic components in household and industrial waste). Secondary products are generated by the decomposition or conversion of organic substances in higher organisms (e.g. the digestion system of animals); these are for example liquid manure and sewage sludge.

2.5.1 Principles

Structure and composition of plants. Plants consist of the stem, leaves and roots. The latter anchors the plant in the ground and enables it to withdraw water and nutrients from the soil. The stem carries the leaves, supplies the plant with water and nutrients from the root, and carries the organic substances generated in the leaves to the root. The leaves absorb the sunlight required for photosynthesis. They enable the gas exchange of carbon dioxide (CO_2), oxygen (O_2) and steam (H_2O) for photosynthesis, breathing and transpiration (Fig. 2.47). For the purpose of multiplication, plants generate blossoms from which the reproductive seeds are growing.

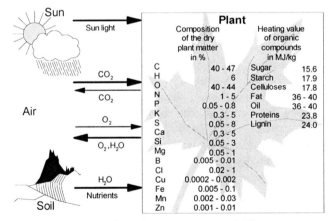

Fig. 2.47 Formation and composition of plant matter (see /2-7/)

Photosynthesis. The most important process for biomass generation is photosynthesis. Using light energy carbon dioxide (CO_2) is absorbed by the process of photosynthesis and carbon (C) is added to the plant substance (assimilation), whereby solar energy is transformed into chemical energy.

Photosynthesis is divided into the processes of light and dark reaction. During the light reaction, the cell produces the energy required for assimilation of carbon dioxide (CO_2) through photo-chemical reactions. Besides oxygen (O_2) the high-energy substances adenosine triphosphate (ATP) and nicotine adenine dinucleotide phosphate (NADPH) as well as hydrogen ions (H^+) are produced. During the process of the dark reaction which takes place without the influence of light, these high-energy substances are used up again for the assimilation of CO_2. The end product of the photosynthesis process is hexoses or sugar ($C_6H_{12}O_6$). The total formula for the entire process is described by Equation (2.27).

$$6\,CO_2 + 6\,H_2O \xrightarrow[Chlorophyll]{Light} C_6H_{12}O_6 + 6\,O_2 \tag{2.27}$$

The conversion of carbon dioxide (CO_2) and water (H_2O) into hexoses takes place in the chloroplasts containing the green, light-absorbing pigment chlorophyll. CO_2 diffuses through the fissure gaps and the cell gaps (intercellular spaces) of the plants to the photosynthetically active cells and binds there to ribulose-1.5-diphosphate. A molecule of 6 carbon atoms is generated which decomposes afterwards. Subsequently, phosphoglycerinaldehyde, a component of carbohydrates (e.g. glucose) is produced consuming ATP and $NADP^+$. These carbohydrates are used by the plants as a source of energy for their metabolic processes as well as elements to generate the plant substance /2-26/.

This photosynthetic path is the so called C_3-type typical for plants growing in temperate climatic zones. The name originates from the first product of photosynthesis, the phosphor-glycerine acid consisting of 3 C-atoms. The so-called CO_2 acceptor of the C_3-plants in charge of binding CO_2 is ribulose-1.5-diphosphate. Some crops, such as maize, sugarcane or miscanthus, mainly originated from subtropical regions, use phosphoenol pyruvic acid as CO_2-acceptor. This substance has a greater affinity to CO_2 than the CO_2-acceptor of the C_3-plants. Firstly, based on this substance compounds of 4 C-atoms are generated. Additionally, for such so-called C_4-plants, the light and dark reaction take place in different types of chloroplasts which are separate from each other. This enables the C_4-plant to bind CO_2 in the chloroplasts of the mesophyll cells at the exterior of the leaves, and to build up a high CO_2 concentration in the chloroplasts of the interior bundled partitions which is advantageous for assimilation.

The efficiency of the photosynthesis process states, what percentage of the radiation hit the plants can be stored in the form of chemical energy through photosynthesis by the plants. In general, the process of photosynthesis utilises 15.9 kJ per gram of assimilated carbohydrate. For individual leaves the utilisation of radiation through photosynthesis can be as high as 15 % under favourable conditions (C_4-grass up to 24 %). In most cases, however, the degree of efficiency is between 5 and 10 % or even lower. With regard to the entire plantation and considering the seasonal and locally changing assimilation conditions, the photosynthetic efficiency of different plant populations fluctuates between 0.04 % in desert areas to up to 1.5 % in rainforests. The efficiency of agricultural crops during their growth period is between 1 and 3 % /2-26/.

CO_2 has a lower energy content compared to organic molecules. This energy difference is used by the plants when breathing, i.e. when decomposing the carbohydrates produced during photosynthesis (dissimilation). The energy is used for metabolic processes, and building up different components of the plant mass, such as proteins, fats and cellulose. Respiration, which increases with an increasing temperature level, leads to a loss in substance. Normally the gain in substance during photosynthesis, which can only take place when light is present, is larger than the loss in substance caused by respiration, which can take place during the day (photorespiration) as well as during the night (dark respiration). The net photosynthesis is the result of the gross photosynthesis minus the respiration losses. For C_3-plants it is up to 30 and for C_4-plants between 50 to 90 mg CO_2 per 100 cm^2 leaf surface and hour /2-27/. One reason for the higher production of organic matter by the C_4-plants is the lower level of photorespiration in comparison to the C_3-plants. This is caused by a more effective binding of CO_2 and the described separation of the different types of chloroplasts. Further energy losses are caused by water evaporation leading to long-wave reflection and heat release in order to keep the temperature at a physiologically acceptable level.

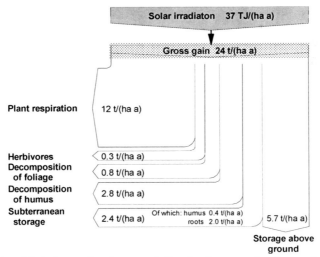

Fig. 2.48 Materials balance of a plant population for the example of a hornbeam forest of temperate climate (see /2-26/)

Fig. 2.48 shows the gain in net biomass of an ecosystem using the example of a hornbeam forest. By utilising 1 % of the incident solar energy, 24 t/(ha a) of biomass (dry matter) are produced. Half of that amount is lost by plant respiration. One part of the remaining biomass is added to the soil as leaves fall and is decomposed by microorganisms. The net storage of biomass per hectare and year is approximately 5.7 t above ground and approximately 2.4 t as roots and humus underground

Influence of various growth factors. The formation of biomass is mainly influenced by irradiation, water, temperature, soil, nutrients, and plant cultivation measures. These parameters are discussed below /2-28/.

Irradiation. Net photosynthesis increases with an increased level of irradiation intensity until it reaches a saturation point. If irradiation is very low, respiration of carbon dioxide exceeds its assimilation. The level of irradiation intensity, at which the respired amount of CO_2 equals the amount assimilated, is called photo compensation point and for most plants it varies between 4 and 12 W/m^2.

Only part of the irradiation incident on a plant is absorbed. The rest is reflected by the plant or penetrates it. The absorption of irradiation in the plant tissue is selective, i.e. it depends on the wavelength. Especially in the infrared range between 0.7 and 1.1 μm a major share of energy penetrates the plant body without being absorbed (Fig. 2.49).

The net incident irradiation is a result of the non-reflected total irradiation and the long-wave reflection. The reflection coefficient represents the ratio of reflected to incident energy. This coefficient mainly depends on the incident angle, the surface texture, and the colour of the plant. For a green plant body, the reflection coefficient is between 0.1 and 0.4.

CO_2-assimilation of individual leaves of different plants increases proportionally to the incident irradiation and the type of photosynthesis. At the same level of irradiation, the assimilation of C_4-plants is higher than that of C_3-plants (see e.g. /2-29/).

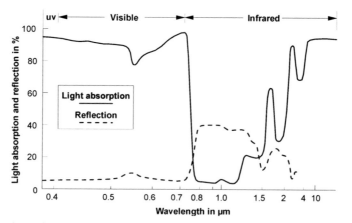

Fig. 2.49 Absorption and reflection spectrum of poplar leaves (see /2-26/, /2-37/)

Water. Green plants consist of approximately 70 to 90 % of water. This water content changes with type and age of the plant organ. It has very important functions within the plant, among others transporting dissolved substances and maintaining hydrostatic pressure keeping the tissue firm. Water is also an important raw material for all metabolic processes such as photosynthesis. Furthermore, most bio-chemical reactions take place in aqueous solution.

Water balance within plants is defined by water absorption, mainly through the root, and the release of water. The latter mainly takes place through transpiration by the leaves. A water deficit occurs if the amount of released water exceeds the absorbed amount. This can happen at a high level of transpiration, low water availability in the ground or an inhibited metabolism in the root. The root absorbs water from the ground by means of the suction power of the root cells. The water absorption capacity stops at the wilting point, where the water content in the ground is so low that the water retaining capacity of the ground exceeds the suction power of the roots.

The plant biomass production is directly linked to their water supply. Each plant requires a specific amount of water to produce organic mass. The transpiration coefficient describes the amount of water required by the plant to produce 1 kg of dry mass. With 220 to 350 l/kg, C_4-plants, such as maize and miscanthus, use water most efficiently and thus show the lowest transpiration coefficients. This is, among other factors, due to their closely arranged photosynthetically active cells and the thus lower transpiration loss. C_3-plants such as wheat and the fast-growing willow require between 500 and 700 l/kg. In general, increased water supply increases the potential biomass productivity of a site (Fig. 2:50, left).

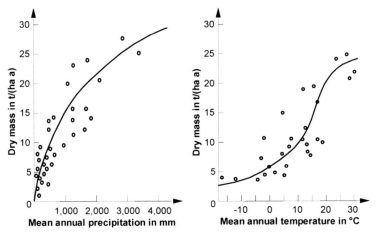

Fig. 2.50 Net productivity of forests depending on the mean annual precipitation (left) and mean annual temperature (right) (see /2-30/)

Temperature. Temperature affects all living processes. This is particularly the case for photosynthesis, respiration and transpiration. In their activity, plants show an optimum temperature range which is typical for the respective species. C_4-plants are thus characterised by a higher temperature optimum (over 30 °C) than C_3-plants (approximately 20 °C).

The bottom margin of the photosynthetic activity, the temperature minimum, is a few degrees below zero for plants in cold and moderate climate zones. With an increase of the annual mean temperature (up to approximately 30 °C) the biomass yield potential of a site also increases if the water supply is sufficient (Fig. 2.50, on the right).

The upper temperature limit for various plants varies between 38 and 60 °C, as proteins are destroyed above this temperature, leading to a reduced enzyme activity and damaging of membranes. This stops the metabolic processes.

Soil and nutrients. Soil is produced by weathering of the earth's crust under the influence of microorganisms (biosphere). It consists of minerals of different kinds and sizes and humus produced from organic substances. Furthermore, it contains water, air and various living organisms. The soil offers the plants space for their roots, anchoring, and the supply with water, nutrients and oxygen.

Growth and development respectively the yield of plants are strongly influenced by the physical, biological, and chemical properties of the soil. Physical attributes are the thickness of the soil, i.e. the depth of the top layer that can be reached by the roots. Other physical properties are the texture or granulation size, the proportion of air-conducting pores and the capacity of the soil to store water and to store and release heat. A sufficiently large root space is important for an optimum plant growth in order to tap nutrients and water. Among the chemical properties are i.e. the nutrient content of the soil and its pH-value. Biological properties of the soil are determined by the presence and activity of microorganisms in the soil. To a large extent these microorganisms live from organic matter released into the ground from dead plants. Microbial activity releases nutrients that are absorbed by plants through their roots.

The non-mineral nutrient elements carbon (C) and oxygen (O_2) are absorbed from the air by the leaves of the plant. In contrast to oxygen which is largely available, carbon dioxide (CO_2), with only 0.03 Vol.-%, is only available in a very low concentration in the air. If irradiation is stronger, the CO_2-supply of the chloroplasts can limit the production rate of a plant body.

The main mineral nutrients nitrogen, phosphorus, potassium, magnesium, and sulphur plus the micronutrients iron, manganese, zinc, copper, molybdenum, chlorine, and boron mainly have to be absorbed from the soil by the plants through their roots.

The larger the root surface, i.e. the better the root system can develop the more nutrients and water can be absorbed by the plant. The rooting decreases with an increasing density of the soil components and the occurrence of compression zones in the soil, which can occur when soil cultivation is performed on wet soil or by too heavy machinery.

Plant cultivation measures. Together with the factors set by the natural location such as temperature of precipitation, an anthropogenic influence on plant growth through plant cultivation measures is possible. These measures can be choosing crops suitable for the individual conditions of the site, the cultivation of the soil, the sowing method, fertilisation, plant husbandry, and harvesting measures. The plant cultivation production technique should support the realisation of the respective yield potential of a plant in a particular location.

The main requirement is the selection of a plant type adapted to the ecological and environmental conditions of the production site. This affects the demands towards the condition of the soil as much as the amount of precipitation and its distribution, the temperature and the temperature distribution curve within the year. Soil cultivation is performed to loosen the soil, to mix remaining crop residues as well as mineral and organic fertilisers (i.e. animal manure) into the soil, to fight weeds, to prepare the soil for sowing and to sow or to plant young plants. Timing and technique of soil cultivation have to be adjusted to the soil condition (e.g. wetness of the soil) and the requirements of the plants.

Crop rotation determines the timely sequence of crops in a field. It is limited by biological factors, as the cultivation of the same or related crops in consecutive years is limited by diseases. Consequently, some crops cannot be produced year after year on the same field. Crop rotation has to be planned in a way that leaves enough time for preparing the soil between harvesting one crop and sowing the next. Therefore crops with early sowing times, such as winter rape and winter barley, cannot be cultivated in the same field that was used to produce late-harvest crops such as maize or sugar beets.

Fertilisation measures are aimed to directly improve the nutrient supply of the plants (e.g. mineral nitrogen fertiliser) and the properties of the soil (e.g. lime treatment or addition of organic substances). The level of fertilising depends on the amount of nutrients withdrawn from the soil by the plants. The strongest influence on the yield is achieved with nitrogen fertilising, as the nitrogen supply in the ground is mostly a limiting factor for yield, and nitrogen mainly supports mass growth. Nitrogen is mainly added to soil in the form of mineral or organic fertiliser, by nitrogen fixation of leguminous plants (i.e. clover), or through rain from the air (i.e. the nitrogen released from anthropogenic sources mainly as NO_x is deposited partly also on agricultural land and acts here as a nitrogen fertilizer). Together with nitrogen, fertilising with phosphorus and potassium is normally also carried out on a regular basis. In addition to its function as a plant nutrient, potassium is also important for the fertility of the soil. It influences the pH-value of the soil and thus its chemical reactions and the availability of various nutrients, and stabilizes the structure of the soil with its bridging role. Apart from magnesium, which can often be found in potassium fertilisers, all further nutrients are in most cases sufficiently available in the soil and are only applied in the event of an obvious deficiency.

Crop protection measures during the vegetation period serve to prevent or fight weeds, diseases and pests. Weeds compete with the crops for growth factors, thus

reducing their growth or supplanting them entirely. This not only leads to a reduced biomass yield in most cases, it also produces lower quality or undesired
properties of the harvested biomass. Diseases and pests living on the photosynthesis products and reserve substances generated by the plants have the same effect.

It depends on the harvesting process and technology which proportion and
which quality of biomass is available for energetic use. The right harvest time and
appropriate technique are particularly important factors for harvesting without
incurring major losses.

2.5.2 Supply characteristics

Spatial supply characteristics. The area-related supply characteristics of biomass
are defined by the combination of soil quality, precipitation level and distribution
plus the annual temperature curve. While precipitation and temperature only vary
to a limited extent across larger areas, the soil quality differs in very small spatial
dimensions. Therefore the biomass yields are characterised by tremendous variations throughout the world.

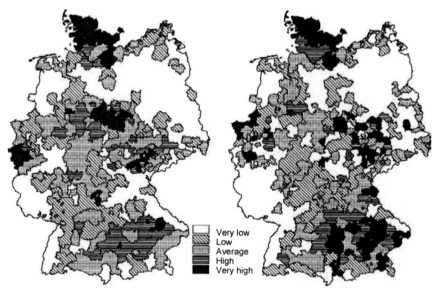

Fig. 2.51 Yield of winter wheat (left) and winter rape (right) exemplarily for Germany

Areas with a high level of biomass productivity are mostly characterised by
high-quality soils and sufficient precipitation amounts. In this respect it has to be
considered that different crops have varying requirements towards soil, temperature and precipitation conditions.

As an example Fig. 2.51 shows the yield level of winter wheat and rape exemplarily in the municipal and regional districts of Germany. If sandy soils coincide

with a low level of water-holding capacity and a low level of precipitation, like for example in Brandenburg (i.e. the mid eastern part of the country), the biomass production potential is low. Areas with a higher level of biomass productivity for winter wheat cultivation are mostly characterised by high-quality soils which occur in areas with loessic soil, or in regions with marshy soil. The far northern part of Germany, Schleswig-Holstein, is for example particularly favoured by soil quality and balanced precipitation. Centres of high soil quality and biomass productivity can also be found for example in the centre of Germany, at the far western side (i.e. Cologne Bay), and in large areas of Central Bavaria (i.e. the south-eastern part of Germany).

Temporal supply characteristics. The increase in biomass is characterised by a daily and annual cycle.

As the process of photosynthesis is dependent on insulated solar energy, the diurnal rhythm of the photosynthesis is determined by the daily course of solar irradiation (Fig. 2.52). The photosynthetic activity increases with increasing incident irradiation, and reaches its peak when the sun is at its highest level at midday and decreases again towards the evening. Reduced irradiation caused by cloudiness reduces photosynthetic activity.

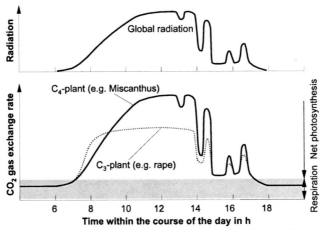

Fig. 2.52 Schematic daily cycle of CO_2-gas exchange (bottom) depending of the insulated solar irradiation (top) (see /2-31/)

The annual cycle of biomass production is determined by the course of the ambient temperature and the length of the day. The lower temperature limit of an increase in biomass by most crops in Central Europe, for instance, is at a daily mean temperature of 5 °C; this might be different in other parts of the world (e.g. tropical areas). In general, the photosynthetic activity and biomass accumulation increase with increasingly long days and higher temperatures, and in Central Europe reach peak levels during the months from May to August (Fig. 2.53).

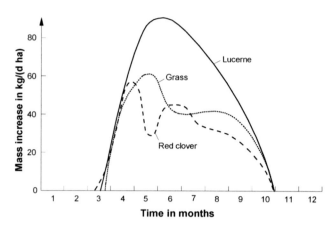

Fig. 2.53 Biomass growth increment of various forage crops in the course of the year exemplarily for sites in Central Europe (see /2-32/)

2.6 Geothermal energy

Apart from the energy resulting from solar energy and the interaction of planet gravitation and planet motion, the heat stored in the earth is another renewable energy source available to human mankind. In the following chapter the principles of this type of energy supply are described and discussed.

2.6.1 Principles

Structure of the earth. Earthquakes cause sound waves occurring as compressions of matter (compression waves) or as movements perpendicular to the direction of propagation (so-called shear waves). They can be measured with receivers (seismometers) distributed throughout the overall globe. By tracing and analysing these sound waves, a layered structure of the earth can be determined.

Table 2.6 Physical properties in the interior of the earth

	Depth in km	Density in kg/dm³	Temperature in °C
Earth's crust	0 - 30	2 - 3	up to 1,000
Earth's mantle	up to 3,000	3 - 5.5	1,000 - 3,000
Earth's core	up to 6,370	10 - 13	3,000 - 5,000

The crust of the earth, or the top layer, reaches below the continents to a depth of approximately 30 km; below the oceans the earth crust only has an average thickness of around 10 km (Table 2.6). The Mohorovicic discontinuity separates the earth's crust from the mantle. By going from the crust to the mantle, the velocity of seismic compression waves increases. The earth's mantle is solid, and reaches down to a depth of up to approximately 3,000 km. It surrounds the core of the earth, which is assumed to be liquid, at least in its external part (approximately 3,000 to 5,100 km). The core shows no shear waves propagation (Fig. 2.54).

The upper crust of the earth up to a depth of approximately 20 km mainly consists of granite types of rocks (approximately 70 % SiO_2, approximately 15 % Al_2O_3 and approximately 8 % K_2O/Na_2O). The lower crust primarily consists of basaltic rocks (approximately 50 % SiO_2, approximately 18 % Al_2O_3, approximately 17 % $FeO/Fe_2O_3/MgO$, and approximately 11 % CaO). The mantle below mainly consists of peridotite with the mineral olivine. It is assumed that the earth' core consists of iron and nickel. The assumptions about the structure of the deep earth are – among other studies – based on spectral analyses of extraterrestrial bodies, composition of volcanic plutonic rocks and modelling of geophysical measurements.

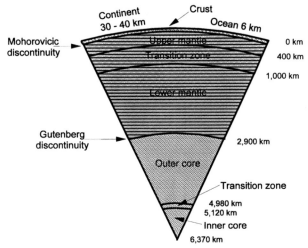

Fig. 2.54 Basic structure of the earth (see /2-4/)

According to the theory of plate tectonics developed over the last decades, the outer shell of the earth's crust consists of mainly solid plates (i.e. of the lithosphere) which are approximately 30 km thick, floating on a "softer" lower level (i.e. the asthenosphere); see Fig. 2.55. This theory of plate tectonics is in line with the layered structure of the earth discussed here. But in addition, it provides an explanation for the dynamic processes such as the formation of new earth crust, the subduction of plates or the volcanism at the boundary of lithosphere plates.

Temperature gradient. The temperature gradient within the outer earth crust which has been measured in deep wells, on average is 30 K/km (Fig. 2.56). In geological old continental crustal shields (e.g. Canada, India, South Africa) lower temperature gradients can be observed (e.g. 10 K/km). In contrast, much higher gradients are measured in tectonically active, young crust areas, e.g. at the boundaries of lithosphere plates (Fig. 2.55; e.g. in Iceland, in Larderello in Italy (approximately 200 K/km)) or in rift regions (e.g. up to 100 K/km in the Rhine rift) (see /2-4/).

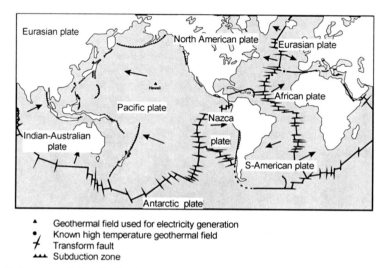

- ▲   Geothermal field used for electricity generation
- •   Known high temperature geothermal field
- ✝   Transform fault
- ▲▲▲ Subduction zone

Fig. 2.55 Important lithospheric plates on earth (see /2-4/)

The temperature gradient in the earth's mantle can be estimated from its geophysical properties. The temperature has to be below the melting point of the mantle's siliceous rocks, even taking into consideration the pressure dependence of the melting temperatures. The maximum temperature gradient in the earth's mantle is therefore estimated to be in the order of 1 K/km.

Heat content and distribution of sources. The temperature profile in the earth's mantle is limited by the melting temperature for iron and nickel in the core of the earth. According to that estimate, temperatures of around 1,000 °C predominate in the upper mantle. For the earth's interior, maximum temperatures of 3,000 to 5,000 °C can be assumed (Table 2.6).

Assuming a mean specific heat of 1 kJ/(kg K) and a mean density of the earth of around 5.5 kg/dm^3, the heat content of the earth can be estimated to be at about 12 to 24 10^{30} J. The heat content of the outer earth crust up to a depth of 10,000 m is approximately 10^{26} J.

The heat stored within the earth results firstly from the gravitational energy at the time of the formation of the earth approximately 4.5 billion years ago due to contractions of gas, dust and rocks. Secondly there is a probably still existing primordial heat. Thirdly the heat existing within our planet has been produced partly during the decay of radioactive isotopes since the formation of the primordial earth. According to current perceptions, the heat-generating isotopes U^{238} and U^{235} of uranium, Th^{232} of thorium and K^{40} of potassium are enriched within the continental crust consisting mainly of granite and basaltic rock. In granite rock, the radiogenic heat production rate is approximately 2.5 $\mu W/m^3$ and in basaltic rock approximately 0.5 $\mu W/m^3$. In addition, heat is also discharged to a small extent by chemical processes in the earth.

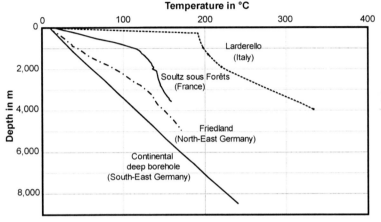

Fig. 2.56 Mean temperature increase with increasing depth (see /2-4/)

Terrestrial heat flow density. The terrestrial heat is conducted within the earth via solid rock (so-called conductive proportion ($\dot{q}_{conductive}$) of the entire heat flow $\dot{q}$) and by being transported in and with liquids (so-called convective proportion of the entire heat flow ($\dot{q}_{convective}$)). This heat flow or heat flow density is defined as an amount of heat flowing through a unit of area within a defined unit of time. The terrestrial heat flow density $\dot{q}_t$, i.e. the heat flow per area in the range of depth of the layer-thickness Δz, consists of these two components and the heat production H, summed up along the range of depth. According to Fourier's heat conductivity equation, the conductive proportion of the heat flow density $\dot{q}_{conductive}$, which normally dominates in the continental crust, results from the product of the temperature gradient $\Delta\theta/\Delta z$ and the thermal conductivity λ of rocks in the upper crust. It can be described according to Equation (2.28).

$$\dot{q}_{conductive} = \lambda \frac{\Delta\theta}{\Delta z} \tag{2.28}$$

The heat conductivity of the rocks in the upper crust varies between 0.5 and 7 W/(m K). This relatively large range of values is mainly caused by the variation of the chemical-mineralogical structure and in the textural differences (e.g. degree of grain contacts, porosity) of the rocks. This results in a mean heat flow density of 65 mW/m² at the earth's surface for the continental earth crust.

The radiogenic heat generation $\dot{q}_{rad}$ (e.g. the proportion of radiogenic heat production in the entire heat flow density) of an area at a particular depth results from the layer thickness Δz and the heat production H according to Equation (2.29).

$$\dot{q}_{rad} = H \, \Delta z \qquad\qquad (2.29)$$

Radiogenic heat production has a minor role in the Earth's mantle due to the lack of radioactive isotopes. By Equation (2.29) it can be estimated that the heat flow density resulting from radioactive decay is approximately 35 mW/m² at a depth of ca. 30 km, where the boundary between the crust and the mantle is located. This represents a mean heat production rate of 1 µW/m³ and thus, a rather common value for the earth's crust. The main part of geothermal heat supplied at the earth's surface is therefore generated in the earth crust during the decomposition of radioactive elements.

The oceanic earth crust mainly consists of basaltic rock with low level of heat generation. Nevertheless the average heat flow density is around 65 mW/m². Heat from the convective surge of hot rock masses from the earth's coat at the rims of the lithosphere play an important role for the heat flow here, as above-average temperatures can be reached due to this process.

Heat balance at the surface of the earth. The heat flux density of 65 mW/m² results in a radiation power of the earth of approximately 33 10¹² W; thus the earth provides energy of 1,000 EJ per year to the atmosphere. In contrast, solar radiation incident on the surface of the earth is around 20,000 times more than the energy released by the terrestrial heat flow. The released and the absorbed heat radiation define the observed temperature balance of approximately 14 °C at the surface of the earth.

This can also be concluded from Fig. 2.57. It has to be assumed that the temperature very close to the surface of the earth (i.e. to a depth of some meters) is dominated by the heat energy radiated from the sun. Among other things, this reveals also the fact that the soil is frozen up to depths of several meters at some places and is heated up to sometimes very high temperatures during the summer on the other side (sometimes up to 50 °C and more at a correspondingly high solar radiation incident). This is exclusively caused by the seasonal differences of the solar radiation supply and the resulting temperature level in the atmospheric layers close to the surface of the earth. Solar radiation influences the temperature course within the earth up to a depth of 10 to 20 m (annual course).

In the top layer of the earth's crust, the proportion of the entire geothermal flow, resulting from the geothermal heat and the incident solar energy, is influenced by a number of very different effects. One main influencing parameter is rain. The resulting surface and groundwater is "heated up" by solar energy, and transports the incident solar energy to shallow layers of the earth. Heated surface waters can therefore locally influence the temperatures in the near-surface layers of the ground up to a depth of around 20 m, in exceptional cases at even deeper levels.

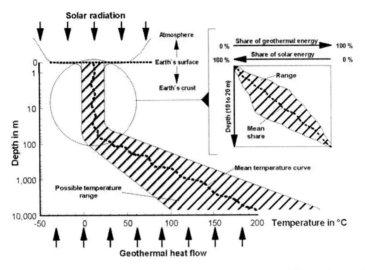

Fig. 2.57 Thermal flows and temperature course in the top layers of the earth crust /2-7/

Geothermal systems and resources. The natural surface heat flow density of approximately 65 mW/m² cannot be used directly at the surface of the earth due to economic reasons. On the other hand, hot water in depths of sometimes only a few hundred meters, with increasing temperatures up to depths of several thousand of meters, hot wells at the surface of the earth, and the well-known volcanic areas demonstrate the existence of such heat, and thus the geothermal energy potential.

Usually several essentially different types of geothermal heat reservoirs can be distinguished.
- Heat in the shallow layers of the earth. These heat reservoirs stretch from the surface up to a few hundred meters below the earth's surface and show temperatures of up to about 20 °C. This geothermal heat is mainly influenced by solar radiation up to a depth of around 10 to 20 m, by the heat conductivity of the soil, and by circulating groundwater heated by solar energy. Although this kind of energy is strongly influenced by solar energy, it is referred to as geothermal energy by definition.
- Hydrothermal low-pressure reservoirs. These reservoirs are further subdivided into warm (i.e. temperatures of up to 100 °C) and hot water (i.e. temperatures above 100 °C) reservoirs, wet steam reservoirs and hot- or dry steam reservoirs characterised by temperatures above 150 °C. These reservoirs are based on heat in water or steam bearing rocks. In Central Europe for example such deposits can be found at depths of up to approximately 3,000 m, with temperatures varying between 60 to 120 °C. At present, some of these hydrothermal energy

reservoirs are already in use. Very high temperatures (150 to 250 °C) of the depth sections of up to 3,000 m only occur in areas with specific tectonic situations, for example in the fracture zones of the earth's crust, where magma rocks are uplifted from the deep underground.

– Hydrothermal high-pressure reservoirs. Such deposits contain hot water which is preloaded and mixed with gas (such as methane) (e.g. in the South of the USA, in the region of the Gulf Coasts of Texas and Louisiana). They are formed by porous isolated rock units which are subducted within a short time into the earth by tectonic processes. The pore waters and gas contents are then exposed to the prevailing pressure and temperature conditions of those depths.

– Hot dry rocks. Really dry rock layers are the exception at earth's crust in depths of up to approximately 10 km to be reached with the currently available drilling techniques. Therefore, also rock formations that do not have enough naturally available water to enable thermal water loops over a longer period of time (several years) are called "dry". Thus, a large spectrum of rock layers with different levels of permeability and percentages of water are summed up under the expression of "hot dry rocks". Such occurrences, however, contain by far the largest potential of geothermal energy accessible with the currently available technology.

– Magma deposits. Close to tectonically active zones, melted rocks (so-called magmas) of temperatures of over 700 °C can be found which often have a lower density compared to the still solid rocks surrounding them. These partially liquefied materials have been uplifted due to their low density from larger depths to depths of 3 to 10 km. Fluid systems with high temperatures can generally be found around such magma chambers. They can be used to supply high-quality geothermal energy. However, developing such systems is a technological challenge.

The use of such geothermal reservoirs largely depends on the energy content and thus on the temperature.

Above 130 °C geothermal reservoirs can be used for electricity generation. Basically the provision of electrical energy is also possible at temperatures below this temperature range; the "coldest" geothermal power plant of the world located in Neustadt-Glewe/Germany operates at a temperature level of approximately 98 °C; but temperatures above 130 °C are necessary to get sufficient efficiency to convert the earth's heat to electrical power. Geothermal energy of temperatures above 150 °C is already utilised in several locations (i.e. Italy, New Zealand). Usually, geothermal power production is independent of the time of the day or the season and the local weather conditions. The reservoirs can be used economically for base load energy provision and in most cases in an environmentally friendly way. However, the appropriate geological requirements have to be met. This is only the case to a limited extent worldwide.

Even at temperatures below 130 °C geothermal heat can be utilised in many ways. Typical examples are outlined as follows:

- Heating stations to supply district heat to private households (i.e. for heating and domestic hot water), small consumers (i.e. heating greenhouses and fish ponds) and industry (i.e. drying wood, heating up dip pools).
- Earth-coupled heat pumps, i.e. to heat single and multi-family homes or for industrial air-conditioning.
- Underground-touched building components with heat exchanger systems for air conditioning (i.e. heating, cooling).
- Material use, i.e. for bathing and curative purposes.

2.6.2 Supply characteristics

Shallow underground. In the shallow layers of the earth located close to the earth's surface, the temperature regime is mainly determined by solar radiation and reflection, precipitation, groundwater movement, and the heat transfer within the underground. The geothermal heat flow in these layers only has negligible influence. As the temperature regime is subject to very high seasonal fluctuations, the temperatures correspondingly fluctuate at these low depths below the surface of the earth.

Within the shallow underground of the earth's crust the different shares in the entire geothermal heat flow resulting from geothermal heat and the incident solar energy are influenced by a number of different effects. For example, precipitation, in the same way as the soil, is affected by the ambient temperature of the environment. Part of the rain water seeps into the underground. This type of heat transport is convective, and takes place through a heat transfer medium, in this case water. Water can have widely varying temperatures when it seeps into the soil. The quicker it reaches the groundwater and the more water seeps into the subsoil, the less the heat condition of the entering water is generally changed, and the more it can have a warming or cooling effect on the groundwater. This is mainly the case for very permeable cover layers and groundwater aquifers. The situation is different if the water remains in the subsoil for a long time before it reaches the groundwater. Under these circumstances the water temperature can then largely be adjusted to that of the surrounding rocks. If the water enters into loose rocks (e.g. sands) the contact area is very large, and thus heat exchange is favoured.

The seasonal influence of the solar radiation can be observed up to a depth of 10 to 20 m below the earth's surface. (Fig. 2.58 and 2.59) /2-4/. These seasonal temperature variations up to depths of roughly 20 m below ground results from the seasonally varying solar radiation supply. Temperature fluctuations in the air layers near to the earth's surface have no direct effect on the temperature of the near-surface soil due to the energy storage capacity of the soil. This has the consequence that the responding temperature profile follows the seasonally changing mean air temperatures with a certain time lag. In 10 to 20 m below the surface the temperature is constant at the annual mean temperature of approximately 9 to

10 °C. (Fig. 2.58). In the top centimetres of the Earth' surface temperature variations occur in the diurnal course due to solar radiation. These variations decrease
with increasing depth and are practically of no importance for the energetic utilisation of shallow geothermal energy.

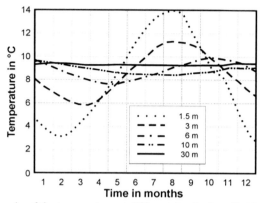

Fig. 2.58 Annual cycle of the temperature at different depths (see /2-4/)

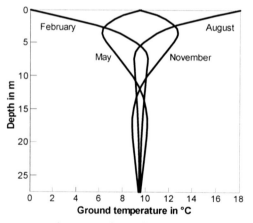

Fig. 2.59 Ground temperature in near-surface soil (see /2-33/)

The depth where no seasonal temperature fluctuation occurs is referred to as
neutral zone. According to DIN 4049 this is the section below the surface of the
earth where the annual temperature course does not fluctuate by more than 0.1 K.
The temperature fluctuations decrease with depth and they are strongly influenced
by the thermal conductivity of rocks and the groundwater flow. This neutral zone
can be identified at depths of about 20 m. Below that, the temperature is
dominantly influenced by the geothermal heat flow. In an approximation, the
temperature within the neutral zone equals the mean long-term (over many years)
annual temperature at the surface of the earth in the respective region (i.e. 9 to

perature at the surface of the earth in the respective region (i.e. 9 to 10 °C in Central Europe) /2-4/.

Deep underground. Different heat transfer processes lead to varying temperatures at the same depth. The local thermal temperature gradient can thus deviate significantly from the regional or the global mean value.

The geothermal heat flow gradient can be influenced by volcanic and tectonic activities. Fig. 2.60 gives a schematic view over the most important sources of geothermal heat with respect to the dynamic behaviour of the earth. High potential of geothermal energy is given at the middle oceanic ridges. Here, the oceanic crust is relatively thin and hot mantle rock is uplifted close to the earth's surface. This is, for example, the case in Iceland where geothermal energy already contributes substantially to the energy demand. Fig. 2.60 also shows that high potentials for using geothermal power are also given where the oceanic crust moves below the continental crust. At the border between the two plates, hot mantle rock might move upwards and become visible in the form of volcanoes.

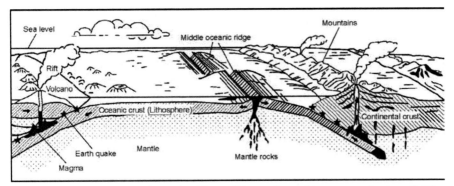

Fig. 2.60 View of the dynamic earth with possible geothermal heat reservoirs (see /4-42)

Typical examples for this type of geothermal activity are the numerous volcanoes located along the Andes in South America. Even when it is difficult to tap the energy of volcanoes directly in geological periods post volcanic areas with relatively high temperatures in shallow depth are potential areas for the use of geothermal energy. The Geysers field in California/USA and the geothermal fields in New Zealand are examples for this type of geothermal resources.

Also, other areas like the continental crust (e.g. the Eurasian plate; see Fig. 2.55) show considerable heat flows. Therefore Fig. 2.61 illustrates the heat flow density in parts of Europe according to the current status of knowledge. According to this map areas are visible which are characterised by a relatively high heat flow density. This is for example valid for the Upper Rhine Valley and for large areas south of Paris. Very high heat flow densities are given in Tuscany; this is the area where the first geothermal power was installed and brought to operation.

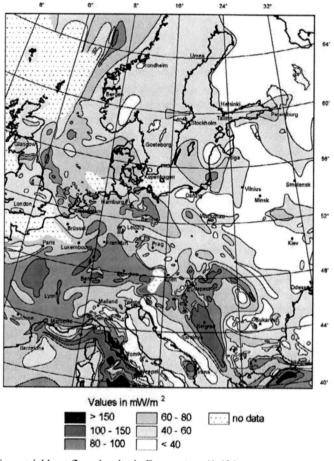

Values in mW/m ^{2}

▉ > 150	▨ 60 - 80	⬚ no data
▨ 100 - 150	▨ 40 - 60	
▨ 80 - 100	☐ < 40	

Fig. 2.61 Terrestrial heat flow density in Europe (see /4-42/)

Temperature maps can be developed based on such maps of the heat flow density and using measured values from deep boreholes. As an example, Fig. 2.62 shows temperatures contours exemplarily for Germany at a depth of 2,000 m. The high temperatures in the Upper Rhine Valley are noticeable; they are influenced by rift-wide groundwater circulations. Further on, high temperatures are given in the area of the Swabian Mountains, south of Stuttgart near Bad Urach and in the North German-Polish Basin; saline structures support such temperature anomalies.

In order to exploit this energy technically, the respective temperature level is nothing more than a parameter. It is also decisive whether the heat is exclusively stored in rocks, or whether it is possibly also a part of fluids existing naturally in rock pores which can be extracted (i.e. carrier rocks with a high level of porosity and permeability). Usually there are wet rocks in depths. In order to directly exploit the energy, the availability of aquifers, carrying enough water and located not too deep in the ground, is the main requirement. These aquifers can be found in the large sediment structures existing at many locations worldwide. Fig. 2.63 shows exemplarily the existing sediment basins. The most interesting area, in terms of its hydrothermal conditions, is the Molasse Basin between the Danube and the Alps, where the water of the Malm (i.e. a geological layer with limestone rocks in the Upper Jurassic), can be used.

Fig. 2.62 Temperature distribution at a depth below NN of 2,000 m within Germany (see /2-34/, /2-35/, /2-36/)

The geothermal heat flow is independent of daily or seasonal influences. Thus, there is no variation in the energy supply in human time dimensions. Only in geological timeframes the geothermal heat flow might change. A slow cooling of the heat stored in the interior of the earth is certainly the most lengthy process. Within shorter periods – still periods of maybe several million of years – the migration of locally confined heat centres within the earth's crust can, for example, lead to a change in the location of the reservoir through tectonic processes. Also, the processes associated with the rising salts in the sediment basin (i.e. the growing of salt domes) also play a certain role. However, all these slow time-related changes in the heat content have no impact on energetic utilisation.

Fig. 2.63 Areas in Germany with hydrothermal and potential hydrothermal energy resources (see /2-4/)

3 Utilisation of Passive Solar Energy

The term "Utilisation of Passive Solar Energy" was first introduced in the seventies of the last century. At that time, the criterion of "adding auxiliary energy" was used to clearly distinguish between active solar energy applications. When auxiliary units (such as fans) were used, the systems were referred to as hybrid systems. However, the delimitation between passive and active systems remained fluid: for a window equipped with automatic shading devices is both passive and hybrid. Only recently, the term "Passive Solar Energy Utilisation" has been defined in a more realistic and precise manner. According to the new definitions passive solar systems convert solar radiation into heat by means of the building structure itself, i.e. by the transparent building envelope and solid storage elements. Utilisation of passive solar energy (often also referred to as passive solar architecture) is thus characterised by the use of the building envelope as absorber and the building structure as heat store. In most cases, solar energy is transferred without any intermediate heat transfer devices. However, also this definition does not always allow for a clear differentiation of active and passive solar utilisation.

3.1 Principles

In a building, several energy flows can be observed (see Fig. 3.1). Energy is primarily supplied by means of space heating systems, secondly, heat is created by people, lighting and household appliances (so-called internal heat gain), and thirdly, there are passive solar heat gains, such as heat created by transparent surfaces (so-called passive solar energy utilisation). Heat losses or heat gains (depending on the ambient temperature) are due to the heat conductivity of the building envelope (i.e. transmission). Further heat losses are attributable to ventilation and infiltration, generally required to maintain a certain air quality and to prevent the system from exceeding the prescribed levels of carbon dioxide (CO_2) and other harmful substances, air humidity and certain odours. Within the building, additional energy can either be absorbed or re-radiated by the available thermal mass in the form of absorbed solar radiation. Thermal mass is also capable of absorbing and intermediately storing heat in case of overheating. Heat is only released if the thermal mass becomes warmer than the room temperature. The following explanations focus entirely on the utilisation of passive solar energy, and thus only apply to one of the energy flows in a building.

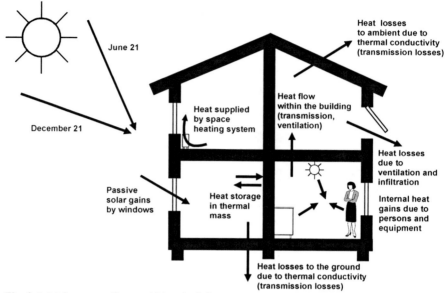

Fig. 3.1 Main energy flows within a building

Passive solar energy utilisation is based on the absorption of short wave solar radiation, either by the building interior, as solar radiation penetrates through the transparent external structural elements, or by the building envelope. The concerned structural elements are warmed up by the absorbed solar energy. The energy is released back to the exterior by convection and long wave radiation. The quantity of absorbed solar energy of surfaces, exposed to radiation, depends on their orientation, shading equipment and the absorption coefficient of the concerned absorber surface (see Fig. 2.20 and Fig. 2.21). The quantity and timing of released energy are determined by thermal conductivity, density, and the specific thermal capacity of the absorbing material itself and the material placed behind, as well as by the difference to surrounding temperature. The seasonal effects of passive solar energy utilisation can be further intensified by an appropriate orientation of the concerned surfaces or shading devices (see Fig. 2.21 and Fig. 3.3).

3.2 Technical description

In the following chapters all system elements involved in the utilisation of passive solar energy are described and illustrated. However, explanations are limited to a selection of the current main aspects. First, a compilation of the fundamental technical terms is provided.

3.2.1 Definitions

Terms. The translucence of walls is often described by the terms opaque, transparent, and translucent, as well as by solar aperture surface.

The opaque building envelope is not permeated by light and includes, for instance, brick walls, or a roof covered with tiles.

Transparent and translucent parts (e.g. windows) of the building are permeable by solar radiation. In general, the word transparent means clear, whereas translucent parts of the building can not be seen through. In terms of solar energy utilisation, the word transparent is also used to describe external parts of the building that can be seen through but that are not clear, in order to express their permeability not only by visible light but also by other components of the solar spectrum.

The term solar aperture surface refers to the translucent envelope surface that is suitable for solar energy utilisation.

Key figures. Following are some definitions of the main key figures used in terms of passive solar energy utilisation.

Transmission coefficient. The transmission coefficient τ_e indicates the share of global radiation incident on the irradiated structural element, which is transmitted through the glazing into the building as short wave radiation. It also considers the invisible wavelengths of solar radiation. If the transmission coefficient refers to the vertical radiation incidence, it is designated as τ_e^*.

Secondary heat flow. The secondary heat flow factor q_i indicates the share of global radiation G_g which is absorbed by a structural element and re-radiated into the building in the form of long wave radiation and convection (see /3-1/). Transparent elements (glazing) also warm up a little by the absorbed incident radiation and thus also present a secondary heat flow.

Energy transmittance factor (g-value). In addition to the energy supplied by radiation transmission (i.e. in addition to transmission coefficient τ_e), the g-value g or energy transmittance factor also includes the secondary heat flow q_i. It has been defined for a vertical radiation incidence and the same temperatures on both sides of the structural element /3-2/. For transparent building parts (glazing), it consists of the transmission coefficient τ_e^*, assuming a vertical radiation incidence, and the secondary heat flow q_i (see Equation (3.1)). q_{in} represents the heat flow added to the structural element. The secondary heat flow q_i is calculated by means of this added heat flow and the global radiation G_g.

$$g = \tau_e^* + q_i \quad \textit{with} \quad q_i = q_{in}/G_g \tag{3.1}$$

Diffuse energy transmittance factor (diffuse g-value). Depending on time and season, solar radiation strikes the transparent building elements from very different angles. On average, solar incidence on transparent surfaces is thus not vertical. Furthermore, moderate climate is characterised by a high share of diffuse radiation, amounting to about 60 % of the total incident solar radiation and presenting an average incidence angle of about 60°. The diffuse g-value $g_{diffuse}$ considers the decreased energy transmittance factor or g-value in case of vertical incidence, amounting to about 10 % /3-3/. When compared to the conventional g-value g, the diffuse g-value $g_{diffuse}$, allows for more realistic results.

Thermal transmittance coefficient (U-value). The U-value or thermal transmittance coefficient is a measure of the heat that is transmitted from the front side of a façade to the inside, assuming an area of 1 m^2 and a temperature difference of 1 K. It consists of heat transfer from air on one side of the element, thermal conductivity within the structural element, and thermal transmission from the other side of element to the air. In case of double-glazing, heat is transmitted by long wave radiation and convection between the two glass panes. For windows, we distinguish the U_G-value, which solely refers to the glazing, and the U_W-value, which also considers heat losses of the window frame, and thus refers to the entire window.

Equivalent thermal transmittance coefficient (equivalent U-value). The equivalent U-value is a measure which describes the difference between the specific thermal losses of a structural element and its specific heat gain by solar radiation. Like the U-and g-values, it also depends on the radiation incidence on the transparent surface and the dynamic behaviour of the building located behind. For its determination only thermal gains during the heating season must be taken into account, as overheating of rooms due to the solar radiation on glazed surfaces is not desirable. A negative equivalent U-value indicates that the heat gained by a transparent surface exceeds its thermal transmission. The Equation (3.2) allows for approximate estimation of the equivalent U-value U_{eq} by means of the U_W-value, referring to the entire window (including frame), the g-value (energy transmittance factor) and a correction factor S_W for window orientation. The correction factor S_W varies between 0.95 for north facing, 1.65 for east and west facing, and 2.4 for south facing orientation.

$$U_{eq} = U_W - S_W\, g \tag{3.2}$$

Transmission losses. As illustrated in Fig. 3.1 thermal losses of a building $\dot{Q}_l$ consist of ventilation and infiltration losses $\dot{Q}_V$ and transmission losses $\dot{Q}_T$. Transmission losses are calculated by means of the U-values of the respective surfaces (i.e. of surfaces A_n) and the temperature difference between internal room temperature θ_i and the corresponding external temperature θ_e of all structural elements of a house (see Equation (3.3)). However, transmission losses must not be mistaken for the transmission coefficient τ_e of transparent structural elements.

$$\dot{Q}_T = \sum_{n=1}^{m} \left(U_n \cdot A_n \cdot \left(\theta_{i,n} - \theta_e \right) \right)$$

(3.3)

3.2.2 System components

Passive solar systems may consist of transparent covers (such as windows, transparent thermal insulation), absorbers, heat stores and/or shading devices. Following is a detailed description of all system components.

Transparent covers. As an example, Fig. 3.2 illustrate the energy flow through double-glazing. Only one part of the incident solar radiation is transferred into the interior, whereas the remaining part is reflected from the outer pane surface. The radiation share directly transferred into the interior through both panes is indicated, in proportion to the radiation incidence on the outer pane surface, by the transmission coefficient τ_e. Another portion of the incident solar radiation is absorbed by the glass panes and heats up the gap between the two panes, and thus leads to further heat transmission into the interior by long wave radiation and convection. The g-value or energy transmittance factor indicates the ratio of total heat transferred into the interior and the incident radiation.

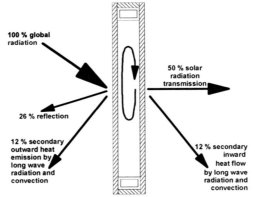

Fig. 3.2 Total energy transmittance factor of an average thermal insulation double-glazing (see /3-1/)

Transparent covers (such as windows) serve to transmit a maximum share of solar radiation to the interior and ensure at the same time utmost insulation from the outside. Typically, these two properties are expressed by the g-value (energy transmittance factor) and the U-value (thermal transmittance coefficient).

Good transparent covers are characterised by high g-values and low U-values. The single glass panes and insulating glazing used in the past offered very high g-values on the one hand, but also very high, and disadvantageous, U-values on the other. Noble gas fillings between the two glass panes, characterised by low thermal conductivity, low specific thermal capacity and high viscosity, help further reduce the thermal transmission by convection between the two glass panes. Furthermore, optimum adjustment of the pane spacing ensures the lowest possible U-values. Table 3.1 illustrates examples of optimum spacing and lists the physical characteristics of some of the most common filling gases.

So-called low ε-coatings help reduce thermal losses due to radiation exchanges in between the two panes. These coatings reduce the emission coefficient ε, for long wave radiation from originally 0.84 to 0.04. For short wave radiation these coatings are highly transparent. Low ε-coated double and triple glazing with noble gas fillings and panes with infrared reflecting coating ensure both: low U-values (thermal transmittance coefficient) and high g-values (energy transmittance factor).

Table 3.1 Optimum spacing between panes and thermodynamic properties of some window filling gases at 10 °C /3-1/

Filling gas	Optimum spacing between panes in mm	Thermal conductivity in W/(m K)	Density in kg/m^3	Dynamic viscosity in Pas	Specific thermal capacity in J/(kg K)
Air	15.5	$2.53\ 10^{-2}$	1.23	$1.75\ 10^{-5}$	1,007
Argon	14.7	$1.648\ 10^{-2}$	1.699	$2.164\ 10^{-5}$	519
Krypton	9.5	$0.9\ 10^{-2}$	3.56	$2.34\ 10^{-5}$	345
SF$_6$	4.6	$1.275\ 10^{-2}$	6.36	$1.459\ 10^{-5}$	614

By developing glazing with a high energy transmittance factor g, and transparent thermal insulation material (TI), transparent covers that offer both a high energy transmittance and good thermal insulation properties, are obtained. Table 3.2 illustrates some examples of g and U-values of some typical glazing types and a selection of transparent insulation systems. To indicate the energy transmittance also the diffuse g-value is considered.

Table 3.3 shows the equivalent U-values corresponding to different glazing types. By selecting state-of-the-art south facing double-glazing with thermal insulation, heat losses can be nearly compensated; triple glazing can achieve energy gains. The heat gain of a high-class north facing triple glazing can even exceed its heat transmission.

Table 3.2 Diffuse g-values and U-values of different glazing types and transparent thermal materials /3-4/

	Diffuse g-value	U-value glazing in W/(m² K)
Insulating glazing (4 + 16 + 4 mm, air)	0.65	3.00
Thermal insulation double-glazing (4 + 14 + 4 mm, argon)	0.60	1.30
Thermal insulation double-glazing (4 + 14 + 4 mm, xenon)	0.58	0.90
Thermal insulation triple-glazing with argon filling	0.44	0.80
Thermal insulation triple-glazing with krypton filling	0.44	0.70
Thermal insulation triple-glazing with xenon filling	0.42	0.40
10 cm plastic capillaries, one cover pane	0.67	0.90
10 cm plastic honeycombs, one cover pane	0.71	0.90
10 cm glass capillaries, two panes	0.65	0.97
2.4 cm granular aerogel, two panes filled with air	0.50	0.90
2 cm evacuated (100 mbar) aerogel plate, two panes	0.60	0.50

The diffuse g-values were measured for a 4 mm front pane, poor in iron, whereas for the U-values an average sample temperature of 10 °C has been assumed.

However, the diffuse g-values indicated in Table 3.3 only apply to the glazing itself. For window calculation the frame thus needs to be deducted from the window surface. For large-surface windows the U-value of a window U_w includes a 30 % frame surface. For smaller windows the U_w-value needs to be recalculated by means of the thermal transmittance coefficient (U-value) for both frame and glass pane, and additional thermal losses due to connecting sections have to be considered.

Table 3.3 Diffuse g-value ($g_{diffuse}$), U-value of the window (U_w) and equivalent U-values (U_{eq}) corresponding to different glazing types (see /3-5/)

	$g_{diffuse}$	U_w	U_{eq} (south)	U_{eq} (east/west)	U_{eq} (north)
			in W/(m² K)		
Simple glazing	0.87	5.8	3.7	4.4	5.0
Double-glazing (air 4 + 12 + 4 mm)	0.78	2.9	1.0	1.6	2.2
Double-glazing with thermal insulation and argon filling (6 + 15 + 6 mm)	0.60	1.5	0.1	0.5	0.9
Triple-glazing with thermal insulation and krypton filling (4 + 8 + 4 + 8 + 4 mm)	0.48	0.9	-0.3	0.1	0.4
Triple-glazing with thermal insulation and xenon filling (4 + 16 + 4 + 16 + 4 mm)	0.46	0.6	-0.5	-0.2	0.2

The g-value (energy transmittance factor) of a glass pane is additionally reduced by dust on the glass pane F_D and possible fixed shading F_S and flexible shading F_C. Even for frequently cleaned surfaces, due to dust, a reduction of the g-value by 5 % has to be assumed /3-6/. The value needs to be further reduced to consider the inclined radiation incidence. This factor is taken into account by the diffuse g-value in Table 3.2 and Table 3.3.

According to Equation (3.4), the solar heat generated within a defined period of time within area Q_S is thus calculated by multiplying the solar global radiation incident on window $G_{g,t,a}$, g-value and reduction factors, such as fixed and flexible shading (F_S and F_C) (Table 3.4 and Table 3.5), contamination F_D, and frame section F_F (Table 3.6).

$$Q_S = F_S\, F_D\, F_F\, F_C\, g\, G_{g,t,a} \tag{3.4}$$

Shading devices. By an appropriate building design, e.g. by balconies and projections, adequate shading protection from high-angle sun in summer can be provided without incurring any additional costs. The advantages of such fixed shading devices are simplicity and permanent function, as the devices lack moving parts which would require special control. However, they are applied most easily in a new building, where they can be incorporated into the original design, and should be south facing to ensure good shading in summertime and high irradiation into the building by the low-angle sun in winter (see Fig. 3.3). Even in summertime orientation to the east and west provides for high irradiation of the building by the low-angle sun, whereas in wintertime irradiation for these orientations is low (see Fig. 2.21). However, fixed shading devices reduce the efficiency of passive solar energy utilisation, as they also provide shading in-between seasons (in spring and fall), when space heating systems are still required.

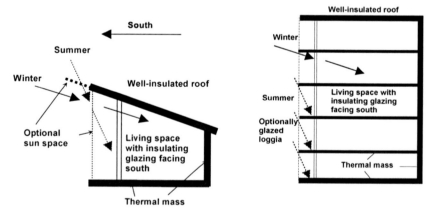

Fig. 3.3 Shading of transparent building surfaces by roof overhangs (left: single family house, right: multi family house)

Shading of buildings thus depends on the following parameters or factors. For a definition of the relevant angles refer to Fig. 3.4.

- Shading by the horizon F_h: determined by the solar position plot (see Fig. 2.20) or according to Table 3.4.
- Shading by projecting structures: overhangs F_o and side overhangs (fins) F_f are distinguished (Table 3.5).

The so-called shading factor F_S covers the entire shading. According to Equation (3.5), it is composed of the horizon shading factor F_h, the shading factors of overhangs F_o and side overhangs F_f. Dynamic building simulations allow for a more precise determination of the total shading of a building than this simplified equation.

$$F_S = F_h F_o F_f \tag{3.5}$$

Table 3.4 Part of shading factor to determine the shading by the horizon F_h for different latitudes, window orientations and horizon angle (S south, E east, W west, N north; see e.g. /3-3/)

Horizon angle	Latitude 45°			Latitude 55°			Latitude 65°		
	S	E/W	N	S	E/W	N	S	E/W	N
0°	1.00	1.00	1.00	1.00	1.00	1.00	1.00	1.00	1.00
10°	0.97	0.95	1.00	0.94	0.92	0.99	0.86	0.89	0.97
20°	0.85	0.82	0.98	0.68	0.75	0.95	0.58	0.68	0.93
30°	0.62	0.70	0.94	0.49	0.62	0.92	0.41	0.54	0.89
40°	0.46	0.61	0.90	0.40	0.56	0.89	0.29	0.49	0,85

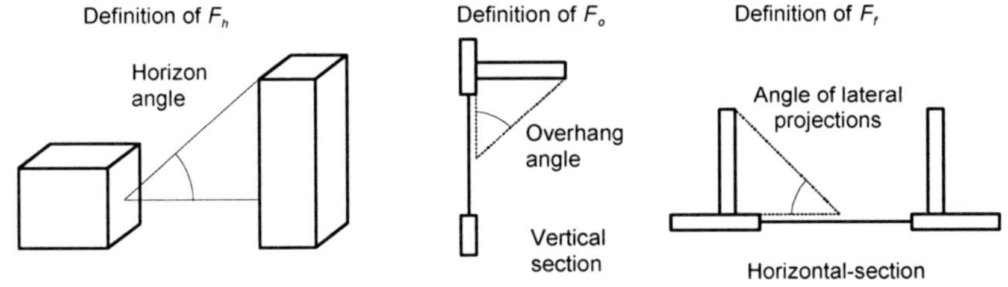

Fig. 3.4 Definition of angles corresponding to different kinds of shading (left: angle corresponding to the horizon shading factor F_h; centre: angle corresponding to the overhang shading factor F_o; right: angle corresponding to the shading factor F_f of side overhangs (fins); see /3-3/)

Besides the described fixed shading elements, also adjustable shading devices are used for passive solar system control. If the solar heat gain exceeds, for instance, the heat demand of a living space to be covered by solar resources, the solar aperture surface can be shaded to prevent overheating. External shading elements, such as window blinds and shutters, re-transmit the absorbed radiant heat to the ambient air, and are thus often more efficient than internal shading equipment. However, internal shading equipment (such as shutters and drapes) does not have to be weatherproof and is thus less demanding in terms of design.

In Table 3.6 are illustrated some examples of shading factors F_C for flexible shadings, corresponding to various types of adjustable shading equipment, along with the corresponding orientation.

Table 3.5 Shading factor corresponding to shading by overhangs (F_o) and by overhangs (fins) (F_f) for different latitudes, window orientations and horizon angles (S south, E east, W west, N north; for angle definition see Fig. 3.4; refer to /3-3/)

Angle	Latitude 45°			Latitude 55°			Latitude 65°		
	S	E/W	N	S	E/W	N	S	E/W	N
Shading by overhangs (F_o)									
0° [a]	1.00	1.00	1.00	1.00	1.00	1.00	1.00	1.00	1.00
30° [a]	0.90	0.89	0.91	0.93	0.91	0.91	0.95	0.92	0.90
45° [a]	0.74	0.76	0.80	0.80	0.79	0.80	0.85	0.81	0.80
60° [a]	0.50	0.58	0.66	0.60	0.61	0.65	0.66	0.65	0.66
Shading by projections (F_f)									
0° [b]	1.00	1.00	1.00	1.00	1.00	1.00	1.00	1.00	1.00
30° [b]	0.94	0.92	1.00	0.94	0.91	0.99	0.94	0.90	0.98
45° [b]	0.84	0.84	1.00	0.86	0.83	0.99	0.85	0.82	0.98
60° [b]	0.72	0.75	1.00	0.74	0.75	0.99	0.73	0.73	0.98

[a] overhang angle; [b] angle of lateral projections

Table 3.6 Shading factors F_C of selected internal and external shading devices /3-3/

Shading devices	Optical properties		Shading factor	
	Absorption	Transmission	Internal shading devices	External shading devices
White window blinds	0.1	0.05	0.25	0.10
		0.1	0.30	0.15
		0.3	0.45	0.35
White cloth	0.1	0.5	0.65	0.55
		0.7	0.80	0.75
		0.9	0.95	0.95
Coloured cloth	0.3	0.1	0.42	0.17
		0.3	0.57	0.37
		0.5	0.77	0.57
Aluminium-coated cloth	0.2	0.05	0.20	0.08

As an example, Fig. 3.5 illustrates the room temperature of a building calculated by dynamic building simulation for both cases, without shading devices and for two different shading methods in the course of one week in summer and ambient temperature θ_e ranging from 12 to 27 °C. Passive solar heat gains have been considered. However, for the example, active air cooling is assumed for room temperatures above 26 °C; room temperature θ_i will thus not increase beyond this value. The figure shows furthermore, that internal window blinds can only slightly reduce the inside temperature, whereas external blinds can reduce the inside temperature by several degrees Kelvin. In the present case, no additional cooling is required when using external window blinds.

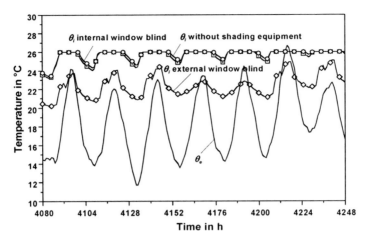

Fig. 3.5 Shading by internal and external window blinds (θ_e ambient temperature, θ_i room temperature) /3-7/

In contrast, reliability is considerably enhanced with shading systems incorporated into the glass pane. These systems include the following operating principles.
– Thermotropic glazing becomes opaque at defined outside or system temperatures, as molecules tend to accumulate to a pane-incorporated gel layer.
– Electrochromic glazing is characterised by a special coating which converts from transparent to opaque at a defined voltage.
– Glazing covered by holographic foils reflect the irradiation from the high-angle sun, so that sun-rays incident at small angles reach the absorber without any obstacle.

Absorber and heat storage. While absorber and heat storage are individual components in active solar systems, they are integrated into the building structure of passive systems.

Within direct gain systems, the room envelopes with solar radiation exposure, serve as absorber surfaces. Passive solar systems should thus offer well-absorbing outer surfaces and a heat-storing building structure which is well-adapted to the solar system.

The "classic" passive energy system is not equipped with any control. The thermal mass of a house, which is heated up by solar radiation, releases the heat back into the internal space with a certain time lag and reduced temperature without any user intervention. It is thus essential to prevent passive accumulators from overheating the rooms. For this purpose, the time lag and heat flow reduction by passive storage need to be known factors. Also, in most cases, additional (active) shading devices need to be provided to reduce energy absorption in summertime.

Indirectly heated thermal mass (e.g. unheated internal walls) can only be used sensibly if the corresponding room temperature variations are permitted. In case of high room temperatures, heat is slowly absorbed by thermal mass which is gradually heated up by the space. If, by contrast, the room temperature falls below the thermal mass surface temperature, the stored heat is released back into the space.

The established heat flow $\dot{q}$ depends on the temperature (θ) difference between the warm and the cold accumulator, the specific thermal capacity c_p, the density ρ_{SM} and the thermal conductivity λ of the storage medium, as well as on the heating and heat dissipation time t. Within a very short period of time, for instance, the accumulator will only heat up at the surface, and the absorbed energy quantity is thus only little.

The heat flow $\dot{q}$ is calculated by Fourier's (one-dimensional) law on heat conduction according to Equation (3.6) and by the heat flow into and from the storage element (storage equation, Equation (3.7)).

$$\dot{q} = -\lambda \frac{\partial \theta}{\partial x} \tag{3.6}$$

$$\frac{\partial \dot{q}}{\partial x} = -\lambda \frac{\partial^2 \theta}{\partial x^2} = \rho_{SM} \, c_p \, \frac{\partial \theta}{\partial t} \tag{3.7}$$

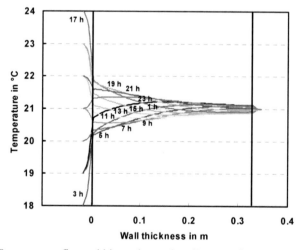

Fig. 3.6 Temperature flow within an internal wall exposed to radiation and varying temperatures on one side (left) /see 3-9/

As an example, Fig. 3.6 illustrates the room and wall temperatures of an internal concrete wall (storage wall) within the course of 24 hours. Within this period the temperature varies by 6 °C on one room side, whereas the stored and released energy amounts to 0.076 kWh/(m² d) under the side conditions assumed for this wall within the given period. The temperature only varies significantly up to a wall thickness of approximately 15 cm. An increased wall thickness does thus not enhance the heat storage capacity.

In most indirect gain systems only the outer wall is used for heat storage, which is thus of solid design, whereas the outer wall surface serves as absorber. For this purpose the surface is either painted in black or covered with black absorber foil.

Only within decoupled systems, absorber and accumulator are separate components, whereas black or selectively coated sheet metal serves as absorber medium. The heat carrier is transferred to the accumulator by means of a channel or a more sophisticated medium. The accumulator itself may also be part of the building structure, for instance, if it is designed as hollow ceiling or double wall masonry. Rock storage, however, are not of double use, as they are not part of the building structure.

Dynamic building simulations permit the determination of energy requirements for heating Q_H (see Equation 3.8) by means of the building heat losses Q_L reduced by usable energy from solar radiation Q_S and internal heat Q_i (i.e. heat created by people and household appliances) as well as utilisation factor η.

$$Q_H = Q_L - \eta(Q_S + Q_i) \tag{3.8}$$

In simplified terms, this is attributable to the determination of the utilisation factor η for the available solar energy Q_S, and internal heat gains Q_i. Equation (3.9) shows the approximate calculation of utilisation factor η for a ratio of heat gain to heat loss γ below 1.6, and common thermal inertia of Equation (3.10) /3-6/.

$$\eta = 1 - 0.3\gamma \tag{3.9}$$

$$\gamma = (Q_S + Q_i)/Q_L \tag{3.10}$$

3.2.3 Functional systems

Depending on their form and arrangement, four different functional systems (i.e. direct gain systems, indirect gain systems, decoupled systems, sunspaces) are distinguished; however, the borderline between the systems remains fluid.

Direct gain systems. Solar radiation penetrates into the living space through transparent external surfaces and is converted into heat at the internal room surfaces. Room temperature and room surface temperature change almost simultaneously. Typical direct gain systems are regular windows and skylights (see Fig. 3.7).

Direct gain systems are characterised by a simple structure, low control requirements as well as low storage losses, as radiation energy is on the spot converted into heat, inside the living space. However, its poor ability to react to phase shifts between radiation and internal temperature may be disadvantageous. Direct gain systems can only be controlled by shading, since the heat which is re-radiated into space by thermal mass cannot be influenced. Hence, to ensure profitable utilisation of solar gains, additional heating systems with low inertia are required.

However, direct gain systems are especially profitable if radiation and heating demand occur simultaneously, as in many office buildings. Direct gain systems may also be combined with daylight systems in order to save the energy required for lighting. Direct gain systems are thus especially suitable to complement indirect gain systems which respond more easily to the phase shifts between incident radiation and heating demand.

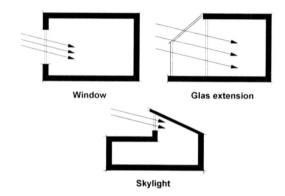

Fig. 3.7 Different types of direct gain systems (according /3-8/)

Indirect gain systems. Within indirect gain systems (solar wall) solar radiation is converted into heat on the side of the storage element opposite to the living space. Energy migrates to the storage element's inner surface (room side) by heat conductivity and is released into the room air (Fig. 3.8). There is thus a certain phase shift between solar radiation and internal temperature, which is influenced by the storage element material and its thickness.

Solar wall systems are characterised by a simple structure, phase shifted room heating and lower room temperature variations in comparison to direct gain systems. However, their higher heat losses to ambient may be disadvantageous. Heat supply can only be regulated by appropriate shading devices. Once the heat has been absorbed by the storage element, heat transfer into the living space can no longer be controlled. For systems supported by convection, also the inner side of the transparent cover needs to be cleaned as room air and heated air tend to mingle.

Solar wall systems or indirect gain systems are suitable to complement direct gain systems, since, when combined, they extend heat distribution into the living space. Combinations of both systems are particularly appropriate for living spaces with continuous heat demand.

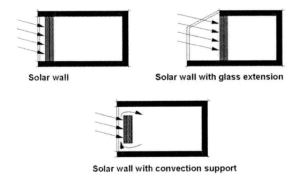

Solar wall   Solar wall with glass extension

Solar wall with convection support

Fig. 3.8 Different types of solar wall systems (see /3-8/)

Transparent thermal insulation. Besides the direct utilisation of solar radiation by windows, within the last few years, also wall systems equipped with transparent thermal insulation (TI) have been developed to enhance passive solar energy gains. They are a special variant of indirect gain systems, as they enable the effective utilisation of passive solar systems in Central Europe thanks to the use of transparent heat insulators.

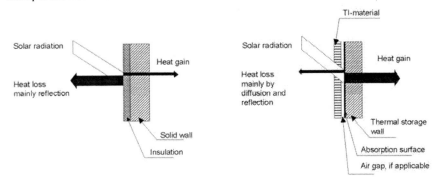

Fig. 3.9 Comparison of opaque (left) and transparent thermal insulation (right)

Only very little of the solar radiation incident on an opaque thermal insulation can be utilised (Fig. 3.9, left). During absorption of solar radiation the outer surface may heat up; however, due to the low thermal conductivity of the insulation layer, also in case of considerable temperature differences (between the inside and the outside of the space) only very little heat will penetrate to the inside.

In contrast, a large amount of solar radiation penetrates through elements with transparent insulation and is converted into heat when striking the black coated element (absorber) (Fig. 3.9, right). Due to the high thermal resistance of the insulating material, a large amount of heat is transferred into the storage wall.

Fig. 3.10 illustrates the temperature flow and the corresponding heat flow densities as well the U-value (thermal transmittance coefficient) and the equivalent U-value (U_{eq}) of an exemplary solar wall with thermal insulation (TI) on a cold and foggy winter's day within the course of 24 hours. To provide for good transfer of the heat created at the absorber (useful heat), and to avoid excessive maximum temperatures, the wall behind the transparent thermal insulation material (TI) must have a high thermal transmission coefficient and good storage properties. However, these properties will result in very low thermal insulation. The total U-values of transparent thermal insulation (TI) are thus often higher than those of a solely insulated wall. At night time, when thermal mass has cooled down, the wall will have higher losses than a solely insulated wall. However, for well-designed walls with thermal insulation heat losses are in most cases overcompensated by heat gains; the equivalent U-values (U_{eq}) (including solar gains) are lower or even negative (net heat gain). In the example illustrated in Fig. 3.10 the U-value amounts to 0.527 W/(m^2 K) and the equivalent U-value (U_{eq}) to 0.267 W/(m^2 K), on this unpleasant day.

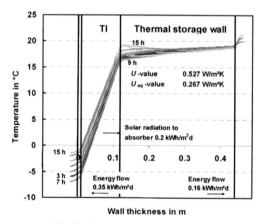

Fig. 3.10 Temperature distribution of a system consisting of glass – transparent thermal insulation (TI) – air – absorber – concrete wall on a cold and foggy winter's day /3-9/

South-facing solar walls with thermal insulation and without shading devices allow for annual savings of useful energy within the range from 350 to 400 MJ/(m^2 a), with regard to the solar aperture area and in comparison to common opaque insulation systems (such as heat-insulation connection system or non-bearing façades ventilated at rear). If the internal temperature is very high even energy savings of above 700 MJ/(m^2 a) can be achieved /3-8/.

Solar systems with convective heat flow. A further variant of indirect gain systems are solar systems with convective heat flow (Fig. 3.11). These systems do not require any shading, as, in summertime, the hot air between absorber and accumulator is evacuated to the outside.

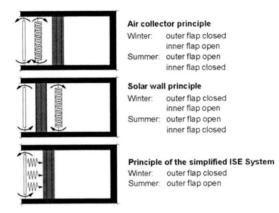

Air collector principle
Winter: outer flap closed
          inner flap open
Summer: outer flap open
          inner flap closed

Solar wall principle
Winter: outer flap closed
          inner flap open
Summer: outer flap open
          inner flap closed

Principle of the simplified ISE System
Winter: outer flap closed
Summer: outer flap open

Fig. 3.11 Different types of solar wall systems with convective heat flows (see /3-8/)

Decoupled systems. Some components of decoupled solar systems (such as heat transfer devices and fans) are not part of the building structure. Since they actually belong to the energy provision system, it is not always possible to clearly distinguish them from active solar systems.

These decoupled systems convert the incident solar radiation on an absorbing surface, thermally insulated from the internal space (Fig. 3.12). Subsequently, solar heat is transferred to a heat accumulator via a channel system, using air as heat carrier. The heat accumulator can either be incorporated into the building structure or be itself an individual technical component (or a combination of both). Hollow ceilings and double wall masonry are examples of accumulators incorporated into the building structure, whereas rock and water storage tanks are systems that are independent of the building structure.

If heat exchange is entirely convectional (thus excluding auxiliary units), and if the accumulator is part of the building structure, the solar system is unambiguously passive. However, if fans are used for air circulation, the respective system is referred to as semi-passive. By means of accumulators equipped with thermal insulation heat distribution throughout the room can be controlled independently from absorber and storage temperatures.

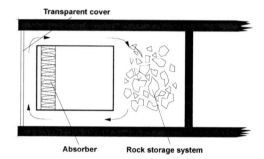

Fig. 3.12 Thermally decoupled solar system attached to the building envelope (see /3-8/)

Decoupled systems are characterised by good controllability. Thanks to the thermal insulation between absorber and the internal space, at night heat losses are very low. However, these systems present the disadvantages of high construction costs, a propensity towards defects (e.g. leakages) and high absorber temperatures.

Thermally decoupled systems are most suitable to compensate considerable time lags between solar radiation and heat demand. They also result advantageous if separate heat accumulators are already available or if they can be easily integrated into the building structure.

Sunspaces. Sunspaces are another variant of functional systems. Most popular are unheated sunspaces, whose connecting doors to the internal living space are left open if heating is needed and the temperature inside the adjacent sunspace becomes warmer. Furthermore, sunspaces of two or more stories also serve for ventilation of houses (Fig. 3.13). In wintertime the minimum temperatures amount to 0 °C, whereas in summertime heat needs to be evacuated to the outside to avoid overheating (temperatures above 50 °C are possible). For this reason slanting windows should be avoided and the roof should be well-insulated. Also, orientation toward the east and west is unfavourable, as the incident solar radiation will be very little in winter, and in summer shading can only be provided by window blinds, but not by roof overhangs.

During the heating season a well designed sunspace, with optimum control, supplies the house with just as much or a little more energy than it receives from the house. Besides the utilisation of passive solar energy, unheated sunspaces also cut the heat load of a building, as the system wall – sunspace – wall usually presents a lower U-value (thermal transmittance coefficient) as the outer wall. A heated sunspace, however, will result in higher heat losses.

Even shaded sunspaces provided with roof overhangs are often overheated in summertime. Fig. 3.14 illustrates the temperature curves of the sunspace (θ_{ss}) shown in Fig. 3.13, the living space temperatures (θ_i), the ambient temperatures (θ_e) and the floor (θ_{fl}) and ceiling temperatures (θ_{ce}) in a house equipped with a floor heating system on three nice days in summer. In spite of a high ventilation rate, temperature rises to above 40 °C. However, the maximum temperature in the adjacent living space only amounts to 30 °C.

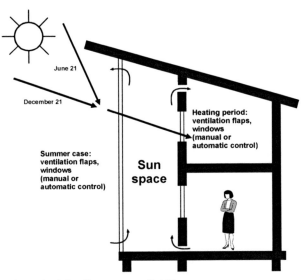

Fig. 3.13 Operation principle of a sunspace /3-10/

In multiple family homes, sunspaces often serve as hallways, as room temperature fluctuations are more easily accepted there than in living spaces. However, temperature distribution needs to be carefully observed in these high spaces, and natural air circulation/air currents need to be taken into account. For this purpose most sunspaces are provided with flaps at the bottom and the top, to let fresh air in and to evacuate overheated air.

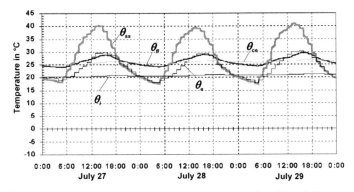

Fig. 3.14 Temperature flow of a sunspace in summertime (for abbreviations see text; e.g. /3-11/)

4 Solar Thermal Heat Utilisation

4.1 Principles

Part of the solar radiation energy can be converted into heat by using absorbers (e.g. solar collectors). The absorbers together with the other necessary components are the solar system. Solar systems are installations converting solar radiation into heat in order to heat swimming pools, produce domestic hot water, cover the demand for space heating or supply other heat consumers. In the following, the physical principles of energy conversion for these forms of solar thermal heat utilisation are described (see /4-1/, /4-2/, /4-3/).

4.1.1 Absorption, emission and transmission

The basic principle of solar thermal utilisation is the conversion of short-wave solar radiation into heat. This energy conversion process can also be described as photo-thermal conversion. If radiation incidences on material a certain part of the radiation is absorbed. A body's capacity to absorb radiation is called absorbing capacity or absorption α, where α reflects the share of absorbed radiation as part of the entire radiation on matter. An ideal black body absorbs radiation at every wavelength and therefore has an absorption coefficient equal to one.

The emission ε represents the power radiated by a body. The relationship between absorption α and emission ε is defined by "Kirchhoff's law". For all bodies the ratio of specific radiation and the absorption coefficient is constant at a given temperature, and in terms of its amount, equal to the specific radiation of the black body at this temperature. This ratio is exclusively a functionality of temperature and wavelength. Matter with a high absorption capacity within a defined wave range also has a high emission capacity within that same wave range.

In addition to absorption and emission, also reflection and transmission play a role. The reflection coefficient ρ describes the ratio of the reflected to the incident radiation. The transmission coefficient τ defines the ratio of the radiation transmitted through a given material to the entire radiation incident, according to Equation (4.1). Thus the sum of absorption, reflection and transmission is one.

$$\alpha + \rho + \tau = 1 \tag{4.1}$$

4.1.2 Optical features of absorbers

Absorbers have to absorb radiation and partially convert it into heat. The absorber is, amongst other things, characterised by being opaque for radiation ($\tau = 0$) as expressed in Equation (4.2); the sum of absorption α and reflection ρ at the absorber area is one.

$$\alpha + \rho = 1 \tag{4.2}$$

An ideal absorber does not reflect any short-wave radiation ($\rho = 0$) and thus – in line with Equation (4.2) – completely absorbs solar radiation within this wave range ($\alpha = 1$). For long-wave radiation above a certain boundary wavelength, the situation is exactly the opposite. Given an ideal absorber, it reflects all of the radiation and does not absorb any at all. Accordingly, the emission in this wave range is zero (Kirchhoff's law). Fig. 4.1 shows the wavelength dependency of the absorption and reflection coefficient in the case of an ideal absorber.

Ideal scenarios cannot be completely recreated in real life. So-called selective surfaces (or selective coatings) are close to the optimal absorber features (Fig. 4.1). Within the spectrum of solar irradiance, the reflection coefficient ρ_{real} is close to zero, in the infrared spectrum (> 3 μm) close to one. The absorption coefficient α_{real} demonstrates exactly the opposite.

Table 4.1 shows the absorption coefficient for various different materials and the transmission and reflection coefficients for the solar irradiance and the infrared range of the solar radiation spectrum. Compared to the non-selective absorber, selective absorber surfaces show high degrees of α_s/ε_l.

Table 4.1 Optical features of absorbers (according to /4-1/)

		Solar irradiance			Infrared-radiation			α_s/ε_l
		α_s (ε_s)	τ_s	ρ_s	α_l (ε_l)	τ_l	ρ_l	
Selective	Black nickel	0.88	0	0.12	0.07	0	0.93	12.57
Absorber	Black chromium	0.87	0	0.13	0.09	0	0.91	9.67
	Aluminium grid	0.70	0	0.30	0.07	0	0.93	10.00
	Titanium-oxide- nitride	0.95	0	0.05	0.05	0	0.95	19.00
Non-selective absorber		0.97	0	0.03	0.97	0	0.03	1.00

α_s is the absorption coefficient in the spectrum of solar irradiance, ε_l is the emission coefficient in the infrared radiation spectrum. Such surfaces are thus also called α/ε-surfaces. For the selective absorbers described in Table 4.1, the ratios α_s/ε_I are between 9 and 19. Titanium oxide with 19 for example shows a particularly high α_s/ε_I-ratio.

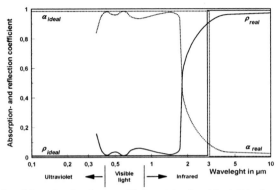

Fig. 4.1 Absorption (α) and reflection coefficient (ρ) of an ideal (*ideal*) and a standard real absorber (*real*)

4.1.3 Optical features of covers

In order to reduce the convective thermal losses of the absorber to the environment (Chapter 4.1.4), in many cases absorbers used in solar thermal systems have a transparent cover. Ideal covers have a transmission coefficient of one in the range of solar radiation, whereas reflection and absorption coefficient equal zero in this spectrum.

Table 4.2 Optical features of covers (according to /4-1/)

	Solar spectrum			Infrared radiation		
	$\alpha_S (\varepsilon_S)$	τ_S	ρ_S	$\alpha_I (\varepsilon_I)$	τ_I	ρ_I
Sheet glass	0.02	0.97	0.01	0.94	0	0.06
Infrared reflecting glass (In$_2$O$_3$)	0.10	0.85	0.05	0.15	0	0.85
Infrared reflecting glass (ZnO$_2$)	0.20	0.79	0.01	0.16	0	0.84

In real life such conditions cannot be achieved. Table 4.2 shows the attributes of the different cover materials. According to that table, glass fulfils the required optical features within the luminous spectrum very well. Infrared light emitted by the collector , however, cannot pass through, but is mainly absorbed. If the degree of absorption is high, the temperature of the glass cover rises and the radiation losses to the environment are correspondingly high according to Kirchhoff's law. These losses can be reduced by vacuum-coating of layers that reflect infrared light.

4.1.4 Energy balance

General energy balance. Equation (4.3) describes the general energy balance of a medium that absorbs radiation and converts it into heat.

$$\dot{G}_{G.abs} = \dot{Q}_{conv.abs} + \dot{Q}_{rad.abs} + \dot{Q}_{refl.abs} + \dot{Q}_{cond..abs} + \dot{Q}_{useful} \tag{4.3}$$

$\dot{G}_{G,abs}$ is the entire global radiation incident on the absorber surface; $\dot{Q}_{useful}$ is the utilisable thermal flow. In addition there are four different loss flows:
- convection losses of the absorber to the ambient air $\dot{Q}_{conv,abs}$,
- long-wave radiation losses of the absorber $\dot{Q}_{rad,abs}$,
- reflection losses of the absorber $\dot{Q}_{refl,abs}$,
- thermal conductivity losses $\dot{Q}_{cond,abs}$.

Energy balance of the collector. In solar thermal systems the absorber is normally part of a collector. Other components of the collector are frame, cover and insulation. Given these conditions, the energy balance will be discussed further in the following.

A collector removes the utilisable heat by a heat transfer medium flowing through the collector (Fig. 4.2). The difference between the energy at the inlet and the outlet of the heat transfer medium is the thermal flow removed by the transfer medium $\dot{Q}_{useful}$ (Equation (4.4)), where c_p is the specific heat capacity, $\dot{m}$ is the mass flow of the transfer medium and θ_{in} and θ_{out} are the inlet and the outlet temperature of the transfer medium flowing into, or out of the collector.

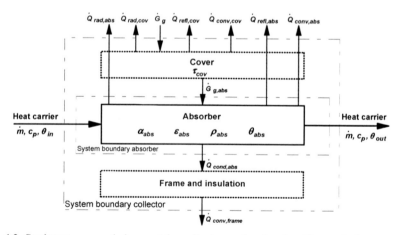

Fig. 4.2 Stationary energy balance at the collector or the absorber (for symbols see text)

$$\dot{Q}_{useful} = c_p \, \dot{m} \left(\theta_{out} - \theta_{in} \right) \tag{4.4}$$

This results in the following energy balance for the absorber of a collector (Equation (4.5)).

$$\dot{G}_{g,abs} = c_p \, \dot{m} \, \theta_{out} - c_p \, \dot{m} \, \theta_{in} + \dot{Q}_{conv,abs} + \dot{Q}_{rad,abs} + \dot{Q}_{refl,abs} + \dot{Q}_{cond,abs} \tag{4.5}$$

The global radiation on the absorber $\dot{G}_{g,abs}$ is defined by the total global radiation $\dot{G}_g$ on the collector cover and the corresponding transmission coefficient τ_{cov} (Equation (4.6)).

$$\dot{G}_{g,abs} = \tau_{cov} \, \dot{G}_g \tag{4.6}$$

The reflection losses of the absorber $\dot{Q}_{refl,abs}$ can be calculated with the radiation on the absorber and the degree of reflection (Equation (4.7)). It is neglected that a small part of the radiation reflected by the absorber is again reflected by the cover back towards the absorber. τ_{cov} is the transmission coefficient of the cover and the reflection coefficient of the absorber is ρ_{abs}.

$$\dot{Q}_{refl,abs} = \tau_{cov} \, \dot{G}_g \, \rho_{abs} \tag{4.7}$$

According to the Stefan-Boltzmann radiation law, the radiation losses $\dot{Q}_{rad,abs}$ result from the degree of emission ε, the difference between the absorber temperature θ_{abs} and the ambient external temperature θ_e, to the fourth power (in Kelvin), plus the Stefan-Boltzmann-constant σ ($5{,}67 \cdot 10^{-8}$ W/(m^2 K^4)) according to Equation (4.8). In addition, they are proportional to the radiating absorber area S_{abs}.

$$\dot{Q}_{rad,abs} = \varepsilon_{abs} \, \sigma \left(\theta_{abs}^4 - \theta_e^4 \right) S_{abs} \tag{4.8}$$

The convective thermal losses of the absorber are initially transferred to the cover plate. In a steady state (i.e. the temperature changes of the cover plate do not change) this thermal flow is then transferred entirely to the environment. This convective thermal flow $\dot{Q}_{conv,abs}$ can be assumed to be approximately linear. It depends on the difference between the absorber temperature θ_{abs} and the ambient air temperature θ_e, and can be described by using the heat transfer coefficient U^*_{coll} that is constant in the first approximation (i.e. temperature-independent heat transfer coefficient). The corresponding equation is as follows (Equation (4.9)).

$$\dot{Q}_{conv,abs} = U^*_{coll} \left(\theta_{abs} - \theta_e \right) S_{abs} \tag{4.9}$$

The thermal flow $\dot{Q}_{cond,abs}$ due to the heat conduction from the absorber to the frame and the insulation is very small compared to the other thermal flows and can be neglected. The energy balance results therefore in Equation (4.10) for the heat $\dot{Q}_{useful}$ transported by the heat transfer medium.

$$\dot{Q}_{useful} = \tau_{cov}\dot{G}_g - \tau_{cov}P_{abs}\dot{G}_g - \tag{4.10}$$

$$U^*_{coll}(\theta_{abs} - \theta_e)S_{abs} - \varepsilon_{abs}\sigma(T^4_{abs} - T^4_e)S_{abs}$$

Considering Equation (4.1), the first two terms of Equation (4.10) can be joined. Furthermore, the absorber normally has low degrees of emission. If the temperature difference between the absorber and the environment is kept low, the last term of Equation (4.10) can be neglected in many cases. The entire heat and radiation losses can be described, in an approximation using a heat transfer coefficient U_{coll}, as linearly dependent on the temperature that takes the entire thermal losses into account. These assumptions result in Equation (4.11).

$$\dot{Q}_{useful} = \tau_{cov}\,\alpha_{abs}\,\dot{G}_g - U_{coll}(\theta_{abs} - \theta_e)S_{abs} \tag{4.11}$$

In some cases neglecting the 4th order dependency can be too big of an omission. The dependency can then be approximated by a 2nd order term. This is described in Equation (4.12). C_1 and C_2 are corresponding auxiliary constants.

$$\dot{Q}_{useful} = \tau_{cov}\,\alpha_{abs}\,\dot{G}_g - C_1(\theta_{abs} - \theta_e)S_{abs} - C_2(\theta_{abs} - \theta_e)^2 S_{abs} \tag{4.12}$$

4.1.5 Efficiency and solar fractional savings

The efficiency η of the conversion of solar radiation energy into useable heat in the collector results from the ratio of the useful thermal flow transported by the heat transfer medium Q_{useful} to the global radiation incident on the collector (Equation (4.13)).

$$\eta = \dot{Q}_{useful} \Big/ \dot{G}_g \tag{4.13}$$

For a collector with given transmission and absorption coefficients, plus a given thermal conductivity coefficient, the efficiency can be calculated combining Equations (4.11) or (4.12) with Equation (4.13) (Equation (4.14) and (4.16) respectively). If the energy balance is drawn for a collector area of one square metre, this results in Equations (4.15) and (4.17) respectively. $\dot{G}_{g,rel}$ is the global radiation on an absorber area of one square metre (net collector area). C_1 and C_2 are auxiliary constants to calculate the utilisable heat of the collector.

With given material parameters, the highest efficiency is achieved at the lowest possible temperature difference between the absorber, the environment and a maximum radiation.

$$\eta = \tau_{cov}\,\alpha_{abs} - \frac{U_{coll}\,(\theta_{abs} - \theta_e)\,S_{abs}}{\dot{G}_g} \qquad (4.14)$$

$$\eta = \tau_{cov}\,\alpha_{abs} - \frac{U_{coll}\,(\theta_{abs} - \theta_e)}{\dot{G}_{g,rel}} \qquad (4.15)$$

$$\eta = \tau_{cov}\,\alpha_{abs} - \frac{C_1\,(\theta_{abs} - \theta_e)\,S_{abs}}{\dot{G}_g} - \frac{C_2\,(\theta_{abs} - \theta_e)^2\,S_{abs}}{\dot{G}_g} \qquad (4.16)$$

$$\eta = \tau_{cov}\,\alpha_{abs} - \frac{C_1\,(\theta_{abs} - \theta_e)}{\dot{G}_{g,rel}} - \frac{C_2\,(\theta_{abs} - \theta_e)^2}{\dot{G}_{g,rel}} \qquad (4.17)$$

In many cases the solar fractional saving F_s is significant. It is defined in different ways by the relevant literature. In this context, it is the ratio between the utilisable emitted energy through conversion of solar radiation by the solar installation ex-storage to the actual demand for heating, domestic warm water or process heat that is to be covered partly or entirely by solar energy (Equation (4.18)). All heat losses of the heat storage are allocated to the solar system when using this definition. It is thus defined as the conventional energy carrier saving $\dot{Q}_{aux}$, in relation to the corresponding demand for heat $\dot{Q}_{demand}$. If the substitution of the conventional energy carrier is the starting point – as it is common – then for this exclusively conventional system no storage, or only a very small storage, is necessary. Thus the storage would only be used by the solar system and the definition in Equation (4.18) is correct.

$$F_s = 1 - \frac{\dot{Q}_{aux}}{\dot{Q}_{demand}} \qquad (4.18)$$

4.2 Technical description

Besides the collector a solar thermal system also consists of other system components. Essential are a liquid or gaseous heat transfer medium and pipes to transport the heat transfer medium. Normally, a heat store with none, one or several heat exchangers plus, for certain designs, pumps with a drive to maintain the heat carrier cycle, sensors and control instruments are required.

4.2.1 Collectors

Collectors are part of solar thermal systems, partly converting solar radiation into heat. Part of this heat is subsequently transported by a heat carrier flowing through the collector. For that purpose, a collector consists of several components described in detail in the following.

Collector components. Fig. 4.3 shows the main components of a liquid-type flat-plate collector. Accordingly, a collector consists of the absorber, the transparent cover, the frame and the heat insulation. Additionally, heat carrier inlet and outlet, plus fixing methods are shown. Depending on the collector design, it does not include all of the illustrated components. The absorber, however, including the appropriate pipes for the heat carrier, is an absolutely essential part. For most designs, the other components described below are also collector parts.

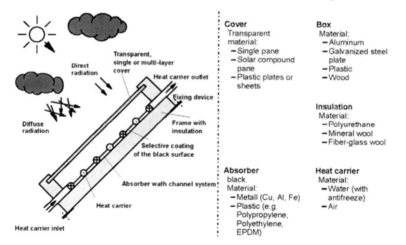

Fig. 4.3 Main components and materials as well as the schematic design of a flat-plate liquid-type collector (according to /4-6/ and various other sources)

Absorber. The absorber converts short-wave radiation into heat (photo-thermal conversion). The function of "radiation absorption" is carried out by a type of absorber material with quite a high absorption capacity within the luminous spectrum. On the other hand, a low absorption and thus emission capacity, is aimed for in the thermal radiation wave spectrum. In addition, the absorber has to enable a good heat transfer to the heat carrier and also be temperature-resistant, as normally temperatures of up to 200 °C occur in an insulated absorber with glass cover and selective coating. In concentrating collectors temperatures are generally even higher.

In line with these requirements, mainly copper and aluminium are used to build absorbers. As there may be a shortcoming of these materials, provided that the market increase of solar thermal collectors is continuing, polymeric materials and steel could become more important in the future. In the simplest case, this basic material is painted black on the side receiving radiation (maximum absorber temperature approximately 130 °C). For a large number of absorbers, this side is also coated selectively (maximum absorber temperature approximately 200 °C).

The heat carrier flows through the channels inside the absorber. The energy proportion of the solar radiation on the absorber converted into heat inside the absorber is partly transported to the heat carrier (by heat transfer). The system of pipes in the absorber can vary in terms of pipe material, pipe cross-section, length and pipe allocation within the collector.

Cover. The transparent cover of collectors ought to be as transparent for solar radiation as possible and retain the long-wave thermal reflection of the absorber. At the same time it has to reduce convective thermal losses to the environment.

Suitable materials are glass sheets, synthetic plates or synthetic foils (e.g. made of polyethylene or Teflon). The high level of material stress often leads to brittle and tarnished synthetic materials. Furthermore, the outer area can also be scratched very easily by atmospheric exposure. Thus transmission values of synthetic covers are often not stable long-term. Therefore for most applications glass is used. For solar collectors mainly safety glass is used that is characterised by a high level of transparency and resistance to hail. Additionally, low iron contents can reduce the absorption capacity in the short-wave spectrum. Thus it is avoided that the glass sheet heats up. Convective thermal losses to the colder environment are reduced. Often infrared-reflecting layers are vacuum coated on the bottom side of the cover to reflect the long-wave heat radiation from the absorber to the cover into the direction of the absorber. Thus losses can be reduced even further.

Collector box. The collector box holds the components required for radiation transmission, absorption, heat conversion and insulation. It can be made of aluminium, galvanised steel plate, synthetic material or wood. It gives the collector mechanical firmness and makes it environment-proof. However, a low level of ventilation has to be ensured in order to reduce high or low pressure caused by temperature fluctuations and remove possible humidity.

Independent of the material, designs can be differentiated as boxes for on-roof installation on top of the roof tiles and boxes for in-roof integration for collectors installed on pitched roofs. Boxes installed externally on the roof have a cover (e.g. made of aluminium) at their back, whereas boxes integrated into the roof do not need such a cover.

Other components. Thermal insulation made of standard insulation material (e.g. polyurethane, glass fibre wool, mineral wool) belongs to the group of other components. On the outside of the box, one inlet pipe for heat carrier charging and one outlet pipe for heat carrier discharging are installed. Furthermore, the necessary components to attach the collector are outside of the box. Often additional components are offered for on-roof installed collectors that enable an on-roof collector installation with a certain angle to the roof slope. In general, the energy output is only marginally increased by doing so. For in-roof installed collectors, sheet metals for plumbing are often delivered alongside. If the temperature needs to be measured inside or outside of the collector, there are drill-holes or other means available.

Installation. Collectors are mainly installed on pitched roofs; in this respect the integration into the roofing or the on-roof installation, on top of the tiles, are common technical solutions. Independent of the type of installation
- the static's of the roof have to carry the collector load (in-roof collectors are often lighter than the tiles that are generally intended to the used),
- the coupling to the roof has to ensure that the collectors are not separated from the roof (e.g. by wind) and
- the heat expansion of the collectors and pipes must not be obstructed.
Integration into the roof is less visible and cheaper than the on-roof installation. It is preferably used for new buildings or larger collector arrays on already existing roofs. Additionally, roofing costs are saved for the parts of the roof where the collectors are installed. If retrofitted, collectors are often installed on top of the roof tiles. This easier form of installation does not damage the roof cladding and consequential damages of the building can be largely ruled out in the event of collector leakages or damaged glazing

Installation of collectors on flat areas (e.g. on flat roofs, in gardens) facilitates optimal adjustment and incline when compared to the installation on pitched roofs. Mainly standardised frames are used to integrate the collector. Frames need to be arranged so that shading is avoided. It can be useful to build the collectors with a comparatively low incline (e.g. 20°). Because of the lower level of internal shading, larger collector areas can be built on the same area. Furthermore, installation costs decrease with smaller frames and lower wind loads. The reduced output of installation due to a flatter incline of the absorber area compared to the optimal installation is insignificant for systems with low solar fractional savings.

Collector designs and practical applications. The different collector designs can be differentiated according to the heat carrier and the way they absorb radiation. According to this method, four basic collector designs can be identified:
- Non-concentrating swimming pool liquid-type collectors
- Non-concentrating glazed flat-plate liquid-type collectors
- Non-concentrating glazed air collectors,
- Radiation-concentrating liquid-type collectors and

– Radiation-concentrating air collectors.

Within these five basic designs there are a number of variations. Fig. 4.4 shows a selection – however, only very few have been successful as standard solutions in the market.

Non-concentrating swimming pool liquid-type collectors. This basic design that is used most frequently in its simplest form consists of an absorber mat with a corresponding system of pipes for the heat carrier (Fig. 4.4, on the top left). This collector design is often referred to as the collector type "absorber". It is preferably used for heating open-air swimming pools. This application needs water at a temperature around ambient temperature. Heat insulation to the ambient is not needed, because there is no driving force (temperature difference) for heat losses. Therefore a transparent cover and an insulation on the back side of the collector are not needed and the optical losses are only due to the reflection coefficient of the absorber ρ_{abs}. The absorber material is mainly EPDM (ethylene-propylen-dien-monomers) which is able to withstand UV radiation and temperatures up to 150 °C. Such high temperatures do not occur due to the lack of insulation. This absorber type is very cheap and results highly efficient for the swimming pool application.

Non-concentrating glazed flat-plate liquid-type collectors. Should higher temperature levels be required, glazed flat-plate collectors are used in many cases (Fig. 4.3 and Fig. 4.4, on the left). They can be built with one or more transparent cover sheets. In order to further reduce the convective thermal losses from the absorber to the cover, the space between the two can be evacuated, which turns the collector into a vacuum flat-plate collector. Due to the pressure difference, the cover sheet has to be supported from the inside in that case. Heat losses to the back of the collector are avoided by insulation material. Absorber, cover and insulation are fixed by a collector case.

The piping in the collector can either be designed with many parallel tubes that are connected by a distributor and a collector in the absorber or by a single bent tube covering the whole collector area. In the former case there is a high total mass flow in the absorber (parallel tubes) but the temperature lift during irradiation is small (high flow principle), in the latter case the total mass flow is low (only one tube) but the temperature lift is high (low flow principle).

Glazed collectors are built from units down to 1 m² up to units with 16 m². The advantage of the collectors with bigger areas is the reduced pipe work on site and therefore reduced possibility of failure. On the other hand such units cannot be mounted manually but need a crane. The collectors can be integrated into the roof, which means that the collector is the rain-tight surface of the roof and roof-tiles can be spared. The frame within the roof can be made out of wood. This is only feasible, if the roof has the right orientation, but most of the solar thermal collectors e.g. in Germany and Austria are mounted this way. The other possibility is to mount the collector on the roof as separate element. In this case the whole frame must be resistant to all weather conditions and is made of aluminium or steel. There is also a tendency to integrate solar collectors into south-facing walls. This is especially interesting for solar combisystems (domestic hot water and space heating) whose collector area for domestic hot water production is too big in summer, but which need as much solar irradiation in the heating season as possible.

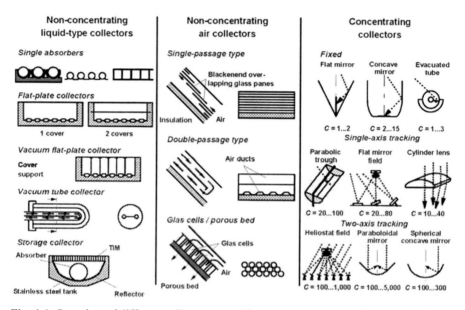

Fig. 4.4 Overview of different collector types (*C* concentration ratio; defined as the ratio of the optically active collector area to the absorber area exposed to radiation; TIM transparent insulation material; see e.g. /4-2/ and various other sources)

The storage collector (Fig. 4.4, left) is a special type of flat-plate collector. It combines the collector and heat accumulation functionality in one component. A pressure-resistant tank is installed in the centre of a radiation-focusing mirror. The tank area is either selectively coated or painted in black. The storage collector is directly connected with the cold and hot water pipes. The radiation on the collector is reflected onto the tank by the mirror; the water flowing through the tank absorbs heat and can be used; sometimes a further heating is realised with other sources of energy. Advantages of this design are the low number of components and the compact design. A major disadvantage is the high convective thermal loss that results in a significant temperature decrease inside the storage during the night or bad weather. If installed on the roof, the weight of the water has to be taken into account for the purpose of roof static's. Furthermore, ordinary tap water normally flows through this collector; therefore there is a risk of frost due to the meteorological conditions in Central and Northern Europe during the winter.

A further special type of liquid-type collector is the heat pipe, using the phase change of a working medium that interchanges between evaporation and condensation. Thus, heat can be transferred at very low temperature differences. In spite of these and further advantages (e.g. self-regulation, no overheating), this concept has not been widely accepted so far due to its comparably costly production.

Non-concentrating air collectors. Fig. 4.4 also shows different design types of non-concentrating air collectors. Due to the low heat transfer coefficient between the absorber and the air, the contact area between absorber and air flow has to be large. This is for example ensured by ribbed absorbers, multi-pass systems or porous absorber structures.

As no frost, overheating or corrosion problems can occur, air collectors have a simpler design when compared to liquid-type collectors, for example. Even leakages of the heat carrier are comparatively uncomplicated. Disadvantages are the large channels and the often significant drive capacities required for fans.

The reason why air collectors are not widely used for the heating of buildings or the supply of domestic hot water in Central and Northern Europe is that heating systems based on hot water distribution networks are commonly applied. Nevertheless they are used in individual cases, e.g. for solar food drying systems and low-energy houses with exhaust air heat recovery that are already equipped with air distribution and collector systems and thus do not require a water heating system.

Concentrating liquid-type or air collectors. These collector types reflect the direct share of solar radiation through mirror areas and thus concentrate the direct radiation on the absorber area. The level of concentration of solar radiation is the concentration ratio or the concentration factor C. It is defined as the ratio of the optically active collector area to the absorber area impinged on by radiation. The maximum theoretical concentration ratio of 46,211 is the result of the distance between sun and earth, and the sun radius. Technically, concentration factors of up to a maximum of 5,000 can be achieved at present /4-1/, /4-2/, /4-3/.

Mainly, the temperature that can be achieved in the absorber depends on the concentration factor (Fig. 4.5). The theoretical maximum absorber temperature just equals the surface temperature of the sun in the case of a maximum concentration ratio (approximately 5,000 K). The temperatures that can be realistically achieved in the absorber are significantly lower. Rotation parabolic mirrors, for example, can achieve absorber temperatures of a maximum of 1,600 °C /4-2/.

Concentrating collectors can be divided into three different groups: fixed, plus one-axis and two-axis tracking systems (Fig. 4.4, right). Fixed concentrating collectors have the lowest concentration ratios, whereas two-axis tracking systems have the highest.

Which heat carrier is used mainly depends on the achievable temperatures. Liquids are preferred in the low temperature range, whereas gaseous media are also used with increasing working temperatures.

As only direct radiation can be concentrated, the use of concentrating collectors from a technical point of view only makes sense in areas with a high level of direct radiation. In Central and Northern Europe they are practically not used at all.

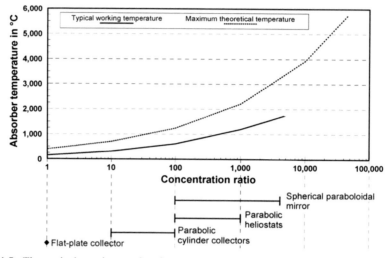

Fig. 4.5 Theoretical maximum absorber temperature and actual temperatures of concentrating collectors (e.g. /4-1/, /4-2/, /4-3/)

Data and characteristic curves. Optical and thermal losses are the decisive factors for the collector efficiency (Chapter 4.1.4). Optical losses are determined by the product of the cover transmission coefficient and the collector absorption coefficient. This loss is only dependent on the material and – approximately – radiation and temperature-independent. Thermal losses are described together with other non-constant losses by a constant heat transition coefficient (Equation (4.11), Chapter 4.1.4). As a first approximation, this loss is linearly dependent on the difference between the absorber and the ambient temperature and inversely proportional to radiation (Equation (4.14), Chapter 4.1.5).

The resulting efficiency curve for a single flat-plate collector is shown in Fig. 4.6. In the case of large temperature differences, assuming the linear dependency on the temperature, an increasing deviation from the real efficiency curve is observed. The reason for this is the non-linear increase in heat radiation within this temperature difference range. Therefore the collector Equation (4.12) or the efficiency Equation (4.17) is applied in many cases – the approximation of heat radiation is performed by using a square term in that case.

Additionally, Fig. 4.6 shows the course of the characteristic curve for the same collector at different levels of radiation. It becomes obvious that the approximation line for the efficiency is getting flatter with an increase in radiation and thus a change in the temperature difference between absorber and the environment has less impact. If the characteristic curve is drawn across the temperature difference related to radiation, the curves for the different radiation intensities almost merge into one. Therefore this form of representation is preferred in many cases (see e.g. /4-1/, /4-2/, /4-3/).

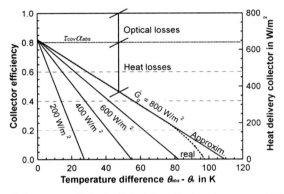

Fig. 4.6 Characteristic curves of single flat-plate collectors ($\tau_{cov}\alpha_{abs} = 0.82$; $\dot{G}_g$ global radiation on horizontal receiving area; Approxim. Approximation; see e.g. /4-1/, /4-2/, /4-3/)

Fig. 4.7 shows the courses of the characteristic curves for a number of different non-concentrating liquid-type collector designs. A single absorber might have a significantly steeper course of the characteristic curve and nevertheless achieve high specific energy yields, if it is only used in cases in which the difference between the absorber and the ambient temperature, on average, is very low. This is for example the case for absorbers for solar open-air swimming pool heating, as they are only run during the summer and the temperature level of the required heat is also low for this kind of application. Due to the missing cover ($\tau_{cov} = 1$) the optical collector efficiency is higher than for the other collector types at these small temperature differences. Collectors used all year round would generally show flatter courses of the characteristic curves as efficiency must not decrease too much with larger temperature differences.

Some typical parameters and important areas of use for non-concentrating liquid-type collectors that are most often used in Central and Northern Europe are shown in Table 4.3. The temperatures of the heat carrier in the collector are – depending on the meteorological conditions and the collector design – between 0 and approximately 100 °C during operation. Solar open-air swimming pool heating and the partial cover of the demand for domestic water are typical areas of application. Coupled solar thermal supply of domestic water and space heating (solar combined system) is also increasingly used. In Austria and Switzerland 50 %, and in Germany 30 % of the collector areas are solar combined systems /4-4/.

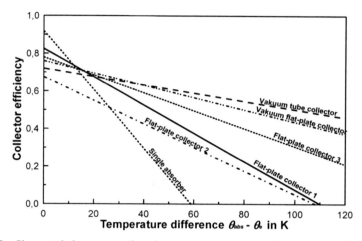

Fig. 4.7 Characteristic curves of various non-concentrating liquid-type collectors at a global radiation of 800 W/m² (e.g. /4-1/, /4-2/, /4-3/)

Table 4.3 Parameters of various non-concentrating liquid-type collector designs (according to /4-1/, /4-2/, /4-3/ and various other sources)

	Optical Efficiency	Thermal loss factor in W/(m² K)	Typ. temperature range[a] in °C	Required effort[g]	Typical application
Single absorber[b]	0.92	12 – 17	0 – 30	small	OASW
Flat-plate collector 1[c]	0.80 – 0.85	5 – 7	20 – 80	medium	DHW
Flat-plate collector 2[d]	0.65 – 0.70	4 – 6	20 – 80	medium	DHW
Flat-plate collector 3[e]	0.75 – 0.81	3.0 – 4.0	20 – 80	medium	DHW, SH
Vacuum flat-plate collector	0.72 – 0.80	2.4 – 2.8	50 – 120	large	DHW, SH, PH
Vacuum-pipe collector	0.64 – 0.80	1.5 – 2.0	50 – 120	very large	DHW, SH, PH
Accumulation collector[f]	ca. 0.55	0.55	20 – 70	very large	DHW

OASW Open-air swimming pool; DHW Domestic hot water; SH Space heating; PH Process heat; Typ typical; [a] medium working temperatures; [b] black, non-selective, not covered; [c] non-selective absorber, single cover; [d] non-selective absorber, double glass and supporting foil; [e] selective absorber, single cover; [f] Prototype ISE; [g] necessary effort for the production of the absorber.

Collector circuit. In most cases several single collectors are linked together. These collectors can be either connected in series or in parallel; mainly combinations of the two forms of connection are used. By connecting in series, the achievable temperature rise in the collector is basically increased and the total mass flow is reduced (Low-Flow). The advantage of being able to supply hot water quickly is opposed by the disadvantage of a higher thermal loss of the absorber to the environment, which is due to the larger temperature difference. The higher pressure loss of collectors connected in series can be overcome by the lower pressure losses in the pipes due to a lower level of total mass flow. The pump output decreases due to the lower level of total mass flow. When connecting in series, there is a more regular flow through the collector areas. The hydraulic layout has to be adjusted to the total mass flow. High-Flow installations with collector areas below 15 m² are mostly combined by internal heat exchangers for heat generation, whereas Low-Flow systems try to load the heat store using stratification units in order to prevent cooling the heated water in the store by mixing.

Collectors are connected to distributing pipes both for charging and discharging. In order to equally distribute the heat carrier to the individual absorbers and thus keep the pressure loss in the connecting pipe and with it, the electricity demand of the circulation pump at a low level, the distributing pipe should be larger in diameter than the absorber pipes. For the same reason, the flow channels in the collectors connected in parallel should also be the same, if at all possible. Inlets and outlets of the pipes should be connected at opposite ends /4-2/. However, together with the number of collectors connected in parallel, the differences in flow and thus the differences in the temperature rise in the collector (θ_{out} - θ_{in}) increase. Therefore, for large collector arrays, the circuits connected in parallel should be adjusted with control valves.

4.2.2 Further system elements

Heat store. A heat store is not required for the physical basic principle of solar thermal heat accumulation. However, heat stores are part of most solar thermal installations. The reason is the general non-correlation between the solar radiation supply and the demand for heat.

The heat store accumulates the heat generated in the collector by solar radiation, and stores it for times when it is required. For that purpose, a heat store has to consist of the heat accumulation medium, a solid cover with insulation material plus heat inlet and outlet devices.

The heat capacity is an important parameter for the heat accumulation medium. It is the amount of heat required to increase the temperature of a specific substance by 1 K. The specific heat capacities (in relation to mass and volume), plus the densities of various heat accumulation media are shown in Table 4.4. Further technical criteria determining a substance to be used as a heat accumulation medium are the availability, the compatibility with other substances (e.g. corrosion risk) and the environmental friendliness.

Several types of heat store designs are used. They can be differentiated according to the type of heat accumulation (chemical, thermal) and the condition of the accumulative substance. In the segment of low-temperature heat storage (up to approximately 80 °C) mainly thermal heat accumulation is used. Liquid heat stores (water heat stores), solid-matter storage and latent heat stores can be distinguished.

Liquid storage (Water storage). This is the form of storage used in most cases. The simplest case is a multi-functional open-air swimming pool. In most cases, separately installed, pressure-free tanks or tanks under pressure are used.

Table 4.4 Heat capacity and specific density of different heat accumulation media at a temperature of 20 °C (see /4-5/)

	in kJ/(kg K)	Heat capacity in kJ/(m³ K)	in kWh/(m³ K)	Density in kg/m³
Water	4.18	4,175	1.16	998
Pebbles, sand	0.71	1,278 – 1,420	0.36 – 0.39	1,800 – 2,000
Granite	0.75	2,063	0.57	2,750
Brick	0.84	1,176 – 1,596	0.33 – 0.44	1,400 – 1,900
Iron	0.47	3,655	1.02	7,860
Oil	1.6 – 1.8	1,360 – 1,620	0.38 – 0.45	850 – 900
Pebbles-water[a]	1.32	2,895	0.80	2,200

[a] 37 vol.-% water

Such storages can be charged directly or indirectly. The forced circulation systems mainly used in Central and Northern Europe normally have a pressurised heat store with a heat carrier for the collector circuit plus a cold water inlet and a hot water outlet. Often, the heat store has a second heat exchanger or an electric immersed heater for auxiliary heating of storage. The storage is normally divided into zones. The solar installation feeds the heat into the system at the lowest and thus coldest point in order to be able to drive the collector with the highest possible efficiency. The volume for the auxiliary heating is at the top end of the storage. Its size is determined by the efficiency and the required minimum running time of auxiliary heater (Fig. 4.8).

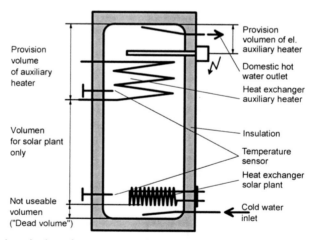

Fig. 4.8 Zonal sectioning of water storage for solar plants (el. Electrical; see e.g. /4-6/)

High-grade or enamelled steel, or steel with a temperature resistant coating (approximately 120 °C) are used as corrosion resistant and long-life tank material. In individual cases also temperature resistant, glass-fibre enhanced synthetic material can be used. The tank is insulated with mineral wool, soft foam and special synthetics on its outside. In order to avoid cold bridges, connecting flanges and fixings have to be especially insulated against thermal losses. If the solar thermal installations for domestic hot water supply are designed properly, the annual average loss including the remaining thermal bridges is between 10 and 15 % of the heat released to the heat store from the collector.

An internal heat exchanger of the collector circuit has to be located at the bottom of the storage due to the temperature layering within the boiler – the heavier cold water is located at the bottom, whereas the specifically lighter warm water is at the top. The temperature profile inside the heat store is also the reason why the cold water inlet is at the bottom and the hot water outlet is at the top of the storage in the case of direct charging and discharging. The heat exchanger for the auxiliary heating is located in the upper part of the storage. Thus, the lower volume can be entirely used by the solar system.

One variation of this type of water storage is the thermo-siphon storage. In contrast to traditional heat stores, the bottom heat exchanger is positioned vertically in an ascending pipe that is open at the bottom. Thus, during operation of the solar system, water can flow past the heat exchanger and the heated water flows upwards. The ascending pipe has specially designed openings all the way through. Depending on the temperature level, the heated water is released through these openings into the storage volume, if rising further is no longer possible due to the higher temperature in the layer above. With this stratification unit the heated water is always released to the storage at a height where both temperatures of storage and heated water are the same. This process is dependent on the collector efficiency and thus dependent on the solar radiation supply. Such an ascending pipe is called a stratification charging unit (see Fig. 4.15, right).

For larger collector areas, an external heat exchanger has to be applied as it is no longer possible to transport heat at the required minimum temperature loss using an internal heat exchanger. This requires an additional pump between the heat exchanger and the storage. The storage is either charged with water heated up by the collector at one or two fixed heights or via a stratification charging unit /4-6/.

Solid storage. Solid matter storages are mainly used in systems with air collectors and are often directly integrated into the building. They are normally fills of pebbles or other rock or mass-intensive parts of the building (e.g. walls, floors, ceilings). To give some examples, the fill can be underneath the basement floor or vertically integrated into the wall of the building. Solid matter storages can also be run with liquids as heat carriers.

In loose rock fills, the hot air from the collector is let in from the top; it releases its heat into the rocks before leaving the storage again at its bottom. Heat release works the other way round. If the parts of the building are used directly as storage, they are called hypocausts. Warm air is transferred through channels to the individual components and heats them up. The components then release the heat with a time lag and at lower amplitude to the building. In contrast to rock storages, hypocausts can only be charged in a regulated way, whereas the heat discharge is not regulated.

As the heat capacity of rock is significantly lower than the capacity of liquids, two to three times larger volumes are required for the same storage capacity. Additionally, the inlet and outlet of heat, at low temperature differences requires large heat transfer areas evenly distributed within the storage. This heat transfer is not required in fills with a direct heat carrier flow and the hypocausts. The disadvantage of the need for more space is set off by the advantage of easier production, as the rock storage is run without pressure. Furthermore there are only a few requirements in terms of tightness; it can also be run at very high temperatures.

Latent heat store. Changing the state of aggregate (phase change) of a material is performing at a constant temperature by charging and discharging energy. During the melting or evaporation processes, heat has to be added; correspondingly, heat is released during the solidification or condensation processes. Melting and solidification temperature, and evaporation and condensation temperature are the same in that case. The heat accumulated or released by the material during this phase change is called latent heat or heat of fusion. If the phase change takes place at higher temperatures than the ambient temperature, the latent heat can be stored by the material. In order to accumulate heat, heat has to be added accordingly in order to increase the temperature to the level of the phase change temperature.

In the case of low-temperature heat storage, only the phase change from solid to liquid is used, as the volume increase during the phase change from liquid to gaseous under normal pressure conditions requires a lot of effort for expansion devices in closed heat stores.

Latent heat stores are defined by a high level of energy density. Heat charging and discharging can occur at almost constant temperature levels. The main disadvantages are the volume changes that occur during phase changes. Some materials can also cool down too much during the release of heat. The varying heat conductivity in a solid and a liquid condition is also problematic. If inorganic salts are used, corrosion is an additional problem.

A special type of latent heat store is the sorption heat store. Silica gel for example can be used as a sorbent. During charging, water is extracted from the silica gel through heating. This can start from temperatures of 60 °C onwards; thus the heat provided by the solar collectors can be used effectively. The dried silica gel is easy to store. In order to extract heat, it is brought in contact with steam, and through an exothermal reaction water is absorbed. The resulting heat can be used. Due to low absolute operating pressures (10 to 100 mbar), steam can be produced by solar collectors during the winter. The energy densities range from 150 to 250 kWh/m^3 /4-7/. Unfortunately, the released heat has nearly the same amount of energy as the steam production, although at a higher temperature. Therefore sorption heat stores can be seen as a kind of heat pump. So far, no latent heat stores for solar installations are available on the market.

Duration of storage. Heat stores can be distinguished as short-term, daily and seasonal heat stores.

Short-term heat stores only store heat for a number of hours. One typical example is the water tank of storage collectors integrated into the collector.

Daily heat stores can store heat for one to several days. This is the classic case of solar thermal domestic hot water installations and solar combined systems (for domestic hot water and space heating) with solar fractional savings of up to approximately 60 %.

Seasonal heat stores are mainly used if a solar thermal installation is supposed to cover almost the entire heating demand. Large storage volumes are required in that case. Water, aquifer and vertically ground coupled storages can be used.
- Water storages can be built above or below ground with a heat-insulated steel or concrete cover or in sealed rock caverns.
- Heat accumulation in aquifers (i.e. water permeable, separated rock formations) takes place by inflow of hot and removal of cold water through a specific arrangement of wells. The heat store is discharged exactly the other way round (see Chapter 9).
- Pebble bed-water storages consist of a sealed tank that is filled with pebbles and water. These storages are self-supporting and can thus be produced cheaply. Heat capacity is lower than in water storages. It is however possible to achieve a similar layering as in water storages. Heat is added to or removed from the different layers of the storage via heat exchangers.
- Probe storages use soil or rocks as the storage media. Vertical probes are bored or rammed into the earth (see Chapter 9). Heat produced by solar energy is added or extracted through corresponding pipes serving as underground heat exchangers. Storage media are mainly rock, loam or clay. It has to be considered that the store cannot be located in an area with a groundwater stream as it would divert the heat.

Sensors and control systems. The number and the type of sensors and control instruments are largely dependent on the concept of the system. Natural circulation systems normally do not require any active regulation instruments. In forced circulation systems, mainly used in Central and Northern Europe, the collector circuit is generally actively controlled by a temperature difference control device. Temperature sensors on or inside the collector, and on or inside the storage, measure the temperature and convert it into electronic signals. The temperature in the storage is measured at the level of the heat carrier releasing the heat from the collector circuit to the storage in the case of internal heat exchangers. If external heat exchangers are used, the temperature is measured slightly above the store outlet position to the heat exchanger. Within the collector, the measurement should be taken at the hottest point near the outlet to the storage. Both temperature measurement signals are compared in the control instrument. If the target collector temperature exceeds the storage temperature, the collector circuit pump is switched on. If the temperature difference sinks below a second set value, the pump is switched off. For common solar thermal domestic hot water systems the set value for the temperature difference when switched on is between 5 and 7 K. A set value for the temperature difference when switched off is normally at approximately 3 K. The control should be exact up to 1 K. Additionally the use of time-lag devices is useful as temperature swings might occur in longer pipes. A recently introduced control strategy uses the pressure rise of the collector loop when the collector is heated up. If a certain pressure rise is detected, the collector loop pump switches on. Two temperature sensors on the hot and cold side of the collector loop are used to switch the pump off again. Such systems can be prefabricated units. No sensors have to be mounted on site /4-8/.

Apart from controlling the circulation pump within the collector circuit in forced circulation systems, maintenance of temperature limits in the storage and the collector circuit must be guaranteed. The storage temperature must not exceed a certain maximum value. In standard tanks of solar thermal domestic hot water supply systems, lime stone deposits can be caused by temperatures over 70 °C. Furthermore, an evaporation of the heat carrier in the collector circuit must definitely be avoided or, the resulting steam has to be condensed by system parts designed for that process (i.e. the heat exchanger in the heat store).

There are several ways /4-6/ of avoiding problems that may occur in the case of a collector standstill. If the maximum allowed storage temperature is exceeded, the circulation pump in the collector circuit can be switched off completely in order to avoid charging the storage with further energy. In that case the collector reaches its standstill temperature, which is significantly above 140 °C for selectively coated collectors. The collector content is evaporated. Due to the increase in volume during the evaporation process, in the best case, the entire liquid content is pressed out of the absorber and captured by an expansion tank designed for that purpose. In the worst case, the entire liquid content of the collector has to be evaporated and condensed again within the system. This normally occurs in the heat exchanger to the storage. In that case the expansion tank has to be able to also absorb the volume of the pipes /4-9/. The evaporation strategy is often used because no auxiliary energy is required. Recently, temperature-resistant heat exchangers are also offered so that there is no risk of premature ageing of the heat exchanger with this type of operation. The circulation pump should only be switched on after a collector standstill if the collector temperature is below 100 °C. Thus, it is ensured that the collector is free of evaporated media.

- The Drain-Back collector system solves the standstill problem by including a gaseous volume (nitrogen or air) into the circuit from the collector to the storage – either into the storage itself or into an intermediary integrated tank. When operating the system, the heat transfer medium flows through the gaseous volume. At collector standstill, the gaseous volume moves into the collector and the collector liquid fills the space filled by the gaseous volume beforehand. This process requires no auxiliary energy. It however requires the ability of the collector to empty itself (falling pipes, no "liquid sacks"). The gas in the collector can heat itself up to standstill temperature now without the heat carrier having to evaporate. When started up again, the circulation pump presses the gaseous volume from the collector to the envisaged tank again. It thus needs to have a larger pressure head than ordinary circulation pumps. If the gaseous volume is dimensioned in a way that all system components exposed to ambient temperature are filled with gas at standstill, the collector circuit can even be run without antifreeze.

- By running the circulation pump during the night, the collector circuit can also be used to cool the storage. As the thermal losses of the collector circuit are much higher than in the storage, the storage cools down below a defined temperature limit during the night. This temperature has to be at a level that prevents the collector from heating the storage up above its maximum temperature if the next day is warm and sunny. The disadvantage is that this type of cooling is dependent on the use of auxiliary energy and thus no heat removal can take place if a power breakdown occurs. Additionally, the decision of how much cooling down the storage needs during the night needs to be based on knowing the weather forecast for the next day.
- A system can also have its own integrated heat removal system that is switched on through the control instrument on demand (e.g. swimming pool, heat exchanger on the roof, auxiliary boiler that is cooled by natural convection through the chimney), but again having to use auxiliary energy is problematic.

Apart from these two tasks – regulating the circulation pump and maintaining the temperature limits – a suitable regulation also has to ensure additional heating in the case of low radiation.

Heat transfer medium. Some of the requirements of a heat transfer medium are:
- high specific heat capacity,
- low viscosity, i.e. good flow capability,
- no freezing or boiling at operating temperature,
- non-corrosion in the conduit system,
- non-flammable and
- non-toxic and biologically degradable.

Water fulfils most of these requirements very well. However, the danger of freezing at temperatures below 0 °C can cause problems. Water without additives can thus only be used in the warmer zones of the earth without the risk of frost.

In Central and Northern Europe mainly mixtures of water and antifreeze are used. Normally, an anticorrosive agent is also added to the antifreeze as mixtures of water and antifreeze are more corrosive than pure water. The most commonly used substances are ethylene glycol and propylene glycol; for domestic hot water supply systems normally the food-safe propylene glycol is used. The disadvantages of this additive are the lower specific heat capacity compared to water, higher viscosity and the reduced area tension. The mixture can thus permeate pores that pure water cannot get through. Furthermore, pressure losses are higher and the heat transfer worse. Thus the main components (pumps, cross-sections of pipes, heat carriers) have to be adjusted to this mix. Recently, heat carriers with a resistance of up to 290 °C in combination with alkylen-glycol and completely desalinated water have been marketed especially for solar systems with standstill operation /4-10/.

Pipes. The collector and the storage are connected by pipes. The size of the system and the absorber material determine the material chosen for these pipes. Mostly hard or soft copper pipes or corrugated stainless steel pipes, and additionally, steel and polyethylene pipes are used. However, if the absorber is made of aluminium, the use of copper piping is not advisable due to the associated danger of corrosion. However if it is the case, at least galvanic isolation has to be applied.

During operation of solar thermal systems providing domestic hot water, flows of 30 to 50 l/h are common per square metre of collector area. For a number of years now, systems with lower flows (10 to 15 l/h per square metre of collector area), so-called Low-Flow concepts, have been used /4-2/, /4-6/. Even one single flow through the collector circuit can heat up the heat transfer medium by the required temperature difference. The advantages are lower pressure losses within the pipes and the quicker supply with hot water from the collector circuit. Disadvantages are higher thermal losses in the collector and thus lower specific energy yield. Furthermore, these systems require longer series connection of the collectors in order to achieve turbulent flows and thus obtain a good heat transfer within the collector. Such Low-Flow concepts are only superior to high flow concepts if external heat exchangers in connection with several charging levels into the storage or stratifying charging units via external or especially built internal heat exchangers are used (see Fig. 4.14, right).

Cross-section and the hydraulic flow scheme determine the pressure loss to be overcome and the mass of the heat transfer medium inherent in the pipes. Large cross-sections reduce the pressure loss but make control more difficult, as the thermal mass of the pipes increases with larger cross-sections. Furthermore, the area of the pipe also increases and the thermal losses increase proportionally.

In order to reduce thermal losses, the pipes of the collector circuit have to be insulated. Materials that can be used are mineral wool, polyurethane-shells and foam rubber. Increasingly, pre-insulated double pipes made of high-grade corrugated steel pipes with an integrated duct for the collector sensor cabling are used.

In standard solar thermal systems for the domestic hot water supply, the thermal losses that still occur in the pipes, in spite of insulation, are 10 to 15 % of the energy released by the collector /4-1/.

Heat exchanger. Heat exchangers serve to transfer heat from one medium to another while separating the media physically. They have to be used if the storage is charged or discharged indirectly. The transferred heat depends on
 – the temperature difference between the two media,
 – the area of the heat exchanger and
 – the heat transfer medium and the flow speed on both sides of the heat exchanger (heat transfer coefficient).

External and internal heat exchangers are used for solar thermal systems.

One advantage of internal heat exchangers is that they do not require a lot of space. A disadvantage is the relatively low heat output, the required larger temperature difference and the limited size. Straight-tube and ribbed-tube heat exchangers are used. Sometimes double-mantle heat exchangers are used.

External heat exchangers are mostly designed with counter current mass flows (counter current heat exchangers). Common designs are shell and tube (for large systems), plate and coaxial heat exchangers. The advantages of using external heat exchangers are the higher heat transfer output at a lower temperature difference plus the possibility to charge the top of the storage with the heated-up water. For that reason, they are preferably used for collector areas larger than 15 to 20 m². Within the storage, a better temperature layering than with internal heat exchangers can be achieved. The higher thermal losses, the requirement for more space, and an additional pump required within the secondary circuit are disadvantageous.

Using a rough average, a ribbed exchange area of approximately 0.4 m² for ribbed tubes, and of around 0.2 m² for bare exchange tubes is required per square metre collector area in hot water systems with an internal heat exchanger /4-11/. Due to a better heat transfer, this exchange area can be reduced to 0.05 up to 0.08 m² /4-6/ for external heat exchangers.

Pumps. In solar thermal systems with a forced circulation a pump is required to operate the collector circuit. For standard domestic solar thermal water systems, volume flow amounts of 30 to 50 $l/(h\,m^2_{collector\ area})$ are common (High-Flow) /4-1/. For Low-Flow systems the volume flows are between 10 and 15 $l//(h\,m^2_{collector\ area})$. The layout of the collector circuit pumps also depends on this volume flow rate.

High-Flow systems always have simple centrifugal pumps, mostly equipped with a manual adjustable speed control. For Low-Flow or Drain-Back systems, however, vane or gear pumps are used that still show good efficiencies at a higher pressure rise and a lower volume flow rate.

The pumps are normally electricity-driven and are generally directly plugged into the public grid. However, they can also be connected with a photovoltaic module of the required power. They then operate as direct current pumps. For this more costly way of pump electricity supply it is advantageous that electrical energy is mainly used for the pumps if the corresponding solar energy supplies are available. Thus, radiation supply and energy demand correlate. Electrical energy storage is not necessary although the pump is driven independently of the grid.

The electrical energy required to drive the pump is between approximately 1 and 2 % for the standard solar thermal systems for domestic hot water supply. This is related to the heat available at the outlet of the solar installation. For larger systems the required electrical energy is even lower due to better pump efficiencies.

4.2.3 Energy conversion chain and losses

Energy conversion chain. A solar thermal system built with the described system components converts solar radiation energy into utilisable heat. Fig. 4.9 shows the entire energy conversion chain of such a solar thermal installation, together with collector, heat carrier and heat store (optional). According to this presentation, photons of solar radiation are absorbed by the absorber and cause the absorber atoms to vibrate. Thus the temperature in the absorber increases and heat is generated. Part of this heat is transported by thermal conduction within the absorber to the absorber pipes that the heat carrier flows through. This heat is released to the heat carrier and transported further. In most cases the heat is then transferred via a heat exchanger to a heat store before it is passed on to the consumer.

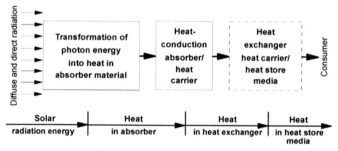

Fig. 4.9 Energy conversion chain of solar thermal heat utilisation (e.g. /4-6/)

Losses. Due to the various loss mechanisms only part of the solar radiation is available as heat to the consumer. Fig. 4.10 shows the energy flow of a solar thermal installation with a flat-plate collector, forced circulation and one to two day storage to support the domestic hot water supply for a private household of 3 to 5 people according to the current state of technology. With a collector area of approximately 6 m^2, the mean annual solar fractional saving is 50 to 60 %. During the summer it is proportionally higher – over 90 % –, and during the winter it sinks to below 15 %.

The relative losses described in Fig. 4.10 are mean annual values. They are typical for Central European meteorological conditions and are related to the radiation on the collector. Large losses of approximately 25 % thus occur due to a collector standstill if the storage has already been heated up to its maximum temperature, or the temperature required to charge the storage has not yet been reached by the collector. The greatest losses, with a total of around 38 %, occur in the collector when converting solar radiation into heat or before transporting it further with the heat transfer medium.

Such solar systems have a total annual system efficiency of around 25 %, starting from solar radiation to the actual utilisable heat of domestic hot water (here all losses of the domestic hot water storage are allocated to the solar installation) or of 32 % up to the release of the collector heat into the domestic hot water storage.

With a radiation at collector level between 3,760 and 4,520 MJ/(m^2 a), this is equivalent to an annual energy yield at the solar system outlet of between 1,200 and 1,450 MJ/(m^2 a) or 330 to 400 kWh/(m^2 a).

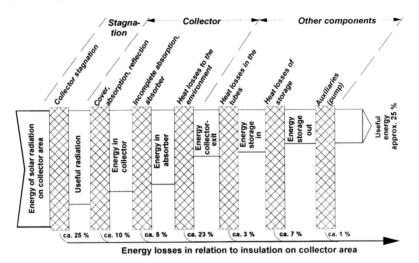

Fig. 4.10 Energy flow of a solar thermal forced-circulation system with flat-plate collector to support the domestic water supply (e.g. /4-6/)

The dimensioning of installations and the coordinated layout of the individual system components are decisive factors for the total system efficiency. The total system efficiency and the solar fractional saving are interdependent. For a given collector area the solar fractional saving increases with the increase of the efficiency of the entire system (e.g. by using better collectors, reducing duct losses or using better storage heat insulation or an increase in the storage volume). If the solar savings are increased within a system that has already been designed, e.g. by a lower demand for domestic hot water, the entire system efficiency is reduced. The reason for this is that during the summer the collector converts too much radiation into heat that cannot be utilised under these circumstances. If, on the other hand, the collector area is enlarged, maintaining the design otherwise, the solar fractional saving increases, but the entire system efficiency is also reduced, as the main part of additional heat is generated during the summer, when the solar fractional saving is already close to 100 %. Thus excess heat is lost during the summer.

4.2.4 System design concepts

The solar system consists of all the system components described above. The variety of different system installations can be described according to the type of heat transfer medium circulation. Thus,

- systems without circulation (storage collector),
- natural circulation systems (Thermo-siphon-systems) and
- forced circulation systems

can be distinguished. If the formation of the solar circuit is used to differentiate,
- open systems and
- closed systems

can be distinguished. On the basis of these criteria, five basic principles of solar systems can be defined. They are shown in Fig. 4.11 which describes the essential system components for the functionality and safe operation.

Systems without circulation (Fig. 4.11, a). In this most basic of all possible principles, heat transfer medium and the liquid actually used by the consumer are the same medium. Within the normal drinking or domestic water circuit, a suitable collector is integrated. When flowing through the collector, the water is heated up and can be used afterwards. This basic principle is used in storage collectors, to give one example.

Open natural circulation systems (Fig. 4.11, b). This most basic of all circulation concepts consists of the collector, the flow and the return pipe and a pressure-free, open storage. The reason for the natural circulation is the decrease in density in a liquid with increasing temperature. To give an example, water density at 20 °C is 998 kg/m^3 and at 80 °C only 972 kg/m^3. These density differences between the hot liquid in the collector and the cold fluid in the storage and the collector flow pipe create circulation within the system if the storage with the colder medium is positioned above the collector.

The driving forces of these differences in density are opposed by the pressure drop caused by pipe friction. Lift pressure and pressure losses caused by flow are the same when stationary. This results in the mass flow of the fluid. If radiation intensity increases, the collector outlet temperature increases, too, and with it the temperature difference between the storage and the collector. The mass flow is increased, an increasing amount of heat transfer medium, and with it heat, are transported to the storage and released to the storage medium. As a result, the temperature in the collector sinks again. Therefore, this is a self-regulating system that, at least in its basic form, can work without sensors and control instruments.

The natural circulation system is open in this case. The same liquid flows through the collector that is directly passed on to the user and is utilised in its heated-up condition. As there is generally no danger of frost in southern countries, and thus the heat transfer medium cannot freeze within the collector circuit, such systems are widespread in those regions. The collector circuit has to be resistant to corrosion as drinking water generally flows through it.

Closed natural circulation systems (Fig. 4.11, c). In order to prevent freezing and corrosion, the collector circuit can be closed in natural circulation systems.

This, however, requires a heat transfer medium that normally releases the heat in the collector circuit to a storage that can distribute the heat further.

As the circuit is sealed off from the environment, it is normally under greater pressure. In order to operate it safely, an expansion tank and a pressure control valve have to be installed within the primary circuit. If such systems are used in areas exposed to frost, frost-resistant heat carriers have to be used and the storage, the cold and the hot water service pipes must be protected against frost.

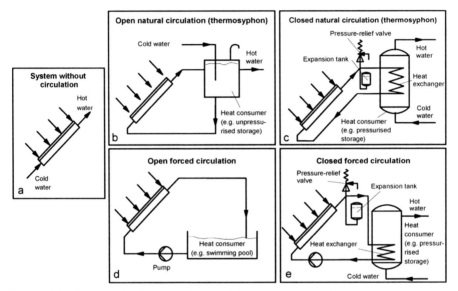

Fig. 4.11 Basic concepts of active solar thermal systems (see e.g. /4-6/)

Open forced circulation systems (Fig. 4.11, d). If the heat sink (e.g. heat store, swimming pool etc.) cannot be installed above the collectors, circulation of the heat transfer medium has to be forced by integration of a pump into the circuit. The given advantage of orientating collectors and heat sink independent from each other is of importance for heating open-air swimming pools, where the collectors are normally positioned on roofs or free spaces above the heat sink.

If the fluid in the collector cools down more quickly than in the collector loop pipes, the circulation might be reversed during the night if the pump is not running. In that case, the cold liquid is pressing downwards from the collector and extracts hot fluid from the storage or the heat exchanger. This can be prevented by e.g. integrating a check valve into the collector return flow pipe.

Closed forced circulation systems (Fig. 4.11, e). For open forced circulation systems the medium flowing through the collector circuit is generally ordinary water. Therefore these systems are exposed to the same frost and corrosion risks as open natural circulation systems. In order to avoid freezing, the forced circuit is sealed off and a frost-resistant liquid flows through. This concept of closed forced circulation is the most practical solution for applications in Central and Northern Europe. If used within buildings, the collector is normally installed on the roof. The heat from the collector circuit is normally transferred to a heat store in the basement. As for closed natural circulation systems, in addition an expansion tank and a pressure control valve are required. Additionally, a check valve, as in the open forced circulation system, also has to be installed.

4.2.5 Applications

Solar heating of open-air swimming pools. One of the best ways of using solar thermal energy is heating open-air swimming pools; where the timing of demand for heat and the available solar radiation more or less correlate. Additionally, an external heat store is not required as the open-air swimming pool filled with water can function as the storage. As the water in the pool only has to be heated up to comparatively low temperatures (a maximum of approximately 28 °C), the use of simple and inexpensive non-covered absorber mats, either installed on the roof of the open-air swimming pool or an adjacent free space, generates high energy output.

Fig. 4.12 shows the diagram and the flows of a solar-heated open-air swimming pool. Whether an additional heating source based on conventional energy carriers is necessary, largely depends on site-specific requirements. Thus, the heat gains of the open-air swimming pool consist of the following: energy $\dot{Q}_{abs}$ released into the pool by the absorbers, heat gains created by radiation impinging on the pool $\dot{Q}_G$, and the release of heat by the pool users $\dot{Q}_{human}$. Convective thermal losses $\dot{Q}_{conv}$, radiation losses $\dot{Q}_{rad}$ and evaporation losses at the water surface $\dot{Q}_{evap}$ as well as the transmission losses into the soil $\dot{Q}_{cond}$ are the inherent losses. Due to the water circulation ($\dot{m}_{in}$ or $\dot{m}_{out}$), a small amount of heat is also lost in the circulation pipes to the ambient.

In an approximation, the total of radiation and convection losses ($\dot{Q}_{rad}$ and $\dot{Q}_{conv}$) is linearly dependent on the difference between the water temperature in the pool and the mean temperature of the air. If the ambient temperature is above the pool water temperature, the convective thermal flow is reversed; the pool water then convectively absorbs ambient heat. The thermal losses caused by evaporation, which are normally the highest ones, depend on the pool area, the wind speed, the atmospheric humidity and the difference between water temperature and ambient temperature. Transmission losses to the ground are low, at around 3 % of the entire heat losses.

If the pool is covered over night, the convection, radiation and evaporation losses can be reduced significantly. Covering the pool for ten hours with the standard absorber materials reduces evaporation losses by approximately 30 % and the radiation and convection losses by approximately 16 %.

The energy gained by the incident solar radiation absorbed by the pool depends on the pool surface area and the degree of absorption of the pool water and the pool floor. The degree of absorption increases as the colour of the pool bottom and walls darken from white over light blue to dark blue, as well as with increasing water depth. Additional energy can be generated depending on the heat released by the swimmers; depending on the swimmer's movements, the heat capacity is between 100 and 400 W per swimmer /4-2/.

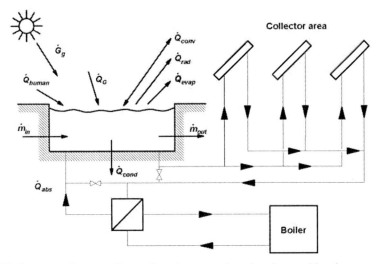

Fig. 4.12 Layout and energy flows of a solar open-air swimming pool heating system (see /4-1/)

Any energy demand that goes beyond this energy generation has to be produced by the absorbers or another fossil fuel (e.g. natural gas) or renewable energy carrier (e.g. wooden pellets) driven heater. Considering a swimming period of 130 days, between 540 and 1,620 MJ are required per square metre of pool area. The absorber area should be between 50 and 70 % of the pool area in order to achieve a mean temperature increase between 3 and 6 °C, depending on the pool cover (e.g. /4-2, 4-3/).

Small systems. In the past, the use of solar thermal systems in domestic households was limited mainly to solar-supported domestic hot water (DHW) heating. The additional space heating support by a solar system, also called solar combined system, is becoming increasingly significant. In Austria and Switzerland, 50 % off all solar systems can be categorised as combined systems /4-12/, /4-13/.

For such systems it is important to consider that the energy demand for domestic hot water is normally at the same level throughout the year; the demand for space heating, however, is generally inversely related with the solar radiation available.

DHW system with closed forced circulation. Fig 4.13 shows a complete layout of a solar thermal system with closed forced circulation to support domestic hot water production. The main criterion for dimensioning the system is the demand for domestic hot water. Demand values for average scenarios are available according to Table 4.5. According to these values, the solar system should cover the entire demand for domestic hot water 70 to 90 % of the time in summer. The volume of the storage is approximately 1.5 to 2.5 times as big as the rated demand for one day. For a household with four people with a daily demand of 50 l per person, a collector area of approximately 7 to 8 m^2 non-selectively coated or 5 to 6 m^2 selectively coated are to be installed, using standard flat-plate collectors. Additionally, a storage with a volume between 250 and 500 l is required /4-1/, /4-2/, /4-3/. If the dimensioning parameters are maintained, approximately 50 to 65 % of the domestic hot water supply can be covered by solar energy.

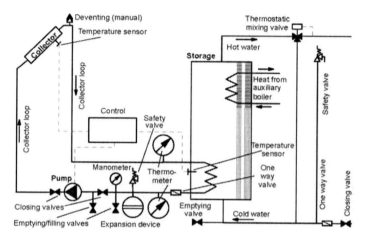

Fig. 4.13 Solar thermal forced flow-through system with flat-plate collector to support the domestic hot water supply (see /4-2/)

Table 4.5 Standard values for the domestic water demand /4-14/

	Domestic hot water l/(Person d)	Utilisable heat in MJ/(Person d)
High demand	70 – 115	10.44 – 16.70
Medium demand	50 – 70	7.31 – 10.44
Low demand	35 – 50	5.22 – 7.31

In order to achieve a higher solar fractional saving of around 70 %, a non-selectively coated collector area of approximately 15 to 18 m^2 or a selectively coated collector area of 10 to 12 m^2 would be required in the described case. The storage volume would have to be around 600 l. During the summer, however, collector standstills can occur on a regular basis.

DHW system with closed natural circulation. Most solar thermal DHW plants in Southern Europe are built with closed natural circulation systems. Fig. 4.14 shows such a system consisting of a metal frame which holds the collector, the water storage (with either horizontal or vertical axis) and all other necessary parts. It is prefabricated and can be easily installed on the flat roof of a house. Only the cold and hot water tubes have to be connected to the house installation. Electricity for pumps or control is not needed. The store is mostly built as mantle tank with the domestic hot water in the inner tank and the collector fluid floating in a mantle around this tank.

One disadvantage is, that the collector loop can not be shut down, when the water store is already at its maximum temperature. If the sun keeps shining, the temperature in the water store rises until it starts boiling. As long as the sun is shining, the boiling carries on and there has to be a pressure-controlled steam release valve incorporated into the water circuit. One simple solution to the problem is to use cheap collectors with a low stagnation temperature (i.e. collectors with non-selective surface).

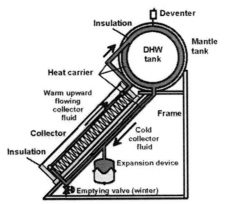

Fig. 4.14 Example of a closed natural circulation system to cover domestic hot water (DHW domestic hot water; see /4-6/)

Solar combined systems. If solar thermal systems have to cover larger portions of the entire heat demand, also space heating has to be partially supplied by solar energy. These plants are called solar combined systems. In general, improving the heat insulation of the building is more efficient than integrating a solar system into the heating system – and less cost-intensive.

There are various ways of integrating solar systems into heating systems. The following parameters have the most significant impact:
- Type of heating boiler (rolling or automatic boiler with on/off operation, solid fuel boiler);
- Type and characteristics of the heating system (high level storage mass, e.g. floor heating) or low level of storage mass (e.g. radiators); high and low-temperature system);

- Solar system (e.g. collector area and efficiency);
- User requirements (constant room temperature or possible temperature fluctuations of several degrees);
- User goal (highest possible efficiency with large effort or good efficiency at low costs) /4-6/.

As examples, Fig. 4.15 shows three types of system design. The left side of the graphic shows a two-storage layout with a power-controlled automatic boiler. One store is used to produce domestic hot water and the second to partly cover the demand for space heating with solar energy. In this case, the water heated in the boiler is directly fed into the space heating network (Fig. 4.15, left). If it is not a power-controlled boiler (e.g. wood pellets) it would have to be integrated via the space heating storage in order to achieve an increased running time and a mass flow decoupled from the mass flow of the space heating system. External heat exchangers are often used to charge the space heating store due to the larger collector areas for solar combined systems.

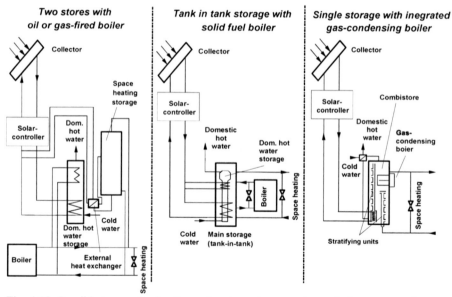

Fig. 4.15 Possible types of solar thermal systems to cover domestic hot water and space heating demands (solar combined systems) (Dom. domestic; see /4-6/, /4-13/)

The system in the middle and the system on the right in Fig. 4.15 are variations of single-storage systems. In terms of installation, they are easier to handle than storage devices that are installed separately. The double heat transfer, however, is a disadvantage (collector/storage and storage/domestic hot water heater).

The system in the middle in Fig. 4.15 is particularly suitable for solar systems in combination with a comparable inert solid fuel boiler (e.g. wood heating boiler). The domestic hot water pressurised tank is integrated into a larger heating tank. Within this double tank, natural convection and vertical temperature layering are utilised. In the top part of the hot water storage there is always enough hot water to fill one bath tub. The higher costs for the double storage are a disadvantage.

The system on the right is an integration of heat store for heating and domestic hot water plus the auxiliary heating device (e.g. a gas-condensing boiler). Such compact heat stores are also called combistores. One advantage is its compact form and the little installation work to be performed on site in order to obtain a solar combined system. The solar system feeds into the heat store via a stratifying charging unit. The burner for the conventional fuel is directly integrated into the storage with a flange. The domestic hot water is generated using a single pass external heat exchanger with controllable mass flow on the store side. Thus, storage of domestic hot water and the possibility of the emergence of legionella is avoided (see /4-6/, /4-13/).

Solar-supported district heating systems. In contrast to systems supplying individual houses with heat by individual solar systems, several heat users join to use heat from a single solar plant. If that is the case, the system is called solar-supported district heating system.

Usually, there are numerous measures of improved heat insulation that are less costly than the supply with district solar heat. In order to technically and economically optimise the overall system, a reduction in space heating in the buildings in question needs to be analysed before installing a heat supply with solar district heating systems. Particularly, low flow and return-flow temperatures of the heat distribution network have a very positive effect (e.g. 80/40 °C).

Solar-supported district heating systems are normally further differentiated as solar-supported district heating systems without and with long-term heat storage (Fig. 4.16). From the collectors installed near the heat store, the heat is transferred to the central storage via pipes and a heat exchanger. In addition, a hot water distribution network in order to distribute the heat from the heating centre to the private households is required. Two-pipe networks and four-pipe networks can be distinguished in this respect.

- The two-pipe network uses decentralised domestic hot water heating – heating is performed via the heating network either with domestic hot water stores or domestic hot water heat exchangers in the individual houses (Fig. 4.16, left); in smaller networks the heater is directly integrated, whereas it is coupled via a heat exchanger in larger networks. In order to keep the thermal losses in low-temperature networks low, heating up the domestic hot water store is done during specific time windows, for example during the night and at the time of highest solar radiation with increased flow and return-flow temperatures of the network. In such two-pipe networks, the domestic hot water circulation pipe that normally leads to thermal losses and destroys the layering, does not have to be used. In addition, the danger of legionella due to small domestic hot water volumes is low.

- In the four-pipe network, the space heating and domestic hot water are distributed separately. The advantage of this separate distribution of heating and domestic hot water (four-pipe-network, Fig. 4.16, right) is the better utilisation of the heat store and the solar system, as domestic hot water is still pre-heated, even at low storage temperatures.

Without long-term storage, solar fractional savings of approximately 10 to 20 % can be achieved for district heating systems with large collector arrays mainly supplying private households. This fractional saving is related to the energy demand for space heating and domestic hot water. Higher savings can be achieved using seasonal long-term storage. If collectors are predominantly installed on the roofs of buildings to save space, fractional savings of a maximum of 50 to 60 % can nowadays be achieved in many Central and Northern European states with their current heat control regulations, the available roof space and long-term heat storage. If enhanced measures for heat insulation are applied, the fractional saving iincreases /4-15/. Only if even higher degrees of fractional savings are aimed at, and in cases where large sunny areas, close to the consumer not used otherwise are available, the collectors are installed on such free areas.

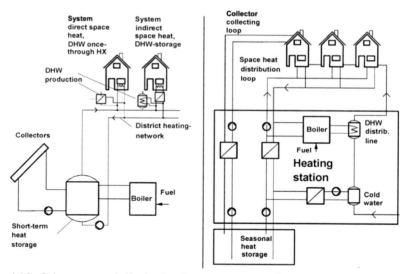

Fig. 4.16 Solar supported district heating systems: two-pipe system with decentralised domestic hot water heating (left) (see /4-18/) and four-pipe system with long-term heat store (right) (DHW Domestic Hot Water, HX heat exchanger, distrib. Distribution; see /4-15/)

For solar supported district systems, the utilisation of highly efficient flat-space collectors with optical efficiencies of almost 80 % and thermal loss factors of approximately 3 W/(m² K) is effective. They can be connected in series and/or in parallel as large collector modules (> 10 m²) with low pressure losses to form large collector arrays. Considering such collector types, for systems with partial solar cover of the domestic hot water demand (i.e. systems without long-term heat storage) collector areas of about 0.9 to 1.2 m^2 per person and a storage volume of 40 to 60 l per square metre of collector area can be expected. In the case of systems with long-term heat storage, the store should have a volume of 2 to 3 m^3 per square metre of collector area. The collector area itself would have to be between 0.4 and 0.7 m^2 per GJ annual demand for heat supply. Systems designed for partial solar coverage of space heating and domestic hot water supply can achieve useful heat gains at the exit of the solar system (i.e. at the outlet of the storage) of around 900 to 1,370 MJ/(m^2 a) or 250 to 380 kWh/(m^2 a) /4-15/. If the solar district heating system is only used to support solar domestic hot water supply, the specific yields are higher since the storage losses are lower in that case (see /4-15/, /4-16/, /4-17/).

Further applications. At the given radiation and temperature levels in Central and Northern Europe, solar thermal heat use is efficient if heat is required at a comparatively low temperature level, and the demand for heat and the availability of solar radiation, either occur simultaneously or at least do not occur at totally different times. In addition to the domestic hot water supply of private households, this is mainly the case in many public facilities. One example is the provision of water for showering in public sports facilities, particularly if the facilities are mainly or exclusively run during the summer (e.g. open air tennis courts). Other examples where a high demand for heat may also occur during the summer include camping sites, small hotels, hospitals and old people's homes as well as nursing homes /4-19/, /4-20/. In addition, there are other applications of solar energy.

– Flat-plate collectors can be integrated into existing networks for district heating at comparatively low cost. The collectors directly feed into the return-flow of a district heating network and can thus cover part of the demand for heat, especially during the summer /4-15/, /4-17/.
– By using highly efficient flat-plate collectors or vacuum pipe collectors, heat at temperatures of more than 90 °C (typically between 90 and 120 °C) can be supplied for industrial use or for BTS (Businesses, Trade and Service Sector, others; mainly small consumers) even under the given radiation conditions in Central and Northern Europe, without having to apply radiation concentration.
– For many industrial uses hot water at temperatures of up to 60 °C (e.g. photo laboratories, washing of parts) is required. These temperatures can be generated with solar energy.
– Hay and grain harvested during the summer can be dried using solar power. Air collectors can be used for this purpose.
– For larger buildings with a requirement for year-round heating and cooling, solar collectors can be used for heating during the summer, at night and during the winter for cooling purposes.
– Solar collectors can also be used to cool rooms via sorption-supported air conditioning processes in the summer.

4.3 Economic and environmental analysis

In the following, a selected number of solar thermal systems with technical parameters that reflect the current market spectrum are analysed under the aspects of cost and selected environmental effects.

4.3.1 Economic analysis

The following five case studies will be looked at, considering the climate of Würzburg (Germany) that could be used to cover the supply requirements outlined in Chapter 1. Their system parameters are shown in Table 4.6.

- Solar thermal system to support domestic water heating (SFH-I) with a collector area of 25 m². It is a solar combined system for a house with a heat load of 5 kW (at an ambient temperature of −12 °C). Approximately 62 % of the average demand for domestic hot water of 200 l/d with a temperature of 45 °C is met by solar energy. This is approximately the demand of a household with four people. The solar fractional saving is around 44 %.
- The same system (i.e. a solar combined system), but for SFH-II with a heat load of 8 kW. The solar fractional saving is around 31 %.
- Solar thermal system to support only domestic hot water heating in a household (SFH-III) with a collector area of 7.4 m². The demand for domestic hot water is the same as in the first example.
- Central solar thermal system with heating support and domestic water heating in multi-family houses (MFH) with a collector area of 60 m² and a two-pipe network directly integrated into the heating system. Of the entire energy demand for 10 flats (approximately 500 GJ/a) around 10.4 % are covered by solar energy. The solar system is combined with a conventional boiler that meets the remaining demand for domestic hot water and space heating in a heating centre for all connected households.
- Solar thermal supported district heating system to meet the demand for space heating and domestic hot water with a short-term storage (DH-I). The solar-supported district heating system supported by solar thermal energy has a boiler capacity of 1 MW and a demand for heat of 9.9 TJ/a on the heat supplier's side. The network is designed with flow and return-flow temperatures of 95/60 °C at −12 °C. The solar fractional saving is 6.2 %. It is a two-pipe network with domestic water heating directly with the consumer.

All collectors apart from the ones in the example SFH-III (decentralised domestic water heating) where copper absorbers with a high-grade steel tray are used, are selective flat-plate collectors (copper absorbers) with a mineral insulation layer and a wooden frame that are installed inside the roof. The solar district heating system (DH-I) additionally uses collector modules with particularly large areas.

The steel tanks to store the heat supplied with solar energy for decentralized domestic water heating (SFH-III) and in solar combined systems (SFH-I and

SFH-II) are located in the basement of the buildings. Apart from the solar-supported domestic hot water system of reference system SFH-III, the storages are charged by an external heat exchanger. The steel buffer storage for the central heating in the multi-family house with the two-pipe system (variation MFH) is located in the heating centre. The domestic hot water storages and the heat transfer stations are decentralised and kept in the flats. They are seen as part of the heating system. For the solar-supported district heating network with short-term storage (DH-I), the storage is kept in the heating station and the collector area is installed on the roof or in the neighbourhood to the heating centre.

Table 4.6 Technical data of the analysed solar thermal systems

System[a]		SFH-I	SFH-II	SFH-III	MFH	DH-I
Demand for space heating	in GJ/a	22	45	108	432	8,000 (+1,900
Demand for domestic hot water	in GJ/a	10.7	10.7	10.7	64.1	network losses)
Solar system						
type of collector		Flat-plate collector inside roof, selective		Collector on roof, selective	Flat-plate collector inside roof, selective	
inst. net collector area	in m²	25	25	7.4	60	620
collector loop length	in m	30	25	20	120	300
technical life time	in a	20	20	20	20	20
collector efficiency	in %[b]	18.6	21.0	28.1	38.4	26.0
spec. collector yield	in kWh/(m² a)[c]	219	248	331	453	312
	in MJ/(m² a)[c]	787	893	1,191	1,632	1,124
useful solar heat	in kWh/(m² a)[d]	161	193	252	327	274/221[g]
	in MJ/(m² a)[d]	578	696	906	1,178	985/795[g]
solar fractional saving	in %[e]	44	31	63 (5.6)[h]	10.4	6.2
system efficiency	in %[f]	14	16	21	28	23/19[g]
Storage						
type of storage		tank	tank	tank	tank	tank
storage volume	in l	2,000	2,000	500	2,000	55,000
heat exchanger		extern	extern	intern	extern	extern
Collector pump						
connection capacity	in W	2 x 50	2 x 50	30	2 x 75	2 x 400
annual running time	in h/a	1,050	1,173	1,435	2,200	1,364

[a] systems SFH-I, SFH-II, MFH and DH-I serve for the solar-supported production of space heating and domestic hot water; the SFH-III system is solely used for solar-supported domestic hot water heating; [b] useful share of solar radiation energy at the storage inlet at a mean annual solar radiation of 1,180 kWh/m² on the collector area (test reference year Würzburg, Germany, pointing South, 45° inclination); [c] heat input into the storage (without storage losses); [d] effective useful solar heat ex storage outlet; [e] for systems SFH-I and -II plus MFH and DH-I related to the demand for space heating and domestic hot water supply, for system SFH-III related to the demand for domestic hot water; [f] share of the useful solar heat at the storage outlet of the solar radiation on the collector area; all storage losses are allocated to the solar system; [g] without/with network losses (15 %) and losses of the house transfer station (5 %); [h] related to the demand for domestic hot water or the entire demand for space heating and domestic hot water.

The collector loop of solar thermal domestic water heating (SFH-III) and the solar combined systems (SFH-I and SFH-II) consist of heat-insulated copper pipes with a total length of 30 m. For a multi-family house (MFH) a connecting tube length of 120 m and for a solar district heating system with short-term storage (DH-I) a pipe length of 300 m is assumed. The hydraulic layout for the solar combined systems (SFH-I and SFH-II) and the multi-family house (MFH) are designed as Low-Flow systems via storage with an external heat exchanger and stratified charging. The solar system of the SFH-III operates according to the High-Flow principle. The specifications in Table 1.2 (Chapter 1) are valid for the district heating system DH-I.

The system efficiency includes all steps of energy transformation, from solar radiation incident on the collector area up to the useful heat at the storage outlet (for the district heating system the values are indicated without and with thermal losses of the district heating system and the house transfer stations). The relatively high system efficiency of the decentralised domestic water heating (SFH-III) and the district heating network (DH-I) are due to the solar summer yields that can be utilised almost entirely. The lower efficiency of the combined systems (SFH-I und SFH-II) is due to the system being unable to use the excess heat during the summer for domestic water heating and space heating because of the large collector area; it often stands still. The high system efficiency of the system MFH is due to the low solar fractional and the fact that solar radiation can be used during the whole summer, low return-flow temperatures from the heat distribution network.

When calculating the assumed degrees of utilisation of the conventional heating boiler that is needed in addition to the solar installation, the seasonal dependency of the boiler efficiency has to be considered to the same extent as the varying fractional saving of the solar plant during the summer and in winter. At an annual fractional saving of around 60 % for solar heating of domestic water (SFH-III) the fractional saving for example is between 80 and 100 % during the summer months, and sometimes drops to even below 20 % in winter. Thus, the efficiency of the fossil fuel boiler for domestic water heating is lower in summer than in winter. Therefore, the mean efficiency of the boiler is lower than the annual average at times when the heat can also be supplied by a solar system. This is why an efficiency of 80 % is assumed for the decentralised solar thermal domestic water heating (SFH-III) and the combined systems (SFH-I und SFH-II), where oil is substituted. For the centralised solar thermal system for multi-family houses (MFH) and the solar-supported district heating network (DH-I) a condensing boiler with a mean annual efficiency of 98 % is assumed.

In order to evaluate the costs generated by solar thermal heat utilisation, first of all the investment and the operating and maintenance costs of solar thermal systems are described. Subsequently, the specific solar heat generation costs plus the specific equivalent fuel costs will be determined on that basis. The latter are the costs of useful solar energy at the storage outlet evaluated with the efficiency of a conventional heating boiler providing heat in connection with the solar system (i.e. the costs for the (fossil) fuel avoided by solar thermal heat generation).

Investments. Investments into systems for solar thermal low-temperature heat generation can vary significantly. In the following only average costs can be indicated. The actual costs can sometimes differ tremendously from these average costs.

Collector. The approximate costs of the collectors currently available are between 50 and 1,200 €/m². The decisive factor is the collector type; simple absorber mats cost between approximately 40 and 80 €/m² and single-glazed flat-plate collectors with black or selective absorbers can cost between 200 and 500 €/m² depending on the plant size. Vacuum pipe collectors, multi-covered flat-plate collectors or collectors improved by transparent heat insulation, can increase costs to more than 700 €/m² and sometimes above.

Apart from the technology, the collector costs also depend on the size of the collector. Collector modules with large areas are cheaper, relative to their size, than small collectors; in some cases large collector modules have been offered at 220 €/m², or even less for very large collector areas (i.e. below 200 €/m²), including installation and piping /4-15/. However, costs are generally slightly above that level.

The collectors are often also available as kit and can assembled by the person/company in charge of operating the system. Costs are then somewhat lower. However, collector kits for self-assembly have lost market share over the last few years as prices for readily installed systems have gone down.

Storage. The costs for the storage depend mainly on the storage volume; investment costs for smaller systems with a storage content between 200 and 500 l including the heat exchanger are between 1.5 and 3 €/l storage volume or 100 to 200 €/m² collector area.

Heat-insulated steel tanks of up to 200 m³ are currently the state-of-the-art of technology. A 100 m³ storage costs between 300 and 400 €/m³. Larger reservoirs in the ground are significantly cheaper. Total costs between 75 and 80 €/m³ were estimated for a ground reservoir with a size of 12,000 m³. This includes the labour and material costs for setting up the building site, ground works and drainage plus steel and concrete works. Other sources quote costs between 50 and 80 €/m³ for heat-insulated ground reservoirs with metal foils for sealing and volumes between 7,000 and 40,000 m³.

Other system components. This includes the pipes, the sensors and control instruments, the pump, the anti-freeze plus all installations related to security-technology (e.g. security and shut-off valves, expansion tank). For decentralised domestic hot water supply systems, normally 20 to 30 m of pipes have to be installed. Thus, the costs for the pipes including the heat insulation are between 40 and 70 €/m² collector area. In total, the investment costs for these components are between 60 and 90 €/m² collector area for decentralised solar thermal systems.

For centralised, solar thermal domestic hot water systems the total costs of other components can vary between 65 and 130 €/m². As a first estimate, this range can also be taken as being representative for larger solar-supported district heating systems.

Installation and operation. Solar thermal systems for domestic hot water heating for households are often partly or entirely self-installed. Costs for the potential people in charge of the system are generally very low. However, if the system is installed by a company, the specific installation costs are between 70 and 250 €/m^2 of collector area. These costs include installation of the collectors, mounting of pipes, connection to the solar storage, installation of the sensors and the control instruments and the pump, connection to the residual heating system plus charging and commissioning of the system. Mounting the pipes accounts for the largest share of the costs. The costs for installing the collectors are approximately 20 to 30 % of the overall installation costs.

For central solar thermal domestic hot water support and the larger solar district heating systems, the specific costs for installation and commissioning of the system are often lower. The installation costs for larger collector arrays are approximately between 10 and 20 % of the overall collector costs or between 30 and 50 €/m^2. The total costs for installation and commissioning of the system are approximately between 50 and 100 €/m^2.

Total investments. There is a large range of total investments for solar thermal systems. Standard domestic hot water systems available in the marketplace usually cost between 5,000 and 6,000 € (e.g. /4-16/ and various other sources). In comparison, self-installation systems exclusively designed for domestic water heating are significantly cheaper; most of these systems cost between 3,000 and 5,000 €. Systems with a larger collector area are proportionally cheaper.

With around 45 % of the total costs, collectors account for the largest share of the total investment, the storage accounts for approximately 20 %, and installation and commissioning are approximately 25 % of the overall costs. Other system components make the smallest contribution with around 10 % /4-16/. The latter also include the costs for piping and antifreezing compound. For larger collector areas, the contribution to the overall cost increases to up to 60 % for a district heating network (DH-I).

Adding everything up, the average total costs for solar domestic hot water heating for a detached family house (SFH-III) according to Table 4.7 can be assumed to be at around 5,200 € (700 €/m^2 of collector area), if installation and commissioning are entirely the responsibility of a specialised business (Table 4.7, SFH-III). If the investment for the conventional domestic hot water storage and the savings on tiles for roof-integrated systems are deducted as well, the costs for a turnkey system are reduced to around 4,400 € (600 €/m^2 of collector area). If the system is self-installed and only charged and commissioned by a specialised company, the total investments are reduced to approximately 3,500 € (550 €/m^2 of collector area).

For solar-supported decentralised space heating systems (SFH-I and SFH-II) the specific investment costs are slightly lower than for smaller systems. This is due to the larger collector area with approximately 500 €/m^2 of collector area (including credit for storage and tiles). For a multi-family house (MFH) the system costs are reduced to around 460 €/m^2 of collector area due to decreasing collector costs and proportionally less piping.

Currently, total costs of approximately 450 €/m^2 for systems of up to 150 m^2 and approximately 350 €/m^2 of collector area for systems of more than 500 m^2 can be assumed (Table 4.7, reference system DH, including credit for storage and tiles) for solar-supported district heating systems for space and domestic hot water heating /4-4/. These costs include the solar thermal part consisting of collectors, collector circuit and the storage. Degrees of solar fractional savings can be between a few and 15 % when integrating a short-term storage. The pro rata investment costs of the district heating network (investment costs weighed with the solar fractional saving) are approximately 60 %.

Table 4.7 Investment and maintenance costs plus specific heat generation costs for the analysed solar thermal reference systems (for technical data see Table 4.6)

System[a]		SFH-I	SFH-II	SFH-III	MFH	DH-I
Collector area	in m^2	25	25	7,4	60	620
Useful solar heat	In GJ/a	14.4	17.4	6.7	70.7	610/490
Investments						
collector	in €	6,100	6,100	2,600	13,430	132,500
storage[b]	in €	3,400	3,400	1,000	3,400	24,000
control	in €	400	400	300	610	6,500
installation, small parts[c]	in €	3,700	3,700	1,300	13,100	54,000
Subtotal	In €	13,600	13,600	5,200	30,540	217,000
credit tiles[d]	in €	-200	-200		-730	-6,500
credit storage[e]	in €	-750	-750	-750	-2,200	
overall solar system	in €	12,650	12,650	4,440	27,600	151,500
	in €/m^2	500	500	600	460	250
District heating network	in €					1,360,000
Solar share of DH	in €					84,200
Heat transfer station (DH)	in €					6,000
Operating costs[f]	in €/a	220	230	86	540	3,500 / 4,800
Total annual costs[g]	in €/a	1,200	1,200	430	2700	19,600 / 27,900[h]
Heat generation costs	in €/GJ	82.8	68.9	63.8	37.7	32.2 / 45.8[h]
	in €/kWh	0.30	0.25	0.23	0.14	0.12 / 0.16[h]
Equivalent fuel costs	in €/GJ	66.2	55.1	51.0	37.0	31.5 / 44.8[h]
	in €/kWh	0.24	0.20	0.18	0.13	0.11 / 0.16[h]

[a] systems SFH-I, SFH-II, MFH and DH-I with solar-supported space heating and domestic water heating, system SFH-III exclusively for solar-supported domestic water heating; [b] solar storage according to Table 4.6; [c] including piping and insulation; [d] SFH-I, SFH-II, MFH and DH-I are roof-integrated systems, SFH-III is a system installed on the roof without tile credit; [e] costs domestic hot water storage without solar system; [f] at an interest rate of 4.5 % and an amortisation period over the technical life-time of the system; [g] operation and maintenance; [h] solar system of the district heating network without/with pro rata costs for the network and the heat transfer station.

Operation costs. During normal operation of a solar thermal system, maintenance costs only occur for the exchange of the heat transfer medium and for small repairs (e.g. exchange of seals). The operation of the solar thermal system also requires auxiliary energy as the heat transfer medium is normally pumped through the collector circuit. The related costs largely depend on the price for electricity. At an electricity price of 0.19 €/kWh and a demand for electricity between 0.008 and 0.03 kWh per provided kilowatt hour of thermal energy, operation costs are around 6 to 10 €/a for decentralised solar thermal systems for domestic hot water heating and between 18 and 25 €/a for solar combined systems. Maintenance costs for most parts of the system are between 1 and 2 % of the overall investment (without installation and commissioning) /4-14/. Thus, the entire annual maintenance and operating costs for solar thermal domestic water heating and the combined systems are at approximately 0.9 to 1.8 % of the overall investment (including installation and commissioning). In relation to the reference system SFH-III, these are annual costs between 60 and 80 € (see Table 4.7, SFH-III).

For a larger solar-supported district heating system, annual total costs of approximately 1 % of the overall investment costs can be assumed for maintenance and miscellaneous costs (e.g. insurance) (excluding installation and commissioning of the system) /4-15/. Costs for the annual electricity consumed by the collector circulation pump, plus, if included in the calculation, the pro rata maintenance costs for the solar-supported district heating network have to be added (see Table 4.7, reference system DH-I).

The costs mentioned in Table 4.7 are average costs for the respective system components at a given configuration of the reference systems. The installation is performed by a company; it also includes the connection of collector, storage and heating boiler. The sensors and the control instruments, the pump plus all devices for security control (e.g. security and shut-off valves, expansion tank) are combined as control costs.

Saving of tiles, plus investment costs for domestic hot water storage without solar support, is credited to the solar system. Thus, only those additional costs are allocated to the solar systems that result from the necessary increase in storage volume compared to conventional domestic hot water storage.

Heat generation costs. The specific energy supply costs can be derived from the absolute investments indicated plus the costs for maintenance and operation. The investments are amortised over the technical lifetime of the system (20 years, Table 4.7). Under the financing assumptions (see Chapter 1) made so far, the results for solar thermal domestic hot water heating and combined systems in private households (reference systems SFH-I, SFH-II and SFH-III, Table 4.7) are costs between approximately 60 and around 85 €/GJ for the heat provided by solar energy. The solar thermal system in the multi-family house is characterised by specific heat generation costs of approximately 38 €/GJ (Table 4.7, reference system MFH). For the solar district heating system, the costs for using short-term storages are at approximately 32 €/GJ excluding a proportion of the solar-supported district heating network and the house transfer station and at 46 €/GJ including the latter (Table 4.7, reference plant DH-I). Doubling the heat generation costs by considering the network and the house transfer station is, on the one hand, determined by the increase in investment costs by 40 % and the operating costs of 30 % and by the degree of efficiency of the district heating network on the other hand. Thus, out of the examples analysed in this chapter the solar thermal support of space heating and domestic hot water heating of multi-family houses and solar-supported district heating with a short-term storage with comparatively low degrees of solar coverage (here 10.4 % and 6.2 %) are the most favourable options in economic terms.

In addition to these solar heat generation costs, namely the costs resulting from investments and operating costs of the solar system, the equivalent fuel costs are also shown in Table 4.7. The costs of useful solar energy at the storage outlet are assessed, considering the degree of utilisation of the conventional heating boiler that supplies heat in combination with the solar installation. The equivalent fuel costs have a major impact on the decision of a house owner whether a solar thermal system will be installed or not, as they allow to immediately calculating the direct annual cost saving for e.g. fossil energy carriers and the expected annual amount of fuel saved. They enable the direct comparison between solar thermal heat supply and the avoided fuel costs for fossil and sometimes also biogenous energy carriers.

Such a comparison with the equivalent fuel costs shown in Table 4.7 reveals that currently all analysed variations of solar thermal heat supply are generally significantly more expensive than conventional space heating or domestic water heating with oil or gas. When comparing domestic water heating with electricity during the summer at an electricity price of approximately 70 €/GJ (household tariff) with specific heat costs, all solar examples are more cost-effective.

However, these values should not be considered as generally valid mean or reference values. In special cases significant deviations can occur under the given marginal and boundary conditions. The price for solar heat for open-air swimming pools, for example, is between 7 and 14 €/GJ. Hence, in many cases solar open-air swimming pool heating is currently already cheaper than conventional heating. The reason is that times with a high level of radiation supply coincide with times of a high demand for low-temperature heat – without a storage system, as the pool water serves as a heat store. Furthermore, the uncovered absorbers used for open-air swimming pools are cheaper than covered collectors.

In order to better assess and evaluate the influence of the different variables, Fig. 4.17 shows a variation of the main sensitive parameters. A decentralised solar thermal domestic water heating system (SFH-III) served as starting point. According to the example, a change in investment costs plus a variation of the interest rate have the most significant impact on the heating costs for a given climate. In the given example, an investment reduced by 30 % cuts the specific heat generation costs from almost 55 to around 41 €/GJ. This shows the importance of economic incentives for the market implementation of solar thermal systems. Furthermore, different climates have been analysed. Starting from Würzburg/Germany with a reference radiation of 100 %, different climates from Genoa with 30 % or more, to Helsinki with 10 % less, irradiation are shown. Costs were estimated to remain the same. It can be observed that the climate of all analysed parameters has by far the highest impact on the heat production cost.

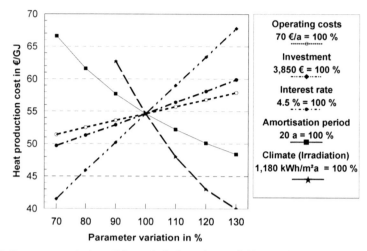

Fig. 4.17 Parameter variation of the main influencing variables on the specific heat generation costs for decentralised domestic water heating (reference system SFH-III, Table 4.7)

4.3.2 Environmental analysis

Solar systems are characterised by noise-free operation without direct substance releases. The following analysis of the local environmental aspects differentiates between construction, normal operation and malfunction, plus end of operation.

Construction. The environmental effects of the construction of solar systems are generally in line with those of the manufacturing industry. Only the production of the absorber is of particular environmental significance. In the past galvanic coating methods were used that required a high level of energy input and produced problematic waste. Recently vacuum coating or sputtering, which is much less problematic in terms of environmental impact during the production process, has increasingly gained importance /4-7/, /4-9/. The anti-reflection glasses that have recently been increasingly used to cover the solar collectors can also be produced following environmental criteria /4-18/.

When producing solar storages, materials that can be produced and processed with little environmental impact have increasingly been used over the last few years. For example polyurethane foams (PU) that can cause environmental problems during production and disposal, have been replaced by polypropylene (PPP) in many cases /4-21/.

More than half of the solar collectors offered in the German Market have been allocated the Blue Angel (RAL-ZU 73). This Blue Angel shows that no halogenated hydrocarbons have been used as heat transfer medium and that the substances used for collector insulation have not been produced using halogenated hydrocarbons.

Thus during the production process of solar systems no environmental effects occur that exceed the general average. If the appropriate environmental protection regulations are adhered to, a very environmentally-friendly production is generally possible.

The rooftop installation of collectors can be dangerous. The risk of dying as a result of falling from the roof during system installation, is comparable to that of a roofer, chimney sweeper or carpenter, and is thus considered low.

Normal operation. As the operation of solar collectors does not release any substances, they can generally be run in a very environmentally-friendly way. Additionally, collectors installed on the roof are relatively similar to roofs in terms of their absorption and reflection behaviour. Thus hardly any negative impacts on the local climate are to be expected in the case of an on-roof installation. The roof areas covered with collectors that can sometimes be seen from far away only have a minor impact on the visual appearance of cities and villages. The space utilisation of solar collectors is also quite low, as generally already existing roof areas are used.

Only if collectors are installed on free areas, a negative impact on the microclimate might be possible. However, it is limited mainly to the shadow area and is negligibly low. In principle, shadowed areas can still be used extensively for farming.

Furthermore, evaporation during collector standstill ought to be prevented by an appropriate system design and thus not be a health risk.

Malfunction. Environmental effects caused by larger failures cannot be expected from solar collector systems. Health risks for human beings or groundwater or soil contamination by a possible leakage of the heat transfer medium containing antifreeze compound are very unlikely due to an advanced technology. Such problems can also be avoided by regular inspections and the use of food-safe heat transfer media (e.g. propylene-glycol-water-mixes).

It is possible that fires can release a limited amount of air transported trace gases into the environment. However, they are not specific for solar thermal systems; furthermore, because of their design, fires at the collectors can only be expected if the entire building on which they are installed is on fire.

Additionally, possible dangers of injuries by falling collectors that have not been correctly installed on the roof can normally be avoided by maintaining the generally valid health and safety standards; the danger potential is at the same level as that of roof tiles.

Legionella can multiply significantly in domestic hot water systems and thus become a danger for human beings if they get in contact with the infected water. However, this is not a problem specific to solar systems, but this problem has also occurred in solar systems in the past. As legionella die quickly at a temperature of approximately 60 °C, this danger can be easily limited by appropriate technical measures. Tests have also shown that extended storage periods, exceeding by far the requirements of DVGW recommendation (German Association for Gas and Water), do not necessarily lead to a multiplication of legionella /4-22/. If the corresponding DVGW regulations are observed, the multiplication of legionella can be safely avoided. These requirements have been fulfilled for all modern solar systems. Out of the system layouts analysed in this chapter, this problem might only occur with regard to the system SFH-III having domestic hot water storage. In all other systems legionella will virtually not occur.

Altogether, the potential environmental impacts of solar thermal heating are also low in case of an accident.

End of operation. In principle, recycling main parts of solar thermal systems (e.g. solar collector, storage) is possible. The producers in Germany, for example, are also committed to take the collectors back after the end of the technical lifetime and recycle the materials as part of the German Blue Angel Agreement. Thus, there are environmental effects common for certain materials being recycled. They are, however, not specific to solar systems.

5 Solar Thermal Power Plants

The term "solar thermal power plant" comprises power plants which first convert solar radiation into heat. The resulting thermal energy is subsequently transformed into mechanical energy by a thermal engine, and then converted into electricity. For thermodynamic reasons high temperatures are required to achieve the utmost efficiency. Such high temperatures are reached by increasing the energy flux density of the solar radiation incident on a collector. In this respect, we refer to concentrated radiation or concentrating collectors. As an alternative, with regard to technical/economic optimisation of the overall system, also lower temperatures, resulting in considerably reduced costs may be desired in some cases. However, such concepts imply the use of large-surface cost-efficient collectors. The above mentioned framework conditions give rise to a whole series of different solar thermal power plant concepts.

According to the type of solar radiation concentration, solar thermal power plants are subdivided into concentrating and non-concentrating systems. The former are further subdivided into point and line focussing systems. In addition, further differentiations can be made, e.g. according to the type of receiver of the solar radiation, the heat transfer media and the heat storage system (if applicable) or the additional firing based on fossil fuel energy; however, to structure this chapter as clear as possible, these differentiations have not been considered here.

The term "concentrating systems" covers primarily the following power plant concepts:
- solar tower power plants (i.e. central receiver systems) as point focussing power plants,
- dish/Stirling systems as point focussing power plants and
- parabolic trough and Fresnel trough power plants as line focussing power plants.

Non-concentrating systems include the concepts of solar updraft tower power plants and solar pond power plants.

The different options are explained and discussed in the following sections. However, emphasis has been laid on those technologies and processes which seem most promising to contribute substantially to cover the given electricity demand on a global scale.

Concentrating collectors can reach temperature levels similar to that of existing fossil-fuel fired thermal power stations (e.g. power plants fired with coal or natural gas). Consequently, the various components required for the actual thermal energy conversion process (including, for instance, turbine and generator) are already state-of-the-art technology. Due to this, within the following sections, only the solar-specific part of such power plants is discussed in detail.

5.1   Principles

Basically, the process of solar thermal power generation is realised within the following steps:
– concentrating solar radiation by means of a collector system;
– increasing radiation flux density (i.e. concentrating of the solar radiation onto a receiver), if applicable;
– absorption of the solar radiation (i.e. conversion of the radiation energy into thermal energy (i.e. heat) inside the receiver);
– transfer of thermal energy to an energy conversion unit;
– conversion of thermal energy into mechanical energy using a thermal engine (e.g. steam turbine);
– conversion of mechanical energy into electrical energy using a generator.

Fig. 5.1 illustrates the general energy conversion chain of such a solar thermal power generation plant. Selected aspects of this kind of solar energy conversion are discussed below.

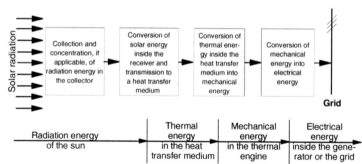

Fig. 5.1 Energy conversion chain of solar thermal power generation

5.1.1   Radiation concentration

Radiation concentration is necessary if higher temperatures than those generated by flat-plate collectors are required. The concentration of solar radiation is described by the concentration ratio. It is defined according to two different methods:
- On the one hand, concentration ratio C can be determined solely geometrically (C_{geom}), describing the ratio of the solar aperture surface A_{ap} to the absorber surface A_{abs} (Equation (5.1)); also, the explanations within this chapter are based

on this definition. Thus, the concentration ratio of a typical parabolic through collector of an aperture width of 5.8 m and an absorber tube diameter of 70 mm amounts to approximately 26. With regard to parabolic through collectors, sometimes the ratio of aperture width to absorber tube diameter is referred to as concentration ratio; this quantity differs from the concentration ratio defined by Equation (5.1) by factor π.

$$C = C_{geom} = \frac{A_{ap}}{A_{abs}} \tag{5.1}$$

- On the other hand, the concentration ratio C can be defined as the ratio of the radiation flux density G_{ap} at the aperture level and the corresponding value G_{abs} of the absorber (C_{flux}, Equation (5.2)). However, this definition is only mentioned here to complete the picture.

$$C = C_{flux} = \frac{G_{ap}}{G_{abs}} \tag{5.2}$$

On the basis of the second fundamental theorem of thermodynamics the maximum possible concentration ratios for two-dimensional (parabolic trough-type) and three-dimensional (e.g. paraboloids of revolution) concentrators can be deduced /5-1/. For this purpose the "acceptance angle" $2\theta_a$ is required. This angle covers the entire angular field of solar beams to be focussed by the collector, without having to move the collector or part of it.

For single-axis concentrators (e.g. parabolic trough), for instance, the maximum concentration ratio $C_{ideal,2D}$ for a given acceptance semi-angle θ_a is calculated according to Equation (5.3).

$$C_{ideal,2D} = \frac{1}{\sin\theta_a} \tag{5.3}$$

For two-axis concentrators (e.g. paraboloids of revolution) the maximum concentration ratio $C_{ideal,3D}$ is calculated according to Equation (5.4).

$$C_{ideal,3D} = \frac{1}{(\sin\theta_a)^2} \tag{5.4}$$

Since on the earth's surface the acceptance angle $2\theta_a$ for the sun amounts to $0.53°$ or 9.3 mrad, maximum ideal concentration factors of 213 are determined for two-dimensional geometries (line focussing parabolic through) and of 45,300 for three-dimensional (point focussing) geometries.

However, in practice the acceptance angle of the concentrator must be increased so that the actual achievable concentration ratio is necessarily considerably reduced. This is due to the following aspects:

- Tracking errors, geometric reflections as well as imperfect orientation of the receiver lead to acceptance angles which are considerably bigger than the aperture angle of the sun.
- The applied mirrors are imperfect and expand the reflected beam.
- Atmospheric scattering expands the efficient aperture angle of the sun far beyond the ideal geometric value of the acceptance semi-angle of approximately 4.7 mrad.

Radiation concentration is aimed at increasing the possible absorber temperature and consequently the exergy of the concentrated heat. In addition, absorber surfaces may be of smaller design due to the concentrated solar radiation. It is thus easier to reduce the inevitable thermal losses due to radiation, convection and heat conduction. In case of absorbers of parabolic through collectors, this is achieved by evacuated cladding tubes and by an absorber coating with a low emission coefficient within the relevant wavelength range. Fig. 5.2 shows the impact of the concentration ratio on the collector efficiency η_{Coll} over the absorber temperature θ_{abs} for two different emission coefficients ε of the absorber (left $\varepsilon_{abs} = 1$, right $\varepsilon_{abs} = 0.08$).

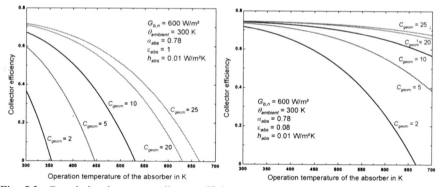

Fig. 5.2 Correlation between collector efficiency η_{Coll} and operation temperature of the absorber θ_{abs} as well as geometric concentration factor C_{geom} (left for an absorber of an emission coefficient $\varepsilon_{abs} = 1$ ("black body radiator"), right for $\varepsilon_{abs} = 0.08$, a practically achievable value; for the sake of simplicity a constant intercept factor (i.e. ratio of incident to reflected radiation) of 0.96 has been assumed; $G_{b,n}$ describes the direct normal radiation; α_{abs} is the absorption coefficient of the absorber; h_{abs} is the thermal loss coefficient of the absorber)

Since direct concentration by diffraction or refraction can only be performed by rigid, transparent materials (e.g. glass lenses), characterised by high costs, this option has not been applied on a large scale for economic reasons. Reflecting surfaces have proven most cost-efficient since they reflect the almost parallel

incident radiation onto a certain point or line. Parabola profiles show these properties (Fig. 5.3 a)). To achieve the highest possible concentration ratio – i.e. the ratio
of reflector surface to absorber surface should be very high for economic reasons
– the profiles are designed as rotational solids (Fig. 5.3 e)). As an alternative, the
reflecting profile can also be extruded, so that the focus is not of point but of line-
shape. These possibilities are outlined in Fig. 5.3 c).

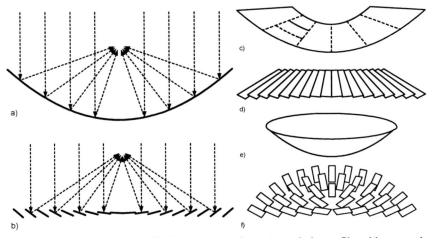

Fig. 5.3 Parabola profiles for radiation concentration (a) parabola profile with ray path,
b) segmented parabola profile (Fresnel) with ray path, c) and d) profiles extruded from
profiles a) and b), e) and f) rotational solids of profiles a) and b))

The more flat the parabola, the more distant is the focal point (or the focal
plane of the image of the sun) from the apex of the parabola. When compared to
steeper parabola profiles, flat profiles are characterised by a lower ratio of reflector surface to aperture surface (i.e. effective collector surface). Specific material
consumption is thus reduced. Nevertheless, a certain depth of the profile and considerable technical effort are inevitable. As an alternative, segmented parabola
profiles – also referred to as Fresnel profiles – are applied. The parabola profile is
subdivided into smaller segments of the same slope at the same point as the profile, but located on the same level. Due to blocking of incident and reflected radiation, reflection efficiencies (i.e. ratio of radiation on aperture surface and concentrated radiation) are generally lower than for realised parabola profiles. Fig. 5.3 d)
shows schematic representations of such segmented profiles for extruded profiles
and Fig. 5.3 f) contains a schematic representation of rotational solids.

Rotation profiles present the advantage of generally higher concentration factors and thus higher process temperatures. However, these concentrators need
two-axis sun-tracking requiring greater technical efforts. Line-focussing systems
only require single-axis tracking and result in lower process temperatures and
efficiencies.

In the following, plants equipped with rotation parabola profiles are referred to as dish/Stirling systems or parabola power plants, and large-scale plants with segmented rotation profiles as solar tower power plants (due to the focal plane being located on a tower). In case of line-focussing plants equipped with extruded parabola profiles the common technical terms are parabolic trough power plants or linear Fresnel collector power plants.

Besides the optical properties of the material applied for the reflector, achievable efficiencies are largely influenced by the geometry of the reflectors and the precision of the sun-tracking system. In practice, mainly optical measuring methods are used to assess the concentrator quality and the performance.

Typical concentration factors and parameters of different solar power generation technologies applying concentrating collectors are summarised in Table 5.1. To allow for a comparison, technical data of non-concentrating solar-thermal power plants have also been added.

Table 5.1 Concentration factors and technical parameters of selected solar thermal power generation technologies

	Solar tower	Dish/Stirling	Parabolic trough	Fresnel reflector	Solar pond	Solar up-draft tower
Typical capacity in MW	$30 - 200$	$0.01 - 1^a$	$10 - 200^c$	$10 - 200^c$	$0.2 - 5$	$30 - 200^c$
Real capacity in MW	10	0.025	80	0.3^d	5	0.05
Concentration factor	$600 - 1,000$	up to 3,000	$50 - 90$	$25 - 50$	1	1
Efficiency^b in %	$10 - 28$	$15 - 25$	$10 - 23$	$9 - 17^d$	1	$0.7 - 1.2$
Operation mode	grid	grid/island	grid	grid	grid	grid
Development status^e	+	+	++	0	+	+

^{a} by interconnection of many individual plants within a farm; ^{b} conversion of radiation energy into electrical energy, annual average is site-specific; ^{c} assuming a solar multiple of 1.0; ^{d} incorporated into a conventional power station; ^{e} 0 successful operation of demonstration plants, + successful continuous operation of demonstration plants, ++ commercial plants in operation.

5.1.2 Radiation absorption

All materials absorb part of the incident solar radiation. The absorbed incident radiation causes atoms of the material to vibrate, whereby heat is generated. This heat is either transferred within the absorbing material by heat conduction and/or released by heat radiation or convection back to the atmosphere.

The major portion of solar radiation consists of visible light (Fig. 2.8); i.e. the short-wave portion of radiation is predominant (Chapter 2.2). The distribution of luminosity of the different wavelengths corresponds approximately to that of a black body radiator of a temperature of approximately 5,700 K. In contrast, with regard to the temperatures relevant to solar thermal plants (approx. 100 to 1,000 °C) bodies radiate mainly medium and short-wave radiation (Wien's Law). When observing only a small spectral range, the absorption coefficient and emission coefficient are identical (Kirchhoff's Law). However, suitable "selective" coatings ensure that short-wave sunlight is well absorbed while (long-wave) heat radiation is inhibited. Such absorber materials are thus characterised by high absorption coefficients α_{abs} with regard to solar radiation and low emission coefficients ε_{abs} in terms of long-wave heat radiation: they are sometimes also referred to as α/ε coatings (see Fig. 5.2).

5.1.3 High-temperature heat storage

Solar radiation is an energy source whose intensity varies deterministically due to the rotation of the earth (day/night) and stochastically as a result of actual meteorological influences (clouds, aerosols, etc.). To compensate for such fluctuations thermal storages can be applied.

In this respect, heat transfer medium storage, mass storage and storage of phase-change material are distinguished.

- In case of storage of the heat transfer medium, it is intermediately stored in thermally insulated containers. However, this implies that the heat transfer medium is inexpensively available and has a high volume-specific heat capacity to minimise container costs. To date, thermal oil and molten salt containers have been applied; however, also water/steam accumulators have been planned. The advantage of this storage mode is the constant temperature of the hot heat transfer medium, which is only reduced by heat losses of the storage tank (and thus is a function of the storage period, the container surface and the insulation).
- In case of mass storage, the heat transfer medium thermally charges a second material of a high heat capacity. For this purpose a good heat transfer (i.e. large surfaces and high heat transmission coefficients) must be provided between the heat transfer medium and the storage material, to ensure the required driving temperature difference and to reduce the ensuing exergy loss of heat transmission. Mass storages are applied if the heat transfer medium itself is too expensive (e.g. synthetic heat transfer oil) or difficult to store (e.g. depressurised air). For mass storages the following combinations are applied: thermal oil/concrete, thermal oil/molten salt, steam/oil-sand and air/ceramics bricks. Mass storage offers the advantage of very inexpensive storage material. However, it also presents the disadvantage that in addition to the common heat loss of the storage tank also an exergy loss occurs during double heat transmission when charging and discharging the storage material.
- Within storage tanks with phase-change material steam is condensed isothermally, so that a storage material (e.g. salts such as NaCl, NaNO$_3$, KOH) is solidified/melted isothermally. Also in this case, an exergy loss occurs due to double heat transmission. Moreover, such phase-change materials are still very expensive.

Table 5.2 shows the thermodynamic data of selected storage media. One characteristic parameter is the thermal penetration coefficient a_{th}. According to Equation

(5.5) a_{th} it is defined as the root of the product of thermal conductivity λ, density of the storage medium ρ_{SM} and specific heat capacity c_p.

$$a_{th} = \sqrt{\lambda \cdot \rho_{SM} \cdot c_p}$$

(5.5)

Table 5.2 Parameters of different storage materials (standard values)

	Maximum temperature	Thermal conductivity	Density	Specific heat capacity	Thermal penetration coefficient
	in °C	in W/(m K)	in kg/m³	in J/(kg K)	in Ws$^{1/2}$/Km²
Silicone oil	400	0.1	970[a]	2,100	450
Mineral oil	300	0.12	900[a]	2,600	530
Molten sodium chloride	450	0.57	927[b]	1,500	890
Insulating bricks	700	0.9[c]	Approx. 1,000[d]	950	925
Reinf. concrete	400	1.5	2,500	850	1785
Construction steel	700	40	7,900	430	11,700

Spec. Specific; Reinf. Reinforced; Approx. approximately; [a] at 20 °C; [b] at melting point; [c] 0.18 to 1.6 W/(mK); [d] between approximately 800 to 1,200 kg/m³.

5.1.4 Thermodynamic cycles

The exergy of heat can be utilised by closed or open cycles. In these processes a working medium undergoes a series of state changes which are either caused by heat exchange or performance of work.

-   If the initial state is identical to the final state, so that the working medium could undergo the same process again, the process is referred to as a "closed cycle" (Fig. 5.4).

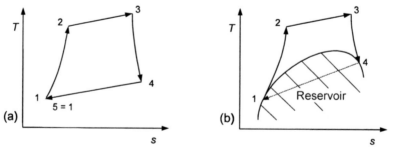

Fig. 5.4 Temperature/entropy-diagram (*T,s*-diagram) of a closed (left, (a)) and an open cycle (right, (b)) (the arrows refer to the direction the cycles performed)

- If the working medium is part of an "inexhaustible" reservoir (e.g. ambient air) and its final state is different from the initial state, the process is referred to as an "open cycle" (Fig. 5.4); yet, strictly speaking, such a process is also closed since the last state change takes place outside of the actual process, namely within the "inexhaustible" reservoir.

In the following such cycles are illustrated by means of temperature/entropy-diagrams. These representations offer the advantage, that both isothermal (i.e. constant temperatures) as well as isentropic (i.e. constant entropy) state changes can be represented as straight lines (Fig. 5.4 (a)) /5-2/.

- Within the Carnot cycle the entire exergy is extracted from the supplied heat so that its full working capacity becomes useful. This cycle consists of isentropic compression/decompression (i.e. performance of pressure change work) and isothermal heat supply and dissipation. The Carnot cycle is an ideal comparative process; however, mainly the isentropic compression/expansion cannot be put into practice (Fig. 5.5 (a)).
- The Ericson cycle represents the first technical approach to an ideal Carnot cycle; isobaric compression and expansion substitute isentropic compression/decompression. Within this cycle addition and evacuation of heat is supported by internal heat transmission (Fig. 5.5 (b)).
- The Stirling cycle is similar to the Ericson cycle. However, compression/decompression is isochore (i.e. density remains constant) (Fig. 5.5 (c)).

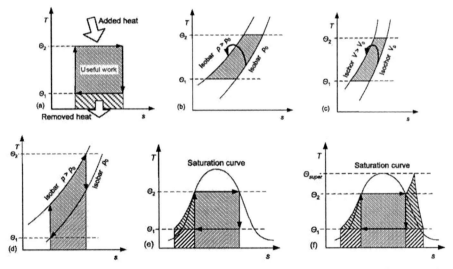

Fig. 5.5 Temperature/entropy diagram (T,s-diagram) of various cycles ((a) Carnot cycle, (b) Ericson cycle, (c) Stirling cycle, (d) Joule cycle, (e) Clausius-Rankine cycle, (f) Clausius-Rankine cycle with superheating) (p pressure, V volume, T,θ Temperature, s entropy)

- The Joule cycle is composed of isentropic compression, isobaric heat addition (combustion), isentropic expansion and isobaric heat dissipation (Fig. 5.5 (d)).

- The Clausius-Rankine cycle (steam power cycle/two phase cycle) makes use of the phase transformation of matters. Such phase transformations correspond to isothermal heat addition and large additions of specific volume. Their technical application is easy (isotropic compression/decompression, isothermal heat addition and dissipation). This is why such processes were first technically applied (Fig. 5.5 (e), (f)).

For current industrial applications Joule and Rankine cycles are most commonly applied.

- For the Joule cycle the working medium "ambient air" is aspirated and compressed prior to adding heat. Heat can either be added by caloric devices or internal combustion (e.g. by combustion of natural gas). For solar applications heat is transferred directly from the absorber to the working medium of the energy conversion process. The volumetric absorber itself has a very large surface to benefit both heat transfer and radiation absorption. Since pressurised air is used as working medium such an absorber must be of closed design. Indirect heat addition, for instance by means of a heat transfer medium, is disadvantageous since the working medium air only has a poor thermal conductivity and thus requires large surfaces for heat transmission.

- The Clausius-Rankine cycle, by contrast, requires a phase change medium to allow for isothermal heat addition. In most cases water is applied, but there are also processes using organic working media for low-temperature applications (so-called Organic Rankine Cycles (ORC)). At the beginning, the liquid working medium is highly pressurised and undergoes a phase change while heat is added. The now gaseous material is subsequently expanded, possibly after further heat has been added. Afterwards condensation is performed under low pressure while heat is dissipated.

All above-mentioned cycles have in common that heat is first applied to increase the volume flow of a gaseous working medium. Subsequently, during its expansion, this volume flow performs mechanical work in pressure engines, which can either be designed as oscillating machines of varying working volume (i.e. reciprocating engines) or as machines with stationary flow (i.e. turbo-machines or turbines). For large-scale power plants dealing with large volume flows almost exclusively turbo-engines are applied.

Turbines are referred to as turbo-engines which first transform the potential energy of a flowing working medium into kinetic energy and afterwards into mechanical energy of the rotating turbine shaft. The medium flows through the turbine either axially or radially, causing it to rotate. A stator whose blades form nozzles causes the working medium to first expand and at the same time accelerates the rotor. Inside the rotor coupled to the turbine shaft the kinetic energy of the working medium is subsequently converted into shaft torque. The combination of rotor and stator is referred to as turbine stage; for instance, in large turbines up to sixty subsequent stages are implemented. Inevitable friction, inconvertible kinetic energy at the turbine exit and so-called gap leakages are considered to measure the efficiency of a turbine; current steam turbines reach efficiencies above 40 %, while those of gas turbines even exceed 55 %.

5.2   Solar tower power stations

Within solar tower power plants (also called "central receiver systems") mirrors tracking the course of the sun in two axes, so-called heliostats (Greek term for "immobile sun"), reflect the direct solar radiation onto a receiver, centrally positioned on a tower. There, radiation energy is converted into heat and transferred to a heat transfer medium (e.g. air, liquid salt, water/steam). This heat drives a conventional thermal engine. To ensure constant parameters and a constant flow of the working medium also at times of varying solar radiation, either a heat storage can be incorporated into the system or additional firing using e.g. fossil fuels (like natural gas) or renewable energy (like biofuels) can be used. Such systems are described in detail below.

5.2.1   Technical description

In the following the technology of solar tower power plants including all related components are described.

5.2.1.1   System components

Heliostats. Heliostats are reflecting surfaces provided with a two-axis tracking system which ensures that the incident sunlight is reflected towards a certain target point throughout the day. In addition, heliostats commonly concentrate sunlight by means of a curved surface or an appropriate orientation of partial areas, so that radiation flux density is increased.

Heliostats consist of the reflector surface (e.g. mirrors, mirror facets, other sunlight-reflecting surfaces), a sun-tracking system provided with drive motors, foundations and control electronics. The individual heliostat's orientation is commonly calculated on the basis of the current position of the sun, the spatial position of the heliostats and the target point. The target value is communicated electronically to the respective drive motors via a communication line. This information is updated every few seconds. The concentrator surface size of currently available heliostats varies between 20 and 150 m²; to date, the largest heliostat surface amounts to 200 m².

The heliostat field accounts for about half the cost of the solar components of such a power plant. This is why tremendous efforts have been made to develop heliostats of good optical quality, high reliability, long technical life and low specific costs. Due to economic considerations there is a tendency to manufacture heliostats with surfaces ranging between 100 and 200 m^2 and possibly beyond. However, there are also approaches to manufacture smaller heliostats to reduce costs by efficient mass-production.

Heliostats are usually centrally controlled and centrally supplied with electrical energy. As an alternative, autonomous heliostats have been developed which are controlled locally. There, the energy required for the control processor and the drives is provided by photovoltaic cells mounted parallel to the reflector surface.

The heliostats are individually controlled in order to control the radiation flux density on the receiver. For this reason not all of the heliostats are focussed on the same point of the receiver; their control rather ensures a smooth flux distribution over the entire receiver surface.

Based on the developments of the last few years, faceted glass/metal heliostats and membrane heliostats are distinguished (Fig. 5.6). These types are described below.

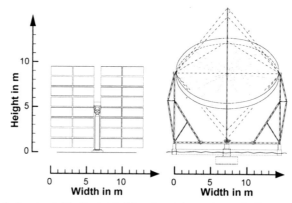

Fig. 5.6 Faceted glass/metal heliostat (left) and metal membrane heliostat (right) /5-3/

Faceted heliostats. Usually, faceted heliostats consist of a certain number of reflecting facets mounted on lattice work which in turn is positioned on a mounting tube. The facets are commonly designed as individual mirrors of sizes between 2 and 4 m^2. The orientation of the individual mirrors on top of the mounting structure (referred to as "canting") is different for every heliostat within the heliostat field, and thus results very expensive. The heliostats are usually tracking in two axes positioned vertically to each other (commonly mounting tube and vertical main axis), according to the desired azimuth and elevation angle. Mostly to reduce canting efforts and the number of individual drives, currently wide facetted heliostats are proposed. The glass/metal heliostat illustrated in Fig. 5.6 (left) as an example has a concentrator width of 12.8 m and a height of 8.94 m. The size of the individual facets is 3 by 1.1 m each. The total weight without foundation amounts to scarcely 5.1 t /5-3/.

Membrane heliostats. To avoid or reduce manufacturing and assembly efforts associated with individual facets and at the same time obtain high optical quality, so-called "stretched membrane" heliostats have been developed. The reflecting surface consists of a "drum", which in turn is composed of a metallic pressure ring with stressed membranes attached to the front and rear side. For this purpose plastic foils or metal membranes are used. In case of metal membranes, characterised by a considerably longer technical lifetime, the front side membrane is covered with thin glass mirrors to achieve the desired reflectivity. Inside the concentrator, a slight vacuum (only a few millibars) is created either by a vacuum blower or a vacuum pump. By this measure the membrane shape is altered so that the even mirror is transformed into a concentrator. Other designs use a central mechanical or hydraulic stamp to deform the membrane. Both configurations are advantageous since the focal length can easily be set and may even be altered during operation. Disadvantageous is the impact of wind on the optical quality of the heliostat and, in the case of using a vacuum blower, the energy consumption of the blower.

Fig. 5.6 shows the example of such a metal membrane heliostat equipped with a simple tubular steel space framework moving with six wheels on a ring foundation for vertical rotation. Two bearings form the horizontal axis. For this type of tracking, forces are introduced into the stable pressure ring far from the rotation axis (approximately 7 m for the illustrated heliostat). The reduced drive torque keeps the gear units small and inexpensive. The concentrator diameter of the heliostat shown (ASM 150) with a mirror surface of 150 m^2 amounts to 14 m. The concentrator thickness is 750 mm and its weight excluding the foundation is approximately 7.5 t.

Heliostat fields and tower. The layout of a heliostat field is determined by technical and economic optimisation. The heliostats located closest to the tower present the lowest shading, while the heliostats placed north on the northern hemisphere (or south on the southern hemisphere) show the lowest cosine losses. Heliostats placed far off the tower, by contrast, require highly precise tracking and, depending on the geographic location, have to be placed farer from the neighbouring heliostats. The cost of the land, the tracking and the orientation precision thus determine the economic size of the field.

The height of the tower, on which the receiver is mounted, is also determined by technical and economic optimisation. Higher towers are generally more favourable, since bigger and denser heliostat fields presenting lower shading losses may be applied. However, this advantage is counteracted by the high requirements in terms of tracking precision placed on the individual heliostats, tower and piping costs as well as pumping and heat losses. Common towers have a height of 80 to 100 m. Lattice as well as concrete towers are applied.

The costs for piping or the technical challenge of a thermal engine mounted on top of the tower can be avoided by a secondary reflector installed on the tower top, which directs incident radiation to a receiver located at the bottom (beam-down principle). Although this measure helps to reduce costs for tower, piping and thermal engine, the overall efficiency of the heliostat field is reduced due to additional optical losses caused by the secondary reflector.

Receiver. Receivers of solar tower power stations serve to transform the radiation energy, diverted and concentrated by the heliostat field, into technical useful energy. Nowadays, common radiation flux densities vary between 600 and 1,000 kW/m^2. Such receivers are further distinguished according to the applied heat transfer medium (e.g. air, molten salt, water/steam, liquid metal) and the receiver geometry (e.g. even, cavity, cylindrical or cone-shaped receivers) /5-4/, /5-5/. In the following, the main technical developments are presented according to the applied heat transfer medium.

Water/steam receiver. The first solar tower power stations (e.g. Solar One in California, CESA-1 in Spain) have been designed with tube receivers. Their design corresponds to a large extent to the salt tube receiver shown in Fig. 5.7. Similar to conventional steam processes, water is vaporised and partly superheated in such a heat exchanger (i.e. tube receiver). Since superheating is prone to unfavourable heat transmission, and due to the fact that start-up operation or part-load operation require complicated controls, this approach is currently not developed further. The above-mentioned difficulties can partly be prevented by avoiding superheating (i.e. saturated steam is generated). However, under these circumstances the power plant process only allows for comparatively low efficiencies due to thermodynamic constraints.

Salt receiver. The difficulties of heat transmission with a vertical tube receiver, exemplarily shown in Fig. 5.7, can partly be avoided by an additional heat transfer medium circuit. The heat transfer medium applied for this secondary circuit should have a high heat capacity and good thermal conduction properties. Molten salt consisting of sodium or potassium nitrate (NaNO$_3$, KNO$_3$) complies with these requirements. For both options, thanks to good thermal conduction properties, the heat transfer medium additionally serves as storage medium and can thus compensate fluctuations of the available radiation. The heat of this heat transfer medium is then incorporated into the thermal process via corresponding heat exchangers.

One disadvantage of all such salt receiver is that the salt must be kept liquid also during idle times when there is no solar radiation. This requires to either heat the whole part of the installation that is filled with salt (including, among other components, tanks, tubes, valves) and thus increases the energy consumption of the plant itself, or to completely flush the salt circuit. The highly corrosive gas phase of the used salts also has a detrimental effect, since, for certain operations, undesired evaporation of small amounts of salt due to local overheating cannot be entirely ruled out.

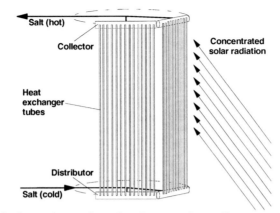

Fig. 5.7 Vertical tube receivers using salt as heat transfer medium

As an alternative to salt tube receivers also salt film receivers may be applied. During this process either molten salt is directly exposed to concentrated solar radiation or a flat or cavity absorber is cooled by an interior salt film /5-6/, /5-7/.

Besides molten salts, generally also liquid metals, such as e.g. sodium (Na) can be applied. However, due to negative experiences made with this heat transfer medium (fire hazard), this approach has been discontinued.

Open volumetric air receiver. Concentrated solar radiation is incident on volumetric absorber material consisting of steel wire or porous ceramics. Such volumetric receivers are characterised by a high ratio of absorbing surface to flow path of the absorbing heat transfer medium air. Ambient air is sucked in by a blower and penetrates the radiated absorber material (Fig. 5.8). The air flow absorbs the heat, so that those absorber areas facing the heliostat field (i.e. illuminated by the solar radiation reflected by the heliostats) are cooled by the inflowing air. Due to this cooling effect the absorber areas radiated by solar radiation are cooler than the interior absorber areas where the heat is transported by the in-flowing air. Therefore the air leaving the absorber shows also a higher temperature compared to the temperature of the absorber areas radiated by solar radiation. This is the reason why this receiver type presents comparatively low thermal losses. Being an open receiver, such power plants operate with ambient pressure. As air is characterised by a relatively low heat capacity, large volume flows and absorber surfaces are required.

Air as heat transfer medium presents the advantages of being non-toxic, non-corrosive, fire-proof, everywhere available and easy to handle. Its disadvantage is its comparatively low heat capacity requiring large heat transmission surfaces which, however, are generally feasible with volumetric receivers. Their lower thermal masses ensure a smooth start-up of the plant.

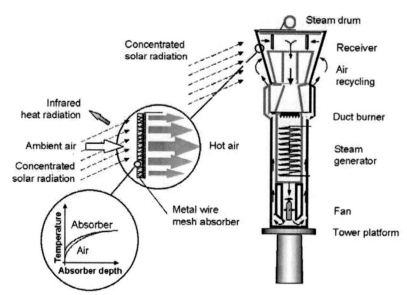

Fig. 5.8 Open volumetric air receiver according to the Phoebus principle (see /5-8/)

Closed (pressurised) air receivers. Receivers of solar tower power plants may also be designed as closed pressurised receivers. The aperture of such receivers is closed by a fused quartz window, so that the working medium air may be heated under overpressure and may, for instance, be directly transferred to the combustor of a gas turbine. To date, for example, a group of closed air receivers of a heat capacity of up to 1,000 kW has been tested at 15 bar. The obtained air outlet temperatures are slightly above 1,000 °C /5-9/. The individual receivers have been designed and interconnected according to their different thermal strain. For commercial applications several module groups may be added (Fig. 5.9).

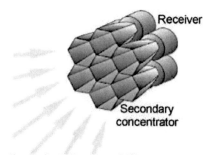

Fig. 5.9 Closed volumetric air receiver cluster equipped with secondary concentrators (see /5-9/)

Power plant cycles. The power plant cycles applied for solar tower power plants are mainly based on conventional power plant components commercially available today; the currently achievable pressures and temperatures of the working media applied for solar tower power plants are in line with the current power plant technology. Solar tower power plants within the capacity range from 5 to 200 MW can thus be designed using commercially available turbines and generators including all required auxiliaries.

5.2.1.2   System concepts

According to the applied heat transfer fluid or working medium, different system concepts are applied. Since open or cavity tube receivers reach working temperatures of approximately 500 to 550 °C, they are predominantly applied for Rankine cycles run by steam. Steam is either generated directly inside the receiver or by the secondary circuit (e.g. molten salt).

Hot air of approximately 700 °C generated by open volumetric receivers can be used within existing steam generators, similar to e.g. heat recovery boilers. The inlet temperature can, for instance, be maintained constant by an incorporated natural gas-fired duct burner, so that this concept is particularly suitable for hybridisation (i.e. application of solar energy in combination with fossil fuels like e.g. natural gas). Outlet gas/air is re-transferred to the receiver by means of a blower so that at least up to approximately 60 % are re-circulated.

Another possibility is the so called inverse gas turbine process. Within such a cycle an open volumetric air receiver is used and the hot air is directly fed into the gas turbine where the air is expanded /5-10/. One advantage compared to a steam cycle is a much simpler design. But so far such cycles have only been analysed theoretically.

In the past, several solar tower power plants have been realised within R&D projects sponsored by public money and industry. Within the following explanations some of these research plants are presented.

Solar One. Solar One is a solar power plant of an electric capacity of 10 MW, which was operated from 1982 to 1988 in the Californian Mojave Desert. This plant proved the general feasibility of solar thermal power generation by tower plants at the megawatt scale. Water was used as heat transfer fluid for the receiver. Among other difficulties, the plant showed the problem of maintaining operation when there are cloud passages.

Fig. 5.10 shows the performance characteristics of the Solar One plant as a typical example of performance characteristics of solar tower power plants. According to this figure, electricity is generated from a daily overall direct radiation of 4 to 5 kWh/(m^2 d) onwards. With increasing direct radiation the electricity output increases approximately linearly. The threshold from which electricity is generated, is mainly determined by the water steam tube receiver technology. This threshold can be lowered by the use of molten salt and volumetric receivers in particular.

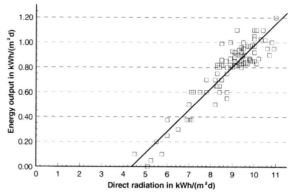

Fig. 5.10 Performance characteristics of Solar One

Solar Two. With the aim to solve the problems encountered with the Solar One plant the latter was remodelled to the Solar Two plant. As heat transfer and heat storage medium, molten salt consisting of 40 % potassium and 60 % sodium nitrate was applied. Thanks to the use of additional thermal energy storage the system is more independent from the available solar radiation.

The functional principle of the Solar Two power plant is shown in Fig. 5.11. Salt is pumped from a "cold" salt storage onto the tower and transferred from there into the receiver, where it is heated by the reflected solar radiation. Subsequently, it reaches the "hot" tank. Hot salt, and thus energy is taken as needed from the storage facility and pumped through a steam generator that generates steam for a conventional steam turbine cycle. Afterwards, the salt cooled inside the steam generator reaches again the "cold" salt storage.

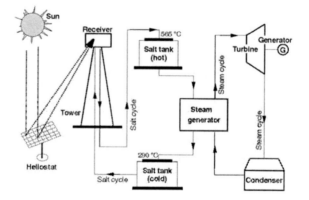

Fig. 5.11 Principle of the Solar Two power plant using molten salt as heat transfer and heat storage medium (see /5-11/)

In principle, this concept allows to generate power not only at daytime but also during the entire course of the day provided that the energy storage and the solar field are sufficiently large. Solar Two shows an electric output of 10 MW which can be maintained up to three hours after sunset thanks to the plant's energy storage.

Phoebus/TSA/Solair. Phoebus/TSA/Solair is a power plant concept with an open volumetric air receiver that provides hot air /5-8/. The hot air is subsequently passed through a steam generator providing superheated steam that can be used to drive a turbine/generator unit. Fig. 5.12 shows the corresponding schematic diagram.

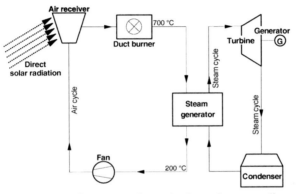

Fig. 5.12 Schematic diagram of an open volumetric air receiver according to the Phoebus-principle equipped with duct burner for additional fossil fuel firing (see /5-12/)

A natural gas-fired duct burner placed in between the receiver and the steam generator adds heat to the air if solar radiation is insufficient to supply the desired steam quantity. The Phoebus plant can thus not only generate power in times of sunshine but also during spells of bad weather and during the night; power generation is thus not exclusively dependent on the available solar radiation.

From 1993 to 1997 such a cycle equipped with an air receiver and a thermal capacity of 3 MW was continuously operated, containing all components of a future Phoebus power plant (so-called TSA system (Technology Program Solar Air Receiver)). Test results have shown the good interaction of components and the low thermal inertia of such systems which enable fast start-up. Further benefits of this technological approach are a simple structure and the unproblematic heat transfer fluid air /5-13/.

PS10. Because of the good experience made with the Phoebus/TSA/Solair System a European consortium, led by a Spanish company, planned the construction and operation of a 10 MW plant named PS 10, equipped with a volumetric air receiver, in the Southwest of Spain in 2004 /5-14/. However, the concept has been changed. The plant is now provided with a tube saturated steam receiver which supplies steam of 40 bar and 250 °C.

The northern heliostat field was erected from 2005 to 2006. It consists of 624 faceted glass/metal heliostats (T type) "Sanlúcar 120" of a mirror surface of 121 m² each. The cavity receiver is mounted on a tower of an approximate height of 100 m, consisting of four 5.36 x 12.0 m tube panels. The thermal storage incorporated into the plant has a useful heat energy of 20 MWh permitting 30 min of operation at 70 % of load /5-15/. Plant commissioning is scheduled for 2007.

Solar Tres. This plant is based on the know-how gathered during the construction and operation of the Solar Two plant (using salt as heat transfer and heat storage medium). This is why the project is called "Solar Tres" (being the Spanish translation for "Solar Three"). This solar tower power plant provided with a molten salt tube receiver and an electric capacity of 15 MW has been exclusively designed for solar operation. The northern heliostat field has 2,494 heliostats of a surface of 96 m² each. The heliostats to be used are of the faceted glass/metal heliostat type (T type) equipped with highly reflecting mirrors in simplified design (solar multiple of 3). It is planned that the receiver will have a heat capacity of 120 MW and be of cylindrical molten salt tube design. The storage (600 MWh) incorporated into the concept is to enable operation using heat from the storage for 16 h /5-15/.

Solgate. Solgate is a pilot solar tower power plant equipped with a closed volumetric receiver, a secondary concentrator and a ceramic absorber of an electric nominal capacity of 250 kW designed for hybrid operation (i.e. combined operation using natural gas and solar radiation). The heliostats arranged in the PSA's CESA-1 field have a mirror surface of 40 m² each (solar multiple of 1). To date, operation conditions have permitted air outlet temperatures of up to 1,050 °C and the direct drive of the gas turbine /5-9/

5.2.2   Economic and environmental analysis

The following sections are aimed at assessing solar tower power plants according to economic and environmental parameters.

Economic analysis. Within the scope of the economic analysis, power generation costs will be calculated for the discussed types of solar thermal power plants. In line with the preceding assessment method, applied throughout this book, the costs for construction and operation are determined and distributed in the form of annuities over the technical lifetime of the power plant. On the basis of these annual amortisation costs and the provided electric energy, the electricity generation costs per kilowatt hour are calculated. If not otherwise indicated, a technical lifetime of 25 years for all machine equipment and an interest rate of 4.5 % have been assumed to allow for a comparison with other power generation options.

Since the installation of solar thermal plants only makes sense in areas with a high share of direct radiation, a reference site with a total annual global radiation on the horizontal surface of 2,300 kWh/m² and a direct radiation total of 2,700 kWh/m² has been defined. Such values can, for instance, be obtained at favourable sites in California or North and South Africa.

Based on these site conditions a 30 MW solar tower power plant provided with an open volumetric receiver is assessed. The corresponding technical data are outlined in Table 5.3. This plant is state-of-the-art in terms of solar tower technology.

Table 5.3 Technical data of the assessed 30 MW solar tower power plant

Nominal capacity	30 MW
Mirror surface	175,000 m²
Full-load hours	2,100 h/a
Storage capacity	0.5 h
Solar share	100 %
Technical lifetime	25 a

Investments. To date, no commercial solar tower power plant has been put into operation. To estimate the power generation costs, we thus need to revert to the figures published by manufacturing and project development companies. On the basis of these data and the adapted specific costs for the heliostats, the overall investment costs for such a power plant amount to roughly 99 Mio. € (Table 5.4). Within the scope of a sensitivity analysis the influence of investment cost variations on the power generation costs are assessed.

Table 5.4 Mean investment and operation costs as well as resulting power generation costs for the reference solar tower power plant

Nominal capacity	30 MW
Investments	
heliostat field	30 Mio. €
receiver and steam generator system	20 Mio. €
tower	15 Mio. €
other components	20 Mio. €
assembly and commissioning	10 Mio. €
design, engineering, consulting, miscellaneous	5 Mio. €
Total	99 Mio. €
Operation and maintenance costs	1.5 Mio. €/a
Power generation costs	0.13 €/kWh

Operation costs. Annual operation costs comprising the costs for operation and maintenance are estimated at 50 €/kW. In the present case, this corresponds to costs of approximately 8.6 €/m² of mirror surface and year (Table 5.4). For the observed plant, the annual overall operation costs thus amount to approximately 1.5 Mio. €.

Electricity generation costs. On the supposition of the above-mentioned invest-ment as well as operation and maintenance costs, the power generation costs re-lated to the solar tower power plant at the reference site amount to approximately 0.13 €/kWh (Table 5.4).

Power generation costs are largely influenced by the number of full-load hours per year, the investment costs and the mean interest rate. A sensitivity analysis conducted on the basis of these parameters reveals the correlations shown in Fig. 5.13. If investments are, for instance, reduced by 30 % power generation costs are cut down to approximately 0.10 €/kWh.

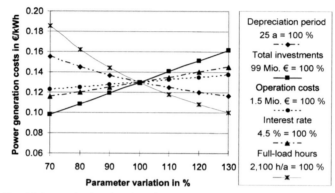

Fig. 5.13 Sensitivity analysis related to the power generation costs of the reference solar tower power plant

However, the sensitivity analysis also shows that for other economic parame-ters (i.e. higher interest rates, short depreciation period) and different conditions for the particular site, power generation costs can vary significantly.

Environmental analysis. The following analyses are aimed at discussing selected environmental effects with regard to plant erection, normal operation, malfunc-tion, and the end of operation.

Manufacture (construction). Environmental effects related to solar thermal plants may already arise during production of the different plant components. They are to a large share the same as for conventional power plants and other industrial production processes. However, the resulting environmental effects are restricted to very limited periods of time, and in many countries they are subject to exten-sive legal requirements. Furthermore, solar thermal power plants are primarily located in deserts and steppes where the population density is relatively low. This is why there is so far only a very limited knowledge on the potential effect on human beings and on the environment.

Normal operation. As for all thermal power plants also solar thermal power plants have certain environmental effects during normal operation. In the following, selected aspects are discussed.

- Land requirements. Solar thermal power plants use solar radiation as a source of energy, i.e. an energy source with a comparatively low energy density. This is why such plants necessarily require large collector areas and thus extensive land areas (Table 5.5). Since the individual collectors must be accessible during operation, the soil, where the collector field is installed, is compacted and levelled during construction. Furthermore, highly grown plants may disturb operation and reduce the technical lifetime of the collectors (due to e.g. humidity, shading, and fire hazard). Thus, at the most, grass vegetation is permitted on the respective terrain of e.g. a solar thermal power plant). The soil is therefore more susceptible to erosions. Since solar thermal plants are generally located in areas with little precipitation (i.e. deserts and steppes) and collector fields must be provided with extensive drainage systems to protect foundations and ensure accessibility, this influence can almost be neglected. Due to the glass reflectors, the collector fields are susceptible to damages caused by e.g. extreme winds. The entire plant is thus generally fenced in, so that during the construction of such a solar thermal power plant it has to be ensured that neither natural habitats of large animals nor typical passages are affected. However, due to the preferred location in desert or steppe areas this requirement can generally be fulfilled without any problem.

Table 5.5 Space requirements of solar thermal plants

Tower power plants	$20 - 35 \text{ m}^2/\text{kW}$
Line-focussing power plants	$10 - 25 \text{ m}^2/\text{kW}$
Parabolic trough power plant	$15 - 30 \text{ m}^2/\text{kW}$
Solar updraft tower power plant	approximately $200 \text{ m}^2/\text{kW}$
Solar pond power plant	approximately $55 \text{ m}^2/\text{kW}$

- Visual impact. Due to their central tower, mainly tower and solar updraft power plants have a non-negligible impact on the appearance of the natural scenery; and the higher the tower the bigger this influence is (approximately 100 m in case of solar tower power plants, and 1,000 m for solar updraft tower power plants). Contrary to wind energy converters, the disturbance of the scenery is static since no moving parts (such as the rotors of wind energy converters) are involved. This is why the impact on the natural scenery is more readily accepted by the spectator. Furthermore, the visual impact can be limited by corresponding shapes and colours as it is presently the case for wind energy converters. At night time, indispensable aviation warning lamps might be perceived as disturbing. Since the tower height amounts to 15 to 25 % of the collector field radius, disturbing shading outside of the plant only occurs with solar altitudes below 15°. Since such plants are preferably built in desert or steppe areas, due to the low population density, negative optical impacts to humans are hardly to be expected.

- Reflections. Solar tower, trough or dish power plants focus solar radiation onto a particular line or point. Provided that power plants are operated properly, i.e. mirrors are precisely tracking, none of the known environmental effects will occur. However, due to the partly very high energy flow densities of the focussed solar radiation, persons and/or assets may be exposed to considerable hazards in case of improper operation. Proper operation and precise radiation focus of the plants must thus be ensured by all means.

- Emissions. Since some solar thermal power plants also apply conventional power plant technology they are also potential sources of airborne emissions. However, greenhouse gas emissions as well as other types of emissions are only released into the atmosphere during hybrid operation involving fossil or biogenous combustibles. Since the application of these types of fuels is generally subject to extensive legal requirements, environmental effects are usually very limited. Additionally, noise may be created by the turbines and pumps. These sound emissions may be kept to a minimum by appropriate noise protection which has become a state-of-the-art requirement, also with regard to approval according to the existing noise protection regulations, slightly different in different countries. Since the power block is furthermore most commonly located in a central position in the collector field, noise emissions at the facility limits are practically negligible. This is why noise emissions issues have not been reported so far. Further emissions are fumes originating from possible re-cooling systems of the condensers and the general thermal load of the thermal circuits. But also in this respect, there are extensive legal and regulatory requirements to be fulfilled in order to obtain permission for plant operation.

Malfunction. In case of malfunction, at the most, the same environmental effects as for conventional power stations fired by fossil or biogenous energy carriers are to be expected. Further malfunctions that may arise with regard to solar farms are operating failures due to heat transfer fluid leakage causing personal and environmental damages.

End of operation. To avoid undesired environmental effects, the plants are properly dismantled and discarded at the end of operation. Discarding of the applied plant components should not cause any major environmental effects. They are mostly similar to those of conventional machine technologies, which are relatively low due to the applicable legal requirements.

5.3   Parabolic trough power plants

The line focussing solar fields of parabolic trough and linear Fresnel collectors reflect the incident radiation on an absorber positioned in the focal line of the concentrator. The collector tracks the sun in one axis (Fig. 5.14). Due to this "one-dimensional concentration" the geometric concentration factors of 15 to 30 are considerably lower than those of two-dimensional collectors discussed above. This is why lower temperatures are achieved when compared to solar tower power plants. However, this disadvantage is compensated by lower specific costs as well as a simpler structure and maintenance.

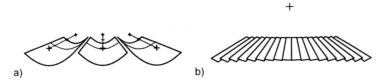

a)   b)

Fig. 5.14 Principles of line focussing collectors (a) parabolic troughs, b) Fresnel troughs)

Line focusing solar power plants have a modular structure. Because of this characteristic and the shape of the solar field, line-focussing solar power plants have in the past also been referred to as "solar farms".

5.3.1 Technical description

In the following, the technology of parabolic trough power plants and Fresnel collector and all related components are described.

5.3.1.1   System components

The system elements composing parabolic trough power plants comprise collector, absorber, heat transfer fluid circuit and power block.

Collectors. The collectors which are typically 100 m, but may nowadays also be 150 m long, are provided with single-axis solar tracking. The annual mean cosine losses of parabolic troughs vary between 10 and 13 %, whereas those of Fresnel concepts are double. After deduction of the optical and thermal losses inside the collector 40 to 70 % of the radiation incident on the mirrors can be used technically. The percentage depends on the design, the field size and on the geographic location of the power plant. In the following, the main collector types are discussed.

Parabolic trough collectors. This collector type is characterised by a parabolic reflector which concentrates the incident radiation onto a tube positioned in the focal line (Fig. 5.14 a), Fig. 5.15, Fig. 5.16).

The reflector itself may either consist of one surface provided with a reflecting layer (metal foil, thin glass mirrors) or of several curved mirror segments arranged in a truss-type structure; the latter variant is commercially applied. Collectors are mounted on a mounting structure and track the sun's diurnal course by a single-axis system following the longitudinal axis.

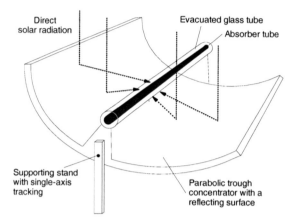

Fig. 5.15 Functional principle of a parabolic trough

The mirror segments typically consist of back-silvered white low iron glass to achieve high reflectivity values in the solar spectrum. When the mirror segments are clean the mean solar reflectivity amounts to approximately 94 %. Since the glass is weatherproof the reflectivity of the cleaned reflector remains practically unchanged.

One collector (SCA = Solar Collector Assembly) is composed of a number of collector elements (SCE = Solar Collector Elements) with a typical length of 12 m each. The largest collector built to date (Skal-ET) consists of 12 SCEs (6 on each side of the central drive pylon). It has an overall length of 150 m and an aperture width of 5.77 m.

Each collector unit is equipped with an angular position sensor to track the position of the collector and optionally additionally with a sun sensor. First electric motors provided with gearboxes or cable winches were used as drives. For the more recent LS-3 and EuroTrough collectors, cost-efficient hydraulic drives have been applied /5-16/.

Fresnel collectors. For this collector type the parabola profile is approximated with individual segments (Fig. 5.14 b)). Individual long rectangular mirror segments of an approximate width of up to 2 m are tracking the sun similar to helio-stat fields, so that they reflect the incident radiation to a common focal line. All segments are mounted at the same level (either near-ground or higher on mounting structures). Due to their lower width they are exposed to lower wind loads compared to parabolic trough collectors. However, the different segments shadow each other. Due to their specific geometry, Fresnel collectors are characterised by lower concentrations and a lower optical efficiency when compared to parabolic trough collectors. Such losses may at least partially be compensated by alternating mirror positions.

Each reflecting segment rotates around its centre of gravity. Segments are actuated either individually or as a group. Fresnel collectors thus require a more sophisticated control system than parabolic trough collectors since a specifically higher number of drives must be used. This is why Fresnel collectors have to date only been tested on a minor scale and have, also on a minor scale, first been commercially applied since 2004 in Liddell, Australia /5-17/.

Absorber / Heat Collecting Element (HCE). Individual horizontal tubes are used as absorbers in the focal line of the collectors; for Fresnel collectors also tube groups may have to be used due to their wider focal line. Today's stainless steel absorber tubes of parabolic trough collectors (i.e. Heat Collecting Element (HCE)) are enclosed in an evacuated glass tube to minimise heat losses (Fig. 5.16). In case of parabolic trough collectors the vacuum also serves to protect the sensitive highly selective coating. Nowadays, such selective coatings remain stable up to temperatures of 450 to 500 °C; solar absorption is above 95 %, and at a temperature of 400 °C emissivity is below 14 % /5-18/.

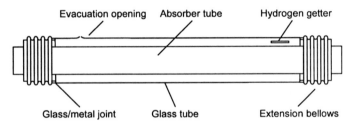

Fig. 5.16 Absorber tube of a parabolic trough collector

In addition to the proven HCE design (Fig. 5.17 a)), Fig. 5.17 b) and c) show two additional variants using a secondary concentrator and a tube bundle receiver. Both options have been proposed to suit the optics of Fresnel collectors.

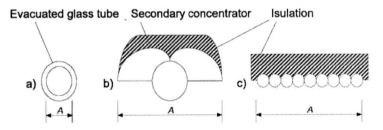

Fig. 5.17 Design principle of receivers of parabolic trough collectors (a); see also Fig. 5.16) and Fresnel collectors b) and c)) (A width of the focal line)

Heat transfer medium. To date, high-boiling, synthetic thermal oil has been applied as heat transfer medium in the absorber tubes. Due to the limited thermal stability of the oil, the maximum working temperature is limited to scarcely 400 °C. This temperature requires to keep the oil pressurised (approximately 12 to 16 bar). This is why collector tubes as well as expansion reservoirs and heat exchangers must be of pressure-resistant design. Relatively high investments are thus required.

Hence, as an alternative, molten salt has been proposed as heat transfer medium. Molten salt is characterised by the advantages of lower specific costs, a higher heat capacity and thus potentially higher working temperature on the one hand, and by the higher viscosity of the medium and a higher melting temperature, requiring trace heating, on the other hand. Due to the higher heat capacity, pumping power requirements are still expected to be lower when compared to thermal oil. To date, only prototypes of this variant have been built.

For this reason, investigations of direct steam generation inside the collector tubes are promoted since great cost saving and efficiency potentials are expected. The advantages are the higher possible working temperature of steam as a working medium and that there is no secondary heat transfer fluid loop required including the necessary heat exchangers. The expected problems related to evaporation of water in horizontal tubes (including two-phase flow and thus different heat transmission) can be solved by available technology (forced-circulation boiler with a relatively high recirculation rate and water/steam separator). It is thus possible to directly generate saturated steam by line focussing collectors. However, the high steam pressure (usually between 50 and 100 bar) requires a relatively high tube wall thickness, so that for very wide collectors tube bundles might be more suitable than the well proven individual tube /5-40/.

Collector fields. Nowadays, collector fields are composed of a certain number of loops of an approximate length of 600 m each. These loops are connected to one feed ("cold header") and one discharge line ("hot header") each. Collectors are north-south oriented to allow for high and constant energy yields.

With regard to the collector field design special emphasis must be laid on the distance between the individual collector rows. The distance determines shading during morning and evening hours and thus the corresponding efficiency reduction of the whole field. Furthermore, land and piping costs as well as thermal and pump losses must be taken into account. Since the effect of shading also depends on the latitude, each field design must be optimised with regard to the site-specific conditions. As a rule of thumb, the distance between lines of parabolic troughs typically amounts to three times the aperture width.

Collectors are positioned horizontally; a slope of several percent may also be admissible. However, more severe unevenness of the site must be compensated or terraced.

The achievable thermal output of the entire field is limited by pressure losses of the heat transfer medium and the piping costs. Currently, the maximum economically sensible thermal capacity of a solar field operated with thermal oil is estimated to be around 600 MW.

5.3.1.2   Plant concepts

The major share of solar thermal electricity is generated by means of parabolic trough plants. In the Mojave Desert in California/USA nine so-called SEGS (Solar Electricity Generation Systems) have been erected, whose concept is detailed as follows. Additionally, further approaches are discussed.

SEGS plants. In the years from 1985 to 1991, nine SEGS plants (Table 5.6) accounting for an overall electric capacity of 354 MW were installed in the Californian Mojave Desert /5-19/. All plants have been operated for power generation on a commercial basis ever since.

Table 5.6   Technical parameters of built parabolic trough power plants (also refer to /5-19/)

	SEGS I	SEGS II	SEGS III	SEGS IV	SEGS V
Year of construction	1985	1986	1987	1987	1988
Capacity in MW[a]	13.8	30.0	30.0	30.0	30.0
Status	in operation	in operation	in operation	in operation	in operation
Collector field					
Collector type	LS1 / LS2	LS1 / LS2	LS2	LS2	LS2 / LS3
Number	608	1,054	980	980	1,024
Overall surface[b] in m²	82,960	190,338	230,300	230,300	250,560
Max. fluid temp. in °C	307	321	349	349	349
Storage capacity in MWh	120				
	SEGS VI	SEGS VII	SEGS VIII	SEGS IX	
Year of construction	1989	1989	1990	1991	
Capacity in MW[a]	30.0	30.0	80.0	80.0	
Status	in operation	in operation	in operation	in operation	
Collector field					
Collector type	LS2	LS2 / LS3	LS3	LS3	
Number	800	584	852	888	
Overall surface[b] in m²	188,000	194,280	464,340	464,340	
Max. fluid temp. In °C	390	390	390	390	

Max. maximum; temp. temperature; [a] net capacity; [b] overall collector surface.

All SEGS plants are operated with thermal oil which is pumped through the solar field. For the first plant (SEGS I) mineral oil has been selected that can be operated at low temperatures but does not require pressurised operation. The superheating required for steam turbine operation is provided by a natural gas-fired boiler which also ensures constant operation of the entire plant. The applied oil was so cheap that a simple thermal storage of 120 MWh could be added.

For the following power plants both the applied heat transfer fluid and the power plant configuration have been modified. The thermal oil, still in use today, allows maximum operation temperatures of scarcely 400 °C, but must be kept under a pressure of at least 12 bar.

From plant SEGS VI onwards, additionally a solar re-heater has been integrated which (together with enhanced steam parameters) increased the thermal efficiency of the power cycle from 30.6 to 37.5 % (Fig. 5.19). Fig. 5.18 shows an example of the performance characteristic of this power plant (i.e. the provided electrical energy as a function of direct radiation).

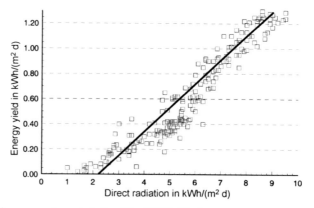

Fig. 5.18 Performance characteristic of the SEGS VI plant

The required steam is either generated directly or indirectly by a secondary circuit (Fig. 5.19). Typical steam parameters are approximately 100 bar / 371 °C for indirect generation (due to the temperature limit of the heat transfer fluid) or 80 bar / 430 °C for direct generation. Compared to conventional steam power plant cycles the indicated values are relatively low. Still, this is to a large extent compensated by an increased technical effort. However, for a plant within this capacity range rather unusual process improvements such as intermediate superheating and multiple-stage internal feed water preheating are required. Consequently, in spite of the rather unfavourable steam parameters, for instance, the 30 MW plants SEGS IV to VI reach thermal efficiencies in the power block of up to 38 %.

Hybridisation is possible by integration of additional firing based on fossil and/or biogenous energy carriers to ensure operation at times of fluctuating or no solar radiation. As an alternative, also parallel steam generators may be applied; this additional technical effort allows for better steam parameters and thus higher electrical efficiencies.

The concept of SEGS plants is also being applied for more recent parabolic trough power plants whose operation is predominantly assured by solar power generation without major additional firing.

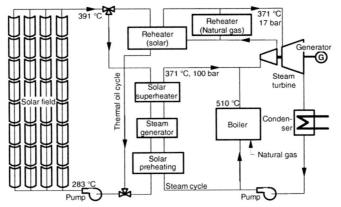

Fig. 5.19 Diagram of a parabolic trough power plant (SEGS VI and VII) (refer to /5-19/)

Integrated Solar Combined Cycle System (ISCCS). To enhance availability, efficiency and controllability, the solar field may be incorporated into a combined cycle power plant. Solar steam is superheated once more up to a temperature of approximately 530 °C in a heat recovery boiler.

If a solar collector field is integrated into the steam cycle of an Integrated Solar Combined Cycle System (ISCCS) the "solar" steam is transferred to the high pressure cycle of the steam generator. The required fossil fuel energy is thus reduced, so that more steam can be generated with the same flue gas stream or less flue gas is required to provide the same quantity of steam. In this operating mode, the gas turbine may be operated at part load; the solar field thus saves fossil fuel energy. The share of solar power generation is between 3 and 10 %.

Integration into conventional power plants. A further possibility to integrate solar heat into conventional power plant processes is to incorporate solar heat into the feed water preheating of conventional steam power plants. For internal feed water preheating, normally turbine extraction steam is required, which is then not available for expansion in the turbine anymore. If solar heat is available for feed water preheating, the steam can be utilised for the turbines.

In summer 2004, the first phase of solar feed water preheating by means of Fresnel collectors was commissioned in Liddell, Australia. For the final configuration it is planned that the last high-pressure pre-heater is exclusively operated by solar heat. In addition, this plant is to be tested with additional steam generation /5-17/.

5.3.2 Economic and environmental analysis

The following explanations are aimed to assess parabolic trough power plants according to economic and environmental parameters.

Economic analysis. Within the following analysis the power generation costs for the outlined solar thermal plants are calculated. In line with the preceding assessment method applied throughout this book, the costs for construction and operation are determined and distributed in the form of annuities over the technical lifetime of the plants. Based on these yearly costs and the provided electrical energy, the electricity generation costs per kilowatt hour are calculated. To allow for a comparison, as usual a technical lifetime of 25 years and an interest rate of 4.5 % are assumed.

Since such plants are only installed in areas with a high share of direct radiation, also a reference site has been assumed, which is characterised by an annual direct radiation total of 2,700 kWh/m^2.

The key data of the assessed 50 MW parabolic trough power plant are outlined in Table 5.7. The plant is defined similar to the AndaSol I power plant /5-15/; differences are due to the selected reference site.

Table 5.7 Key data of the assessed 50 MW parabolic trough power plant equipped with integrated thermal storage

Nominal capacity	50 MW
Mirror surface[a]	510,000 m^2
Full-load hours	3 680 h/a
Storage for 7 h	Molten salt
Solar share	100 %
Technical lifetime	25 years

[a] sufficient for a 80 MW power plant without storage.

Investments. Investment costs for such a power plant vary between 220 and 300 Mio. €. Here a mean of approximately 260 Mio. € is assumed. Specific investment costs therefore amount to 5,200 €/kW, including storage. Table 5.8 shows the approximate distribution of the investment costs.

Table 5.8 Estimation of the investment costs of a parabolic trough solar power plant

Power plant (including thermal balancing)	60 Mio. €
Solar field including heat transfer fluid loop	155 Mio. €
Solar field preparation (i.e. levelling, fencing, access roads)	5 Mio. €
Subtotal (without energy storage)	220 Mio. €
Thermal energy storage (7 h storage)	40 Mio. €
Total	260 Mio. €

Operation costs. Annual operation costs of the collector field are estimated at 5 €/m^2. In addition, there are costs for the remainder of the power plants. All in all, the operation costs for plants of such a size are estimated at 10 €/m^2 of mirror surface and year.

Electricity generation costs. Based on the above-mentioned investment, operation and maintenance costs as well as the possible electricity generation, for the reference site power generation costs are calculated with 0.12 €/kWh for the parabolic trough power plant with integrated thermal energy storage (Table 5.9).

Table 5.9 Estimation of the power generation costs of a parabolic trough power plant with thermal molten salt storage

Nominal capacity	50 MW
Investment costs[a]	260 Mio. €[a]
Operation and maintenance costs	5.1 Mio. €/a
Electricity generation costs	0.12 €/kWh

[a] including storage.

As for solar tower power plants, not only the full-load hours and the assumed mean interest rate can influence the power generation costs significantly. Therefore a sensitivity analysis is conducted on the basis of these and other parameters. The result of such an analysis show very similar correlations as for the reference solar tower power plant (Fig. 5.13). Accordingly different electricity generation costs are obtained depending on the economic frame conditions and/or the assumed technical parameters.

Environmental analysis. The environmental effects of parabolic trough power plants are very similar to those of solar tower power plants. They are thus discussed in the corresponding chapter on solar tower power plants (see Chapter 5.2.2).

5.4 Dish/Stirling systems

Dish/Stirling systems mainly consist of the parabolically shaped concentrator (dish), a solar receiver and a Stirling motor as thermal engine with interconnected generator.

The parabolic concentrator is tracking the sun in two axes, so that it reflects the direct solar radiation onto a receiver positioned in the focus of the concentrator. The radiation energy transformed into heat within the receiver is transferred to the Stirling motor, which, being a thermal engine, converts the thermal energy into mechanical energy. A generator is directly coupled to the Stirling motor shaft, which converts the mechanical energy into the desired electrical energy (Fig. 5.20). For hybrid operation, the system may be heated in parallel or in addition by a gas burner (operating e.g. by natural gas or biogas).

In the following, the main components of such systems are discussed. Subsequently, the corresponding complete systems are assessed.

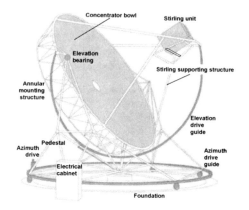

Fig. 5.20 Dish/Stirling systems (Distal II)

5.4.1   Technical description

Within the following explanations, the technology of dish/Stirling power plants including all related components is discussed.

5.4.1.1   System components

Parabolic concentrator (dish). The parabolically shaped concave mirror (dish) concentrates sunlight onto a focal spot. The size of this spot depends on concentrator precision, surface condition and focal distance. Common concentrators achieve concentration ratios between 1,500 and 4,000. Common maximum diameters amount to 25 m.

With regard to concentrator design facetted paraboloids (i.e. consisting of individual segments) and full-surface paraboloids are distinguished.
- For facetted concentrators several mirror segments are mounted on a mounting structure. The segments are supported and oriented individually. Such mirror segments may either consist of glass mirrors or media covered with reflecting foil or thin-glass mirrors.
- For full-surface concentrators the entire concentrator surface is shaped parabolically by a forming process. For instance, a pre-stressed metallic or plastic membrane is attached on both sides onto a stable ring (stretched membrane technology). Subsequently, it is transformed into the desired shape via a forming process (e.g. by water load) and stabilised via a specific vacuum. Such low-weight metal membrane designs provide full-surface concentrators with high rigidity and high optical quality. As an alternative, the facets may also consist of sandwich elements made of fibre-glass reinforced epoxy resin with thin-glass mirrors glued onto them /5-20/.

Mounting structure. The mounting structure of parabolic concentrators is necessarily determined by the shape of the reflector segments or the full-surface concentrator. There is a great variety of technical solutions. However, there is a certain tendency to turntables which at the same time serve as a drive ring. Turntables permit to minimise material consumption and the drive torque.

Solar tracking system. Point-focussing parabolic concentrators must be continuously tracking the sun's path to ensure that solar radiation is always parallel to the optical concentrator axis. Solar tracking systems are further distinguished into azimuth/elevation and polar tracking systems.
- For azimuth/elevation solar tracking, the concentrator is moved parallel to the earth's surface on one axis (elevation axis) and vertically to the earth's surface (azimuth) on a second axis.
- For polar (or parallactic) solar tracking one axis is parallel to the earth's rotation axis (polar axis) and the other vertically to the first (axis of declination).
Both systems are available fully automated. As reference parameter for the control system serve either the solar position, calculated on the basis of date and time of day, or the signal of a solar sensor.

Receiver. The receiver absorbs solar radiation reflected by the concentrator and converts it into technically useful heat. Either the working medium itself or a heat transfer medium may undergo a temperature rise and/or phase change.

Thus, the highest temperatures of the system occur at the receiver. For systems which directly heat the working medium, currently common operation temperatures vary between 600 and 800 °C, whereas pressures are between 40 and 200 bar. Intensity distribution of the focussed radiation within the focal spot cannot be entirely homogenous due to inevitable mirror errors. This is why in addition, large temperature gradients may occur on the absorber surface /5-21/.

Out of the multitude of available receiver technologies, in the following, two different systems are discussed.

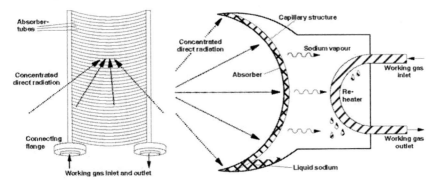

Fig. 5.21 Different receiver types for dish/Stirling systems (left: tube receiver, right: heat pipe receiver, schematic diagrams according to /5-21/)

Tube receiver. The directly radiated tube receiver is the simplest type of a solar receiver suitable for operation with a Stirling motor (Fig. 5.21, left). The heating tubes of the Stirling motor, through which the working medium flows, serve as absorber surface. Thus the receiver tubes are directly heated by the concentrated solar radiation. The volume of the tubes filled with the working medium should be as low as possible to keep the dead volume of the engine low. The shape of the receiver must fit to the geometry of the focal spot produced by the concentrator.

Heat pipe receiver. For heat pipe receivers (Fig. 5.22, right) a phase-change heat transfer medium (e.g. sodium) is applied. Since this heat transfer medium undergoes an evaporation and condensation cycle, the latent evaporation heat is transferred from the radiated absorber surface to the heater and from there to the working medium of the Stirling motor, while the temperature is almost kept constant. Subsequently, the condensate is re-transferred to the heating zone via a capillary structure. Due to the heat pipe principle this structure requires comparatively high efforts in terms of production engineering. However, this concept offers the advantage that high or extremely different heat flow densities may be homogenously transferred onto the Stirling heater thanks to good heat transmission. It is also beneficial that the heat pipe receiver may comparatively easily be combined with other types of operation; i.e. in addition to solar radiation it can also be operated by liquid or gaseous, fossil or biogenous fuels /5-22/.

Such receivers are most commonly designed as cavity receivers. The concentrated radiation passes through a small aperture and impinges on a cavity. The actual absorbing surface, which is subject to temperature rise due to the incident radiation, is positioned behind the focal spot. Because of this geometric position, the absorber surface is bigger than the aperture; the radiation intensity which impinges on the receiver is thus reduced. Yet, with regard to cavity receivers heat losses are relatively low since only a small portion of the diffuse radiation emitted by the absorber is lost by the aperture and convection losses, caused for instance by wind.

Stirling motor. Thermal energy provided by concentrated solar radiation can be converted into electrical energy using a Stirling motor with coupled generator. Stirling motors belong to the group of hot-gas machines and use a closed system; i.e. within the working cycle always the same working gas is used /5-23/. Contrary to Otto or Diesel engines, energy is provided by external heat supply, so that Stirling motors are also suitable for solar operation.

The basic principle of a Stirling motor is based on the effect that gas performs a certain volume change work in case of a temperature change. The process is based on isothermal compression of the cold and isothermal expansion of the hot medium at a constantly low volume, in case of heat supply, and, at a constantly large volume (isochorous), in case of heat removal (Fig. 5.5 (c)). Periodic temperature change – and thus continuous operation – can be ensured by moving the working gas between two chambers of constantly high and constantly low temperature.

For the technical realisation a compression piston is moved to the closed side, so that the cold working gas flows to the warm space, passing through a regenerator. The regenerator transmits the previously absorbed heat to the working gas (isochorous heating phase (1); Fig. 5.22). The gas is warmed up to the temperature of the warm space while the regenerator cools down to the temperature of the cold space. Subsequently, the working gas inside the warm space expands isothermally and absorbs the heat from the warm space (isothermal expansion phase (2); Fig. 5.22).

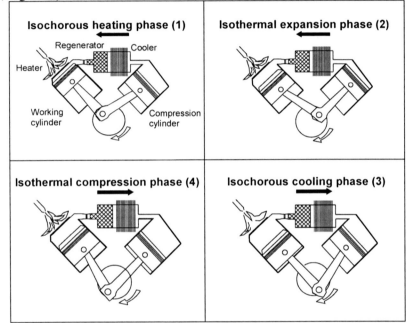

Fig. 5.22 Working principle of a Stirling motor (see /5-24/)

The expanding working gas moves the working piston to the open side and performs work. If the working piston passes the lower dead centre and is thus moved to the closed side, the hot working gas is forced to pass the regenerator and to move into the cold room. Heat is transferred isochorously from the working gas to the regenerator (isochorous cooling phase (3); Fig. 5.22). The gas is cooled down to the temperature of the cold space while the regenerator is warmed up to the temperature of the warm space. The working gas is subsequently compressed isothermally and transmits the generated heat to the cold space (isothermal compression phase (4); Fig. 5.22).

The basic system components thus include the heated working cylinder, the cooled compression cylinder and a regenerator which serves for intermediate energy storage. In most cases, the regenerator is a highly porous body of a high heat capacity; this porous body has a considerably larger mass than the gas mass flowing through the body. The more complete the alternating heat transmission is performed inside the regenerator, the bigger the mean temperature difference between working and compression cylinder and thus the efficiency of the Stirling motor.

If the displacing piston is coupled to the working piston in the appropriate phase angle via a driving mechanism or a vibratory system, the whole system can serve as thermal engine.

In terms of mechanical design, single and double-acting machines are distinguished. In single-acting machines only one side of the compression or expansion piston undergoes pressure fluctuations inside the working space, while the pressure of the working gas is effective on both sides of the piston of double-acting machines; in the latter case, they simultaneously work as compression and expansion piston /5-21/, /5-24/.

Stirling machines can also be distinguished into kinematic and free piston Stirling engines.

- Kinematic Stirling engines perform power transmission via a link mechanism. A generator can be coupled to this gear by a shaft, leading to the exterior.
- Free piston Stirling engines lack mechanical interlinkage between the working piston, the displacement device and the environment. Both pistons move freely. The converted energy can be transferred to the exterior by an axial generator, for instance. Mechanical interlinkage is replaced by an interior spring damping system; this is why only two movable parts are required. The machine is hermetically sealed, so that tightening issues are avoided. Free piston Stirling machines present the theoretical benefits of a simple structure and a high reliability, but are currently far behind in terms of development when compared to kinematic machines.

The machines applied for dish/Stirling systems use helium or hydrogen between 600 and 800 °C as working gas temperatures. Power output of the Stirling motor is controlled by varying the working gas mean pressure.

5.4.1.2   Plant concepts

Because of their size and space requirements individual dish/Stirling systems are suitable to supply power to small and medium grids (micro and mini grids). When combined with batteries and/or additional generators operated by fossil or biogenous combustibles, they are suitable for the energy supply of rural communities. Since, in this respect, they have to compete with a multitude of other renewable sources, current developments concentrate on automated operation and cost-cutting.

Alternatively, dish/Stirling power plants can be interconnected to provide larger quantities of heat and power. The biggest park was commissioned in 1984 in California and consisted of 700 individual collectors and a central thermal engine of an electric overall capacity of almost 5 MW.

In the last years, various dish/Stirling prototypes have been developed and tested. Of some types, several units have been built. Table 5.10 shows the main parameters of the current dish/Stirling plants /5-25/.

Table 5.10 Solar thermal trial and pilot plants (see also /5-25/)

	MDAC	SES/ Boeing	SAIC/ STM	WGA ADDS	SBP	EuroDish
Year of operation	84 – 88	since 98	since 94	since 99	1990 – 2000	since 2000
Capacity in kW$_{net}$	25	25	22	9	9	10
Efficiency in %	29 – 30[a]	27	20	22	18 – 21	22
Number	6	3	5	2	9	7
Operating hours in h	12,000	25,000	6,400	5,000	40,000	10,000
Availability in %	40 – 84	94			50 – 90	80 – 95
Status	Terminated	Trial run	Trial run	Trial run	Terminated	Trial run
Concentrator						
Diameter in m	10.57	10.57	12.25	7.5	7.5 – 8.5	8.5
Design	1[b]	1[b]	2[c]	3[d]	2[c]	3[d]
Number of facets	82	82	16	24	1	12
Facet size in cm	91 x 122	91 x 122	∅ 300		∅750 – 850	
Mirror support/	glass/	glass/	glass/	glass/	glass/	glass/
Reflector	silver	silver	silver	silver	silver	silver
Reflectivity in %	91	>90	>90	94	94	94 (new)
Concentration factor	2,800	2,800			3,000	2,500
Operation hours in h	175,000	30,000	18,000	54,000	100,000	10,000
Efficiency in %	88.1		90 (design)		88	88
Machine						
Manufacturer	USAB	USAB/SES	STM Corp.	SOLO	SOLO	SOLO
Capacity in kW$_{el}$	25	25	22	10	9	10
Working gas	H$_2$	H$_2$	H$_2$	H$_2$	He	He or H$_2$
Pressure in MPa$_{max}$	20	20	12	15	15	15
Max. gas temp. in °C	720	720	720	650	650	650
Operating hours in h	8, 000	35,000		80,000	350,000	100,000
Efficiency in %	38.5		33.2	33	30 – 32	30 – 33
Receiver						
Type	tube	tube	tube	tube	tube	tube
Aperture diameter in cm	20	20	22	15	25 – 15	15
Tube temperature in °C	810	810	800	850	850	850
Efficiency in %	90			90	90	90

Max. maximum; temp. temperature; [a] at a gas temperature of 760 °C; [b] facetted glass mirror; [c] stretched membrane; [d] sandwich structure.

Fig. 5.23 shows the example of a typical characteristic power curve for the 10 kW EuroDish plants which have been operated since 2001 in continuous operation (sunrise to sunset) for a total of approximately 10,000 hours (energy and capacity data standardised according to IEA directive /5-21/, /5-25/). According to these figures, the Stirling machine has a shaft power of 11 kW. The tube receiver is cooled by a fan in the overload range at high radiation of over 800 W/m².

The concentrator consists of sandwich elements made of fibre-glass reinforced epoxy resin. The segments are assembled to form a closed shell supported by a ring truss assembly which is characterised by high stiffness and contour accuracy. The front side of the shell is covered with thin-glass mirror to constantly achieve a high reflectivity of approximately 94 %.

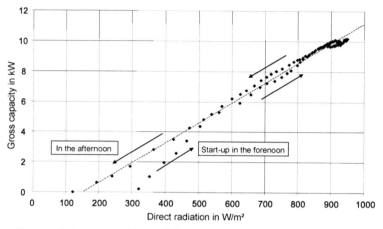

Fig. 5.23 Characteristic curve of 10 kW dish/Stirling systems

5.4.2 Economic and environmental analysis

The following explanations are aimed to assess dish/Stirling systems according to economic and environmental parameters.

Economic analysis. The power generation costs for the described dish/Stirling plants are calculated in line with the assessment method applied throughout this book. According to this, the costs for construction and operation are calculated and distributed in the form of annuities over the technical lifetime of the plant. Specific power generation costs are calculated on the basis of these annual depreciations and the generated electric energy. Unless otherwise indicated, a technical lifetime of 20 years and an interest rate of 4.5 % have been assumed.

Since such plants are exclusively installed in areas of a high share of direct radiation a reference site of an annual direct radiation total of 2,700 kWh/m^2 has been assumed.

The main technical data of the assessed 10 kW dish/Stirling system of the EuroDish type are indicated in Table 5.11.

Investments. Dish/Stirling systems have been tested in continuous operation in the USA and in Spain for more than 15 years. Nevertheless, still they are not commercially available. To approximately quantify the investments, published data of suppliers are used. By doing this we need to bear in mind that the investment costs of trial plants differ tremendously from the values of commercial systems manufactured in series. Thus, if serial production of 1,000 systems per year (i.e. 10 MW/a of installed capacity) is assumed, the unit costs for the entire system amount to approximately 45,000 € or 4,500 €/kW plus shipping, assembly, commissioning, planning, engineering and consulting /5-39/. The concentrator including support structure, drive units and foundation contributes for the major share (Table 5.12).

Table 5.11 Technical data of the assessed 10 kW dish/Stirling system (also see /5-38/)

Nominal capacity	10 kW
Concentrator diameter	8.5 m
Full-load hours	2,400 h/a
Storage capacity	none
Solar share	100 %
Solar multiple	1.3
Technical lifetime	20 years

Operation costs. Operation and maintenance costs are estimated to account for approximately 1.5 % of the investment costs (Table 5.12).

Electricity generation costs. Table 5.12 also illustrates the resulting power generation costs. Due to the solar multiple of 1.3 (i.e. concentrator dimensions are sufficiently large that nominal capacity is already reached at mean radiation) the full load hours of the dish/Stirling system of 2,400 h/a are comparatively high in spite of its design without storage. For the outlined reference system specific power generation costs thus amount to 0.18 €/kWh, provided that the investment cost reductions for the assumed 1,000 plants can be reached.

Table 5.12 Investments, operation and maintenance costs as well as power generation costs of 10 kW dish/Stirling power plants

Investments	
mirror, structure, drives and foundation	25,000 €
Stirling motor incl. receiver	11,000 €
transport, assembly and commissioning	4,000 €
planning, engineering, consulting, miscellaneous	6,000 €
contingencies	4,000 €
Total	50,000 €
Operation and maintenance costs	750 €/a
Power generation costs	0.18 €/kWh

As for tower and farm parabolic trough plants, power generation costs are mainly influenced by the achievable number of full load hours, the investments and the assumed mean interest rate.

Environmental analysis. Since the environmental effects of dish/Stirling plants are very similar to those of solar tower power plants, they are discussed within the scope of the corresponding chapters on solar tower power plants (see Chapter 5.2.2).

5.5   Solar updraft tower power plants

For a solar updraft tower power plant the three components of glass roof collector, chimney and turbine are combined. The use of this combination for power generation was already described more than 70 years ago /5-26/.

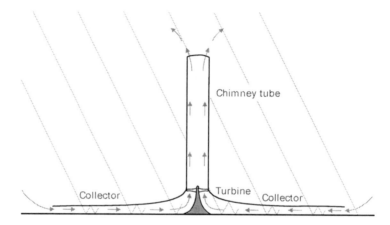

Fig. 5.24 Function principle of a solar updraft tower power plant

The principle, on which a solar updraft tower power plant is based, is shown in Fig. 5.24. The incident direct and diffuse solar radiation warms the air below a flat, circular glass roof, open at the circumference, which, in conjunction with the bottom underneath, forms an air collector. The middle of the roof is equipped with a vertical chimney provided with big openings for air supply. The roof is connected air-tight to the chimney bottom. Since warm air is of lower density than cold air, it rises to the top of the chimney tube. At the same time, the pull of the chimney makes warm air flow from the collector inside the chimney, so that cold ambient air flows inside the collector. Thus, solar radiation ensures continuous updraft inside the chimney. The energy contained in the air flow can be converted into mechanical energy using pressure-staged turbines, located at the bottom of the chimney. Eventually the energy is transformed into electrical energy by means of generators.

The achievable electrical output power P_{el} of a solar updraft tower power plant can be calculated according to Equation (5.6), namely on the basis of the supplied solar energy $\dot{G}_{g,abs}$ and the power plant efficiency η_{PP}. The latter in turn consists of the efficiency of the components collector η_{Coll}, chimney tube η_{Tower} and turbine(s) $\eta_{Turbine}$ (see /5-29/, /5-30/).

$$P_{el} = \dot{G}_{g.abs} \cdot \eta_{PP} = \dot{G}_{g.abs} \cdot \eta_{Coll} \cdot \eta_{Tower} \cdot \eta_{Turbine} \qquad (5.6)$$

The solar radiation supplied to the plant $\dot{G}_{g.abs}$ is calculated according to Equation (5.7) on the basis of the meteorological global radiation incident on the horizontal collector surface $\dot{G}_g$ and the horizontal collector surface S_{abs}.

$$\dot{G}_{g.abs} = \dot{G}_g \cdot S_{abs} \qquad (5.7)$$

The impinging global radiation is characterised by a clear diurnal course which also has an effect on the capacity cycle of the solar updraft tower power plant. A balancing of the electric power output of such a power plant – which might be necessary to allow for a more easy integration within an electricity provision system – is possible by an intermediate storage of the solar energy. This is technically possible by black hoses or bags filled with water, which, placed at the bottom of a solar updraft tower power plant, serve as an intermediate storage. During the day the water is heated up inside these storage elements and the stored energy is again released during the night (Fig. 5.25). This measure allows a continuous updraft inside the tower in spite of the fluctuating solar radiation throughout the day, and thus guarantees a continuous power supply (also refer to /5-27/, /5-28/).

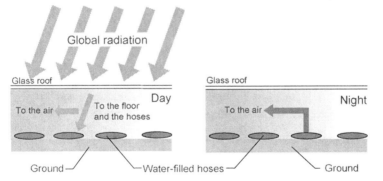

Fig. 5.25 Heat storage by means of water-filled hoses (left: situation during the day; right: situation during the night)

The collector efficiency η_{Coll} is primarily determined by the radiation absorption by the collector and the thermal losses (see Chapter 4.1).

The turbine efficiency $\eta_{Turbine}$ is determined among other parameters by the design of the turbine blades as well as their speed. Profile and tip losses should be minimized. Using multi-bladed turbines with inlet guide vanes (total-to-static pressure) efficiencies of about 80 % can be reached.

Determination of the main parameters of the tower efficiency η_{Tower}, by contrast, is much more complex. For this purpose it is assumed that the tower converts the heat energy supplied by the collector into kinetic energy (convection flow) and into potential energy (pressure drop at the turbine). The difference between the air density inside the tower $\rho_{Air,Tower}$ and the ambient air $\rho_{Air,amb}$ acts as the driving force. The lighter air column of warm air inside a tower of a height h_T is connected to the neighbouring atmosphere at the tower bottom (collector exit and thus tower entry) and consequently gets buoyancy. There is thus a pressure difference Δp between the tower bottom and the environment which is described by Equation (5.8); the pressure difference increases proportionally with tower height. g stands for acceleration of gravity.

$$\Delta p = g \int_0^{h_T} (\rho_{Air,amb} - \rho_{Air,Tower}) \, dh_T \qquad (5.8)$$

The pressure difference Δp consists of a static Δp_s and a dynamic Δp_d component. The static portion of the pressure difference drops at the turbine whereas the dynamic component describes the kinetic energy of the flow. The distribution of the pressure difference depends on the amount of energy which the turbine withdraws from the flow.

The power contained in the flow P_{Flow} can be described according to Equation (5.9) by the pressure difference Δp, calculated according to Equation (5.8), the flow speed of the air inside the tower v_{Air} and the tower diameter d_T.

$$P_{Flow} = \Delta p \cdot v_{Air} \cdot d_T \qquad (5.9)$$

On this basis, the tower efficiency η_{Tower} is calculated as the ratio of the power contained in the flow P_{Flow} (Equation (5.9)) and the power supplied by the collector P_{Abs}, which is calculated by the solar energy supplied to the plant $\dot{G}_{g,abs}$, reduced by the collector efficiency η_{Coll} (Equation (5.10)).

$$\eta_{Tower} = \frac{P_{Flow}}{P_{Abs}} = \frac{P_{Flow}}{\dot{G}_{g,abs} \cdot \eta_{Coll}} \qquad (5.10)$$

Inside the collector the flowing air is warmed up by a certain temperature $\Delta\theta_{Air}$ in dependence of the mass flow $\dot{m}_{Air}$ and the specific heat capacity of the air $c_{p,Air}$. On this basis, the thermal power output of the collector P_{Abs} can be calculated according to Equation (5.11).

$$P_{Abs} = \dot{m}_{Air} \, c_{p,Air} \, \Delta\theta_{Air} \qquad (5.11)$$

Without the energy extraction of the turbines a maximum air flow $v_{Air,max}$ of the flowing air masses $\dot{m}_{Air}$ is created. Under these circumstances the entire pressure difference Δp is converted into kinetic energy, i.e. the air flow is accelerated. The power contained in the air flow P_{Flow} is calculated according to Equation (5.12).

$$P_{Flow} = \frac{1}{2}\dot{m}_{Air} \cdot v_{Air,max}^2 \qquad (5.12)$$

Under the simplifying supposition that temperature profiles inside the tower and the environment are more or less parallel, the flow speed created at free convection can be expressed by Torricelli's temperature-modified equation (Equation (5.13)); g stands for gravity acceleration, θ_{Air} for ambient temperature of the air and $\Delta\theta_{Air}$ for the temperature rise from the temperature at the collector exit or tower entry.

$$v_{Air,max} = \sqrt{2\,g\,h_T\,\frac{\Delta\theta_{Air}}{\theta_{Air}}} \qquad (5.13)$$

Based on the discussed correlations, the tower efficiency η_{Tower} can also be determined by Equation (5.14).

$$\eta_{Tower} = \frac{g \cdot h_T}{c_{p,Air} \cdot \theta_{Air}} \qquad (5.14)$$

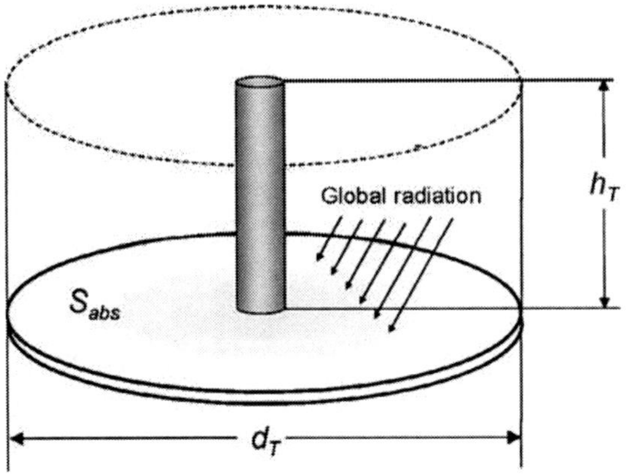

Fig. 5.26 Basic relations of a solar updraft tower power plant

The simplified presentation according to Equation (5.14) reveals that the tower efficiency mainly depends on the tower height. All in all, the power output of the solar updraft tower power plant is thus proportional to the collector surface and to the tower height (i.e. proportional to the volume of the cylinder shown in Fig. 5.26).

Since the electrical output of a solar updraft tower power plant is thus proportional to the volume of the cylinder created by the tower height and the collector surface, a certain capacity can either be achieved by a high tower in combination with a small collector or by a big collector and a smaller tower. There is thus a "classic" technical/economic optimisation problem to be solved.

5.5.1   Technical description

In the following, the technology of solar updraft tower power plants, including all related components, is described.

5.5.1.1   System components

In the following, the individual components of such a power plant are presented and discussed.

Collector. The hot air required for the operation of a solar updraft tower power plant is created by a simple air collector. The latter consists of a horizontal translucent glass or plastic roof located approximately 2 to 6 m above ground (Fig. 5.27).

The translucent roof is permeable by solar radiation, but impermeable by the long-wave heat radiation emitted by the collector bottom, which is heated up by the sun. This is why the bottom underneath the roof is heated strongly and transmits heat to the air, flowing radially from the exterior to the tower, thereby warming the air.

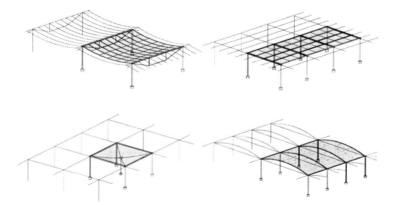

Fig. 5.27 Collector variants suitable for solar updraft tower power plants /5-27/, /5-28/

The height of the air-type collector increases towards the tower. According to this, the flow speed is not increased too much, so that friction losses are kept low. Additionally the losses during the direction change of the air from the horizontal into the vertical direction are minimised.

Storage. If a less pronounced electricity generation peak is desired for the early afternoon, while higher power generation is foreseen for the evening hours, the solar energy can be stored intermediately. For this purpose, water-filled hoses or cushions can be used, which, placed on the collector bottom, considerably enhance the already existing natural heat storage capacity of the ground.

Since already for very low water flow speeds, due to natural convection inside the hoses, heat transfer between the hoses and the water is considerably higher than between the radiation-absorbing surface of the earth (and the soil layers located underneath) below the collector, and also because the heat capacity of water is about five times higher than that of soil, the water inside the hoses stores part of the incident solar radiation. This heat is only released during the night when air temperatures inside the collector are below the water temperature inside the hoses. This is why solar updraft tower power plants can be operated day and night, solely driven by the sun.

Hoses are only filled once and remain sealed afterwards so that no water is evaporated. Depending on the desired performance characteristic the water quantity inside the hoses should correspond to a mean water depth below the collector of 5 to 20 cm (Fig. 5.28).

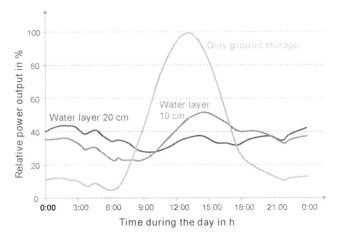

Fig. 5.28 Effect of heat-storing water hoses, located below the collector roof, on the chronological sequence of power provision (simulation results)

Tower. The tower or chimney represents the actual thermal engine of a solar updraft tower power plant. In a first approximation, the updraft of the air heated inside the collector is proportional to the air temperature rise obtained inside the collector and to the height of the chimney. For instance, in case of a large solar updraft tower power plant ambient air temperature is typically increased by 35 K, so that an air flow speed of approximately 15 m/s is created inside the chimney. Technically speaking, the chimneys of solar updraft tower power plants are very big atmospheric cooling towers.

Towers of a height of 1,000 m represent a great challenge, which is well controlled nowadays. For instance, the high-rise building Burj Dubai, currently under construction, is to be over 700 m high, and for Shanghai a high-rise building of over 800 m is being planned. For a solar updraft tower power plant only a simple hollow cylinder is required. This cylinder does not have a very slim shape and the requirements are considerably lower compared to residential buildings.

Such towers can be built using different technologies; besides free-standing reinforced concrete tubes also steel towers or guyed tower tube designs with sheet or membrane cladding are possible. Studies have shown that virtually for all considered sites reinforced concrete represents the most durable and cost-effective alternative.

For such a tower of a height of 1,000 m the wall thickness is slightly above 1 m at the bottom. This thickness would decrease to approximately 0.3 m at half of the height and remain constant afterwards. Yet, such thin tunnels are deformed by the wind load to an oval cross-section ("ovalisation"). This is particularly true for the suction flanks, represented in Fig. 5.29. Meridional stress becomes very high, so that stiffness is reduced because of cracking and there is also the danger of buckling. Ovalisation can effectively be avoided by bundles of strands in the form of lying spoke wheels stretched across the tower cross-section. They have the same stiffening effect as diaphragms, but reduce the updraft only minimally.

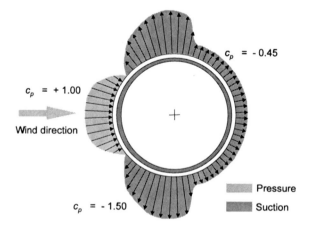

Fig. 5.29 Typical distribution of pressure/suction for the flow around a circular cylinder (c_p coefficient of pressure)

Turbines. Energy is extracted from the air flow by means of turbines. Such turbines to be used for solar updraft tower power plants are not velocity-staged, as free-running wind energy converters (see Chapter 7.1). They are pressure-staged as cased wind turbo generator sets, and extract static pressure similar to a hydro-electric power plant. Achievable efficiencies are thus higher than for free-running wind turbines.

Air speed in front and behind the turbine is almost the same. The extracted power is proportional to the product of volume flow and pressure drop at the turbine. Turbine control aims at maximising this product for all possible operation conditions.

The pressure drop and thus the flow speed and air flow inside the plant are controlled by the blade adjustment mechanism of the turbine. If, in an extreme case, the blades are at a right angle to the air flow, the power generation is zero. If, for the other extreme position, the air flow passes the rotor unhindered, then the pressure drop at the turbine equals zero; also under these circumstances no power is generated by the rotor. The optimum blade position is between these two positions.

For turbine design, we can revert to the experience gathered with hydroelectric power plants, wind energy converters, cooling tower technology and wind tunnel fans. In this respect, the vertical-axis turbine seems to be the most evident solution. As an alternative, also a larger number of horizontal-axis turbines can be used, placed concentrically in between the collector and the tower, so that cost-effective turbines of common dimensions can be applied.

5.5.1.2   Plant concepts

For research purposes, to date, several very small solar updraft tower power experimental facilities of heights of a few metres have been built (e.g. in the USA, South Africa, Iran and China). However, only one single plant of larger dimensions in Spain has been built and operated for power generation during several years in the 1980s.

Prototype located in the vicinity of Manzanares, Spain. In the years 1981/82, a pilot solar updraft tower power plant of a peak capacity 50 kW was built in Manzanares/Spain (about 150 km south of Madrid/Spain) /5-27/, /5-31/, /5-32/.

This research project was aimed at verifying the theoretical approaches and assessing the impact of the individual components on the capacity and the efficiency of the power plant under realistic technical and meteorological conditions. For this purpose a tower (chimney) of a height of 195 m and a diameter of 10 m surrounded by a collector of a diameter of 240 m was built (Table 5.13).

The chimney comprises a guyed tube of trapezoidal sheets (gauge 1.25 mm, knuckle depth 150 mm). The tube stands on a supporting ring 10 m above ground; this ring was carried by 8 thin tubular columns, so that the warm air can flow in practically unhindered at the base of the chimney. A pre-stressed membrane of plastic-coated fabric, with good flow characteristics, forms the transition between the roof and the chimney.

Table 5.13 Technical data of the prototype in Manzanares/Spain

Tower height	194.6 m
Tower radius	5.08 m
Mean collector radius	122.0 m
Mean roof height	1.85 m
Number of turbine blades	4
Turbine blade profile	FX W-151-A
Tip-speed ratio	10
Operation mode	grid-connected or off-grid
Typical temperature rise in the collector	20 K
Installed name-plate capacity	50 kW
Plastic membrane – collector surface	40,000 m^2
Glass roof – collector surface	6,000 m^2

A variety of types of plastic sheet, as well as glass, were selected in order to establish which was the best – and in the long term, most cost effective – material. It was found that glass is able to withstand even severe storms over many years without damage and proved to be self-cleaning; occasional rain showers were sufficient.

Completion of the construction phase in 1982 was followed by an experimental phase, the purpose of which was to demonstrate the operating principle of a solar updraft tower power plant. The goals of this phase of the project are (1) to obtain data on the efficiency of the technology developed, (2) to demonstrate fully automatic, power-plant-like operation with a high degree of reliability, and (3) to record and analyse operational behaviour and physical relationships on the basis of long-term measurements.

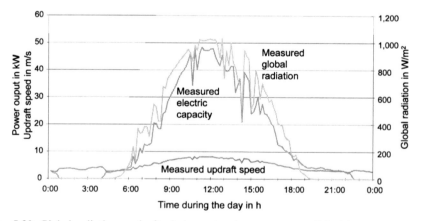

Fig. 5.30 Global radiation, updraft wind speed and power output of the Manzanares prototype (example: measured data at 8th June 1987)

Fig. 5.30 shows selected operational data from Manzanares for a typical day. The data clearly show that for this plant without additional thermal storage the electrical output during the day closely correlates with solar radiation. Neverthe-less, there is a certain updraft that can be utilised for power generation even dur-ing some night hours. This becomes also obvious in Fig. 5.31 showing the charac-teristic curve of such a plant exemplarily for the same day shown in Fig. 5.30. This graphic also outlines that the updraft speed as well as the electrical power provision is directly proportional to global radiation.

In the year 1987, the plant was operated for a total of 3,197 h; this corresponds to a mean daily operating time of 8.8 h. This was achieved by a fully automated plant management which ensured automatic start-up of the plant and synchronisa-tion with the power grid, once the flow speed exceeded a certain value (typically 2.5 m/s).

In spite of the absolutely positive operating results, confirming the calculated data, the test plant was completely dismantled after a storm at the end of the 1980s.

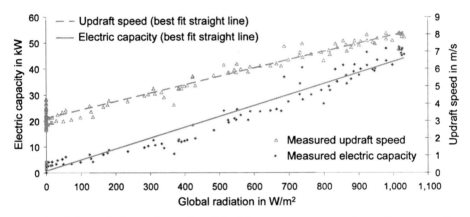

Fig. 5.31 Solar radiation and capacity of the Manzanares prototype (example: measured data at 8th June 1987)

Large solar updraft tower power plants. In spite of the big differences between dimensions of the pilot plant in Manzanares and projected 200 MW plants, ther-modynamic parameters are quite similar: Taking, for instance, the temperature rise and the flow speed inside the collector, the Manzanares plant has a temperature rise of up to 17 K and a speed of up to 12 m/s, whereas the calculated mean values for a 200 MW plant are 18 K or 11 m/s /5-33/.

Fig. 5.32 illustrates the results of a simulation calculation of such a 200 MW plant for a site with pronounced seasons. It shows a period of four days for every season. This plant thus also operates night and day without additional heat stor-age, even though the output power is reduced during the night, especially in win-ter.

Although various big solar updraft tower power plant projects have been de-veloped, e.g. in India and Australia, to date no commercial plant has been built.

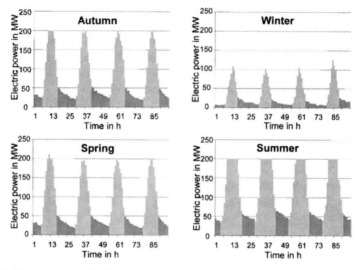

Fig. 5.32 Simulation results for power provision by a 200 MW solar updraft tower power plant without thermal storage

5.5.2   Economic and environmental analysis

The following considerations are aimed at assessing solar updraft tower power plants according to economic and environmental parameters.

Economic analysis. Power generation costs for solar updraft tower power plants are also calculated in line with the method applied throughout this book. Following this, the costs for construction and operation are determined and distributed in the form of annuities over the technical lifetime of the power plant. Power generation costs are calculated on the basis of these depreciations and the generated electrical energy. For this purpose, a technical lifetime of 25 years and an interest rate of 4.5 % have been assumed. In practice, solar updraft towers are designed for longer technical lifetimes (e.g. the tower for 60 years).

Since such plants are only installed in areas with a high share of global radiation, also for this purpose, a reference site characterised by an annual total global radiation on the horizontal plane of 2,300 kWh/m^2 has been assumed.

The energy output of a solar updraft tower power plant is proportional to global radiation, collector surface and tower height. In this respect, there is no physical optimum; thus it is necessary to optimise dimensions according to component costs (collector, tower, turbine) as well as the land costs. This is why plants of different dimensions can be built – at minimised costs – to suit the local conditions at various sites. If the collector surface is cheap and reinforced concrete is expensive, then a large collector and a comparatively small tower will be built.

However, if the collector is expensive, a smaller collector and a bigger tower will be built.

Table 5.14 provides an overview on the typical dimensions of solar updraft tower power plants. The indicated figures are based on internationally common material and construction costs.

Table 5.14 Typical dimensions and power provision of selected solar updraft tower power plants at the reference site

Nominal capacity	MW	5	30	50	100	200
Tower height	m	550	750	750	1,000	1,000
Tower diameter	m	45	70	90	110	120
Collector diameter	m	1,250	2,950	3,750	4,300	7,000
Energy provision	GWh/a	14	87	153	320	680

Investments. Investment costs are determined on the basis of the specific costs and dimensions indicated in Table 5.15. According to this, the investments vary according to the current status of knowledge between 43 Mio. € for a 5 MW plant according to the technical data defined within Table 5.14 and 634 Mio. € for a 200 MW plant (see Table 5.14). However, it needs to be considered that the investment of such plants are afflicted with much more uncertainties than, for instance, conventional power plants, since, to date, no plant of such a size has been built.

Table 5.15 Investment, operation and maintenance costs as well as power generation costs of solar updraft tower power plants

Nominal capacity	MW	5	30	50	100	200
Tower costs	Mio. €	19	49	64	156	170
Collector costs	Mio. €	11	54	87	117	287
Turbine costs	Mio. €	8	32	48	75	133
Engineering, tests, miscellaneous	Mio. €	5	17	24	41	44
Total	Mio. €	43	152	223	389	634
Annuity of investment costs	Mio. €/a	2.3	8.2	12.1	21.2	34.4
Operation and maintenance cost	Mio. €/a	0.2	0.7	1.0	1.7	3.0
Power generation costs	€/kWh	0.18	0.10	0.09	0.07	0.06

Operation costs. The annual operation costs are globally estimated at a lump-sum of 0.5 % of the investment costs. Similar to hydroelectric power plants, operation costs are comparatively low, since the turbines are the only movable components. There are no components that are exposed to high pressures or temperatures. The majority of power plant components is highly durable.

Electricity generation costs. Based on the annual electricity yields determined with simulation models, power generation costs can be calculated; they are within the range indicated in Table 5.15. Following this, they vary between 0.18 €/kWh for the 5 MW solar updraft tower power plant and the order of magnitude of

0.06 €/kWh for the 200 MW plant. Thus, the power generation costs decrease significantly with increasing plant size.

The sensitivity analysis shows similar results as the analysis conducted for a solar tower plant (Fig. 5.13).

Environmental analysis. Environmental effects of solar updraft tower power plants are very similar to that of solar tower power plants. They are therefore discussed within the scope of Chapter 5.2.2 on solar tower power plants.

5.6   Solar pond power plants

Solar ponds are power plants that utilise the effect of water stratification as a basis for the collector. A basin filled with brine (i.e. a water/salt mixture) functions as collector and heat storage. The water at the bottom of the solar pond serves as primary heat storage from which heat is withdrawn. The deeper water layers and the bottom of the solar pond itself serve as absorber for the impinging direct and diffuse solar radiation. Due to the distribution of the salt concentration within the basin, which increases towards the bottom of the basin, natural convection and the ensuing heat loss at the surface due to evaporation, convection and radiation is minimised. This is why heat of an approximate temperature between 80 and 90 °C (approximate stagnation temperature 100 °C) can be extracted from the bottom. Thanks to suitable thermodynamic cycles (e.g. ORC process) heat can then be used for power generation.

5.6.1 Technical description

In the following, the technology of solar pond power plants, including all related components, is explained.

5.6.1.1   System components

As follows the main system components of a solar pond power plant are described in detail.

Pond collector. Pond collectors are either natural or artificial lakes, ponds or basins that act as a flat-plate collector because of the different salt contents of water layers due to stratification. The upper water layers of relatively low salt content are often provided with plastic covers to inhibit waves. This upper mixing zone of such pond collectors usually is approximately 0.5 m thick. The adjacent transition zone has a thickness of 1 to 2 m, and the lower storage zone is of 1.5 to 5 m thickness.

If deeper layers of a common pond or lake are heated by the sun, the heated water rises up to the surface since warm water has a lower density than cold water. The heat supplied by the sun is returned to the atmosphere at the water surface. This is why, in most cases, the mean water temperature approximately equals ambient temperature. In a solar pond heat transmission to the atmosphere is prevented by the salt dissolved in deeper layers, since, due to the salt, water density at the bottom of the pond is that high, that the water cannot rise to the surface, even if the sun heats up the water to temperatures that are close to the boiling point.

The salt concentration of the different layers must thus increase with increasing depth (Fig. 5.33). In a first phase, this ensures stable water stratification. The upper, almost salt-less layer only acts as transparent, heat-insulating cover for the cooling, heat-storing deeper layers at the pond bottom /5-41/.

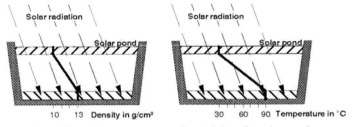

Fig. 5.33 Density (left) and temperature gradient (right) of a solar pond

To ensure stable stratification of a solar pond, with increasing depth the temperature increase must not exceed density increase (i.e. salt content). This is why all relevant parameters must be continuously monitored in order to take appropriate measures (e.g. heat withdrawal, salt supply) in due time.

To achieve the utmost collector efficiency, a high portion of the solar radiation must reach the absorption zone. Yet, this can only be achieved, if the top layers are of sufficient transmission capability.

During the operation of a solar pond, the transmissivity, the salt content and the temperature must be regularly monitored. The timely course of these parameters must be measured from the water surface to the ground in order to determine the heat quantity that can be withdrawn from the pond or to determine the measures to maintain the respective required salt concentration and the water quality (prevention of turbidity due to particulate matter, algae or bacteria).

Diffusion ensures permanent equalisation of the salt concentration in a solar pond which is even intensified by wave motion due to wind near the surface. This is why salt needs to be withdrawn from the surface water and added to deeper layers. For this purpose surface water is evaporated in separate flat basins (salines). Subsequently, the extracted salt is added to deeper zones.

Heat exchangers. Basically, there are two methods to withdraw heat from a solar pond.

- The working fluid of the thermal engine flows through tube bundle heat exchangers installed within the storage zone of the solar pond, and is thereby heated up.
- The hot brine can also be pumped from the storage zone by means of an intake diffuser, subsequently be transmitted to the working fluid of the thermal engine and eventually be re-supplied to greater depths of the pond by another diffuser, once the brine has cooled down. The technical approach allows adjusting the position of the intake diffuser to the depth of the highest temperature. Secondly, heat losses by the pond bottom are reduced, since the cooled water is recycled to the pond near the bottom.

As a matter of principle, a sufficiently dimensioned heat exchanger unit results indispensable for the successful operation of a solar pond. Especially in times of high radiation (i.e. at noon) it has to be ensured that heat can securely be withdrawn from the pond, to prevent phase changes and/or make stratification instable.

Thermal engine. To convert solar thermal energy into mechanical and afterwards in electrical energy, usually ORC processes are applied (see Chapter 5.1.4 and 10.3). These are basically steam cycles which utilise a low-boiling, generally organic, cycle fluid. Such processes permit to provide electrical energy also at low useful temperature differences.

5.6.1.2   Plant concepts

Fig. 5.34 shows the general structure of a solar pond power generation plant. According to this graphic, the water absorbs the incident direct and diffuse radiation, similar to the absorber of a conventional solar collector, and is heated up. The technically adjusted salt concentration prevents natural convection and the resulting heat loss at the surface due to evaporation, convection and radiation.

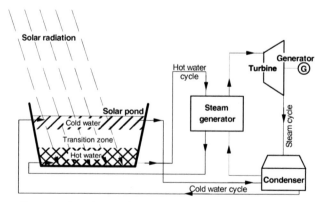

Fig. 5.34 Plant diagram of a solar pond power plant

Water can thus be withdrawn from the storage zone at the bottom at an approximate temperature of 80 to 90 °C. This heat can subsequently be used for power generation by an ORC process.

Solar pond power plants of electric capacities from a few ten kW up to a few MW have been built in Israel, the US (Texas), Australia and India (for process heat provision /5-34/), among other countries. With approximately one percent, solar thermal efficiencies are low; the mean specific capacities range from 5 to 10 W/m² depending on radiation, salt content and maximum temperature. For the short-term, also higher capacities can be withdrawn; however, in such a case the solar pond would cool down much faster. Table 5.16 shows typical examples (see also /5-35/, /5-36/, /5-37/).

Table 5.16 Data of selected solar pond power plants

	El Paso Texas, USA	Beit Ha'Arava Israel	Pyramid Hill Australia
Capacity	300 kW$_{th}$ 70 kW$_{el}$	5 MW$_{el}$ max. 570 kW$_{el}$ (average)	60 kW$_{th}$
Pond surface	3,350 m²	250,000 m²	3,000 m²

5.6.2 Economic and environmental analysis

The following explanations are aimed at assessing solar pond power plants according to economic and environmental parameters.

Economic analysis. In line with the preceding assessment method, in the following, the power generation costs are calculated for solar pond power plants. According to this, the costs for construction and operation are determined and assessed in the form of annuities. On this basis and the produced electrical energy, the power generation costs are calculated. For this purpose, a technical lifetime of 25 years and an interest rate of 4.5 % have been assumed.

Since such plants are only installed in areas with a high share of solar radiation, a reference site has been assumed which is characterised by an annual total global radiation on the horizontal axis of 2,300 kWh/m².

Table 5.17 outlines the main parameters of the assessed solar pond. The solar pond power plant investigated here has a capacity of 5 MW.

Investment costs. Since there are only a few solar ponds, all of them being unique, there are virtually no market prices available that could serve as a basis for these analyses. This is why the following cost estimations have been based on literature values. For this purpose, specific investment costs of 40 €/m² of pond surface have been assumed. Literature values are based on cost estimations for civil engineering works and geomembranes; other sources indicate values between 15 and

75 US\$/m^2 /5-37/. For the overall plant, in total, specific investment costs of approximately 2,000 €/kW are assumed (Table 5.18).

Table 5.17 Parameters of the assessed solar pond

Nominal capacity	5 MW^a (peak capacity)
Collector surface	250,000 m^2
Heat exchanger	external; the brine is pumped from the pond and recycled to the pond in cooled condition
Full-load hours	1,150 h/a
Storage	deepest water layer of the pond serves as thermal storage
Solar share	100 %
Net efficiency^b	1 %

^{a} The average capacity amounts to approximately 650 kW; at short-term higher capacities are possible so that higher revenues can be achieved (peak-load power); ^{b} efficiency between incident solar radiation and produced electric energy.

Operation costs. For this purpose, operation costs are estimated at a lump-sum of 1 % of the investment costs. Besides the maintenance of the heat exchangers and the ORC plant further measures are required since the salt that diffuses from the bottom to the surface has to be retrieved. Compared to conventional steam power plants, the required water quantity is thus many times higher.

Table 5.18 Estimation of the power generation costs of a 5 MW solar pond provided with external heat exchanger

Specific investment costs	2,000 €/kW
Operation costs	100,000 €/a
Power generation costs	0.14 €/kWh

Electricity generation costs. According to Table 5.18, the power generation costs for a solar pond of a surface of 250,000 m^2 and annual operation costs of approximately 100,000 €, amount to approximately 0.14 €/kWh. However, it has been assumed that brine or appropriate salt are available at the site free of charge, so that only transportation measures are required.

For an economic assessment within the scope of operational optimisation, it also has to be considered that electric capacity must be available at any time of the day. Solar ponds can thus generally also serve as peak-load power stations.

Environmental analysis. The environmental effects of solar pond power plants are largely similar to those of solar tower power plants; they are thus discussed in Chapter 5.2.2 on solar tower power plants. Additionally, the salt brine might cause environmental effects when polluting the surroundings of a solar pond. Also, the use of fresh water during the operation of such plants might be considerable. In areas with a shortage of water this could lead to environmental effects.

6 Photovoltaic Power Generation

6.1 Principles

Besides solar thermal heat and power generation photovoltaic power generation is a further possibility to directly utilise solar radiation energy. However, in contrast to solar thermal electricity generation, solar energy is directly converted into electrical energy. In the following the main physical principles of this energy conversion method are outlined (also refer to /6-1/, /6-2/, /6-3/, /6-4/, /6-5/, /6-6/, 6-7/, /6-8/, /6-9/).

6.1.1 Energy gap model

Besides the positively charged protons and the uncharged neutrons inside the nucleus an atom is composed of the negatively charged electrons that assume discrete energy levels (such as "shells" or "orbitales") around the nucleus. There is a limited number of electrons that can occupy a certain energy level; according to the so-called Pauli exclusion principle any possible energy level may only be occupied by a maximum of two electrons. These two electrons are only allowed if they again differ from each other by their "spin" (i.e. self angular momentum).

If several atoms form a crystal, the different energy levels of the individual atoms overlap each other and stretch to form energy bands. Between these "allowed" energy bands there are energy gaps (i.e. "forbidden" bands). There are narrow permitted bands for the inner electrons, closely bound to the nucleus, and wide permitted bands for the outer electrons. The width of the forbidden bands varies in the opposite way; forbidden bands are wide close to the nucleus and decrease with increasing energy level, so that outer bands overlap. The energetic distances of permitted bands and the width of energy gaps, respectively, and the distribution of electrons to the permitted bands determine the electric and optic properties of a crystal.

Also within these bands the number of energy levels to be occupied by electrons is limited (i.e. the number of spaces is restricted). There is thus a "finite energy state density". The inner shells of atoms and the energy bands of solids with low energy levels, respectively, are almost entirely covered with electrons. The electrons are unable to move freely here; they are only able to change places.

These electrons do not produce any conductivity. The most energy-rich energy band, fully occupied with electrons, is referred to as valence band; the electrons it contains determine the chemical bond type of the material.

A solid with electrical conductivity requires freely moving electrons. However, electrons are only able to move freely if they are located in an energy band that is not fully occupied. For energy reasons, this is only true for the energy band located above the valence band. This energy band is thus referred as the conduction band.

The energy gap E_g between the valence band and conduction band is termed as "band gap" (Fig. 6.1). This energy gap exactly equals the minimum amount of energy required to transfer one electron from the valence band into the conduction band.

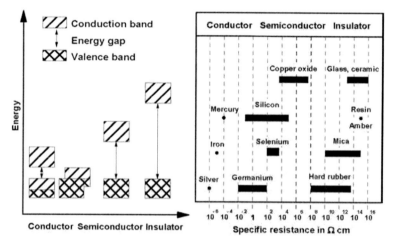

Fig. 6.1 Valence and conduction bands as well as energy gap (left) and specific resistance (right) of conductors, semiconductors and insulators (also refer to /6-2/,/6-10/)

6.1.2 Conductors, semiconductors and insulators

Conductors, semiconductors and insulators are different in terms of their band structure and the occupation of their bands with electrons (Fig. 6.1).

Conductors. Within conductors (e.g. metals and their alloys) two different conditions might occur.
- The most energy-rich band (i.e. conduction band) occupied by electrons is not entirely occupied.
- The most energy-rich band fully occupied with electrons (i.e. valence band) and the conduction band located on top overlap, so that also a partly covered band (conduction band) is formed.

Current is thus transferred by freely moving electrons, abundantly available within the crystal lattice regardless of the respective temperature of the material. Due to this, electrical conductors (like metals) are characterised by a low specific resistance. With rising temperature, the increasing thermal oscillation of the atomic cores impedes the movement of the electrons. This is why the specific resistance of metals increases with a rising temperature.

Insulators. Insulators (e.g. rubber, ceramics) are characterised by a valence band fully filled with electrons, a wide energy gap ($E_g > 3$ eV) and an empty conduction band. Hence, insulators possess virtually no freely moving electrons. Only at very high temperatures (strong "thermal excitation") are a small number of electrons able to overcome the energy gap. Thus, ceramics, for instance, show conductivity only at very high temperatures.

Semiconductors. In principle, semiconductors (e.g. silicon, germanium, gallium-arsenide) are insulators with a relatively narrow energy gap (0.1 eV $< E_g < 3$ eV). Therefore, at low temperatures, a chemically pure semiconductor acts as an insulator. Only by adding thermal energy, electrons are released from their chemical bond, and lifted to the conduction band. This is the reason why semiconductors become conductive with increasing temperatures. This is the other way round compared to metals, where conductivity decreases with rising temperatures. Regarding specific resistance, semiconductors are in-between conductors and insulators. Within the transition area between semiconductors and conductors, in case of very narrow energy gaps (0 eV $< E_g < 0.1$ eV), such elements are also referred to as metalloids or semi-metals as they may show similar conductivity as metals. However, unlike "real" metals they are characterised by a reduced conductivity with decreasing temperatures.

6.1.3 Conduction mechanisms of semiconductors

Intrinsic conductivity. Semiconductors are conductive beyond a certain temperature level as valence electrons are released from their chemical bonds with increasing temperatures and thus reach the conduction band (intrinsic conductivity). They become conduction electrons that are able to move freely through the crystal lattice (i.e. electron conduction).

On the other hand, also the resulting hole inside the valence band can move through the semiconductor material, since a neighbouring electron can advance to the hole. Holes thus contribute equally to conductivity (hole conduction). Since every free electron creates a hole within undisturbed pure semiconductor crystals both types of charge carriers equally exists.

Intrinsic conductivity is counteracted by recombination, namely the recombination of a free electron and a positive hole. Despite this recombination the number of holes and free electrons remains equal since at a certain temperature level always the same number of electron-hole-pairs are formed as recombine. For every temperature, there thus exists an equilibrium state with a certain number of free holes and free electrons. The number of free electron-hole-pairs increases with rising temperature.

If an external voltage is applied to such a crystal lattice from outside, electrons move to the positive pole while the holes move to the negative pole. The mechanism of intrinsic conduction inside semiconductors can also be described by the energy gap model (Fig. 6.2, left).

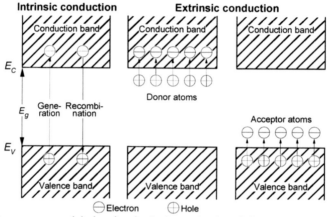

Fig. 6.2 Energy gap model showing intrinsic conduction (left) and extrinsic conduction (centre, right) (E_V valence band energy level, E_C conduction band energy level, E_g energy gap; according to e.g. /6-1/, /6-2/)

Extrinsic conduction. In addition to the – low – intrinsic conduction of pure crystal lattices extrinsic conduction is created by intentional incorporation of foreign atoms ("doping"). Such impurities are effective if their number of valence electrons differs from that of the base material. If for instance the valence electron number of the incorporated impurities exceeds that of the lattice atom (e.g. in pentavalent arsenic (As) incorporated into tetravalent silicon (Si); Fig. 6.3), the excess electron is only weakly bound to the impurity atom. It thus separates easily from the impurity atom due to thermal movements within the lattice and increases the conductivity of the crystal lattice as a freely moving electron. Such foreign atoms which increase the number of electrons are referred to as donor atoms. By this the number of electrons exceeds by far that of the holes. In this case electrons are called majority carriers, whereas the holes constitute the minority carriers. Since conductivity is mainly created by negatively charged particles, this type of conduction is referred to as n-conduction.

If the impurities incorporated into the semiconductor material are by contrast provided with less valence electrons (e.g. trivalent boron (B) or aluminium (Al) incorporated into tetravalent silicon (Si); Fig. 6.3), these doping atoms tend to absorb one additional electron from the valence band of the basic material. Such foreign atoms are thus referred to as acceptor atoms. They increase the number of holes (quasi positive charge carriers) and create p-conductivity. Under these conditions deficit electrons (i.e. holes) are majority carriers whereas electrons act as minority carriers.

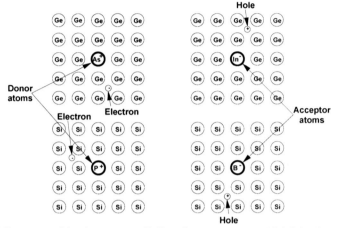

Fig. 6.3 Effects caused by donor atoms (left) and acceptor atoms (right) (various sources)

Also the above-mentioned context can be shown within the energy gap model (Fig. 6.2). Within the undoped semiconductor material (Fig. 6.2, left) a certain equilibrium concentration of free mobile charge carriers is formed due to the recurrent processes of electron-hole-pair creation and recombination. During this process the number of holes and electrons are equal. Besides temperature, charge carrier density at equilibrium concentration is determined by the minimum energy required to release one electron from the valence band, and is thus described by the energy gap E_g. For example for germanium (Ge) the energy gap amounts to 0.75 eV and to 1.12 eV for silicon. Since electron and hole densities remain relatively low, also conductivity of the undoped semi-conductor material is low.

The number of electrons within the conduction band is considerably increased by the addition of donors (Fig. 6.2, centre). Within the energy gap model, donors correspond to energy levels scarcely below the conduction band. Adding acceptors increases the number of holes inside a valence band (Fig. 6.2, right) accordingly. Within the energy gap model, energy levels of acceptors are a little above the valence band.

By doping the semiconductor with acceptors (p-doping) and donors (n-doping) conductivity of semiconductor materials can be controlled across several orders of magnitudes. However, the product of electron density and hole density of a certain material is a temperature-dependent material constant. Hence, if for instance the electron density is increased by the incorporation of donors, the hole density is automatically reduced. Nevertheless, the conductivity increases. However, both kinds of doping must not be applied simultaneously, since the effects of acceptors and donors cancel each other (also refer to /6-3/, /6-4/).

Semiconductors are further distinguished into "direct" and "indirect" semiconductors. While for direct semiconductors only energy is required to transfer charge carriers from the valence band to the conduction band, for indirect semiconductors also a momentum needs to be transferred to the charge carrier. This is mainly due to the band structure and has a tremendous impact on the appropriateness of semiconductor materials for solar cells. While for a direct semiconductor absorption of an incident photon is sufficient to lift the charge carrier up to the conduction band (i.e. exclusive energy transfer by the photon), for indirect semiconductors an appropriate momentum additionally needs to be transferred. This process requires three particles: the charge carrier (1st particle) simultaneously receives sufficient energy quantities from the photon (2nd particle) and the required momentum from a phonon (3rd particle); quanta of a crystal momentum are referred to as phonons. Only if all three particles meet simultaneously (i.e. three particle process), charge carriers are lifted into the conduction band. In comparison to direct semiconductors (two body process) these conditions are much rarer. This is why in case of indirect semiconductors the photon inside the semiconductor material has to travel a much longer distance until it is absorbed.

Crystalline silicon is such an indirect semiconductor, and silicon cells must thus be relatively thick and/or contain an appropriate light-trapping scheme to generate a prolonged optical path length. Amorphous silicon, CdTe or CIS (see Chapter 6.2.1), are in contrast direct semiconductors. Solar cells made of these materials can thus have a thickness clearly below 10 µm, while the thickness of crystalline silicon solar cells typically stretches from 200 to 300 µm. Thinner crystalline silicon cells are under development, but must be provided with the discussed optical properties, resulting in increased manufacturing expenditure.

6.1.4 Photo effect

The term "photo effect" refers to the energy transfer from photons (i.e. quantum of electromagnetic radiation) to electrons contained inside material. Photon energy is thereby converted into potential and kinetic energy of electrons. The electron absorbs the entire quantum energy of the photon defined as the product of Planck's quantum and the photon frequency. External and internal photo effect is distinguished.

External photo effect. If electromagnetic radiation hits the surface of a solid body within the ultraviolet range, electrons can absorb energy from the photon. Then they are able to surmount the required work function to escape from the solid body, provided that there is sufficient photon energy. This process is referred to as the external photo effect.

Within the energy gap model electrons are energetically lifted beyond the conduction band so that they are no longer considered as a part of the solid body. For individual atoms the corresponding energetic borderline is the ionisation energy; whereas it is referred to as vacuum level for solid bodies. It is defined as the energetic border at which the energy of the electron detached from the solid body, inside the vacuum, equals zero. The vacuum level is thus identical to the upper edge of the conduction band or the upper edge of all bands above the valence band.

Internal photo effect. The internal photo effect describes also absorption of electromagnetic radiation within a solid body. The electrons are in this case not detached from the solid body. They are only lifted from the valence band up to the conduction band. Therefore, electron-hole-pairs are created which enhance the electric conductivity of the solid body (Chapter 6.1.3)

The internal photo effect is the basis for the photovoltaic effect and thus of the solar cell. However, the photovoltaic effect requires an additional boundary layer, for instance a metal-semiconductor junction, a p-n-junction or a p-n-heterojunction (i.e. an interface between two different materials with different types of conductivity; Chapter 6.1.6).

6.1.5 P-n-junction

By well defined addition of donors and acceptors (diffusion, alloying, ion implantation) adjacent p- and n-regions are created inside a semiconductor crystal (Fig. 6.4). Especially abrupt transitions from one type of conductivity to the other one are obtained by epitaxy. Here, the layer by layer growth of a semiconductor enables a transition within almost one atomic layer to the subsequent one.

If p- and n-doped materials brought into contact, holes from the p-doped side diffuse into the n-type region and vice versa. First a strong concentration gradient is formed at the p-n-junction, consisting of electrons inside the conduction band and holes inside the valence band. Due to this concentration, gradient holes from the p-region diffuse into the n-region while electrons diffuse from the n- to the p-area. Due to the diffusion, the number of majority carriers are reduced on both sides of the p-n-junction. The charge attached to the stationary donors or acceptors then creates a negative space charge on the p-side of the transition area and a positive space charge on the n-side.

As a result of the equilibrated concentration of free charge carriers an electrical field is built up across the border interface (p-n-junction). The described process creates a depletion layer in which diffusion flow and reverse current compensate each other. The no longer compensated stationary charges of donors and acceptors define a depletion layer whose width is dependent on the doping concentration (Fig. 6.4, Fig. 6.5).

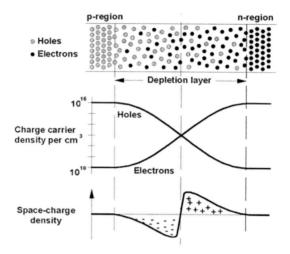

Fig. 6.4 Creation of a depletion layer within the p-n-junction (various sources)

Fig. 6.5 shows idealised conditions. Simplified, it has been assumed that majority carrier density is negligible over the entire space-charge region and that depletion layer density (Fig. 6.5 c) remains constant up to the edges of the depletion zone. This also implies that the respective doping concentration is constant up to the p-n-border; p-n-transition is thus abrupt. Fig. 6.5 d shows the corresponding potential curve of a positively charged particle and the diffusion voltage created within the depletion layer.

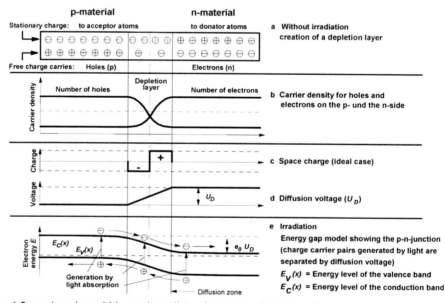

Fig. 6.5 p-n-junction within a solar cell (e_0 elementary charge) (various sources)

Since within the energy gap model, the energy of the electron is represented on the y-axis, in Fig. 6.5 c diffusion voltage is shown in opposite direction and does therefore not match the distribution function shown above (Fig. 6.5 b).

6.1.6 Photovoltaic effect

If photons, the quantum's of light energy, hit and penetrate into a semiconductor, they can transfer their energy to an electron from the valance band (Fig. 6.5 e). If such a photon is absorbed within the depletion layer, the region's electrical field directly separates the created charge carrier pair. The electron moves towards the n-region, whereas the hole moves to the p-region. If, during such light absorption, electron-hole-pairs are created outside of the depletion region within the p- or n-region (i.e. outside of the electrical field), they may also reach the space-charge region by diffusion due to thermal movements (i.e. without the direction being predetermined by an electrical field). At this point the respective minority carriers (i.e. the electrons within the p-region and the holes in the n-region) are collected by the electrical field of the space-charge region and are transferred to the opposite side. The potential barrier of the depletion layer, in contrast, reflects the respective majority carriers.

Finally, the p-side becomes charged positively while the n-side is charged negatively. Both, photons absorbed within, and outside, of the depletion layer contribute to this charging. This process of light-induced charge separation is referred as p-n-photo effect or as photovoltaic effect.

Thus, the photovoltaic effect only occurs if one of the two charge carriers created during light absorption passes the p-n-junction. This is only likely to occur when the electron-hole-pair are generated within the depletion layer. Outside of this electrical field there is an increasing likelihood that charge carrier pairs created by light get lost by recombination. This is more likely the greater the distance is between the location of the generation of the electron-hole-pair and the depletion layer. This is quantified by the "diffusion length" of the charge carriers inside the semiconductor material. The term "diffusion length" refers to the average path lengths to be overcome by electrons or holes within the area without an electrical field before recombination takes place. This diffusion length is determined by the semiconductor material and, in case of the identical material, highly depends on the impurity content – and thus also on doping (the more doping the lower the diffusion length) – and on crystal perfection. For silicon the diffusion length varies from approximately 10 up to several 100 μm. If the diffusion length is less than the charge carrier's distance to the p-n-junction most electrons or holes recombine (mathematically spoken: after having overcome a certain diffusion length the number of light-induced charge carriers is reduced to $1/e$; after having covered two diffusion lengths it is reduced to $1/e^2$ etc. To achieve an effective charge carrier separation the diffusion length should be a multiple of the absorption length of the solar radiation incident on a photovoltaic cell.

Due to the charge separation during irradiation, electrons accumulate within the n-region, whereas holes accumulate in the p-region. Electrons and holes will accumulate until the repelling forces of the accumulated charges start to impede additional accumulation; i.e. until the electrical potential created by the accumulation of electrons and holes is balanced by the diffusion potential of the p-n-junction. Then the open-circuit voltage of the solar cell is reached. The time to achieve these conditions is almost immeasurably short.

If p- and n-sides are short-circuited by an external connection, the short-circuit current is measured. In this operating mode the diffusion voltage at the p-n-junction is restored. According to the operating principle of a solar cell, short-circuit current increase is proportional and almost linear to solar irradiance (see also e.g. /6-3/, /6-10/).

6.2 Technical description

The technical basics of photovoltaic power generation are outlined within the following chapters. All explanations, including the indicated key figures, reflect state-of-the art technology. High-end laboratory cells or modules might have a better performance.

6.2.1 Photovoltaic cell and module

Structure. Fig. 6.6 shows the basic structure of a photovoltaic cell consisting of p-conducting base material and an n-conducting layer on the topside. The entire cell rear side is covered with a metallic contact while the irradiated side is equipped with a finger-type contact system to minimise shading losses. Also full cover, transparent conductive layers are used. To reduce reflection losses the cell surface may additionally be provided with an anti-reflecting coating. A silicon solar cell with such construction usually has a blue colour. By the incorporation of inverse pyramids into the surface reflection losses are further reduced. The inclination of the pyramid surfaces is such that photons are reflected onto another pyramid surface, and thus considerably enhance the possibility of photon penetration into the crystal. Absorption of the solar light by these cells is almost complete, the cells appear black.

Current-voltage characteristic and equivalent circuit. An illuminated solar cell ideally can be considered as a current source provided with a parallel diode. The photocurrent I_{Ph} is assumed to be proportional to the photon flow incident on the cell. The Shockley equation for ideal diodes (Equation (6.1) /6-3/) describes the interdependence of current and voltage (current-voltage characteristic) of a solar cell.

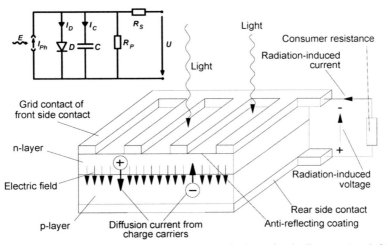

Fig. 6.6 Structure of a typical solar cell and an equivalent circuit diagram (top left) (see also /6-1/, /6-10/; for symbols see text)

$$I = I_{Ph} - I_0 \left(e^{\frac{e_0 U}{kT}} - 1 \right)$$

$$U = \frac{k}{e_0} \cdot T \cdot ln\left(1 - \frac{I - I_{Ph}}{I_0} \right)$$

(6.1)

I stands for the current flowing through the terminals, I_{Ph} for the photocurrent and I_0 for the saturation current of the diode, whereas e_0 represents the elementary charge (1.6021 10^{-19} As), U the cell voltage and k the Boltzmann constant (1.3806 10^{-23} J/K), and θ stands for the temperature. However in Equation (6.1) the sign for current I have been inverted compared to the conventional notation. This is why the characteristic curves (Fig. 6.7) are not located in the forth but in the first quadrant. However, this kind of representation has become common practice.

Under realistic conditions, the performance of a solar cell can be described as shown in the equivalent circuit diagram illustrated in Fig. 6.6, left top. Without irradiation, the solar cell is equal to an ordinary semiconductor diode whose effect is also maintained at the incidence of light. This is why diode D has been connected in parallel to the photovoltaic cell in the equivalent circuit diagram. Each p-n-junction also has a certain depletion layer capacitance, which is, however, typically neglected for modelling of solar cells. At increased inverse voltage the depletion layer becomes wider so that the capacitance is reduced similar to stretching the electrodes of a plate capacitor. Thus, solar cells represent variable capacitances whose magnitude depends on the present voltage. This effect is considered by the capacitor C located in parallel to the diode. Series resistance R_S consists of the resistance of contacts and cables as well as of the resistance of the semiconductor material itself. To minimise losses, cables should be provided with a maximum cross-section.

Parallel or shunt resistance R_P includes the "leakage currents" at the photo-voltaic cell edges at which the ideal shunt reaction of the p-n-junction may be reduced. However, for good mono-crystalline solar cells shunt resistance usually is within the kΩ region and thus has almost no effect on the current-voltage characteristic.

Fig. 6.7 shows the typical shape of a current-voltage curve for various operating modes (i.e. changing irradiation and temperature). At the intercept points of the curve and the axes the short-circuit current I_{SC} (which is in 1st order approximation equal to I_{Ph}) is supplied at $U = 0$ and the open-circuit voltage U_{OC} at $I = 0$. Starting with the short-circuit current, the cell current is at first only slightly reduced and decreases over-proportionally shortly before reaching open-circuit voltage when increasing the cell voltage continuously. These effects result in the characteristic shape (see /6-2/, /6-5/).

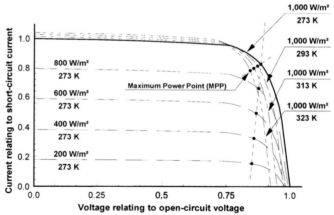

Fig. 6.7 Influence of radiation and temperature on the characteristic current-voltage curve assuming standard test conditions (typical curve shape for a silicon solar cell according to e.g. /6-11/)

Electric power is defined as the product of voltage and current. Thus at a certain point of the characteristic curve the maximum power of the solar cell is reached. This operating point is referred to as MPP (Maximum Power Point). The characteristic curve, and thus also the MPP, are a function of solar radiation and the temperature of the photovoltaic cell.

- Photo-current or short-circuit current increases almost linearly with increasing irradiance of the photovoltaic cell. Also, open-circuit voltage is increased according to Equation (6.1); however, the increase is logarithmic. The current-voltage curve moves thus parallel to the vertical axis with increasing solar ra-

diation. The corresponding power of the solar cell increases altogether over-proportionally with increasing irradiation; in Fig. 6.7 this effect is expressed by the inclined curve connecting the different MPP's.
- This correlation is only true if the temperature of the solar cell is kept constant. If the temperature is increased the diffusion voltage within the p-n-junction is reduced. The open-circuit voltage of a silicon solar cell is, for instance, changed by approximately -2.1 mV/K. In parallel, the short-circuit current increases by approximately 0.01 %/K due to the enhanced mobility of charge carriers within the semiconductor. Thus, at increased temperatures the characteristic current-voltage curve of a commercially available silicon solar cell is characterised by a slightly increasing short-circuit current and relatively strong decreasing open-circuit voltage (Fig. 6.7). Cell power is therefore reduced with increasing temperatures (i.e. MPP shift in the set of characteristic curves shown in Fig. 6.7 for 1,000 W/m² irradiation).

The relation between the maximum power (product of current I_{MPP} and voltage U_{MPP} within the MPP) and the product of open-circuit voltage U_{OC} and short-circuit current I_{SC} is referred to as fill factor FF (Equation (6.2)).

$$FF = \frac{I_{MPP} U_{MPP}}{I_{SC} U_{OC}} \tag{6.2}$$

The fill factor serves as an index for the "quality" of the photovoltaic cell. High values are achieved with good rectifying properties of the p-n-junction (i.e. for a low saturation/latching current I_0, at a low series resistance R_S and a high parallel resistance R_P).

Efficiencies and losses. In order to raise an electron from the valence to the conduction band a material-related, precisely defined minimum energy quantity defined by the energy gap E_g is required. Photons characterised by energy below the energy gap are inappropriate to initiate this process since their energy is insufficient to lift the electrons into the conduction band. However, of photons whose energy quantity exceeds that of E_g, only precisely the energy E_g can be used for electrical energy generation. Any energy that exceeds E_g is directly transmitted to the crystal in the form of heat. Thus, only one electron-hole pair per photon is generated (see Fig. 6.2) in conventional solar cells.

Solar radiation is characterised by wide spectral distribution (see Fig. 2.8), i.e. it contains photons of very different energy quantities. Therefore a solar cell should convert, thus adsorb, as many photons as possible on the one hand, and should transform the available photon energy as good as possible.

The smaller the energy gap between valence and conduction gap of the applied semiconductor, the better the first requirement is fulfilled. For instance, the energy gap E_g of silicon amounts to approximately 1.1 eV and can thus absorb the major portion of the solar spectrum. Since the photocurrent is proportional to the number of absorbed photons per time unit, the photocurrent of a solar cell increases with a decreasing energy gap.

However, the energy gap also determines the upper limit of the potential barrier within the p-n-junction (see diffusion voltage in Fig. 6.5). A small energy gap is thus always associated with a small open-circuit voltage. Since power is the product of current and voltage, very small energy gaps only have small efficiencies. Large energy gaps create high open-circuit voltage, but only allow absorbing a very limited portion of the solar spectrum. Photocurrent thus only has small values, and finally the product of current and voltage is small.

This analysis of extreme cases reveals that there is an optimum energy gap with regard to the choice of semiconductor material for photovoltaic application. Fig. 6.8 shows the corresponding calculation of the theoretical solar cell efficiency in relation to the energy gap E_g of the semiconductor material for a average solar spectrum /6-12/. Depending on the respective applied material, simple solar cells (i.e. no tandem solar cell or other type of combined cell) can achieve maximum theoretical efficiencies of approximately 30 %.

Due to other effects, the efficiencies of real solar cells are much lower than the indicated theoretical efficiencies (also refer to /6-28/). This is, among other factors, mainly attributed to the following mechanisms.

– Part of the incident light is reflected by the finger-type contact system or conducting grid mounted on the front side (see Fig. 6.6). By choosing small grid contacts with maximum spacing in between reflection losses are kept to a minimum. Yet, for a low-impedance transition resistance between semiconductor layer and grid contact maximum contact areas are required. Also the spacing between the grid contacts must not exceed inadmissible limits to minimise the resistance losses of the charge carriers on their way through the semiconductor.

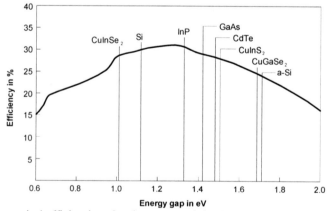

Fig. 6.8 Theoretical efficiencies of various types of simple solar cells under average conditions (see /6-12/)

- Due to the different refractive indices, reflection losses occur when radiation is transmitted from air to the semiconductor material. Anti-reflecting coatings and structured cell surfaces considerably reduce these losses.
- Short-wavelength light usually does not penetrate as deep into the semiconductor material as long-wavelength light. To utilise short-wavelength light the structuring of the upper semiconductor layer properties are of major importance. The higher the layer doping, the thinner the layer should be, since charge carriers tend to re-combine very quickly in such layers. The adsorbed light thus contributes only very little to the photocurrent of the solar cell.
- High short-circuit currents, open-circuit voltages and fill factors imply maximum diffusion lengths. However, charge carriers tend to re-combine at imperfections and impurities of the crystal lattice. Thus the bulk material must be of good crystallographic quality and must meet maximum purity requirements.
- Also the surface of the semiconductor material (i.e. the photovoltaic cell) is a large-surface imperfection of the crystal lattice. There are various techniques available to passivate such surface imperfections and to reduce the resulting efficiency losses.
- Further losses occur when transferring energy from the solar cell. Resistance losses occur as charge carriers move towards the contacts and as they are transferred through the connecting cables. Manufacturing imperfections may cause local short-circuit between the front and rear side of the solar cell.

For highly efficient laboratory silicon solar cells these losses amount to approximately 10 %. Under otherwise optimum conditions, the theoretical maximum efficiency of a solar cell of 28 % (Fig. 6.8) is thus reduced to an actual efficiency of 25 % (maximum laboratory values achieved, see Table 6.1).

The efficiencies indicated for photovoltaic cells usually only apply for determined standardised measuring conditions, since the power output of a solar cell depends on spectral light composition, temperature and irradiation intensity. The standardised conditions mentioned above generally refer to so-called "Standard Test Conditions" (STC): radiation 1,000 W/m^2, solar cell temperature 25 °C, spectral distribution of the irradiation according to AM (air mass) = 1.5 (AM = 1.5 implies an effective atmosphere thickness of 1.5 times the vertical light penetration; spectral distribution of solar irradiation is thus changed in a characteristic manner mainly due to the absorption of photons at certain frequencies within the atmosphere; the AM 1.5 spectrum has been standardised and the light used for solar cell or module calibration has to comply with this spectrum). The power generated by solar cells under these conditions is referred to as peak power.

However, standard test conditions (STC) occur only very rarely in practice, precisely spoken, almost never. In Europe, for example, at a radiation of 1,000 W/m^2 modules heat up by 20 to 50 K above ambient temperature, depending on the concept of module mounting or integration into the building environment. STC temperature and radiation thus only occur under ideal conditions in winter when ambient temperature amounts to 0 °C or below. But due to the low-angle of the sun AM values increase in winter and lead to a shift of the solar spectrum. Nevertheless, highest solar module efficiencies are achieved at clear cold days during wintertime.

Table 6.1 Efficiencies of solar cells (also refer to /6-16/; only cells with a cell surface larger than 1 cm² have been considered)

Material	Type	Efficiency		State
		Lab.	Manufac. in %	of tech-nology[a]
Silicon	monocrystalline	24.7	14.0 – 18.0	1
Polysilicon, simple	polycrystalline	19.8	13.0 – 15.5	1
MIS inversion layer (silicon)	monocrystalline	17.9	16.0	2
Concentrator solar cell (silicon)	monocrystalline	26.8	25.0	2
Silicon on glass substrate	transfer technol.	16.6		3
Amorphous silicon, simple	thin film	13.0	8.0	1
Tandem 2 layers, amorphous silicon	thin film	13.0	8.8	2
Tandem 3 layers, amorphous silicon	thin film	14.6	10.4	1
Gallium indium phosphate / Gallium arsenide[b]	tandem cell	30.3	21.0	2
Cadmium-telluride[c]	thin film	16.5	10.7	2
Copper indium di-selenium[d]	thin film	18.4	12.0	2

Lab. Laboratory; Manufac. Industrial manufacturing; technol. technology; [a] 1 large scale production, 2 small scale production, 3 pilot production, 4 development on a laboratory scale; [b] GaInP/GaAs; [c] CdTe; [c] CuInSe₂

    To assess the power output of a photovoltaic module under the site-specific meteorological conditions the so-called annual efficiency concept has been developed. Actual module temperatures, solar irradiation, and solar spectra are assessed according to the frequency of their occurrence and according to the product-specific parameters of the efficiency dependence on temperature, radiation, and spectrum. Based on this approach evaluation of the power output of various solar modules may thus differ from the efficiencies determined under STC conditions. However, for the plant operator in the end only the annual efficiency is of importance since it determines the energy yield /6-13/.

Cell types. Due to the energy gap, shown in Fig. 6.8, crystalline silicon is not regarded as an ideal semiconductor material for photovoltaic cells. Furthermore, silicon is a so-called indirect semiconductor whose absorption coefficient for solar radiation shows relatively low values. Solar cells made of such semiconductor material must thus be relatively thick; a conventional crystalline silicon cell of simple planar structure, as shown in Fig. 6.6, must have a layer thickness of at least 50 μm to nearly completely absorb the incident sunlight. High layer thickness implies high material consumption and thus high costs. Nevertheless, crystalline silicon is commonly used for photovoltaic cells. The main reason is that silicon is the semiconductor material that shows the widest market penetration, that has been theoretically best understood, and that is most easily controlled.

Already in the sixties of the last century, a multitude of research and development activities have been conducted to develop cost-efficient thin film solar cells (see /6-14/). For this purpose "direct" semiconductors are required. This substance category mainly includes II-VI, III-V and I-III-VI$_2$ compounds. Also amorphous silicon (a-Si), discovered in the 1970's within the scope of photovoltaic projects, is a direct semiconductor. It is characterised by good absorption properties and seemed suitable as base material for thin film solar cells.

Yet, due to still unresolved problems with regard to the competing semiconductor materials or technologies, crystalline silicon (includes crystalline and polycrystalline technologies) will continue to be predominantly used as base material within the years to come.

Since solar cells are (still) relatively expensive, there is a tendency to concentrate solar radiation and thus to reduce the required surface of the photovoltaic cells. Furthermore, efficiencies of photovoltaic cells tend to increase with increased irradiance – if cell temperature remains constant. For concentrating systems, more expensive but more efficient solar cell technologies may be applied cost-efficiently. For instance, mirror and lens systems are used to concentrate solar radiation. But under these circumstances tracking systems are additionally needed, helping to enhance the energy yield per unit surface. Such concentrating systems are most suitable for direct radiation (only direct radiation can be focused) and thus for regions throughout the world where the solar radiation is determined by direct irradiance (like in deserts). Therefore their application in Middle Europe, with a ratio of direct to diffuse radiation of approximately One, based on the annual radiation (see Fig. 2.9), is not appropriate in most cases.

State-of-the-art technologies in terms of solar cell development, with regard to laboratory research and manufacturing, have been summarised in Table 6.1. In the following the various solar cell type technologies are briefly outlined.

Solar cells from crystalline silicon. This cell technology is mainly based on processes applied within the semiconductor industry (see /6-7/, /6-8/, /6-9/, /6-15/). With regard to crystalline cell manufacturing three steps are distinguished:
– production of high-purity silicon as base material,
– manufacture of wafers or thin films and
– solar cell production.
Silica sand (SiO$_2$) serves as base material for high-purity silicon. By means of a specific reduction method (melting electrolysis) silica sand is transformed into "metallurgical grade silicon" characterised by a maximum purity of 99 %. However, this purity is still insufficient for solar cell production.

Thus, further expensive purification steps are required for the production of silicon used by the semiconductor industry, since for semiconductor silicon (semiconductor grade silicon; SeG-Si) the impurity content must not exceed 10^{-9}. The required silicon purification is worldwide in most cases performed by the Siemens process. This purification method starts with the conversion of metallurgical grade silicon into trichlorosilane using hydrochloric acid. The subsequent fractional distillation ensures compliance with the extreme purity requirements. Afterwards silicon is again obtained by pyrolysis of the purified trichlorosilane. In appropriate pyrolysis reactors within a reducing atmosphere the trichlorosilane is decomposed at hot bars. Elementary silicon is separated as polycrystalline material. The obtained "poly-silicon" meets the requirements of "SeG-Si" (semiconductor grade silicon) and shows grit sizes within the μm range.

Up to now the photovoltaic industry was able to use silicon of polysilicon quality for the production of standard products. This polysilicon quality did not meet the requirements of the semiconductor industry, but is still sufficient for solar cell manufacturing. Due to the decreasing growth rates of the semiconductor industry on the one hand and the strongly increasing production of the photovoltaic industry on the other this kind of "off grade" material is expected to become more and more scarce in the years to come. Therefore some countries already develop alternative purification methods for metallurgical grade silicon with the aim to produce cost-efficient "solar silicon". Such developments have already been conducted on a global scale in the early eighties of the last century; however, they had been stopped due to the competition of "off grade silicon".

Polycrystalline silicon serves as base material for the provision of silicon mono-crystals. The standard process applied for the production of such mono-crystals is the Czochralski process (Cz process). Within a shielding gas atmosphere polysilicon is melted down in a crucible (see Fig. 6.9). A seed crystal is dipped into the molten silicon and is again removed slowly by continuous turning, while precisely controlling the temperature gradients so that cylindrical mono-crystal bars are obtained. By cutting these bars with the help of a wire saw thin (250 to 300 μm) mono-crystal silicon wafers are obtained. However, sawing (standard technology) wastes up to 50 % of the (expensive) material. The semiconductor industry uses such silicon wafers to produce integrated electrical circuits and subsequently subdivides the individual wafers into different functional "chips" to be used in computers and other electronic devices. Within the photovoltaic industry these wafers are used to manufacture mono-crystalline silicon solar cells. Therefore the circular wafers are additionally trimmed to obtain square plates to allow for a better space utilisation and thus for enhanced surface specific module efficiencies.

The Czochralski process produces inappropriate wafers to manufacture solar cells with record efficiencies of 25 %, since too many crystal imperfections and impurities occur. Such high tech solar cells require mono-crystals manufactured by the float zone process, which is much more sophisticated and more costly compared to the Czochralski method. The float zone process is thus unsuitable for a mass production of solar cell.

Besides single-crystal or mono-crystalline wafers also "poly-crystalline" wafers are successfully used by the photovoltaic industry. For this purpose, polysilicon is melted down and cast into ingot moulds to solidify slowly and adjusted. These poly-crystalline blocks (with grain sizes ranging from the mm to the cm range) are cut into quadratic poly-crystalline plates by sawing. However, the cheaper manufacture and improved mass utilisation of poly-crystalline compared to mono-crystalline material is counteracted by lower efficiencies since the numerous crystal boundaries form recombination centres that reduce the diffusion length of the minority carriers despite of passivation measures.

Since the mid 1960's studies have been conducted to directly manufacture silicon wafers in the form of bands or cast or sintered plates for photovoltaic purposes, thus avoiding crystal growth or casting an ingot and posterior sawing. Within the scope of these investigations more than 20 different processes have been investigated and tested with regard to manufacturing technology /6-17/. However, to date only the "Edge-defined Film-fed Growth (EFG-ribbon)" process has been applied in practice for solar cell manufacturing successfully /6-18/. By this process hexagon silicon tubes are obtained which are subsequently cut into bands and plates by laser. Efficiencies of solar cells made of this kind of band material amount to about 15 %.

For the actual photovoltaic cell manufacturing from silicon wafers only a few process steps must be performed (Fig. 6.9). As base material, poly-crystalline or mono-crystalline wafers, usually already p-doped, are used. First, chemical etching purifies the wafer surface. Subsequently, the p-n-junction is obtained by diffusion of phosphorus within the material surface aiming at building-up an n-doped surface layer (i.e. p-doping of the silicon wafer is overcompensated by diffusion of phosphorus atoms up to a maximum depth of 0.2 to 0.5 µm). N-doping must then be removed from the wafer edges by plasma etching. In addition, during the diffusion of the phosphorus a phosphorus glass builds up at the silicon wafer surface, which also has to be removed prior to the following process steps.

To remove the n-doping at the rear side of the wafer, first an aluminium coating is applied by silk-screen process printing, and dried prior to applying an additional rear side metallization. Subsequent sintering ensures that aluminium atoms diffuse into the silicon wafer from the rear side and thereby overcompensate the undesired n-doping at the rear side of the wafer. At the end of the day as p-n-junction only the phosphorus-diffused layer at the front side of the wafer remains.

Subsequently, the front side contact is printed on the wafer in the form of a grid and dried afterwards. After applying an anti-reflex coating to enhance light trapping a last sintering step is performed prior to conclusive electrical gauging of the photovoltaic cell.

Virtually all mono-crystalline and poly-crystalline silicon wafer manufacturers apply the described technology as a standard technology. Cells manufactured as mono-crystalline silicon wafers based on the Czochralski process allow achievement of efficiencies from 14 to 18 %, while the efficiencies of poly-crystalline wafers stretch from approximately 13 to 15.5 %. Thus, the described process sequence is a compromise between a simple and cost-efficient process design and wafer materials on the one hand, and minimised efficiency losses due to simplicity, on the other.

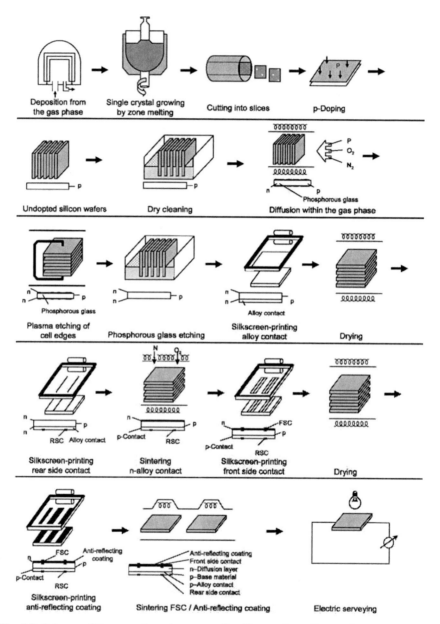

Fig. 6.9 Scheme of the manufacturing steps of a silicon solar cell manufactured according to the silk-screen process printing (FSC front side contact; RSC rear side contact; according to /6-44/)

But both, mono-crystalline and poly-crystalline silicon allow achievement of considerably higher efficiencies compared with those obtained by industrial manufacturing today (Table 6.1). However, to exploit such enhanced efficiencies additional process steps are required, mainly to improve light trapping and the electric properties of surfaces and boundary areas of the photovoltaic cell. For instance, for high-efficiency solar cells made of mono-crystalline silicon multi-stage phosphorus diffusion is applied for the p-n-junction and for structuring of the rear side contact (either point or line-shaped contact) /6-7/, /6-8/. Since the high-performance processes mentioned above imply a multitude of additional complex and expensive process steps (e.g. photolithography), they are currently not cost-efficient despite the enhanced cell efficiencies. Only if the specific costs with regard to the overall photovoltaic system decrease significantly, high-efficiency cells are expected to be manufactured on a broader commercial scale.

Another concept to increase cell efficiency is the metal insulator semiconductor (MIS) solar cell. The name of this solar cell type stems from the impact of a layer of fixed positive charges, located at the surface of a p-doped layer. This layer is referred to as inversion layer since the part of the p-layer close to the surface virtually acts as an n-layer, due to the electrical field created by the fixed surface charges; the p-layer close to the surface is thus quasi inverted. The advantage of these cells is that they only require six manufacturing steps at a relatively low temperature level. In large-scale production the electrical cell efficiency amounts to roughly 16 %.

For a similar concept, aimed at a significant simplification of the manufacturing process /6-19/, the so-called Hetero-junction with Intrinsic Thin-layer (HIT) structure, the rectifying front contact to a mono-crystalline silicon wafer (n-doped) is created by deposition of a double layer consisting of un-doped (intrinsic) and p-doped amorphous silicon. Thus, a rectifying p-n-hetero-contact between n-conducting crystalline and p-conducting amorphous silicon is created. The overall thickness of both layers only amounts to a few 10 nm, so that amorphous silicon does not contribute to the photocurrent. The actual photovoltaic absorber material still is the mono-crystalline silicon wafer. The sophisticated and energy-consuming manufacturing process of the p-n-junction by diffusion is replaced by the comparatively simple and energy-saving deposition of the amorphous silicon double layer. This new technology allows obtaining electrical cell efficiencies of above 20 % (in the laboratory).

Above all, current research is aimed at reducing the costs of solar cell production while maintaining maximum electrical cell efficiencies. For this purpose the applied materials as well as the overall manufacturing process chain are examined over and over again for cost reduction potentials. Besides the new concepts mentioned above, current investigations focus on the use of thinner silicon wafers (thicknesses as small as 70 µm). Such thin wafers have already obtained good efficiencies in the laboratory. However, disadvantages are the expensive manufacturing of thin wafers, on the one hand, and their poor stability within the industrial manufacturing process on the other.

Thin-layer amorphous silicon (a-Si:H) solar cells. In the mid 1970's hydrogen-passivated amorphous silicon (a-Si:H) was first applied as a base material for photovoltaic cells. This material is directly derived from decomposed silane (SiH₄) at temperatures between 80 and 200 °C by plasma-supported enhanced chemical deposition from the gas-phase. Due to the fact that amorphous silicon forms a direct semiconductor, very thin active layers in the range of 1 μm are required. Therefore very little material is needed. Additionally this process is characterised by very low deposition temperatures and thus small energy consumption. As a consequence the costs of solar cell manufacturing are reduced tremendously compared with crystalline silicon solar cells.

The assembly of a a-Si:H-solar cell is completely different compared to a crystalline silicon photovoltaic cell. Instead of a p-n-junction, p-i-n-structures are used; i.e. the major portion of the photovoltaically active layer of a thickness of several 100 nm consists of un-doped (intrinsic) hydrogen-passivated amorphous silicon (a-Si:H) with n-doped layers of a few 10 nm on top and underneath. The electric field of such a structure thus covers the entire absorber region and ensures the separation of electron-hole-pairs created by the absorbed solar irradiation at all places.

Fig. 6.10 shows the layer sequence of typical a-Si:H solar cells. According to this, different substrate technologies are distinguished. The layers are deposited in the direction from the shaded side to the light-exposed side (Fig. 6.10, left). Starting on top of a conductive (non-transparent) substrate, such as e.g. a stainless-steel foil, a layer sequence consisting of n-doped, un-doped and p-doped hydrogen-passivated amorphous silicon (a-Si:H) is deposited from the gas phase. Eventually, a transparent conductive oxide (TCO) acts as a contact on the light-exposed side. Superstrate technologies start with deposition on the light-exposed side; i.e. first the conductive oxide acting as a transparent contact and subsequently the layer sequence consisting of hydrogen-passivated silicon (a-Si:H) and lastly the metallic rear side contact need to be deposited (Fig. 6.10, centre).

Besides solar cells provided with individual p-i-n-junctions, also tandem cells and even triple cells are in use. For these applications two or three p-i-n-layers are piled on top of each other, whereas the junction between highly n-doped and highly p-doped material bypasses both layers (so-called tunnel contact). The voltages of the two or three piled p-i-n-layers add up. For an optimum utilisation of the solar spectrum frequently the energy gap of one of these p-i-n-structures is enhanced or reduced by alloying the hydrogen-passivated amorphous silicon (a-Si:H) with amorphous carbon or amorphous germanium. The cell that is closest to the light-exposed side should show the largest energy gap. As this cell only absorbs the short-wave radiation of the solar spectrum it better exploits the photon energy. Such a cell thus supplies higher voltages with regard to its energy gap. The cell located on the side shielded from the sun light has the smallest energy gap and can thus still use part of the low-energy photons that have not been absorbed by the first cell. Fig. 6.10, right, shows the example of a tandem cell where two p-i-n-structures of hydrogen-passivated amorphous silicon (a-Si:H) and a a-SiGe:H alloy have been combined.

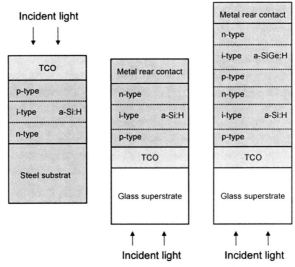

Fig. 6.10 Layer sequence of various p-i-n cell structures of amorphous silicon (a-Si:H) (left: stainless steel substrate cell, centre: superstrate cell on glass, right: tandem cell made of a-Si:H and a-SiGe:H on glass superstrate; for all cells the deposition sequence starts with the bottom layer and ends with the top layer) (TCO transparent conductive oxide)

Within small consumer electronics applications, amorphous silicon has gained an undefeated monopoly position worldwide (watches, calculators etc.). However, due to the poor stability of its physical properties, it is inappropriate for applications where a higher installed power is needed (like in grid-connected photovoltaic systems). When applied outdoors, electrical efficiencies are in some cases considerably reduced within the first months of operation. To date, efficiency reduction is still clearly above one forth so that the efficiency is markedly reduced to below 10 % (degradation; Staebler-Wronski effect) /6-20/. Nevertheless it is necessary to note, that all long-term monitoring programs show a saturation of the degradation effect, latest after two years of operation. Power ratings of amorphous silicon cells are thus typically based on the stabilised power after degradation. Therefore customers can compare the power related costs on a fair basis. However, in the medium term, more stable amorphous silicon cells are expected to be manufactured; latest findings have shown that efficiencies of 14.6 % (tandem cell with three p-i-n-layers) can be achieved and that degradation comes to a standstill at 13 % /6-21/.

Thin film solar cells based on chalcogenides and chalcopyrits, particularly CdTe and CuInSe$_2$ ("CIS"). The advantages of the thin layer technology in case of hydrogen-passivated amorphous silicon (a-Si:H) are counteracted by relatively low electrical cell efficiencies, when compared to crystalline silicon. In contrast polycrystalline thin films made of direct semiconductors, such as cadmium telluride (CdTe) and copper indium di-selenium (CuInSe$_2$), have at least on a laboratory scale reached efficiencies between 16 and 18 %; that is approximately 75 % of the theoretical efficiency of the crystalline silicon technology. Both materials can physically be deposited onto glass at a temperature of 600 °C. Since both materials are direct semiconductors, photovoltaically active layers of a few μm thickness are sufficient to absorb all photons of the solar spectrum with energy superior to the energy gap E_g of the respective absorber material. The energy gap of cadmium telluride (CdTe) amounts to approximately 1.45 eV and that of copper indium di-selenium (CuInSe$_2$) to about 1.04 eV. However, for the latest generation of chalcopyrit solar cells instead of pure CuInSe$_2$, Cu(In,Ga)Se$_2$ alloy with a gallium share of 20 to 30 % in relation to the total indium (In) and gallium (Ga) content is used. With an energy gap of 1.12 to 1.2 eV the alloy is closer to the theoretically achievable optimum of the efficiency (see Fig. 6.8; see /6-21/, /6-22/).

Thin films of good electronic quality of cadmium telluride (CdTe) and copper indium di-selenium (CuInSe$_2$) can only be manufactured p-doped. Thus, a second n-doped material is required for solar cell manufacturing, which can be combined with the first material to form a semiconductor p-n-hetero-junction. In both cases, n-doped cadmium sulfide (CdS) is used. Such semiconductor hetero-structures present the "slight" disadvantage that at the boundary surface between both materials, an increased recombination of photogenerated charge carriers occurs. However, this disadvantage is compensated by the advantage that the top layer may be designed as a "window", thanks to the possibility to choose a semiconductor with a high energy gap (such as CdS: $E_g = 2.4$ eV); hence, the described top layer only absorbs a very limited share of the solar spectrum, which is then lost for the photocurrent. After the transmission of the remaining irradiation through this window layer the major share of the incident radiation is absorbed very close to the p-n-junction – thus at the point of the maximum electrical field strength. The resulting separation of the photogenerated charge carriers is thus highly efficient.

Fig. 6.11, left, shows the layer sequence of a CdS/CdTe hetero-structure solar cell. This cell technology is a superstrate structure; i.e. the transparent front electrode exposed to the sunlight and made of indium stannous oxide (ITO) is usually applied first by means of sputtering. Subsequently, cadmium sulfide (CdS) is deposited as window or buffer layer followed by the actual photovoltaically active absorber layer consisting of cadmium telluride (CdTe). Usually, both layers (i.e. the window layer with a thickness between 0.1 and 0.2 μm and the absorber layer with a thickness of approximately 3 μm) are superimposed using the same technology (e.g. by the sublimation-condensation method or silk-screen process printing). In order to obtain layers of a photovoltaically sufficient quality, an activation step based on a temperature treatment in the presence of cadmium chloride (CdCl$_2$) needs to be performed after deposition. Cell manufacturing is accomplished by depositing a metal rear electrode made of graphite, copper (Cu) or a mixture of both. On a laboratory scale, cadmium telluride (CdTe) solar cells have reached peak efficiencies of nearly 17 % when applied to small surfaces. Over the last few years, several pilot production lines for large-surface CdTe modules of a surface of 0.5 m^2 and efficiencies ranging from 8 to 10 % have been commissioned.

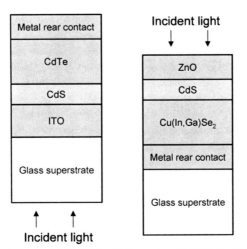

Fig. 6.11   Layer order of a CdS/CdTe (left) and a CdS/Cu(In,Ga)Se$_2$ (right) hetero-structure solar cell (ITO indium stannous oxide)

With efficiencies above 18 %, CdS/Cu(In,Ga)Se$_2$ hetero-structure solar cells achieve the highest electrical efficiencies of all thin film solar cells. Fig. 6.11, right, illustrates the layer sequence of such hetero-structure solar cells. Manufacturing of this solar cell type starts with depositing a molybdenum rear contact, followed by the deposition of the photovoltaically active Cu(In,Ga)Se$_2$ layer of a thickness below 2 μm. The following two deposition methods are suitable for this purpose on an industrial scale.
- Firstly, physical co-evaporation of all elements (i.e. Cu, In, Ga and Se) onto the heated substrate, so that the Cu(In,Ga)Se$_2$ compound is already built up during vapor deposition.
- Secondly, deposition of all elements onto the unheated substrate (e.g. by sputtering). Subsequently, a second heating step, called selenization, is performed to obtain the Cu(In,Ga)Se$_2$ compound.

Subsequent to absorber manufacturing by one of the two described methods, a CdS layer of an approximate thickness 0.05 μm is deposited from the chemical bath prior to depositing the ZnO front electrode by sputtering.

Pilot projects and small scale manufacturing plants have been installed and are operated for Cu(In,Ga)Se$_2$ solar modules. Modules with efficiencies of up to 12 % are commercially available. Both technologies (i.e. CdTe and Cu(In,Ga)Se$_2$) will have to prove within the next years whether they can sustain themselves in the marketplace. Generally speaking, for crystalline silicon solar cells the manufacturing process is quite reasonable and well understood based on the experience from the semiconductor industry. But the material costs are high due to the thick wafers. For the above-discussed thin-film technologies it is the other way round. Material costs are low. Production processes are complex and expensive. Therefore thin-film technologies can compete on an economic basis only if large productions sites are operated.

Thin film solar cells made of crystalline silicon. There are also attempts to utilise the economic and process-related advantages of thin film solar cell technology, such as low material consumption, integrated module manufacturing by structuring of individual layers during manufacturing, for crystalline silicon. Due to the indirect energy gap of crystalline silicon layer thicknesses of at least 20 μm are required to sufficiently absorb the incident solar radiation. "Light-trapping", however, allows further reduction of the layer thickness. If a light beam reflecting diffuser, or an inclined reflecting structure is superimposed on the reverse side of the solar cell, or if a pyramid-type substrate is coated with a thin silicon film, also for crystalline silicon, layers of only a few μm thickness are sufficient to nearly entirely absorb the incident radiation. For silicon layers of only 2 μm thickness and optimised "light-trapping", the efficiency potential has been calculated at approximately 15 %. In practice, various methods of film deposition and posterior treatment are being investigated with the aim to manufacture such silicon cells under commercial circumstances.

A variation of the deposition parameters of the plasma-supported enhanced chemical deposition from the gas phase allows depositing microcrystalline silicon. Although the silicon crystals of this kind of material have a size of only a few 10 nm (and are therefore also referred to as nanocrystalline silicon) based on p-i-n-structures electrical efficiencies above 10 % are achieved in the laboratory scale. Since the deposition conditions and the deposition temperature of nanocrystalline silicon, ranging between 200 and 300 °C, are very similar to those of amorphous silicon both materials may be combined as tandem cells that yield efficiencies of over 10 % on a laboratory scale /6-23/. For silicon thin film solar cells of higher efficiencies, silicon needs to be deposed at temperatures above 700 °C, so that cheap glass-substrates are inappropriate /6-24/. The higher growth temperatures allow achieving grain sizes of up to 100 μm and thus a better photovoltaic quality of the polycrystalline silicon layers. The so-called transfer technologies are a promising alternative to manufacture even mono-crystalline thin film solar cells (see also /5-25/, /6-26/). Typically, a mono-crystalline silicon layer of 20 to 50 μm thickness is manufactured on a pre-treated mono-crystalline silicon substrate which is subsequently detached and transferred to any kind of foreign substrate. The silicon substrate may afterwards be used again for this purpose. With 16.6 %, on a laboratory scale, solar cells made of mono-crystalline transferred silicon permit to achieve the highest efficiencies of thin film silicon on foreign substrates.

Thin film solar cells with integrated serial circuit. The individual cells of a solar module must be series-connected if the module voltage is to be superior to that of its composing cells. Usually, voltages of 12 or 24 V are desired. A major advantage of all thin film technologies is that the serial connection of the individual cells to a module may be combined with the actual cell manufacturing. While the process of serial connection is totally independent from the manufacturing of wafer-based solar cells and is performed a posteriori, thin film modules consist of lamellar arrangements of individual cells placed on a single substrate or superstrate (Fig. 6.12).

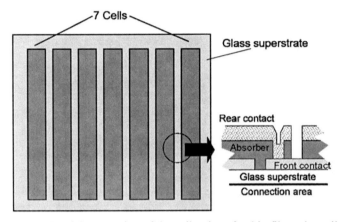

Fig. 6.12 Integrated serial connection of the cell strips of a thin film solar cell on the substrate (i.e. the serial connection links the bottom electrode of a cell strip to the top electrode of the next strip)

For the series connection of the cell strips, three structuring steps must be performed during cell manufacturing. Fig. 6.13 illustrates the production principle of such an integrated serial connection taking a hydrogen-passivated amorphous silicon (a-Si:H) thin film solar cell as an example. Firstly, the glass superstrate is coated with a conductive transparent layer of indium stannous oxide. What later on will become the front contact, is removed with regular intervals during the first structuring step. Structuring may either be performed by laser removal or by simple mechanical scraping off. Within the following preparation step, the photovoltaically active layer ("absorber") is deposed. For a solar cell made of hydrogen-passivated amorphous silicon (a-Si:H) this manufacturing step encompasses the deposition of three layers (p-doped, i-doped and n-doped a-Si:H) in case of a simple p-i-n-type cell or a sequence of six layers for manufacturing of a tandem cell. After the deposition of the absorber a second, a slightly parallel located structuring step is performed, which removes the absorber layer but not the front contact. Then the rear contact is deposited, so that by the two preceding structuring steps the front contact of a cell strip is connected to the rear contact of the next strip. Short-circuiting of individual cell strips by the rear contact is prevented by a third structuring step. The approach is analogous for substrate-based technologies.

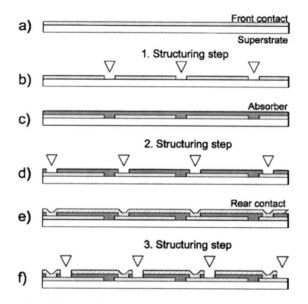

Fig. 6.13 Layer deposition and structure sequence of a photovoltaic thin film module (a) deposition of the transparent front electrode on the glass superstrate; b) first structuring step; c) deposition of the photovoltaic active absorber layer; d) second structuring step; e) deposition of the rear electrode; f) third structuring step)

Solar cells for concentrating photovoltaic systems. Solar cells of concentrating photovoltaic systems are illuminated up to 500 times more at standard test conditions (STC) compared to fixed mounted cells. However, at higher radiation concentrations the serial resistance constitutes a major problem due to the high currents. This is why concentrator cells must be especially highly doped and be provided with low-loss contacts /6-27/.

Terrestrial concentrating photovoltaic systems are almost exclusively provided with silicon-based solar cells, whose structure is similar to that of the highly efficient silicon solar cells mentioned above. On a laboratory scale, they reach electrical efficiencies of up to 29 % at 140-fold concentration of the radiation.

Furthermore, concentrator cells based on gallium arsenic (GaAs) and ternary III-V alloys, partly assembled as tandem structures have been investigated. With regard to epitactically grown mono-crystalline tandem structures, efficiencies of up to 34 % have been reported for 100 to 300-fold concentrations.

For such concentrator systems it is of particular importance to avoid high temperatures, which will cause power losses. Further more it has to be taken into consideration that high concentration factors, in the range of several 100's, do need a two axis tracking system and only direct radiation can be used.

Dye solar cells made of nano-porous titan oxide (TiO2). Electrochemical solar cells made of nano-porous titan oxide (TiO_2) use a TiO_2 particles layer of a typical particle size of 10 to 20 nm. The blocking contact to these nano-particles is ensured by a liquid electrolyte, usually the redox pair J_3^-/J^-. The photovoltaic activity of this kind of solar cell is given due to a monomolecular layer of a rubidium dye adsorbed at the TiO_2 particle surface. Due to the porous sponge-type structure of the titan oxide (TiO_2), its surface is about 1,000 times bigger than the cell surface. Absorption of sunlight by the dye is only possible due to this area enlargement.

The photon irradiated on the surface of this cell lifts one electron inside the dye from the basic state into an excited status. The linkage between the adsorbed dye to TiO_2 is that strong that the excited electron is injected into TiO_2 within only a few pico seconds, while the dye is regenerated by the electrolyte; i.e. one electron is delivered in addition to the basic state of the dye.

Fig. 6.14 gives an overview on the design of such a solar cell, as well as a simplified energy scheme for the primary photovoltaic activity. Primary charge separation thus involves a three-step-process.

1. Excitation of the dye.
2. Injection of the electron from the excited status of the dye into the conduction band of TiO_2.
3. Regeneration of the dye from the electrolyte.

Charge separation is eventually accomplished by the diffusion of the photogenerated electron through the TiO_2 network to the front contact, while the electrolyte is regenerated at the opposite platinum (Pt) rear electrode.

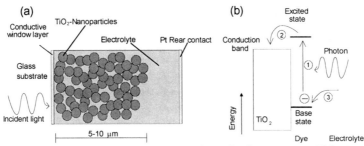

Fig. 6.14 (a) Schematic illustration of a dye solar cell of nano-porous TiO_2 (not shown: single-molecular dye layer adsorbed by TiO_2 nanoparticles of a thickness of approximately 20 nm); (b) simplified energy scheme illustrating the primary charge carrier separation by a three-step process: 1 excitation of dye; 2 injection of the electron from the excited status of the dye into the conduction band of TiO_2; 3 regeneration of dye from the electrolyte

On the one hand, this new solar cell technology is attractive as material costs are low and the production process is very simple, thus allowing for significant cost saving. On the other hand, the physics of dye solar cells is very different from that of all other (solid) solar cells and has not yet been exhaustively investigated, and as a result, has not been fully understood yet. In fact, primary charge carrier transfer into dye solar cells is similar to the charge carrier transfer processes realised within the photosynthesis process.

In laboratories, dye solar cells have reached efficiencies above 10 %, while first small modules have reached efficiencies of up to 5 %. Currently, this solar cell type is being investigated as to its long-term stability /6-28/. Major emphasis is put on the replacement of a liquid electrolyte by gel or a solid electrolyte to avoid the risk of leakage of the modules. Further more it is necessary to understand, that these types of solar cells include electrochemical processes which are much more affected by ageing mechanisms than processes within a solid state.

Solar module. Individual photovoltaic cells are combined to a photovoltaic module forming the basic unit of a solar generator. Usually, such a module consists of electrically interconnected photovoltaic cells, embedding materials including a front pane of glass and a rear side cover, electrical connecting cables or a connection box and partly also a frame, usually made of plastic or aluminium. However, also frameless modules are increasingly applied which require special edge sealing measures. The embedding of the individual cells into a module helps to protect the single cells against atmospheric impacts, ensure a defined upper voltage level and maximum amperage respectively, and thus enables the installation of photovoltaic generators with user-defined current-voltage characteristics.

Embedding of the cells, as well as edge sealing, is subject to very high requirements. For instance, the cell surface which is subject to temperature fluctuations ranging from about -40 °C to approximately +80 °C within the course of one year must be protected against any kind of humidity (such as rain and condensation) throughout the overall technical lifetime of 20 to 30 years and even longer. Furthermore, mechanical deterioration due to hailstones of up to several centimeters in diameter or due to the wind loads with gusts of 50 m/s and more need to be prevented. Moreover, high insulation strength needs to be ensured and the used materials must not be afflicted by bacteria nor be eroded by animals (e.g. birds). The commercially available photovoltaic modules meet all these requirements and ensure a safe operation throughout the overall technical lifetime.

Due to great variety of possible applications, solar modules of different power ratings are available on the market. The nominal open-circuit voltage of the module is determined by the number of serial connected cells and by the nominal open-circuit voltage of the individual cells (Fig. 6.15). The nominal short-circuit current of the module is determined by the number of cell strings connected in parallel and by the nominal short-circuit current of the individual cell. This nominal short-circuit current depends on cell technology, the quality of the used material, the manufacturing process, as well as mainly the size of the cell. As the short-circuit current is linearly proportional to the solar irradiance it is also linearly proportional to the effective surface of the cell. Accordingly, the characteristic current-voltage curve of the entire module changes – compared to that of a single cell – depending on the interconnection of the cells.

The rated power of the photovoltaic module depends on the total number of cells. Power ratings of common photovoltaic applications range from approximately 50 to 75 W for 36 serial connected silicon cells with a cell surface of approximately 100 cm^2. However, particularly for grid-connected installations also large-surface solar modules of capacities of up to 300 W are available. Cell surfaces of 225 cm^2 on a single wafer are offered on the market today.

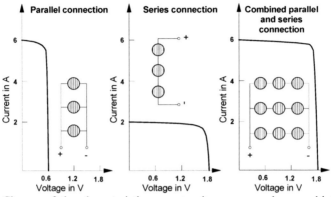

Fig. 6.15 Change of the characteristic current-voltage curve when combining various photovoltaic cells to a module exemplarily for cells of 2 A short-circuit current and 0.6 V open-circuit voltage (according to various sources)

If individual cells of an operating module are shaded or if their original power rating is impaired by defects, they no longer function as generators within the interconnected cells but act as a load. According to the type of connection, they will either be operated in reverse-biasing (wrong voltage direction) or at voltages exceeding their open-circuit voltage (wrong current direction). Under unfavourable conditions they may heat up stronger than the neighbouring cells ("hot spot" effect).

Independent of the described detrimental effects, also partial shading of serial connected cells causes considerable losses. As a first order approximation, the current in a string of serial connected cells is determined by the current of its weakest cell. Thus losses due to partial shading are considerably higher compared to the ratio of the shaded area to the total area. For strings or cells connected in parallel, losses are only proportional to the shaded area.

Particular consideration of the shading effects for building-integrated photovoltaic cells is necessary, as, for instance, façade elements, frames, and windows very often cause partial shading. This is why designs need to be carefully analysed to prevent excessive losses.

Available protection measures for a connection of several modules are outlined in Fig. 6.16; in principle, they also apply to the connection of several cells composing one module. Bypass diodes (freewheeling diodes), arranged in parallel to the cell strings, prevent overheating of possibly shaded solar cells. Incorporated blocking diodes prevent equalising current over cell strings in the wrong direction, if the voltages are lower compared to those of the neighbouring strings due to partial shading or property changes (i.e. damaged cells).

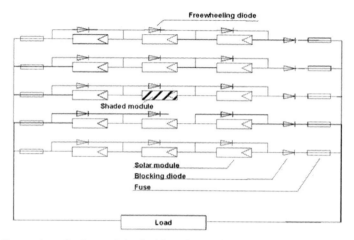

Fig. 6.16 Connection of solar modules inside a photovoltaic generator /6-10/

However, experiences have shown that neither bypass diodes within the individual cell strings of a module, nor blocking diodes at the end of a module are required. These kinds of safety measures create increased energy losses and thus increased costs. Furthermore, diodes show a shorter technical lifetime than cells and modules. However, the case is different if several modules are connected to larger units (arrays, array fields, generators) since partial shading is much more likely under these circumstances (e.g. by clouds floating across the sky, shading caused by building, trees etc. within the course of a day or components of the photovoltaic generator). In this case, the effects discussed above need to be considered for modules instead of individual cells. Each module is therefore bridged by a freewheeling diode, occasionally provided and incorporated into the module by the manufacturer. Nowadays, blocking diodes are mostly omitted, since equalising currents do not impair safety. Additionally, fuses are attached to the terminals of the module strings to prevent overload of modules and cables in case of short-circuit within a module string.

The losses due to the serial connection of individual cells to photovoltaic modules amount to approximately 2 to 3 % of the electrical energy available at terminal clamps or connecting cables of the individual cells. If several modules compose larger generator units, further losses of this magnitude, depending on the generator field size, must be considered. However, pre-selecting each cell for a certain module type by their electrical performance parameters by the manufacturer could reduce these losses.

6.2.2 Further system components

Inverters. Solar generators as well as battery storage systems principally deliver direct voltage or direct current (DC). Many small appliances (such as watches and calculators) are designed for DC supply. However, most appliances require alternating current (AC) of 230 V and a frequency of 50 Hz (in some cases 120 V at 60 Hz, as in the US, for instance). Even, for stand-alone photovoltaic systems without grid-connection, frequently inverters are used to convert direct current (DC) into the appropriate alternating current (AC) power required by commercially available appliances. However, inverters are by all means required to transfer the properties of the electrical energy produced by grid-connected photovoltaic generators into properties similar to these of the grid. A special type are the so-called pump inverters that convert the supplied direct current (DC) from the photovoltaic generator into alternating current (AC) of adjustable voltage and frequency, which is suitable for the rotation-variable operation of water pumps /6-29/, /6-30/.

The power range of inverters used in conjunction with photovoltaic appliances stretches from about 100 W to several 100 kW, applying a great variety of circuitry topologies and components. A continuously growing market, as well as new findings lead to new inverter concepts and new products /6-9/, so that in the following, only the basic principles of inverter designs and the main requirements are outlined.

Island inverters. While the low voltage grid supplies sine-shaped electrical energy of 230 V/50 Hz in Europe, island inverters are subdivided into three groups according to their voltage shape: rectangular, trapezoid, and sine inverters (Fig. 6.17). Within the small power range, e.g. for the local supply of individual AC consumers within a DC grid, frequently rectangular or trapezoid inverters are used. For larger systems (more than 1 kW), by contrast, sine-wave inverters are most commonly applied. Recently there is also a tendency to use sine-wave inverters for small scale applications.

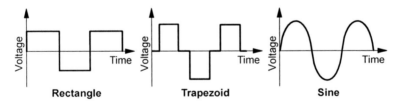

Fig. 6.17 Typical output voltage curves of rectangular, trapezoid, and sine inverters

The rectangular inverter is characterised by a very simple structure. In the example illustrated in Fig. 6.18 a battery voltage of 12 or 24 V is applied to the primary side of the transformer at a rhythm of 50 Hz and with alternating polarities passing through the bridge circuit consisting of switches S1 to S4. The switches S1 and S2, usually bipolar transistors or MOS field effect transistors, are closed during the first phase. The same applies to S3 and S4 during the second phase. This "cut up" direct voltage is subsequently transformed to the required output voltage by the transformer /6-29/.

This concept has the disadvantage that the output voltage level is directly proportional to the fluctuations of the battery voltage between, for instance, a final discharge voltage of 11 V and a gassing voltage of 15.5 V. For a 12 V lead accumulator and assuming a constant transmission ratio for the transformer, the output voltage covers a range from 210 to 297 V.

The so-called trapezoid or "quasi-sine" inverter, based on the same circuit principle, avoids this disadvantage. However, its output voltage contains a blanking interval. An appropriate controlling circuit ensures that the width of the blanking interval is adjusted, so that even for different input voltages an almost constant actual output voltage can be achieved.

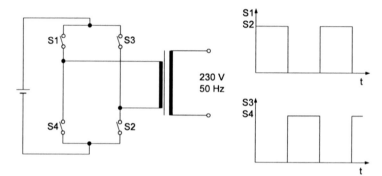

Fig. 6.18 Circuit principle of a rectangular inverter

During operation of such inverters, it always has to be verified whether the intended consumers, generally designed for sine-shaped voltages, can perform reliably under the given voltage conditions. Generally, the supply of electrical power with such properties for light bulbs, irons or other simple consumers, such as drilling machines, does not cause any problems. However, if such consumers are equipped with transformers or capacitive voltage dividers at the power entry, noise, considerable additional losses or even deterioration of the consumers may occur. Some electronic appliances (e.g. washing machines) should not be directly connected to these inverters, since they require the defined zero-crossing of sine-shaped voltages for their internal control system.

The output voltage of the sine inverter, by contrast, is identical to that of the public supply system, so that all consumers can be provided with power useable without any problems. Out of the great variety of topologies or circuits Fig. 6.19 shows the principle of a pulse width modulated (PWM) inverter.

Unlike rectangular inverters the input voltage is "chopped" at much higher frequencies (some 10 kHz to a few 100 kHz). For the positive half-wave of the sine oscillation, shown in Fig. 19 switch S1 is continuously closed while switch S2 is activated and deactivated at high frequencies with a variable pulse/pause ratio (pulse width modulation, PWM). An appropriate pulse pattern ensures a rectangular voltage at the bridge circuit exit, whose timely average curve has a sine shape.

A downstream filter suppresses all high-frequency portions of this signal, so that, at the exit, only the desired sine-shaped voltage is present. A downstream transformer adjusts the voltage to the required 230 V. This system component can be omitted in case of sufficiently high input voltage (> 350 V). This allows for inverter-related savings as well as for considerably higher efficiencies especially within the lower partial-load range. Inverters without transformer are also significantly reduced in weight.

For stand-alone systems or isolated grids, such high input voltages are currently only applied for large plants. This is mainly due to the fact that also batteries would need to be connected in series to obtain such high voltages. However, due to the risk of individualisation of different cells, it is much more difficult to operate a battery at a high voltage level than at low voltages. For plants of installed capacities of up to 10 kW input voltages usually vary between 48 and 60 V.

To avoid these shortcomings another typology includes a DC/DC converter between battery and the inverter that supplies the appropriate voltage level. This concept allows designing transformerless inverters for any type of photovoltaic system. A disadvantage of all typology without transformers is the fact that they have no galvanic interruption and therefore some additional safety features must be integrated.

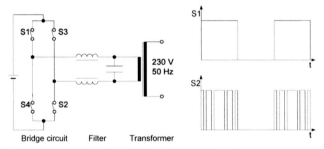

Fig. 6.19 Circuit principle of a pulse-width modulated inverter

Depending on the kind of application the following requirements are relevant for island inverters /6-29/.

– High efficiency. The efficiency of island inverters should be as high as possible and should be high already within the lower section of the partial-load range. However, if the inverter is only occasionally turned on within a DC grid to supply an assigned AC consumer, the internal power consumption and the efficiency of the inverter are of minor importance. Yet, if the inverter is continuously operated, to provide, for instance, power to the grid of a dwelling house, its self-consumption is a critical variable. Every percent of inverter internal consumption reduces the mean annual efficiency by approximately 10 %. Thus, inverter self-consumption should be below 1 % of the nominal output power, corresponding to an efficiency of above 90 % at 10 % of the nominal output power. However, an efficiency of 85 to 90 % at rated power is sufficient since the inverter is only operated at rated power for a small fraction of its operational time. Out of the two efficiency curves represented in Fig. 6.20 the curve "ideal efficiency curve" represents the more appropriate efficiency curve for photovoltaic appliances with the higher annual efficiency.

– Low self-consumption. Energy losses due to high self-consumption can be reduced if the inverter is only switched on when needed, as its stand-by consumption during the remaining time (sleeping mode) is much lower. However, it has to be ensured that even small loads, such as compact fluorescent lamps, are always securely detected and the inverter is turned on. The master / slave process represents such a possibility for a securely load detection: a small inverter (master) ensures permanent power supply whereas the slave is only switched on and operated if additional power is required.

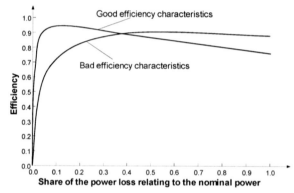

Fig. 6.20 Favourable and unfavourable characteristic efficiency curves of island inverters (see /6-29/)

– Stable operation behaviour. The output voltage of island inverters should be as stable as possible regarding frequency and amplitude (voltage source). This applies in particular to larger scale plants, if many consumers are served simultaneously. During the start-up of larger consumers (e.g. washing machines, refrigerators) the voltage level should not fail and cause, for instance, a simultaneously operated computer to crash.

– Sine-shaped output voltage without direct current bias. The output voltage should be sine-shaped (i.e. few overtones or distortions). An indicator is the distortion factor, which should be below 5 %. In addition, the output voltage should not have a direct current bias, since it may pre-magnetise and deteriorate trans-formers and electric motors. Furthermore, the inverter should be capable of operating inductive as well as capacitive loads (e.g. fluorescent lamps, alternating current motors) and asymmetric loads such as e.g. hair dryers where only one half of the sine wave is used. This feature is described by the admissible power factor. For motor start-up (e.g. of refrigerators, washing machines) it should be able to withstand 2 to 3-fold overload for short periods.

- Coverage of the entire voltage range. The inverter entry side should cover the entire voltage range of the battery energy storage of -10 to +30 % of the nominal voltage. In the event of falling short of a certain minimum input voltage it should be disconnected, either automatically or via a control input to protect the battery from deep discharge.

In the future, island inverters will almost exclusively be designed as sine-wave inverters and an increasing number of devices will meet the discussed requirements. Additionally there are further cost saving potentials due to a large-scale production and the use of modern semiconductor components.

For larger stand-alone hybrid systems increasingly bi-directional inverters are used. They allow battery charging from additional generators, such as e.g. wind generators, diesel generators and hydropower without additional charging equipment. Although the system is simplified by the above-mentioned measures, high requirements are placed on the inverter /6-9/.

Grid-connected inverters. To feed solar power into the grid generally an inverter is required to convert the direct current (DC) power generated by the photovoltaic system into alternating current (AC) power compatible with the mains /6-29/. Unlike island inverters, which are mainly provided with direct current by a battery, grid-connected inverters are directly connected to the photovoltaic system without additional storage systems.

Most of the grid-connected inverters of the 1980's were modified thyristor inverters, which had been applied numerously for electrical drives. However, these devices, optimally operated at nominal load, are often inappropriate for partial load operation typical for photovoltaic generators. Due to this they only achieve very low operational efficiencies.

Modern semiconductor components, such as MOS-FET's (Metal Oxide Semiconductor – field effect transistors) or IGBT's (Insulated Gate Bipolar Transistors), in conjunction with optimised circuit typologies, triggered the development of special solar inverters with significantly improved properties at the early nineties of the last century. Especially, the system's self-consumption had been tremendously reduced, so that the required efficiency of over 90 % could be reached at a nominal output power of 10 %. While island inverters typically convert the major energy share at approximately 20 % of the nominal power, grid-connected inverters under Central European irradiation conditions have a relative evenly distributed loading over the entire power range. Besides the low self-consumption, grid-connected inverters are also characterised by high efficiencies at rated power. In the meantime, a wide range of devices within the power range of 10 W up to several 100 kW has become available on the market. Usually, devices within the MW-range consist of several inverters (e.g. 300 kW devices operated in master/slave mode (often with a rotating master)). For this purpose, a wide range of functional principles is applied. Some are discussed below.

– Grid-commutated inverters. Due to their design, grid-commutated inverters need a strong electrical power grid to function. Thyristors, characterised by robustness and cost-efficiency, serve as basic electronic system elements. However such inverters generally grossly distort (overtones) the supplied output voltage and cause phase shifts with regard to the mains voltage. Since this disadvantage requires additional filter and compensation measures, other concepts within the low power range result as more cost-efficient. But thyristor inverters are still applied for the power range of several 100 kW and above.

– Self-commutated inverters. Self-commutated inverters are based on disengageable power switch and thus do not require an external grid for normal operation. A widely used function principle is pulse-width modulation offering a wide range of circuit topologies. In comparison to island inverters, grid-connected inverters need to be synchronised with the grid; a continuous operation of the inverters in case of a power failure within the grid must by all means be prevented for safety reasons. To minimise the risk of creating isolated grids while the overall mains is disconnected, e.g. German regulations prescribe single-phase grid supervision for solar inverters, abbreviated ENS. The supervision aims at preventing that areas of the mains considered to be disconnected remain live due to the inverters, so to ensure that maintenance work could be performed at the grid under safe conditions. Many other countries lack such regulations.

Provided that the input voltage is sufficiently high, inverters can feed directly into the power grid and do not require any transformer for voltage adjustment. Besides reducing costs, weight, and volume, direct feeding into the grid markedly reduces the inverter's self-consumption. The latter, in turn, contributes to an improved efficiency in the partial-load range. There is thus a tendency towards concepts without transformers.

Devices provided with transformers are further distinguished into systems equipped with a 50 Hz transformer and systems provided with high-frequency transformers. The latter offers weight and volume advantages on the one hand, but is also prone to higher losses and failure probability due to a more complex overall circuit layout. With regard to devices, there is a tendency towards more simple and sound concepts, whereby a favourable efficiency curve is obtained by applying top quality components, especially for the 50 Hz transformer /6-29/.

The first generation of grid-connected devices had been provided with central inverters. Therefore the photovoltaic modules were first connected in series (strings) to obtain the required voltage. To guarantee the required power, the strings are subsequently operated in parallel. The so interconnected photovoltaic generator then feed an individual inverter. Occasionally, the respective power is distributed among several central inverters operating in master/slave mode.

Decentralised modular inverters increasingly gain importance. Such devices are further distinguished into inverters for small strings (string concept) and inverters that are integrated with individual modules or even directly incorporated into their terminal block (module-oriented inverters). These concepts offer a series of benefits.

- Each module or module group is optimally operated in its maximum power point (MPP).
- Losses due to different characteristic curves of the modules (mismatch losses) are reduced.
- Partial shadings only affect individual modules or the module groups.
- Inverter failure only impairs the corresponding generator component.
- Cabling of modules or module groups among each other is only performed on the alternating current (AC) side, thus lowering the hazard potential of a conventional technique (direct-current (DC) arc).
- Mass production of numerous small identical units permits cost reduction.

However, besides the benefits mentioned above also some disadvantages must be expected that have to be compensated by a more sophisticated design of the individual devices.

- Especially module-integrated inverters are subject to high thermal stress that requires the use of appropriately designed components. For some modules warranties of up to 25 years are granted. However, regarding the technical lifetime of module-integrated inverters and the corresponding modules still a big gap needs to be closed.
- The replacement of module-integrated inverters, e.g. in facades, is extremely expensive.
- The function of any small inverter has to be controllable from a central place. This requirement can, for instance, be fulfilled by data transfer over the power line without any additional wiring, but requires special efforts and investment into the communication interfaces.
- At decreasing nominal power of energy converters it is increasingly difficult to obtain an appropriate efficiency curve for photovoltaic appliances, since the own energy consumption cannot be equally reduced. Hence, for very small units a decreased overall efficiency needs to be expected.

Especially, for the large-scale plants currently installed, a tendency towards larger inverter units, ranging from some 5 kW up to several 100 kW, can be observed. Large inverter units permit to considerably reduce inverter costs, both initial investment costs as well as running expenses for maintenance, supervision and repair.

On the whole, grid-connected inverters should meet the following requirements /6-9/, /6-29/.

- The output current is synchronous to the mains. Unlike island inverters supposed to supply basically constant output voltages (voltage source), grid-connected inverters act as power source whose amperage depends on the current input power level.
- The output current should be of sine shape. Distortions and thus the level of harmonics must not exceed the prescribed limits (VDE 0838, EN 60555).

- The output current should not have any direct current bias, since it premagnetises transformers within the grid and may also impair the function of earth leakage protection switches.
- Feed-in current and grid voltage should not show any phase shift (cos $\varphi = 1$), to prevent reactive power from oscillating between the grid and the inverter that may cause additional losses. Future inverter generations should allow for active reactive power compensation to enhance supply quality and reduce transmission losses. The operation of such a plant thus creates an additional value for the network operator.
- In case of abnormal operating conditions (such as missing or excessive mains voltage, strong deviations from the target frequency, short-circuits or isolation errors) the inverter must automatically disconnect from the grid. To monitor parameters characterising the grid, such as voltage and frequency, the monitoring of all three phases has requested also for single-phase inverters in the past. Since the launch of the ENS (device for grid monitoring with allocated series-connected switching) photovoltaic applications up to 5 kW have been considerably simplified. By measuring grid impedance, dynamic grid impedance changes as well as grid voltage and frequency, this system detects grid failure and disconnects the inverter from the mains using two independent switching devices. For safety purposes monitoring devices must be of redundant design. According to the plant concept such an ENS may either be provided individually for every inverter or one central ENS for several inverters.
- Further safety components such as isolation or earth leakage protection switches suitable for AC and DC currents must be provided in accordance with the inverter concept.
- Ripple control signals integrated into the grid voltage by the power supply companies must not be distorted by the inverter nor disturb its operation.
- The entry side should be well adapted to the solar generator, e.g. by Maximum Power Point Tracking (MPPT). The commonly applied MPPT algorithms determine the maximum power point of the photovoltaic generator by performing search functions at regular intervals, e.g. every few seconds or minutes. For this purpose the working voltage of the solar generator is modified by a small quantity; if, following this operation, the output power of the inverter is increased, the search direction is maintained during the next search function, otherwise it is inverted. The optimum voltage value determined by this procedure will be maintained until the next search. Due to this methodological approach the working voltage is subject to fluctuations of a certain range around the actual maximum power point (MPP). Other MPPT processes pass within regular intervals through a certain section of the characteristic curve of the photovoltaic generator to determine the maximum power point, which is also maintained until the next search function.
- Input voltage fluctuations (voltage ripple) should be low (< 3 %) for single-phase inverters that feed energy into the grid at 50 Hz to enable inverter opera-

tion at the optimal operating point. For this purpose a sufficiently dimensioned buffer capacitor is required at the inverter entry.

– Excess voltages, for instance caused by idle solar generators at low temperatures and high solar radiation, but also by distant lightning strikes, must not cause any defects.

– Inverters are generally designed for a slightly lower nominal power than that of the photovoltaic generator (e.g. factor varying from 0.8 to 0.9). This is due the fact that solar generators only rarely reach their nominal power because during full solar radiation the resulting temperature increase of the modules reduces the overall cell efficiency. Additionally, smaller inverters show a lower self-consumption. Therefore an optimum can be found regarding the overall efficiency rate. Nevertheless, if overloading of the inverter does occur, the inverter power input must be precisely limited by displacing the operating point towards the open-circuit voltage. In the ideal case, the admissible power is adapted to the current passive cooling element temperature. Another, however not optimal solution, is to disconnect and periodically restart the inverter in case of over-load.

– Grid-connected inverters should be supplied with energy by the solar generator itself, so that no energy from the grid is consumed at night-time. Furthermore, the inverter should already start and perform reliably at very low solar radiation levels.

– High conversion efficiency should already be achieved for small capacities (> 90 % at 10 % nominal power). The so-called "European Efficiency" allows for a simple comparison of different types of inverters by considering the typical distribution of energy generated by solar generators according to the typical Central European climate. For this purpose inverter efficiency is weighted differently for six different power levels. Medium capacity levels are taken with a higher contribution into consideration because the efficiency curve should yield high values within this section. European efficiencies of smaller inverters (< 1 kW) should be above 90 % and between approximately 95 and 97 % for larger inverters.

– Grid-connected inverters should be provided with integrated self-monitoring systems equipped with user-friendly displays and interfaces to a communication system, if required. The latter allows for a permanent monitoring and remote diagnosis, which cannot necessarily be provided by the average user.

Mounting systems. Energy yields of photovoltaic modules are proportional to radiated solar energy. This is why the orientation of module surfaces towards the sun is of major importance. In this respect, fixed mounting systems and one or two axes tracking systems are distinguished. In general such tracking systems allow increasing the electricity generation compared to installations without a tracking device. For sun-concentrator photovoltaic systems such tracking systems are indispensable because these installations allow only the use of direct radiation.

In this context, energy losses due to excessive temperature rise attributable to sub-optimal installation must be considered (e.g. /6-31/). For instance, a photo-voltaic module mounted at a distance of only 10 cm from a slanting roof generates additional losses due to heating between 1.5 and 2.5 % of the annual energy yield compared to a completely detached solar generator of the same size and orientation. If the modules are fully integrated into the roof without rear ventilation, losses are between 4 and 5 % compared to a completely detached system; for a fully facade incorporated generator the losses amount to 7 to 10 % assuming Central European weather conditions.

Mounting and installations systems must be adapted to the respective site conditions concerning material (hot-galvanized steel, wood etc.) and soil conditions in case of free-standing installation (concrete foundations, ground anchors, pile-driven profiles, foundation-less installation etc.). Regarding roof and facade installation special mounting systems are available. In case of roof installation rack systems may be installed at a certain distance to the roof cladding; however, also integration into the roof is technically possible. The latter option is often preferred for architectural and esthetical reasons and does not require any conventional roof cover. However, it makes convective cooling of modules more difficult and thus leads to increased energy losses as mentioned above.

In terms of stationary systems module orientation towards the south allows to maximise energy yields. Deviations of below 30° to the east or to the west are in most cases of negligible effect since they only decrease the energy yield by less than 5 % /6-32/. Optimal inclination of solar modules is primarily determined by the latitude. If an inclination angle is selected vertical to the mean midday solar altitude this corresponds exactly to the latitude of the respective site. If the maximum annual energy yield is to be achieved, in summertime, due to higher solar radiation during this time of the year, solar modules are to be installed with a lower inclination angle. For Central European latitudes, inclination angles between 25 and 45° allow for the highest energy yields in terms of grid-connected modules.

Yet, it is more difficult to determine the optimal inclination of stand-alone modules. Systems lacking of an additional power generator (e.g. Diesel generator), and that are to provide about the same energy quantity over the whole year, should be set to a much steeper angle of about 60°. While the energy yield needs to be optimised within such systems in wintertime, in summer excess energy usually cannot be utilised. For stand-alone photovoltaic systems provided with additional generators supplying 20 % or more of the annual energy demand inclination angles between 35 and 45° are most suitable for Central European conditions. Seasonal tracking of the inclination is also technically feasible and requires only little effort (summer: flat angle; winter: steep angle).

Tracking of solar modules to the actual solar altitude increases the energy yield. The following tracking systems are distinguished
- single-axis tracking around the horizontal rotation axis,
- single-axis tracking around the polar axis,

- single-axis tracking around the vertical rotation axis for inclined mounted modules, and
- two-axes tracking.

All in all, single-axis tracking systems are less demanding in terms of engineering
expenditure than double-axes systems. According to tracking mode and the site
specific conditions, single-axis tracking systems permit the increase of energy
yields by 20 to 30 %. Two-axes systems achieve the highest energy yields, however, the yields of latter two single-axis options are only slightly lower. Tracking
systems only require little energy varying between approximately 0.03 to 3 % of
the annual energy yield. When the solar radiation is characterised by a high share
of direct radiation, the energy demand of the tracking system is close to the lower
value of the above mentioned interval. If the available solar energy dominated by
diffuse radiation the situation is vice versa.

Small grid-connected photovoltaic systems are usually fixed mounted on house
roofs, and are usually not equipped with tracking systems for civil engineering or
esthetical reasons. For flat roofs – e.g. on garages and industrial buildings – the
situation is different, as advanced tracking systems could be favoured with regard
to the cost-to-benefit ratio. Also passive tracking systems that are currently being
investigated have shown favourable perspectives.

However, it must be observed that tracking systems tend to become less attractive with increased cost reductions for photovoltaic modules. Compared to fixed
systems, tracking systems need more space, have higher maintenance costs, require more expensive mounting equipment and only offer low cost reduction potentials under Central European meteorological conditions.

Under Central European conditions, even for solar power stations with high installed electrical power (i.e. MW-range), also mainly fixed mounting systems are
used. When comparing the achievable additional energy output to the additional
costs required for tracking systems compared to a fixed installation, tracking systems are, so far, always connected with higher costs. However, this situation may
be different for sites with a high share of direct radiation.

Especially high concentrating systems provided with lenses require very precise two-axes tracking systems. As for all concentrating systems, only direct radiation can be concentrated. Thus, this technology is hardly applied in Central
Europe in view of the annual diffuse radiation share of approximately 50 % but is
of interest e.g. in Spain or Northern Africa with a share of direct radiation of more
than 80 % (see Chapter 2.2).

Batteries and charge controllers. While only in very special cases grid-
connected photovoltaic systems are connected with batteries, batteries are integral
components of stand-alone power supply systems. Batteries store the electric energy provided by the photovoltaic modules or additional generators, and provide it
to consumers when they need it. Typically, prior to consumption, between 70 and
100 % of energy is intermediately stored within such systems. Batteries also permit to connect consumers with higher power requirement than the corresponding

nominal power of the photovoltaic generator, since the batteries can provide very high power according to their sizing.

Yet, batteries have a limited technical lifetime and have to be replaced several times, over the entire lifetime of the photovoltaic system (which is determined by the technical lifetime of the photovoltaic generator of approximately 25 years). Calculated over the lifetime of the system, the summarised battery costs usually account for 20 to 40 % of the entire lifetime costs and thus represent the major cost factor, ranking even before those of the photovoltaic generator. Since the technical lifetime of the battery is largely dependent on the stress profile and the operation strategy (i.e. battery management), batteries need special attention both in terms of planning as well as operation of the battery coupled photovoltaic system /6-9/, /6-29/. Especially under economic aspects it has to be mentioned that the lead- based batteries available today are a mass product without a significant "economy of scale" effect. Additionally, due to the upcoming shortage on the world resource markets and the resulting price increase it is expected that lead-based batteries will become more expensive.

For devices (like watches, calculators) directly supplied by photovoltaics, primarily nickel-cadmium-batteries are used. Additionally, nickel-metal-hydride-batteries, lead-acid batteries, lithium-based battery systems, and capacitor (so called bilayer capacitor or SuperCaps) are in operation. Photovoltaic supplied small scale systems and hybrid systems are usually equipped with conventional lead batteries.

To date, except for the small consumer appliances mentioned above, only lead-acid batteries have gained importance /6-9/, /6-33/. However, such lead-acid batteries only have a poor gravimetric energy density of 20 to 30 Wh/kg. Yet, this disadvantage is of minor importance for the use in photovoltaic power supply systems, since, unlike for example in electrically powered cars, batteries are operated stationary and are not moved.

Such lead-acid batteries (accumulators) store electric energy in the form of chemical energy, which is re-converted into electric energy during discharge. Chemical energy is stored in two electrodes (positive and negative) between which there is a potential difference. Fig. 6.21 shows a schematic representation of a lead-acid battery. When fully charged the positive electrode is composed of porous lead dioxide (PbO_2), whereas the negative electrode consists of porous, spongy lead (Pb). The porosity of both electrodes is well above 50 %, and active masses must have a fine crystalline structure to provide a large active surface. The electrodes are submerged into ion-conducting electrolytes of diluted sulphuric acid (H_2SO_4). The electrodes are separated from each other by an ion-permeable separator to prevent short-circuits. The electrochemical process of charging and discharging triggers the conversion of electronic current into ionic current.

During discharge both active masses are converted into lead sulphate ($PbSO_4$) by consumption of sulphuric acid (Equation (6.3)); this process reduces electrolyte concentration. As a result, the physical and chemical electrolyte properties (including freezing point, conductivity, aggressiveness in terms of corrosion and solubility of lead sulphate) are changed. The electrical behaviour of a lead-acid battery deteriorates with increased depth of discharge.

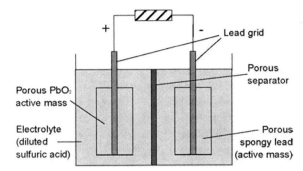

Fig. 6.21 Scheme of a lead-acid battery

$$PbO_2 + Pb + 2H_2SO_4 \quad \xrightarrow[\text{Charge}]{\text{Discharge}} \quad 2PbSO_4 + 2H_2O \tag{6.3}$$

There is a great variety of different designs of lead-acid batteries available on the market. It can be distinguished between batteries designed for high capacities (such as starter batteries for motor vehicles), batteries for a long technical lifetime and few cycles (mainly for uninterrupted power supplies) and batteries designed for strong cyclic operation (e.g. for electrically powered vehicles, fork lifts or wheelchairs). According to the respective requirements, different electrode designs and geometry are used. The main difference can be seen between "flat plate electrode" and "tubular plate electrode", illustrated in Fig. 6.22. Recently also batteries with wound electrodes are available. They show in some application areas very promising results.

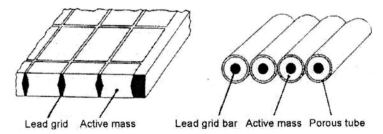

Fig. 6.22 Suitable electrode types for photovoltaic batteries (left: flat plate electrode, right: tubular plate electrode)

For flat plate electrodes a grid from hard lead (lead alloy with antimony or calcium and further additives) carries the active material which is pasted into and onto the grid. This design offers the advantages of a cost-efficient production and high power densities. For tubular plate electrodes the active material is filled into porous tubes around a central hard lead rod. This plate technology which is mainly applied for the positive electrode and whose production costs are higher, allows for considerably longer cycles lifetimes thanks to the good cohesion properties of the active mass. This electrode type is thus ideal for hybrid systems (PV generator plus additional generators) with high energy throughput.

A further distinctive feature is the electrolyte status. The classic lead-acid battery is provided with a liquid electrolyte (so-called flooded batteries). The gases produced within side reactions are directly emitted by the battery due to water electrolysis (oxygen is emitted at the positive electrode, hydrogen at the negative electrode; Fig. 6.23, left). This process consumes water, which must be regularly refilled. In addition, battery rooms have to fulfil high requirements. Such rooms need favourable natural or active ventilation to avoid critical hydrogen or oxygen gas concentrations. Furthermore, electronic components and appliances must be protected from the gases, since the latter have a corrosive effect when they are moisturised by sulphuric acid.

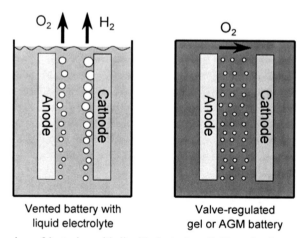

Vented battery with   Valve-regulated
liquid electrolyte   gel or AGM battery

Fig. 6.23 Gassing of batteries with liquid electrolyte (left) and valve-regulated batteries with gel or AGM (absorptive glass mat) electrolyte (right)

As an alternative also so-called valve-regulated gel or AGM (absorptive glass mat) batteries are available, which – instead of liquid sulphuric acid – contain gel or AGM to adhere or to absorb the acid. These kinds of batteries enable the diffusion of oxygen gas created at the positive electrode to the negative electrode, passing through micro-pores inside the gel or the glass mat separator. Since oxygen is again reduced to water at the negative electrode no hydrogen is created here. However, this only applies if all reactions are well balanced. If this is not the case, an excessive pressure inside the cell can occur. Therefore the gas is released through a relief valve in case of excessive pressure. However, water that is lost as a result of this safety measure cannot be refilled. The benefits of this battery technology are reduced gassing, placing considerably lower requirements on the battery compartment, higher installation variability and prevention of electrolyte leakage. However, the disadvantages are higher costs, higher requirements in terms of overcharge protection and – in several applications – shorter technical lifetimes compared to liquid electrolytes. However, particularly high-quality gel batteries provided with tubular electrodes have shown satisfactory durability when combined with suitable charging processes. AGM batteries showed better lifetime in cyclic applications compared with conventional starter batteries for cars /6-34/.

The energy content of such batteries is indicated by their capacities and their nominal voltage. The capacity is defined as the current quantity emitted by the accumulator until reaching a certain end-of-discharge voltage and measured in Ampere hours (Ah) (i.e. amperage (Ampere, A) multiplied by time (hours, h)). The capacity is dependent on the discharge current, the temperature and the defined end-of-discharge voltage. The nominal voltage is defined by the materials participating within the electrochemical reactions and can be calculated for the currentless equilibrium conditions based on thermodynamic coherence. While NiCd and NiMH batteries show a nominal voltage of 1.2 V/cell lead batteries are characterised by 2.0 V/cell and lithium-ionic batteries by 3.6 V/cell; the latter are therefore used often for mobile applications. The nominal voltage defines also how many cells have to be connected in series to obtain the voltage level needed by the consumer.

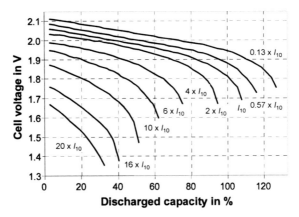

Fig. 6.24 Typical characteristic discharge curves of a battery with tubular electrodes as a function of the discharged capacity (standardised to nominal capacity of ten hours discharge current I_{10}, end-of-discharge voltage 1.8 V/cell)

The useful energy of a battery is largely dependent on the relative discharge current. Currents are expressed as units of the nominal current. The higher the discharge rate is, the faster the voltage decreases. The extractable capacity and consequently also the useful energy are reduced. Fig. 6.24 shows the characteristic discharge curves for different discharge currents. Amperage I is expressed in units of the ten hours discharge current (I_{10}), whereas the extracted capacity is expressed in relation to the capacity available as a result of the ten hours current. According to this rationale, only approximately 50 % of the capacity is available if the current is increased tenfold; if, by contrast, the current is reduced by a factor ten (in relation to the ten hours discharge current), about 30 % more capacity becomes available. But in general a battery fully discharged with a high current can be additionally discharged with a much smaller current afterwards.

Capacities of lead-acid batteries are increased by approximately 0.6 % per K of temperature increase and it is reduced accordingly for decreasing temperatures. The nominal capacity is defined according to the manufacturer definition typically for a reference temperature between 20 and 27 °C. However, the effects of wear and tear and self-discharge multiply with increasing temperature. The optimal operating temperature of lead-acid batteries applied within photovoltaic systems is thus at about 10 °C.

Such batteries store the added electric energy at an Ah-efficiency (Coulomb efficiency) of approximately 95 to 98 %. If the battery is run in cyclic operation in partial state of charge (< 80 % state of charge (SOC)) the Ah-efficiency amounts to almost 100 %. The ratio of received to added energy (typical value 80 to 90 %) is referred to as Wh-efficiency which results from the higher charging voltage when compared to the discharging voltage. At 25 °C self-discharge amounts to 2 to 3 %/month. The rate is approximately doubling with each 10 K temperature increase.

Charge controllers are of major importance for a safe and reliable battery operation. They must protect the battery against deep discharge (according to battery technology, discharged capacity use should not exceed 60 to 80 %) to prevent premature ageing. In addition, charge controllers are responsible for the charging strategy. For this purpose, voltage is limited in that way to enable quick battery charging on the one hand, and to protect the battery against deterioration by gassing or corrosion on the other hand. This is why the voltage needs to be limited. Especially for batteries provided with gel or AGM electrolytes, excessive voltages may create excessive gases which cannot be internally recombined and may thus get lost through the relief valve. It is also of major importance to adapt the maximum charging voltage to the battery temperature. The maximum voltage is lowered with increasing battery temperature to prevent excessive gassing and corrosion.

For this reason high requirements are placed on charge controllers in terms of reliability and the algorithms applied for deep discharge and overcharge protection. However, importance should also be attached to criteria such as self-consumption, user-display to show the battery status and reverse polarity circuit protection. Good charge controllers should allow for additional determination of the battery's state of charge /6-9/, /6-30/, /6-33/, /6-34/.

Further system components. Further photovoltaic system components, that are by no means negligible cost factors, are direct current connecting cables between modules, batteries and inverters. Additionally, usually also fuses, grounding, lightning protection, energy meters as well as low voltage or overvoltage monitoring protection is required. Some components are required to comply with legal prescriptions, whereas others, such as lightning protection, are highly recommended depending on the plant exposure. Photovoltaic power plants also require a transformer to feed the electric energy into the grid with the required power characteristics. Overall losses of these additional system components accounts for 5 to 12 % of the electric energy fed into the mains.

6.2.3 Grid-independent systems

Grid-independent applications are further distinguished into stand-alone and off-grid applications /6-29/ (for examples see Table 6.2).
- In industrialised countries applications are referred to as stand-alone applications if photovoltaic energy supply is applied alternatively to power supply by the power grid for reasons of cost-efficiency, handling, safety or environmental protection; although the mains is available at close hand. If such stand-alone systems are used indoors (at daylight or artificial light) they are referred to as indoor applications; a classic example is the photovoltaic powered calculator. Such systems may be further subdivided into consumer applications (e.g. photovoltaic garden lamps) and industrial or professional applications (e.g. parking ticket vending machines, city furnishings). The solar generator capacity of these applications stretches from a few mW to several 100 W.
- Photovoltaic energy supply systems are referred to as off-grid systems, if the power grid is inaccessible for technical or economic reasons because of the distance to the closest grid connection point (e.g. alpine huts or households located in regions with a missing infrastructure for electricity supply). For larger systems also the terms of "stand-alone applications" or "autonomous energy supply systems" are commonly used. Also in this respect, the systems are further subdivided into industrial applications (the greatest market share accounting for wire-less telecommunication systems) on the one hand, and household supply on the other. Especially for developing countries, the borderlines are loose since economic activities are often only permitted by the power supply to households.

Photovoltaic pump systems receive an exceptional position within grid-independent systems. Their main characteristic feature is their lack of electric energy storage units. Pumped water is stored in an elevated tank when there is sufficient solar radiation for pump operation available. Although there are also pumps operated by direct current, the majority of such systems are equipped with a special inverter, which enables optimum pump operation.

Table 6.2 Selection of typical areas for photovoltaic applications (various sources)

Application area	Application example	Common nominal power in W
Small appliances	Watches, calculators	0.001 to 1
Consumer goods	Radios, vehicle ventilation, tools, lighting	0.5 to 100
Traffic engineering	Buoys, catoptric light, SOS-telephones, solar mobiles, traffic lights, construction site sign-posting, information panels	20 to 500
Foods/Health care	Cooling technology, irrigation	50 to 5,000
Astronautics	Satellite energy supply	500 to 5,000
Communications engineering	Relays stations, repeater, emitter, mobile radio	10 to 7,000
Water engineering	Pumps for surface to subterranean waters, water treatment and ventilation	400 to 6,000
Environmental engineering	Remote measuring stations, sewage-treatment plant	10 to 200,000
Agriculture	Pasture fences, outdoor milking stations, livestock waterers, fish pond ventilation	5 to 20,000
Further stand-alone systems	Cottages and houses, medical care stations, small companies, remote energy supply	40 to 50,000
Grid-connected decentralized applications	One or multiple families home, industrial and municipal buildings	1,000 to 20,000
Photovoltaic power plants	Individual power plants, hybrid systems, hydrogen production	1,000 to more than 1,000,000

System concepts. Autonomous power supply systems are distinguished into
- photovoltaic systems with battery storage and
- photovoltaic systems with battery storage and additional power generator (so-called hybrid systems).

Pure photovoltaic systems can only supply as much energy as supplied by the photovoltaic plant due to the solar radiation. Useful energy quantities are thus subject to fluctuations due to changing seasons and varying weather situations. Hybrid systems, by contrast, ensure a more even energy supply.

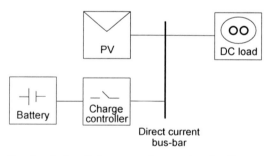

Fig. 6.25 Schematic of a photovoltaic system for supply of a direct current load or direct current consumer application (PV photovoltaic generator, DC direct current)

Photovoltaic systems, applied, for instance, for household or small consumer supply, generally consist of a solar generator, a charge controller and an energy storage which are interconnected by a direct current bus-bar (Fig. 6.25). In most cases such systems directly supply direct current consumers. They are typically applied for camping and leisure as well as for motor vehicle operation; furthermore, solar home systems are applied for basic power supply in rural areas of developing and emerging countries. Yet, there are also systems which are provided with an inverter to supply standard alternating current consumers.

Hybrid systems are applied if a reliable energy supply of large energy quantities independent of season and weather conditions is required. This is especially true for latitudes, such as Central Europe, characterised by high radiation differences between the summer and winter months. For this purpose different power generators are coupled complement to each other in terms of time and energy. The most commonly applied additional power generator is a Diesel engine equipped with a generator (called a Diesel generator). Such Diesel generators are turned on when the battery state of charge is too low due to insufficient solar radiation or excessive power demand. The Diesel generator is operated based on the fossil energy resource. This has the advantage that Diesel fuel can easily be stored. According to the respective system concept batteries may either be recharged or direct current from the Diesel generator can be directly provided to consumers. Fig. 6.26 shows the example of a block diagram including a motor generator (besides Diesel engines also (bio-)gas or bio fuel-fired engines may be used) and a wind energy converter besides the photovoltaic generator. Wind energy converters and solar energy systems complement each other well at many sites with regard to seasonal and weather-related fluctuations. For instance, in Central Europe wind resources complement solar radiation in fall and winter. Furthermore, during prolonged periods of bad weather, often increased wind resources are available. Motor generators which are only turned on in times of high demand offer additional flexibility in terms of system operation, ensure more smooth operation and thus a prolonged service life of the battery.

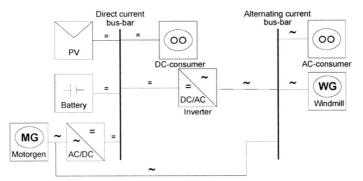

Fig. 6.26 Block diagram of a hybrid system with direct current and alternating current bus-bar (PV photovoltaic generator, MG motor generator, WG wind generator, DC direct current, AC alternating current)

Direct current (DC) generators may be connected to the DC bus-bar of such power supply systems or may be connected via inverters directly to the alternating current AC bus-bar. For hybrid systems, for instance, also mini-hydropower stations or even fuel cells may be applied besides motor or wind generators. Whereas Diesel generator operation only achieves medium efficiencies (scarcely above 15 % measured on an annual basis) and has high maintenance requirements, fuel cells are characterised by low noise emissions, zero-emission local, higher efficiencies, particularly when operated at partial load, and the availability of virtually any nominal power; however, their still high cost and the too low lifetimes are their major disadvantage.

For system designs as shown in Fig. 6.26 a unidirectional or a bi-directional inverter can be applied. Within the former design the motor generator feeds via an inverter the bus-bar; by this the battery is loaded. According to the system layout shown in Fig. 6.26 the windmill can provide only as much power as directly used by the AC-consumers. If instead a bi-directional inverter is used all systems can be created that are provided with only one alternating current bus-bar. Under these circumstances the inverter attached to the motor generator is not needed any more. Additionally excess power can be used for recharging the battery. The integration of bi-directional inverters allow also for the design of systems with only one alternating current bus-bar and no direct current bus-bar. For such systems also batteries and photovoltaic generators are directly connected to the alternating current bus-bar via an individual inverter (unidirectional for the photovoltaic system, bi-directional for the battery).

The voltage on the direct current side is mainly determined by the electrical power of the connected consumers and the power supply from the generators. The resulting currents determine in turn the requirements in terms of cabling, and, in particular, of the required direct current switches and fuses. Especially for high amperage, the latter components are highly sophisticated and expensive. For instance, for systems supplied with alternating current amperage of 100 A at the direct current side represents a reasonable upper limit.

Examples. As follows, some typical applications of autonomous, grid-independent photovoltaic power supply systems, currently in use, are discussed.
– House number illumination. A lot of information should also be legible at night. For instance, emergency rescue services, the police, and emergency relief organisations have for a long time been claiming lighted street names and house number panels. For this purpose, house number illumination supplied by photovoltaic power has been developed. Energy generated during the day is stored in a NiCd or lead-acid accumulator. Such a system allows attaching the house number to virtually any spot of the building without having to lay outdoor cables. House number illumination turns on automatically at the onset of twilight and turns off again the next morning. The entire sequential control as well as load status monitoring of the accumulator is performed electronically, so that the system is nearly maintenance-free. The illuminated surface consists

of a fluorescent collector that also ensures excellent legibility throughout the day by its light collecting capacity based on total internal reflection. Light is emitted only in areas that are printed on rear side (i.e. only the numbers are illuminated). A light-emitting diode serves as light source. Such illumination systems are also suitable for large-surface information panels, to illuminate street nameplates and also suitable for advertising purposes.

− Information panels for stops of public transportation vehicles (like buses, trams). Often stops for public transportation vehicles have no power supply for light or information systems because of the high costs for the cables. Photovoltaic power supply systems allow operating information panels that receive information on schedule deviations by radio. Current departures or delays may be displayed online to the waiting passengers on LCD basis. In addition, in the dark, energy-saving LED light guiding plates illuminate the displays. In conjunction with a motion detector, illumination may be controlled as required in times of low radiation. Larger PV systems also allow illuminating advertisements and make thus an additional income accessible.

− Repeater and base stations for mobile networks. The most important market for commercial, industrial photovoltaic off-grid applications is telecommunications. In view of the ongoing proliferation of cellular phones and other wire-free telecommunication services, and the simultaneous demand for unlimited network availability, there is an increasing demand for autonomous power supply regarding telecommunication network infrastructure; placing the highest requirements on the availability of such supply systems. Oversized photovoltaic generators and battery accumulators meet such high level requirements. For reliable and cost-efficient operation, such power generation systems are equipped with algorithms for telediagnosis and telemonitoring, so that maintenance can be performed on demand.

− Solar home systems. High investment costs for the power distribution infrastructure in conjunction with a low power demand hinder grid connection of remote, scarcely populated areas, particularly in developing countries. This applies to roughly 2 billion people who do not have any access to the grid, and this figure is not likely to diminish within the next decades. This is why stand-alone systems gain increasing importance. Therefore solar home systems represent a technically and often also economically favourable solution for basic electrification of rural areas in developing and industrialising countries. Such electrification is intended to cover the basic requirements for lighting and information of households. Solar home systems normally consist of a 40 to 70 W solar module, a 12 V lead-acid battery of an approximate capacity of 60 to 120 Ah, a charge controller, and the consumers. Typical consumers are energy saving lamps, radios and black and white television systems. In addition, there may be centrally located communal appliances such as water pumps, cooling facilities, e.g. for medication, for video sets for education and advanced training purposes, for medical appliances etc. For such system most commonly direct current appliances are used, offering the advantage of high energy efficiency. However, also inverters of nominal powers from 150 to 500 W are used. The benefits are that virtually any kind of commercial consumer can be connected and that batteries are effectively protected against misuse, since direct connection of consumers to the battery is impossible. The technology of such applications is largely available and reliable, provided that high quality components are used. The spread of these technologies is rather impeded by socio-economic and socio-technical problems hindering the wider use of such systems.

– Village power supply systems. So-called village power supply systems represent an alternative to solar home systems. A central power supply systems provides electric energy to households that are connected to a mini-grid. Central systems are easier to maintain and each household can be supplied with more power while the energy consumption is similar to that of decentralised power supply systems. Central power supply systems are usually hybrid systems that include often all possibilities promising for a site-specific power generation (e.g. micro-hydro, wind, Diesel generator).

– Energy supply of residential buildings and service stations in recreation areas. Even in highly industrialised countries there is still a series of residential building, service stations for leisure activities and alpine huts that are still not connected to the mains, due to the long distance and the related connection costs. Within the EU-15 countries, for instance, about 300,000 houses do not have access to the grid. To date, the only solution available is the installation of a generator (i.e. Diesel generator) for direct power supply of consumers. Diesel generators are thus often operated at partial load, and at disconnecting times, electric energy is usually unavailable. Continuous operation is virtually impossible, not only because of the problems of low partial-load efficiencies, noise and exhaust gases, but also due to the limited technical lifetime of combustion engines under such operating conditions. The example of the "Rotwandhaus", a service station in the European Alps, has shown that a hybrid system, consisting of a photovoltaic generator (5 kW), a wind energy converter (20 kW) and a Diesel generator (20 kW) can provide reliable power supply of approximately 11 MWh/a at an installed capacity of 10 kW to this mountain hut equipped with restaurant and sleeping facilities for 100 people, open year-round. When the available wind speed is sufficient, the wind energy converter contributes to power generation. As soon as the battery is recharged to the maximum voltage, the power output of the wind turbine, the photovoltaic generator and the Diesel generator must be reduced accordingly. An inverter serves for converting the direct current (DC) of the battery into alternating current (AC) of 230 V, so that common electric appliances may be used. The Diesel generator ensures uninterrupted service even in case of extremely unfavourable weather conditions. To utilise the available energy resources as well as possible, a computer monitors and controls the overall system.

6.2.4 Grid-connected systems

Photovoltaic systems feed the generated electric energy into the power grid using inverters to adapt the direct current from the photovoltaic system to the characteristic of the mains (for examples see Table 6.2). The basic structure of such systems is illustrated in Fig. 6.27.

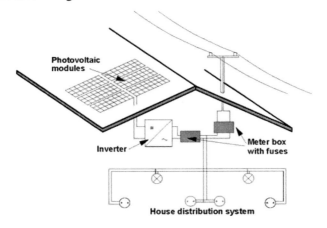

Fig. 6.27 Roof mounted photovoltaic generator directly feeding into the public grid

In terms of grid-connection, the following concepts of photovoltaic systems are available (Fig. 6.28).

- For so-called decentralised systems, that most commonly have the photovoltaic module installed on house roofs, relatively small photovoltaic generators of only a few kW are connected to the mains via an inverter adapted to the photovoltaic generator capacity. They most commonly feed into the low voltage grid. The difference between photovoltaic generator energy provision and the current energy demand of the respective household is balanced by the grid.
- "Quasi centralized" systems (Fig. 6.28) are a very rare mixture of small scale systems and large scale photovoltaic power plants. Within such a system configuration the photovoltaic modules can also be mounted on available support structures (e.g. roofs). Yet, unlike decentralised systems, the individual solar generators are combined to larger units on the direct current (DC) side with an electrical capacity ranging between some 100 kW up to several MW. The systems are then connected to the respective electricity supply grid by larger inverters. For realising such systems a technical-economic optimum between the distance to be covered and the related transportation losses, on the one hand, and the lower inverter losses associated with higher installed capacities, on the other hand, needs to be found. As the electric energy is fed into the medium-voltage power grid additionally a transformer is needed. However, quasi centralised systems have not yet been put into practice on a large scale.

Decentralised systems   Quasi centralised systems   Centralised systems

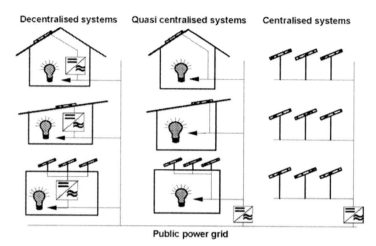

Public power grid

Fig. 6.28 Concepts for grid-connected photovoltaic plants (see also /6-10/)

– Centralised systems with several 100 kW or few MW (Fig. 6.28) are typically mounted on the ground or on very large roofs such as those of fairs. Solar modules may either be stationary mounted or tracked to the current solar altitude by single-axis or two-axis tracking systems. The energy generated by photovoltaics is fed into the low or medium voltage grid by means of one or several inverters and a transformer. Photovoltaic plants of this type show currently electric capacities between some 100 kW and up to 5 MW. However, even higher capacities are achievable from a technical point of view without any problems.

6.2.5 Energy conversion chain, losses, and characteristic power curve

Energy conversion chain. Grid-connected photovoltaic power generation is aimed at the supply of grid-compatible alternating current (AC), which is provided from radiated solar energy at several energy conversion levels illustrated in Fig. 6.29.

As shown in the diagram, solar radiation energy (i.e. diffuse and direct radiation), and thus the photons energy content, is first converted into potential energy of the electrodes of the semi-conductor material; they are now able to move freely within the crystal lattice. If recombination does not immediately take place, and thus no energy in the form of heat is released into the crystal lattice, photovoltaic cells supply this energy as direct current (DC). This direct current is then transformed within grid-connected photovoltaic generators into alternating current (AC) by means of a post-connected inverter customising to properties of the solar energy to the relevant specifications to allow the feeding of the energy into the power grid. Whereas small plants directly feed into the low-voltage grid, larger systems feed into the medium voltage grid.

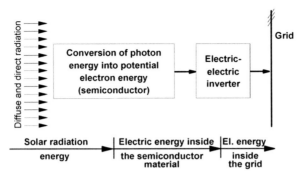

Fig. 6.29 Energy conversion chain of photovoltaic power generation (el. electric; /6-11/)

Losses. Due to the described loss mechanisms only a small part of the solar radiated energy can be fed into the power grid at the connection point. Fig. 6.30 shows the most important losses throughout the overall energy flow of a photovoltaic plant and also indicates the respective magnitude of these losses. The losses indicated in Fig. 6.30 are average figures which may be higher or lower in practical operation; they refer to solar radiation onto the module surface.

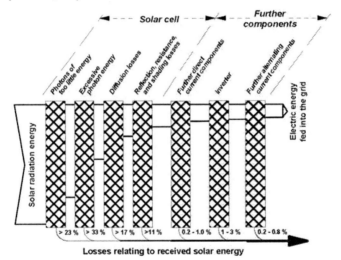

Fig. 6.30 Energy flow of a photovoltaic generator under Central European conditions (solar cell losses have been assumed as minimum losses under standard test conditions (STC); see /6-11/)

According to the diagram, the losses incurred within the actual photovoltaic cell during the conversion of the solar radiation energy into direct electric current account for the major share by far (see /6-2/, /6-10/). For the shown example the approximate efficiency of the solar cell amounts to 16 % referring to the solar radiation; yet, this only corresponds to an approximate efficiency on an annual average of between 13 and 14 %.

Losses incurred outside of the cell are mainly composed by Ohmic losses within the direct current (DC) cabling, the inverter and the required alternating current (AC) cabling. In relation to the radiated solar energy these losses are low and in most cases in the order of magnitude of a few percent. Under the assumed standard test conditions (STC) they result in system efficiencies between 11 and 14 %. The annual average overall system efficiencies of silicon solar cells are thus between 10 and 12 %. It has to be stated that modern photovoltaic plants can show significant better overall efficiencies even on an annual basis.

Characteristic power curve. Radiated solar energy is transformed into electric energy by the described conversion chain. There is a defined correlation between the solar energy radiated onto the cell material within a given period of time and the electric energy effectively provided by the photovoltaic cell or the inverter. But the specific output power or the cell efficiency is reduced by approximately 0.5 %/K due to an increasing module temperature as a result of increasing solar radiation. Fig. 6.31 shows the corresponding characteristic power curves for two different cell types and inverter designs. Such characteristic curves are created by timely summation of solar radiation (kWh/m^2) and supplied alternating current (AC) power generation (kWh/m^2). Fig. 6.31 illustrates the respective daily totals of solar radiation along with the corresponding DC and AC power generation.

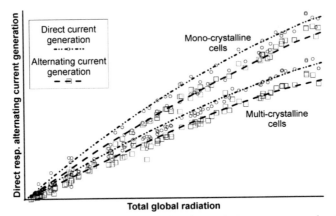

Fig. 6.31 Characteristic performance curve for photovoltaic power generation for different cell types and inverter designs (see /6-11/)

Considering the typical correlation between high radiation and high ambient temperature, the diagram reveals why efficiencies during time periods with high solar radiation (primarily during the summer months) in average are clearly below those achieved in periods with low radiation (especially during the winter months). Fig. 6.31 makes this obvious by the slightly deviating characteristic performance. However, it must also be observed that efficiencies are considerably enhanced starting from low radiation with increasing irradiance due to the open-circuit voltage that increases logarithmically with irradiance. Fig. 6.31 also clearly illustrates this fact within the section of low radiation sums.

According to Fig. 6.31 for the multi-crystalline cells the arising solar radiation onto the module surface generates an increase of the direct current (DC). Yet, doubled irradiance does not exactly double the direct current amount due to the discussed cell temperature rise. The conditions are similar for the mono-crystalline cells also illustrated in Fig. 6.31; however, in the latter case the area-specific power generation is higher due to the higher cell efficiencies.

Fig. 6.31 reveals that for a given radiation the respective area-specific alternating current (AC) power generation is slightly lower when compared to the corresponding direct current (DC) generation. This is due to the losses within the inverter.

6.3 Economic and environmental analysis

Within the following sections the specific power production costs and selected environmental burdens are explained and discussed for state-of-the-art grid connected photovoltaic generators.

Yet, the economic and environmental assessment of stand-alone photovoltaic systems is much more difficult and largely depends on the respective site conditions. For instance, the economic and environmental parameters of hybrid systems are largely influenced by the distribution of the generated power among the individual power generating units. Furthermore, there is no fixed easily determinable comparison facilitating the economic and environmental comparison. Whereas for grid-connected systems, the parameters mentioned above are generally compared to power plant alternatives, the assessment of stand-alone systems is much more difficult. For instance, light generated by photovoltaic solar home systems is often used in replacement of candles, kerosene lamps and lead batteries. Additionally, stand-alone systems are generally only applied in regions where they have an economic edge over grid extension. The economic assessment quickly reveals that not so much the power production costs but rather the power distribution cost account for the major share of the consumer end price. For this reason, such systems will not be discussed in more detail throughout the following sections.

6.3.1 Economic analysis

Currently grid-connected photovoltaic power generation is mainly performed by means of roof-mounted systems as well with an increasing importance by means of so called photovoltaic power plants. Hence, a typical photovoltaic system of an electric nominal capacity of 3 kW, installed on a slanting roof, will be analysed.

Additionally a system located on a horizontal roof of an industrial building with an installed capacity of 20 kW is taken into consideration. To cover the overall market, the analysis will additionally be performed for a 2,000 kW photovoltaic power plant mounted on a steel frame on the ground.

Out of the wide range of solar cell technologies, currently available on the market, exclusively multi-crystalline silicon solar cells with cell efficiencies of 16 %, will be analysed, assuming standard test conditions (in literature referred to as STC).

Under Central European climatic circumstances and current technical boundary conditions, for full-load hours of the analysed systems amount to approximately 800 h/a (Site I) for sites in North to Central Europe, to approximately 1,000 h/a (Site II) for sites in Central to South Europe, and to approximately 1,200 h/a (Site III) for promising sites in South Europe to North Africa. The lifetime of solar modules is estimated to be 30 years. The technical data of the reference plants is summarised in Table 6.3.

Table 6.3 Technical data of the analysed photovoltaic systems

		System I	System II	System III
Nominal system capacity	in kW	3	20	2,000
Basic material		Silicon	Silicon	Silicon
Solar cell type		multi	multi	multi
Efficiency[a]	in %	16	16	16
Technical lifetime	in a	20	20	20
Full-load hours Site 1	in h/a	800	800	900
Site 2	in h/a	1,000	1,000	1,100
Site 3	in h/a	1,200	1,200	1,300

[a] under standard test conditions (STC) conditions.

For these systems a technical availability of 99 % is assumed; i.e. only for 1 % of the year power generation is unavailable due to failures or maintenance. This is realistic, as maintenance work can partly be performed when no electric energy can be provided due to the lack of solar radiation (i.e. during the night).

In the following the variable and fixed costs, as well as the electricity production costs are discussed. Depending on plant size and applied technology, expenditures may vary tremendously. Therefore the costs discussed below can give only a rough indication of the costs based on average conditions.

Investments. The installation costs of photovoltaic systems generally include module and inverter costs, costs for frames, design and mounting as well as further expenditures (including e.g. costs for building permits). Table 6.4 illustrates the cost structure pertaining to the defined photovoltaic systems.

Generally, specific costs decrease with increasing plant size. For instance, the overall specific plant investment costs of a complete 1 kW plant manufactured on the basis of multi-crystalline silicon stretch in average from 4,900 to 6,800 €/kW (excluding value-added tax). For the 3 kW plant defined in Table 6.3 they vary between 4,100 and 5,600 €/kW for the same side conditions. For a 10 kW system, the overall investment costs for the plant, also based on multi-crystalline photovoltaic cells, are around 4,000 to 5,000 €/kW, and for the analysed 2,000 kW plant they roughly vary between 3,500 and 4,100 €/kW (also refer to /6-35/).

Besides reduced module prices, in case of higher sales quantities, cost digression is also given due to decreasing inverter costs for increased installed capacities as well as by the reduction of other specific costs (including electric facilities, planning and mounting). These cost advantages apply even more to larger plants. However, due to higher specific expenditures for pedestals and electric installations required for ground-mounted modules, cost advantages are partly compensated.

Table 6.4 Mean investment and operating costs as well as power production costs of photovoltaic generators (for a definition of the analyzed reference plants see Table 6.3)

			System I	System II	System III
Nominal system capacity		in kW	3	20	2,000
Full load hours	Site 1	in h/a	800	800	900
	Site 2	in h/a	1,000	1,000	1,100
	Site 3	in h/a	1,200	1,200	1,300
Investments					
Modules		in k€	7.8	46.3	4,134
Inverter		in k€	1.1	7.8	741
Further components		in k€	1.2	7.9	872
Miscellaneous		in k€	2.9	16.0	1,667
Total		in k€	13.0	78.0	7,414
Operating costs[a]		in k€/a	0.03	0.8	108
Power generation costs	Site 1	in €/kWh	0.42	0.41	0.36
	Site 2	in €/kWh	0.34	0.33	0.30
	Site 3	in €/kWh	0.28	0.27	0.25

[a] operation, maintenance, miscellaneous.

The major share of the expenditures accounts for module costs. For monocrystalline modules the expenditures are currently roughly between 2,000 and 3,300 €/kW. The prices for multi-crystalline photovoltaic modules are slightly below this order of magnitude; they vary roughly between 1,900 and 3,200 €/kW. According to this the costs for the modules contribute with 55 to 65 % of the overall investments required for an entire photovoltaic generator /6-36/.

A major share of the overall costs accounts for the inverter costs indicated in Table 6.4. They are currently varying roughly between 300 and 450 €/kW. They account for a share of approximately 7 to 12 % of the overall investments of a photovoltaic system.

Besides the expenses mentioned above for photovoltaic modules and inverters, mounting frames account for 10 to 15 % of the overall investments, depending on the required technology (installation on slanting or flat roofs). Additional costs occur for the installation of the photovoltaic module. The figures indicated in Table 6.4 include the complete roof installation as well as the overall electrical installations, such as meter box, meter box installation and connection to the grid. Assuming sufficient experience of the contractor, design costs can be estimated at a maximum of 2 % of the overall plant investment costs (also refer to /6-35/, /6-36/).

Operation costs. Operation costs include maintenance and servicing costs as well as further expenses (e.g. repairs, module cleaning, meter rent, insurance). According to the installation type and the plant size the annual operating costs are between 5 and 30 €/kW. For the analysed reference plants running expenses (Table 6.4) amount to approximately 30 €/a for the 3 kW plant, 800 €/a for the 20 kW plant and to 108,000 €/a for the 2,000 kW plant.

Power production costs. By means of the annuity method electricity generation costs can be calculated on the basis of overall investments and annual operating costs. For this purpose a real interest rate of 4.5 % and depreciation period over a technical lifetime of 30 years is assumed. Table 6.4 shows the calculated power production costs pertaining to the analysed reference models.

Fig. 6.32 shows the resulting specific electricity generation costs for photovoltaically generated electrical energy exemplary for Site 1. According to this diagram the provision costs decrease with increasing installed plant capacity; this applies in particular to capacities between 1 and 5 kW. For example for a 1.5 kW plant provided with multi-crystalline photovoltaic modules power production costs are roughly between 0.50 and 0.60 €/kWh. By installing a 3 kW generator these costs can be reduced to approximately 0.40 to 0.55 €/kWh. With continuing increasing plant capacities specific production costs are even further reduced. For a 2,000 kW plant they are still in the range of 0.34 to 0.38 €/kWh.

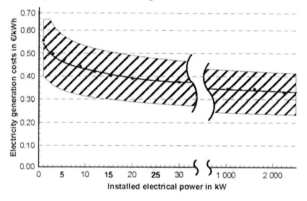

Fig. 6.32 Average specific power production costs of current multi-crystalline photovoltaic generators assuming Middle European radiation conditions (Site 1)

For a system with roof-installed multi-crystalline modules the mean real power generation costs are thus expected to be between 0.40 and 0.55 €/kWh (Site 1). The illustrated range is due to regional radiation variations in Middle Europe and to different investments for plants of the same capacity due to different technical solutions available on the market with different costs.

With increasing capacities of the photovoltaic power station specific power production costs are further reduced, however, the decrease is considerably lower. For a 2,000 kW photovoltaic power plant mean power production costs vary between approximately 0.25 and 0.40 €/kWh under the circumstances discussed above (Site 1). For even higher installed electrical capacities a further reduction of only a few percent can be expected. The partly tremendously lower specific power production costs of photovoltaic power plants in comparison to roof-mounted solar systems are primarily attributable to lower specific overall investment costs, generally higher mean annual inverter efficiencies as well as the optimum inclination and orientation that can be assumed for ground-mounted plants. Thanks to the latter two reasons, an increased number of full-load hours can be achieved at the same site.

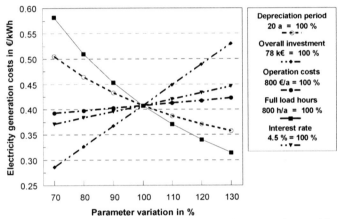

Fig. 6.33 Parameter variation of the main influencing variables on the specific power production costs of the 20 kW photovoltaic multi-crystalline generator at Site 1 indicated in Tables 6.3 and 6.4

Yet, in particular cases and under different site conditions, power production costs may differ tremendously from the magnitudes indicated in Table 6.4. To estimate the impact of such influences on the power production costs, Fig. 6.33 illustrates a sensitivity analysis of predominant parameters within a variation of +/-30 %, using the example of the 20 kW photovoltaic system at Site 1 based on multi-crystalline modules taken from Table 6.4. Thus, besides plant investment costs, the number of full-load hours calculated on the basis of the system utilisation ratio and the effective radiated solar energy within the course of a year and the site conditions, has the most important impact on the power production costs. As a consequence, the costs of photovoltaic power generation can be noticeably decreased by a reduced investment volume, e.g. due to more cost-efficient cell manufacturing methods or enhanced plant efficiency, for instance, achieved by progress beyond the current state of technology. Operating costs and interest rate have, by contrast, only a minor impact on photovoltaic power generation costs. Solely the depreciation period still has certain effect on the specific production costs of photovoltaic power generation.

The indicated figures reveal that the power production costs within the sun belts of the earth (e.g. Southern Europe, North Africa), due to a solar radiation doubled compared to Middle Europe, can be reduced significantly. Under such favourable conditions costs of electricity production with small-scale photovoltaic systems are already below 0.30 €/kWh. For photovoltaic power plants the production costs can even be at 0.25 €/kWh and below.

6.3.2 Environmental analysis

Photovoltaic electricity provision is characterised by noiseless operation without any release of toxic substances or particles by the actual conversion plant. Nevertheless, the following environmental effects may occur.

Construction. Environmental effects related to the manufacture of photovoltaic plants especially occur during the production of the solar cells. In recent years they have been discussed primarily in the context of consumption of scarce mineral resources and toxicity.

Mono-crystalline and multi-crystalline as well as amorphous silicon solar cells are generally characterised by a low consumption of scarce resources, whereas cadmium telluride (CdTe) and CIS cell technologies show medium mineral resource consumption. The application of germanium (Ge) appears to be particularly problematic for amorphous silicon cell production; the same applies to indium (In) with regard to CIS cells and tellurium for CdTe cells. According to current knowledge only limited quantities of these elements are available on earth /6-37/.

In terms of toxicity only low environmental effects are expected for crystalline silicon technologies. However, CdTe and CIS cell technologies are considered more problematic due to their high content of cadmium (Cd), selenium (Se), tellurium (Te) and copper (Cu). In addition, during manufacture of CIS modules gaseous toxic substances (e.g. hydrogen selenide (H_2Se)) may be produced which are generally associated with a certain environmental hazard potential.

On the whole, the environmental effects related to solar cell manufacture are equivalent to those of the overall semiconductor industry. However, the described environmental effects are relatively low due to the challenging legal environmental protection regulations. This is also true due to the required material purity during solar cell manufacture. On the other hand there might exist a manufacture-related hazard potential in case of malfunction.

Normal operation. During the operation of roof-mounted photovoltaic modules no noise is created and no substances are released. Only the inverters currently available on the market are characterised by a low noise development to be minimised with special design measures. This allows a priori for very environmental-friendly power generation. Furthermore, photovoltaic modules are very similar to roofs in terms of absorption and reflection properties. Thus, no major impacts on the locale climate have to be expected. Yet, modules mounted on slanting and flat roofs are in some cases visible from long distances. This might impact the appearance of cities and villages. But on the other side, such an installation does not require any additional (scarce) space.

Ground mounted photovoltaic generators (i.e. photovoltaic power plants installed on e.g. former agricultural areas or on land used formerly for opencast mining) partly or entirely inhibit the use of the ground for other purposes. However, only a very small part of the ground is lost for other purposes (i.e. only around the foundations of the support frames of the solar modules). The major remaining part can still be greened or extensively be used for cultivation (e.g. sheep pasture). In comparison to intensive plant cultivation ecological conditions may even be improved, for instance by biotope creation /6-38/, /6-39/.

Due to the relatively large covered surfaces and the highly divergent absorption and reflection conditions when compared to the agricultural cropland impacts on the microclimate are possible. Yet, these kinds of environmental effects are only relevant in case of intensive photovoltaic utilisation, which is very unlikely due to economic reasons.

The operation of photovoltaic generators is also related to the transmittance of electromagnetic radiation (aspect of electromagnetic compatibility). Unlike common power generation plants, photovoltaic plants are generally provided with extensive direct current cabling and with regard to the solar generator a correspondingly large radiating surface; furthermore, they are partly installed in the vicinity of residential area /6-40/. However, during the installation of such plants it is generally ensured that the wiring loops, acting as antennas, are kept as small as possible. This is a protective measure against both irradiance and receipt of electromagnetic radiation. The latter is particularly critical with regard to lightning strikes in the vicinity of solar modules and could create excess voltages and excess currents in case of a too large receipt area. The destroying of electric components could be a result. However, the low-frequency magnetic fields emitted by photovoltaic components are not higher than those of household appliances; emissions are considerably lower than those of e.g. television sets. The efforts of manufacturers in terms of module design will further reduce emissions, so that no major impacts have to be expected.

Malfunction. To prevent hazards to humans and the environment due to operational malfunctions of photovoltaic generators, generator failures and inadmissible fault currents must be reliably identified and signalized. The inverter and photovoltaic plant design must allow for power disconnection detection and auto shut-off. Photovoltaic systems must only be connected to strong grids. Modern inverters usually include the corresponding safeguarding equipment, so that the above-defined requirements are usually met.

Fires in buildings, causing solar modules and the building envelope to burn, may cause evaporation of certain components contained in the solar cells. For instance, with regard to cadmium telluride and CIS thin-film solar cells critical amounts of cadmium (Cd), tellurium (Te) and selenium (Se) may be released; for instance combustion trials of one hour of duration have shown the release of 4 g/h selenium (Se), 8 g/h cadmium (Cd) and tellurium (Te) /6-37/, /6-43/. Yet, the release of these substances is below the harmful cut-off values defined for these substances. Due to the low concentrations it is expected that even in case of complete cadmium (Cd) release harmful cadmium concentrations to the surrounding air masses can only be reached from plant capacities of 100 kW onwards /6-41/. Roof-mounted modules of such capacities are only applied in exceptional cases (e.g. for factory buildings). In case of fires at electrical plant components (e.g. cables, inverters) additional amounts of harmful substances may be released to the environment; yet, they are not specific for photovoltaic plants.

Furthermore, experience has shown that in case of extreme, hardly realistic, elutriation (e.g. due to rain or modules being submerged into brooks or rivers) the limits of the potable water prescription act are not exceeded.

Injury hazards due to falling solar modules, improperly mounted onto roof panels or facades, or in consequence of electrical voltages between electrical connections, may be largely excluded by adhering to the applicable standards in terms of construction and operation of electro-technical plants.

All in all, photovoltaic power generation has a very low propensity towards malfunctions, and malfunctions are always limited to a certain location. Provided that the modules are appropriately installed and operated, hardly any significant environmental impacts have to be expected.

End of operation. According to current knowledge extensive recycling of solar modules is possible. For instance, extensive recycling of glass components is possible with only little effort. For the recycling of the other module components, by contrast, highly sophisticated chemical separation processes are required. Amorphous frameless modules are best suited for recycling, as they may be transferred to hollow glass recycling without any pre-treatment. Possible recycling methods suitable for "classic" photovoltaic modules include acid separation of solar wafers from the bond, transfer of frameless modules into ferrosilicon suitable for steel production, as well as complete separation of the modules into glass, metals and silicon wafers /6-37/. Yet, cadmium tellurium (CdTe) and CIS technologies need to be further assessed in order to determine whether their heavy metal content precludes or requires further processing /6-42/. The ensuing environmental effects largely correspond to the common impacts of this industrial branch on the natural environment. However, since the recycling of photovoltaic systems is still at its infancy the related environmental effects will be possibly reduced in the future.

7 Wind Power Generation

7.1 Principles

Wind energy converters (WEC) harness the kinetic energy contained in flowing air masses. In the following, the fundamental physical principles of this type of energy conversion are explained. However, explanations do not include wind energy utilisation by sailboats, for instance.

Energy extraction from wind, by wind energy converters, is always related to a certain time difference as wind and operational conditions are usually subject to constant changes. This is why in most cases the instantaneous energy value (power) is determined in order to calculate its useful energy contribution (work) by summation over time (i.e. integration). Kinetic wind power P_{Wi} is thus determined by air density ρ_{Wi}, wind passage area S, and wind velocity $v_{Wi,1}$. By means of wind energy converter the wind power station extracts part of the wind power by reducing the wind speed.

Most modern wind energy converters are equipped with rotors to extract wind power, and consist of one or several rotor blades. The extracted wind power generates rotation and is thereby converted into mechanical power P_{Rot} at the rotor shaft. Mechanical power is taken up at the shaft in the form of a moment at a certain rotation and is transferred to a machine (such as a generator or a pump). The entire wind power station thus consists of a wind energy converter (rotor), a mechanical gear and a generator.

It is physically impossible to technically exploit the entire wind energy, as in this case air flow would come to a standstill. In this case, air would fail to enter the swept rotor area, and wind power would no longer be available.

There are two different physical principles to extract power from wind. The less efficient airfoil drag method is based on the wind drag force incident on a wind-blown surface. The second principle – also referred to as aerodynamic or airfoil lift principle, which is based on flow deviation inside the rotor – is predominantly applied for wind energy conversion. When compared to the drag principle, double or triple the power output is achieved for a given cross-section area. Chapter 7.2 thus focuses entirely on plants based on the airfoil lift principle. Nevertheless, both principles are outlined throughout the following sections to explain the main differences following a discussion on the maximum achievable wind power output by means of an ideal wind energy converter (using the example of a rotor).

7.1.1 Idealised wind energy converter

The considerations outlined throughout this section with regard to assessing the theoretical total capacity extractable by means of wind energy converters (such as a rotor) are based on the following ideal conditions and assumptions:
- frictionless, stationary wind flow,
- constant, shear-free wind flow (i.e. wind speed is the same at every point of the energy extracting surface (e.g. circular rotor surface S_{Rot}) and flows into shaft direction),
- rotation-free flow (i.e. no wind deviation into circumferential direction),
- incompressible flow ($\rho_{Wi} \approx$ const. = 1.22 kg/m^3) and
- free wind flow around the wind energy converter (no external impacts on wind flow).

On the basis of the above conditions the maximum physically achievable wind conversion can be derived by a theoretical model that is independent from the technical construction of a wind power station.

For this purpose an imaginary wind driven air package with air particles is assumed whose flow filaments are bordered by a fictitious stream-tube. The stream-tube is examined at three characteristic cross-sections (S_1 – way before the wind converter rotor, S_{Rot} – at the circular rotor surface, S_2 – way behind the wind converter rotor). Investigations result in the stream-tube illustrated in Fig. 7.1.

According to the mass conservation law, air throughput (i.e. mass flow $\dot{m}_{Wi}$) has to be the same for every cross-section i of the stream-tube (whereby $i = 1$ far before the rotor, $i = Rot$ within the rotor plane and $i = 2$ far behind the rotor). This also applies to the surfaces S_1, S_{Rot} and S_2. Under these circumstances the continuity Equation (7.1) may be applied.

$$\dot{m}_{Wi} = \rho_{Wi} S_i v_{Wi,i} = const.$$

(7.1)

According to Bernoulli's law, the power contained in every point i of the air flow $P_{Wi,i}$ consists of kinetic capacity ($1/2\,(\dot{m}_{Wi,i}\,v_{Wi,i}^2)$), pressure capacity (($\dot{m}_{Wi,i}\,p_{Wi,i})/\rho_{Wi}$) and potential capacity which is in this case negligible by approximation. With regard to continuity the wind capacity balance at any location i far before (e.g. S_1) and far behind the rotor (e.g. S_2) reads as expressed in Equation (7.2). $P_{Rot,th}$ describes the theoretical power at the rotor shaft of the wind turbine.

$$P_{Wi,i} = const. = \frac{1}{2}\dot{m}_{Wi}\,v_{Wi,1}^2 + \frac{\dot{m}_{Wi}\,p_{Wi,1}}{\rho_{Wi}} = \frac{1}{2}\dot{m}_{Wi}\,v_{Wi,2}^2 + \frac{\dot{m}_{Wi}\,p_{Wi,2}}{\rho_{Wi}} + P_{Rot,th}$$

(7.2)

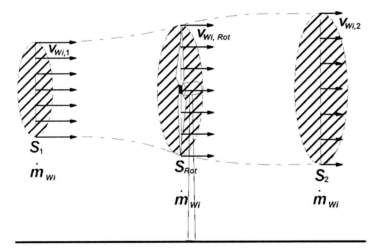

Fig. 7.1 Flow through an idealised wind energy converter (for an explanation of symbols see text; see /7-1/)

Regardless of the actual wind energy converter conditions, the same wind pressure ($p_{Wi,1} = p_{Wi,2}$) and the same density ($\rho_{Wi,1} = \rho_{Wi,2}$) are assumed long before and far behind the wind energy converter. If this assumption is considered for Equation (7.2), the wind capacity $P_{Wi,ext}$ extracted by the wind energy converter which, according to the law of power conservation corresponds to the theoretical rotor capacity $P_{Rot,th,}$ is equal to the difference between the wind capacity before ($P_{Wi,1}$) and behind ($P_{Wi,2}$) the wind energy converter.

$$P_{Wi,ext} = P_{Rot,th} = P_{Wi,1} - P_{Wi,2} = \frac{1}{2}\dot{m}_{Wi}\,v_{Wi,1}^2 - \frac{1}{2}\dot{m}_{Wi}\,v_{Wi,2}^2 = \frac{1}{2}\dot{m}_{Wi}(v_{Wi,1}^2 - v_{Wi,2}^2) \quad (7.3)$$

Equation (7.3) reveals that from free wind flow, energy is only yielded if wind speed is reduced. Hence, only kinetic wind energy can be harnessed.

According to the mass conservation law (Equation (7.1)) the stream-tube must be extended during power extraction – as illustrated in Fig. 7.1 – due to the slow-down of wind speed and power extraction. The stream-tube cross-section is steadily enlarged since wind speed cannot be reduced abruptly.

For the free flow through the rotor surface S_{Rot}, the kinetic wind capacity is calculated according to Equation (7.4). $\dot{m}_{Wi,free}$ refers to the mass flow rate of the free blown stream-tube without any energy extraction ($\dot{m}_{Wi,free} = \rho_{Wi}\,S_{Rot}\,v_{Wi,1}$) and assumes that the surface S_{Rot} is subject to an undisturbed wind speed $v_{Wi,1}$.

$$P_{Wi} = \frac{1}{2}\dot{m}_{Wi,free}\,v_{Wi,1}^2 = \frac{1}{2}\rho_{Wi}\,S_{Rot}\,v_{Wi,1}^3 \qquad (7.4)$$

The above equation reveals that wind capacity depends on the wind speed's third power ($P_{Wi} \sim v_{Wi}^3$); which is of major importance when it comes to selecting a particular site and for wind power plant technology.

When continuously observing the wind flow long before the rotor (S_1) and far behind the rotor (S_2) (Fig. 7.1 and 7.2), wind velocity v_{Wi} is supposed to decrease steadily according to Equation (7.2). On the other hand wind pressure must increase accordingly. At the actual rotor plane S_{Rot} the theoretical rotor capacity is almost abruptly extracted from the wind flow. However, at this point wind velocity cannot change discontinuously; hence, power extraction requires a sudden pressure change Δp_{Wi} (Fig. 7.2). Regardless of these circumstances the wind pressure $p_{Wi,0}$ far before and behind the wind turbine must be taken into account. Wind pressure is subject to weather changes.

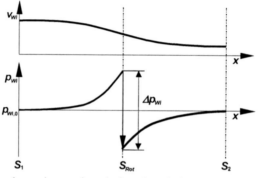

Fig. 7.2 Pressure and speed curve long before the wind energy converter, at the wind energy converter level (rotor plane), and far behind the wind energy converter (for an explanation of symbols see text)

Since according to Newton's law action equals reaction, the dynamic effect which the wind has on the wind energy converter ($F_{Wi,WEC}$) must be equal to the force given by the wind energy converter, which slows down the wind flow ($F_{Wi,slow}$) (Equation (7.5)).

$$F_{Wi,WEC} = F_{Wi,slow} = \dot{m}_{Wi}(v_{Wi,1} - v_{Wi,2}) \tag{7.5}$$

Within the rotor plain S_{Rot}, the wind force $F_{Wi,WEC}$ together with the wind velocity at rotor level $v_{Wi,Rot}$ must be equal to the theoretical rotor power $P_{Rot,th}$ or the power extracted by the rotor $P_{Wi,ext}$ (Equation (7.6)); whereby $P = F\,v$ and power P, force F and velocity v.

$$P_{Wi,ext} = P_{Rot,th} = F_{Wi,WEC} \cdot v_{Wi,Rot} = \dot{m}_{Wi}(v_{Wi,1} - v_{Wi,2})v_{Wi,Rot} \tag{7.6}$$

By equating the relations of (7.3) and (7.6), wind velocity within the rotor level is derived as an arithmetical mean from $v_{Wi,1}$ and $v_{Wi,2}$ (Froude Rankin's theorem;

$v_{Wi,Rot} = (v_{Wi,1} + v_{Wi,2})/2)$. The mass flow rate $\dot{m}_{Wi}$ determined according to Equation (7.1) with regard to the rotor level allows the calculation of the theoretical rotor power $P_{Rot,th}$ or the power extracted by the rotor $P_{Wi,ext}$ long before ($v_{Wi,1}$) and far behind the rotor ($v_{Wi,2}$), the rotor circular surface S_{Rot} and the air density ρ_{Wi}.

$$P_{Wi,ext} = P_{Rot,th} = \frac{1}{2}\rho_{Wi}\left(\frac{v_{Wi,1} + v_{Wi,2}}{2}\right)S_{Rot}(v_{Wi,1}^2 - v_{Wi,2}^2) \tag{7.7}$$

The theoretical power coefficient $c_{p,th}$ expresses the respective maximum physical conversion from wind into rotor power and thus the ratio of the maximum power to the power contained in undisturbed wind. It is defined as the ratio of extractable power ($P_{Wi,ext}$; Equation (7.7)) to the theoretical maximum wind power (P_{Wi}; Equation (7.4)). The power coefficient is calculated by Equation (7.8) on the basis of the wind velocities long before and far behind the wind energy converter (rotor). In this context, the velocity ratio ($v_{Wi,2}/v_{Wi,1}$) is referred to as wind velocity reduction factor.

$$c_{p,th} = \frac{P_{Wi,ext}}{P_{Wi}} = \left(\frac{v_{Wi,1} + v_{Wi,2}}{2v_{Wi,1}}\right)\left(\frac{v_{Wi,1}^2 - v_{Wi,2}^2}{v_{Wi,1}^2}\right) = \frac{1}{2}\left(1 + \frac{v_{Wi,2}}{v_{Wi,1}}\right)\left(1 - \frac{v_{Wi,2}^2}{v_{Wi,1}^2}\right) \tag{7.8}$$

Wind power exploitation aims at extracting the maximum share of wind power. Due to physical restrictions wind masses flowing through rotor level cannot be entirely slowed down, as a complete slowdown would "clog" the rotor and impede power extraction. On the other hand, wind velocity must be decreased if power is to be extracted from flowing air masses. Consequently, there must exist a certain ratio between the speed long before and far behind the rotor that corresponds to the maximum power coefficient $c_{p,th}$.

To determine the maximum wind power that can be extracted from the wind by means of a wind energy converter $P_{Wi,ext}$, Equation (7.7) needs to be differentiated with respect to $v_{Wi,2}$ and zeroed (Equation (7.9)).

$$\frac{d\left(\frac{1}{2}\rho_{Wi}\left(\frac{v_{Wi,1} + v_{Wi,2}}{2}\right)S_{Rot}(v_{Wi,1}^2 - v_{Wi,2}^2)\right)}{d\,v_{Wi,2}} \overset{!}{=} 0 \tag{7.9}$$

When resolving Equation (7.9) with respect to the rotor, the energetically most favourable wind velocity appears to be obtained at one third of the wind speed before the rotor $v_{Wi,1}$. The function shown in Fig. 7.3 also reveals this context. According to the curve the maximum ($c_{p,ideal}$) of the theoretical power coefficient $c_{p,th}$ is achieved at a ratio of wind velocities of one third behind and before the rotor.

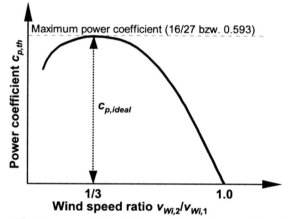

Fig. 7.3 Power coefficient curve relative to the speed ratio between the wind speed in front $v_{Wi,1}$ and behind $v_{Wi,2}$ the rotor (i.e. relative to the reduction factor according to Equation (7.8))

According to Equation (7.8) a maximum theoretical power coefficient of 16/27 is obtained at this ratio. This means that, theoretically, the biggest power extraction from wind amounts to almost 60 % of the theoretical wind power calculated according to Equation (7.4). However, due to general physical restrictions, at least 40 % of the wind power of the undisturbed flow that incident on the surface range of a wind energy converter, is generally unavailable for wind exploitation. One forth of this loss is due to the incomplete wind slow-down at rotor level (1-$(v_{Wi,1}^2/v_{Wi,2}^2)$) and to the stream-tube expansion between S_1 and S_2.

However, even for an ideal rotor, wind slow-down to one third of the original wind velocity presupposes optimum operation conditions (including the rotor speed in relation to wind speed); this power coefficient is referred to as ideal power coefficient ($c_{p,ideal} = 0.593$).

In the 1920s, Albert Betz first published the above theoretical derivation of the maximum extractable wind power, which is entirely independent from the wind energy converter type (see /7-2/ amongst other references). On the basis of Betz' law the described ideal conditions and assumptions have been investigated further within the scope of extended theories and have continuously improved the description of real conditions. For instance, impacts of rotational losses (i.e. kinetic energy that is unavailable due to turbulence of the airflow) which reduces capacities particularly at low tip speed ratios may be considered.

The tip speed ratio λ refers to the ratio of blade tip speed ($v_u = d_{Rot} \pi\, n$; whereby n is the number of rotor revolutions and d_{Rot} the rotor diameter) to the wind velocity at rotor level $v_{Wi,Rot}$ (Equation (7.10)). The lower the number of rotor blades the higher the tip speed ratio.

Especially for modern wind energy converters of few rotor blades (so-called high speed converters) the power coefficient is determined by the rotor blade's angular momentum and friction. Within the optimum operation range, the ideal power coefficient of modern three-blade wind turbines nowadays amounts to values up to 0.47.

$$\lambda = \frac{v_u}{v_{Wi,Rot}} \tag{7.10}$$

Mechanical losses due to bearing and gear friction as well as all losses of the electrical plant components are considered by the corresponding efficiency $\eta_{mech.-elec.}$, which generally amounts to about 90 % for the currently available plants. The useful power of a wind power station P_{WEC} can thus be calculated according to Equation (7.11). P_{Wi} describes the wind power and c_p the power coefficient.

$$P_{WEC} = c_p \eta_{mech.-elec.} P_{Wi} \tag{7.11}$$

7.1.2 Drag and lift principles

There are two different principles available for technically exploiting moving airflow by rotating wind energy converters, which can also be combined under certain conditions. Energy can be extracted from flowing air masses either by the lift or drag method. In the following both principles are explained.

Lift principle. According to the lift principle, wind is deviated to generate peripheral force inside the rotor. For high-speed propeller-type converters, in most cases rotor blades are evaluated according to the wing theory. If a rotor blade (represented schematically as flat profile in Fig. 7.4) is hit symmetrically by airflow at velocity v_{Wi} (angle of inflow $\alpha = 0$), a force referred to as drag force F_D is built up in flow direction due to its shape and frictional drag (i.e. resistance principle; Fig. 7.4, top, and Fig. 7.9). However, there will be little resistance force if the rotor blades are of airflow-friendly design.

Only if the rotor blade is hit asymmetrically (i.e. if the rotor blade profile is inclined at a certain angle to the airflow; flow angle $\alpha > 0$) the streamlines above and below the profile have to cover different lengths (Fig. 7.4, centre). When considering the above framework conditions (i.e. $\alpha > 0$) with regard to a laminar flow pattern, free from losses (not subject to any swirls nor friction), the air particles need to reunify after having passed the cross-section. Consequently, those air particles that need to cover longer distances (see Fig. 7.4, centre, on the topside of the schematic profile exposed to airflow) need to move faster. Energy yield considerations according to Equation (7.2) (i.e. according to Bernoulli's law without considering potential power ($\rho_{Wi} v_{Wi}^2/2 + p_{Wi} = const.$) without power extraction from wind reveal that the faster air particles generate less pressure than the slow particles. There is thus higher pressure below (pressure side) the cross-section exposed to moving air than on the topside (suction side) which produces a force vertically to the flow angle (lifting force F_L). Besides the lifting force also drag force F_D incidents on the cross-section which is higher as in the case of symmetrical blow (blow angle $\alpha > 0$) (Fig. 7.4).

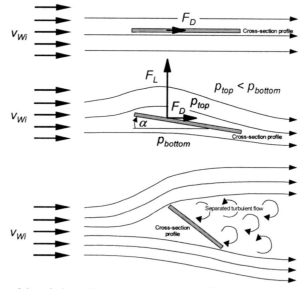

Fig. 7.4 Effect of the wind speed on a cross-section profile exposed to airflow (top: orientation towards flow direction; centre: orientation at a certain flow angle relative to wind speed while laminar flow is present; bottom: orientation at a certain inflow angle relative to wind speed with separated turbulent flow; for an explanation of symbols see text)

The lifting force F_L (Equation (7.12) or Fig. 7.5), that can be sub-divided into a tangential component in circumferential direction $F_{L,t}$ and an axial component $F_{L,a}$ regarding wind velocity direction and the drag force F_D (Equation (7.13) or Fig. 7.5), that can also be sub-divided into a tangential $F_{D,t}$ and an axial component $F_{D,a}$ are dependent on air density ρ_{Wi}, inflow velocity v_l and the cross-section surface projected to the wind attack surface (in case of two-dimensional treatment of the profile length l and, ideally, of an infinitely thin profile thickness b (Fig. 7.6, right)) as well as on the lift coefficient c_l and drag coefficients c_d.

$$F_L = \frac{1}{2} \rho_{Wi} \, v_l^2 \, l \, c_l(\alpha) \, b \tag{7.12}$$

$$F_D = \frac{1}{2} \rho_{Wi} \, v_l^2 \, l \, c_d(\alpha) \, b \tag{7.13}$$

For a rotating (power generating) wind energy converter, inflow velocity v_I at rotor level $v_{Wi,Rot}$ and circumferential speed of the respective cross-section profile add up vectorially (Fig. 7.5). Angle γ is determined by vector v_I and rotor direction v_R. The inflow angle α represents the difference between angle γ and the angle of the profile δ. If the inflow angle is supposed to be almost the same over the entire rotor blade, the angle of attack of the profile must increase steadily from blade tip to hub as the circumferential speed decreases towards the rotation axis (rotor blade torsion).

Fig. 7.5 additionally illustrates the entire force incident on the rotor blade F_R as the vectorial total of F_D and F_L. The figure also reveals that the tangential force F_T, effective on the rotor blade, is calculated by means of the difference between the tangential components of the drag force $F_{D,t}$ and the lift force $F_{L,t}$.

Coefficients c_l and c_d of Equation (7.12) and (7.13) are predetermined by the rotor profile (shape, surface). Furthermore, they depend on the inflow angle α. The described correlation can also be represented graphically by the so-called Lilienthal polars (Fig. 7.6, left). The figure also illustrates the isolated polar (Fig. 7.6, centre). According to this figure, at a certain inflow angle $\alpha_{operation}$ the combination of both polars serves to determine the lift c_l and drag coefficients c_d. Fig 7.6, left and centre, contains a corresponding schematic representation of these facts.

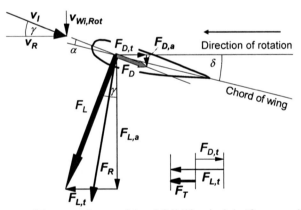

Fig. 7.5 Flow conditions and forces of the airfoil lift principle (for an explanation of symbols see text)

If, by contrast, a vaulted profile shape (Fig. 7.6, right) is used instead of a symmetrical rotor blade profile, lift force F_L, and lift coefficient $c_{l,0}$, are already created for an inflow angle of 0°. Deviating from profile symmetry to increase flow diversion thus increases the lift effect which shifts the profile polar curve and increases the lift coefficient.

Lift coefficient c_l can, for instance, be calculated according to Equation (7.14) for a circular arc profile of arch f and length l (Fig. 7.6, right). The lift coefficient is thus dependent on inflow angle α and angle β between the chord and the arc of the circle.

The relation expressed by the simplified Equation (7.14) applies well to the common profiles if inflow angles are not too elevated. Lift increase – and thus also force F_R resulting from lift and drag forces – is thus linear to the angle of attack and the relative curvature f/l.

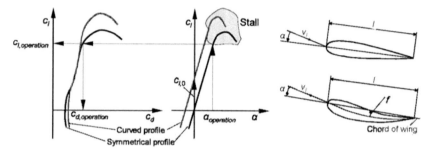

Fig. 7.6 Lilienthal polar curves (left) and isolated profile polar curves (centre) of a symmetrical (right, top) and a curved profile (right, bottom) (for an explanation of symbols see text)

$$c_l = 2\pi \sin(\alpha + \beta/2) \approx 2\pi(\alpha + 2f/l) \qquad (7.14)$$

For wind energy converters, usually almost symmetrical profiles are applied, as they normally produce low drag for low angles of attack ($\alpha - 0°$) and thus almost no lift force. Lift and drag coefficients – and consequently lift and drag force- increase with an increasing angle of attack ($\alpha > 0°$). From a certain point onwards lift is no longer increased but disrupts eventually (so-called "Stall effect"; Fig. 7.4, bottom, and Fig. 7.6, centre). This reveals that the flow incident on the profile breaks down as streamlines can no longer follow the rotor blade contour (Fig. 7.4, bottom). The lift break down involves considerable mechanical strain (i.e. strong shaking) of the rotor – and of all other rotor components, and leads to high material stress and may cause mechanical failures.

Fig. 7.7 on the left once more illustrates the correlation between the angle of attack α and the lift and drag coefficients (c_l respectively c_d) using exact figures. According to the given example the lift coefficient – and thus the lift force – increases up to an angle of attack of approximately 13°, reaches its peak at about 15° and subsequently decreases due to flow break down on the topside of the profile. The drag coefficient, by contrast, reaches its minimum at an angle of attack of –4° and increases almost squarely to both sides.

Fig. 7.7, right shows the polar curves of the profile pertaining to Fig. 7.7, left. The ratio of lift to drag coefficient (i.e. c_d to c_l) is abbreviated as L/D ratio ε. The more flow favourable the profile design, the higher the lift coefficient in comparison to the drag coefficient. The optimum angle of attack is reached at a minimum

L/D ratio and corresponds to an inflow angle of about 0° for the profile under consideration.

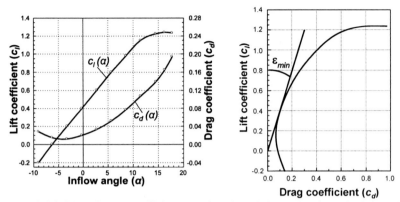

Fig. 7.7 Airfoil lift and drag coefficient as a function of the angle of attack (left) and profile polar curve with marked minimum profile lift/drag ratio (right) (see /7-23/)

The rotor blade receives its driving force F_T in circumferential (tangential) direction from the lift and drag force component ($F_{L,t}$ and $F_{D,t}$) (Fig. 7.5). The rotor torque can be calculated with $F_T = F_{L,t} - F_{D,t}$ multiplied with the effective radius. (Fig. 7.8).

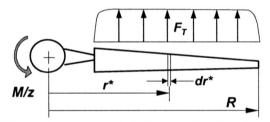

Fig. 7.8 Torque determination at a rotor blade (for an explanation of symbols see text)

By Equation (7.15) the drive torque M may be calculated for a given number of rotor blades z, whereas R is the rotor radius and r a certain point within the radius.

$$M = z \int_{r^*=0}^{R} F_T(r^*) r^* \, dr^* \qquad (7.15)$$

By means of the drive torque M and the number of rotor revolutions n rotor power P_{Rot} is eventually calculated according to Equation (7.16), whereby efficiency η_{Rot} represents twisting and friction losses in comparison to ideal rotor power $P_{Rot,th}$.

$$P_{Rot} = 2\pi n M = P_{Rot,th}\, \eta_{Rot}$$ (7.16)

The tangential force F_T that ultimately creates the power is influenced by the tangential component size of the drag force $F_{D,t}$. Since, especially within the effective outer blade range, the angle γ usually amounts to values below 20°, almost the entire drag force is consumed to reduce the tangential component (Fig. 7.5).

Hence, design and manufacture of wind power stations of favourable power coefficients have to ensure that profile shape and surface roughness produce low drag forces. Modern wind power stations are thus equipped with drop-shaped profiles that are optimised with regard to their L/D ratio ($\varepsilon \approx tan\,\varepsilon = c_d/c_l = F_D/F_L$). Lift/drag ratios ε of aerodynamically favourable profiles amount to 0.02 to 0.08. For aerodynamically inferior profiles (e.g. $\varepsilon = 0.1$) the optimum tip speed ratios and the maximum power coefficient are reduced by approximately 50 %.

The described lift principle almost reaches the Betz power coefficient. This is why practically all commercially available wind energy converters operate according to this principle.

Drag principle. For conversion of the wind energy according to the drag principle (Fig. 7.9) air hits a wind-blown surface S at velocity $v_{Wi,Rot}$ (i.e. projected wind attack surface). The power received P_{Wi} by the wind-blown surface is calculated by means of the drag force F_D and the velocity v_S, at which the wind-blown surface moves (Equation (7.17)). The values of the drag force F_D and the reaction force F_R that slows down the airflow, are equal (Fig. 7.9).

$$P_{Wi} = F_D v_S$$ (7.17)

The relative speed between wind velocity v_{Wi} and the speed incident on the wind-blown surface S ($v_{Wi} - v_S$), have a deciding effect on the air drag. By means of the air drag coefficient c_d, drag force F_D is determined by Equation (7.18). F_D thus largely depends on the square of the effective approach/rotor tip velocity ($v_{Wi} - v_S$).

$$F_D = c_d \frac{\rho_{Wi}}{2}(v_{Wi} - v_S)^2 S$$ (7.18)

The power extracted from the wind $P_{Wi,ext}$ according to the drag principle is calculated by Equation (7.19).

$$P_{Wi,ext} = F_D v_S = c_d \frac{\rho_{Wi}}{2}(v_{Wi} - v_S)^2 S v_S$$ (7.19)

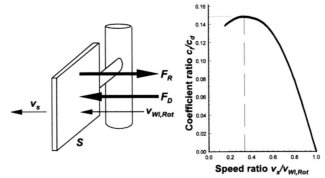

Fig. 7.9 Flow conditions and forces of airfoil drag (left) and shape of the corresponding power coefficient (right) (for an explanation of symbols see text; see /7-4/, /7-5/, /7-6/)

The relation of power extracted from moving air masses to the power contained in the airflow P_{Wi} is expressed by Equation (7.4). Power coefficient c_p is thus deduced from Equation (7.20) (i.e. the relation of extracted power to power contained in the wind).

$$c_p = \frac{P_{Wi,ext}}{P_{Wi}} = \frac{c_d \left(v_{Wi} - v_S\right)^2 v_S}{v_{Wi}^3} \qquad (7.20)$$

Differentiating and zeroing the formula with respect to v_S enables to determine the maximum extractable wind power according to the drag principle. Also these mathematical operations reveal that the maximum power coefficient c_p is achieved if the wind-blown surface is moved with one third of the wind velocity (Fig. 7.9), since, according to Betz, derivation is independent of the wind converter type. For this speed relation a maximum power coefficient $c_{p,max}$ of 14.8 % of the drag coefficient c_d is determined by Equation (7.21).

$$c_{p,max} = \frac{4}{27} c_d \qquad (7.21)$$

For instance, an infinitely large plate has a drag coefficient of 2.01; under these circumstances the maximum power coefficient $c_{p,max}$ would amount to about 0.3. For the rotor blades of drag-type wind turbines maximum drag coefficients of 1.3 are achieved, whereas the corresponding maximum power coefficient amounts to 0.2 or 20 %. The drag principle thus only allows for exploitation of roughly one third of the ideal Betz value of 0.593.

This theoretical principle is even more restricted in practice due to special design requirements. Generally, rotations are required to drive a generator. Consequently, the wind-blown surface needs to rotate around an axis (Fig. 7.9). Usually, several star-type rotor blades are applied. However, wind rotation only advances about half of the rotor, whereas the other half moves in the opposite direction. This is why the opposed half either has to be shielded or has to be designed for a considerably lower drag coefficient than the driven rotor blade side. Only under these conditions a forward force and the resulting driving torque are created. Power conversion (i.e. c_p value) is further decreased. The drag principle thus only applies to very few applications (e.g. cup anemometer).

7.2 Technical description

On the basis of the physical principles outlined for wind energy exploitation, in the following the technical fundamentals of wind power generation are explained. Explanations consider state-of-the-art technology.

7.2.1 Wind turbine design

There is a wide range of different types of wind turbines. The most important features of various concepts include /7-3/:
- rotor axis position (horizontal, vertical),
- number of rotor blades (one, two, three or multiple blade rotors),
- speed (high and low speed energy converters),
- number of rotor revolutions (constant or variable),
- upwind or downwind rotors,
- power control (stall or pitch control),
- wind resisting strength (wind shielding or blade adjustment),
- gearbox (converters equipped with gearbox or gearless converters),
- generator type (synchronous, asynchronous or direct current generator),
- grid connection for power generation plants (direct connection or connection via an intermediate direct current circuit).

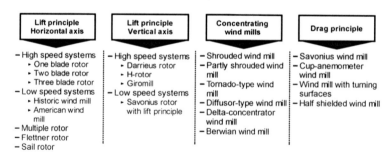

Fig. 7.10 Classification of selected wind power converters (see /7-3/)

The various types of wind energy converters (WEC) can be subdivided into four different groups (Fig. 7.10). Wind turbines operating according to the lift principle are sub-divided into horizontal axis converters (Fig. 7.10, e.g. one, two, or three-blade rotors designed as fast speed or multiple blade converters) and vertical axis machines (Fig. 7.10, e.g. Darrieus rotor or H rotor). Furthermore, there are wind power stations concentrating the airflow (Fig. 7.10, e.g. shrouded wind turbines), as well as wind power stations operating according to the drag principle (Fig. 7.10, e.g. Savonius, cup anemometer).

7.2.2 System elements

Currently and in the near future, almost exclusively grid-connected horizontal (three-blade rotors and, to a very limited extent, also two-blade rotors) hold a predominant market position. The principle plant design is illustrated in Fig. 7.11. A grid-connected wind power station thus consists of rotor blades, rotor hub, gearbox, if applicable, generator, tower, foundation and grid connection. Depending on the respective wind energy converter type further components may be added. Fig. 7.11 also shows the difference between converters with and without gearbox; whereas converters equipped with a gearbox convert the rotor rotations into a higher number of revolutions to apply standardised and less expensive generators, for the gearless converters special generator types must be used that directly operate at a given number of revolutions.

In the past, wind energy converters have mainly been installed on the mainland. However, as average offshore wind speeds are generally higher in comparison to those achieved by mainland installations, offshore installation slowly starts to develop. Since the converter design needs to withstand unfavourable environmental conditions to prevent malfunction, the framework conditions place different requirements on offshore plant technology. For instance, maintenance is more difficult and also more expensive. However, the development of offshore wind energy converters is still very much in its infancy, so that final conclusions cannot yet be drawn.

The following explanations thus focus primarily on mainland wind energy converter technology. However, current developments of offshore wind energy converter installations are also considered.

Rotor. The system component of a modern wind energy converter that transforms the energy contained in the wind into mechanical rotations is referred to as rotor. It consists of one or several rotor blades and the rotor hub (see Fig. 7.11). The rotor blades extract part of the kinetic energy from the moving air masses according to the lift principle. The current maximum efficiency of the kinetic energy of the free flow in relation to the rotor surface (see Equation (7.8)) amounts to 50 %; usually, the so-called aerodynamic efficiency of state-of-the-art rotors amounts to between 42 and 48 % at the turbine design point.

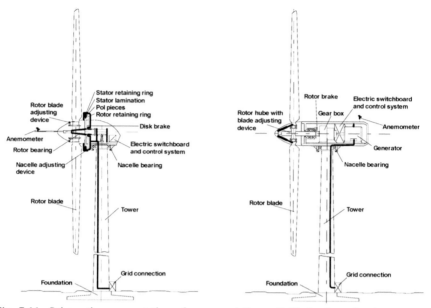

Fig. 7.11 Schematic representation of commercially available horizontal axis converters equipped with gearbox (right) and gearless (left) (see /7-1/, /7-3/)

To prevent exaggerated gear transmission, the number and shape of rotor blades are designed for a relatively high numbers of revolutions to suit the common high-speed generators. Their revolutions generally vary between 1,000 and 1,500 r/min. Alternatively, also low-speed generators directly driven by the rotor (i.e. gearless converters), especially designed for wind energy converters may be applied.

There is an optimum number of revolutions (i.e. turbine design point) for every wind speed at which maximum rotor power is achieved; wind power decreases at lower and higher numbers of revolutions (Fig. 7.12 and 7.16). To maximise useful power and thus the extraction of wind power at a high number of revolutions (ratio of rotor blade tip speed to incident wind speed) (see Fig. 7.16) rotors equipped with a few and narrow-shaped blades are required. They also help keep torsion and friction losses to a minimum. This is why modern wind energy converters are equipped with one to three rotor blades; whereas rotors of more than three blades are virtually not applied for grid-connected power generation. Besides a high number of revolutions (Fig. 7.12) converters provided with only a few rotor blades also save material.

– Three is the lowest number of rotor blades that are still dynamically controllable. Thanks to favourable mass distribution less vibration-dynamic problems are encountered with three-blade rotors than with two or one-blade rotors. Due to tip speed ratios between 6 and 10, blade tip speeds are not particularly high and exaggerated noise is prevented. The smooth running properties also help facilitate the general acceptance of wind energy converters by the population significantly. More than 95 % of the current wind energy converters are three-blade rotors. Even with regard to the current market development, also including the multi-megawatt converters under consideration for offshore installation, the number of rotor blades is not expected to change in the foreseeable future.

– In comparison to three-blade rotors, two-blade rotors save one blade and thus material and costs. On the other hand, the rotor hub has to perform more work, as two blade rotor dynamics are much more difficult to control. Due to the unfavourable mass distribution additional twisting and bending occurs that may be transmitted to the entire converter and may result in higher dynamic stress. Although teetered hubs (see below) are capable of reducing the above disadvantages they require a more sophisticated design and thus lead to additional costs. When compared to three-blade rotors two-blade rotors are characterised by a marginally higher tip speed ratios, varying between 8 and 14, and thus present high blade tip speeds; nevertheless the noise at the blade tips can be kept within certain limits and no longer represents a major problem. Although two-blade rotors are scarcely available in the marketplace they are only of minor importance with regard to the entire available converter range. In Germany, for example, probably only a low percentage of converters is equipped with two rotor blades. Two blade rotors are not expected to be used more widely in the future.

– Although a minimum of material is consumed for one-blade rotors, they require an additional counterweight and a robust rotor hub to compensate for rotor eccentricity. Due to dynamic effects, rotor eccentricity places particularly high requirements on the design and thus increases the cost and propensity toward repairs. However, if the additional technical requirements are fulfilled, the motion of such a rotor is still comparatively uneasy and impairs the scenery. At maximum tip speed ratios of 14 to 16 the blade tip speed is very high and creates high noise emissions. For the described reasons one-blade rotors have not succeeded in the marketplace up to now. It is expected that one-blade rotors will continue to be of very low importance due to the high mechanical stress of the rotor hub and the corresponding requirements placed on the machine design. One-blade rotors also present marginally lower efficiencies when compared to two or three-blade rotors (Fig. 7.16).

Rotor blades. Rotor blades (Fig. 7.11) are usually made of plastic, in individual cases also steel or wood are applied. As plastics in general fibre reinforced material containing glass, coal or aramide fibres is used /7-3/. Up to now, usually glass fibre reinforced plastics (FRP) have been applied. Yet, with increasing plant size, there is a tendency to use coal fibre reinforced plastics. The predominant criterion for material selection is fatigue strength, but also the specific weight, admissible stress, modulus of elasticity and breaking strength. Deciding factors are also the development, material and manufacturing costs resulting from these technical key factors.

Depending on the installed plant capacity rotor blades of common wind energy converters used for power generation usually have lengths of about 5 m, in case of very small wind energy converters, and of about 60 m and above for multi-megawatt wind power stations, for instance required for potential offshore installations. The respective rotors thus cover surfaces ranging from 80 to above 10,000 m^2. In exceptional cases, the surface size may fall below or exceed the above range.

Rotor hub. The rotor hub connects the rotor blades to the rotor shaft (Fig. 7.11). For wind energy converters provided with a blade adjustment mechanism the hub also contains the corresponding mechanics and the blade bearing. For the hub and the pertaining construction besides welded steel sheet constructions primarily cast-steel bodies and forged pieces are applied. The following three different hub designs are distinguished.

– Rigid or hingeless hub. This type of hub is used for three and partly also for two-blade rotors and represents the typical hub for stall-controlled converters (see below). Advantages are low manufacturing and maintenance costs as well as low wear. Disadvantages are relatively high stress of rotor blades and following machine components due to the rigid junction.

– Teetered hub. The teetered hub is a semi-rigid hub design which is partly applied for two-blade rotors. For this hub type rotor blades are suspended in teetered position (gimbal-mounted); they may thus teeter around the (rigid) rotor shaft, considerably reducing asymmetrical rotor loads. Especially in case of a standstill or at very low rotor revolutions, mechanic or hydraulic dampening prevents the blades from teetering too much; however, dampening elements must be designed to securely absorb the enormous strain that may occur at certain wind velocities. This kind of rotor blade suspension also better compensates and reduces the strain resulting from increased twisting and bending.

– Flap and/or lag hinge hub. Rotor blades can also be individually connected to the hub by means of flap hinges. This kind of design is suitable for high-capacity one-blade systems and small multiple-blade converters. Rotor blades are suspended relatively independently from each other and their clamping in shock direction is free from any bending moments. The hinge only needs to absorb centrifugal forces and transmit the torque. Disadvantages of flap and lag hinges are relatively high manufacturing and maintenance costs. This suspension is principally independent from the rotor blade number. However, due to the high design costs and the hardly controllable dynamic behaviour, this kind of hub design is hardly applied in practice.

Blade adjustment mechanism. For common rotor and hub designs, systems provided with rigid and adjustable blades are distinguished. To ensure power and revolution control (Fig. 7.11) wind energy converters of capacities of several 100 kW and above are nowadays usually equipped with a blade adjustment mechanism.

Besides power control the blade adjustment mechanism also transfers the rotor blades into feathered pitch (position without any tangential force), evacuating them from the wind in case of a forced rotor standstill. Since, for the sake of safety, wind energy converters are provided with redundant safety systems (thus equipped with braking features), the blade adjustment mechanism can also be applied for breaking in addition to the mechanical brakes (see also /7-3/, /7-4/, /7-5/, /6-6/).

The main components of the blade adjustment mechanism are rotor blade bearing, blade adjustment drive, energy supply and an emergency adjustment system, if required.

- Rotor blade bearing. Generally, rotor blade bearings hinged at the rotor blade root of the hub. In principle, trunnion and moment bearings are distinguished. Unlike for other typical bearings, this bearing does not need to be optimised with regard to contortions, but regarding static and dynamic strains. Some blade designs only allow for adjustment of the outer blade range (so-called blade tip adjustment). In this case the bearing is located in the outer blade sections.
- Blade adjustment drive. Rotor blades are either adjusted electro-mechanically or hydraulically. For hydraulic systems, for instance, centering actuators are installed inside the rotor hub, which enable rotation either directly or by means of corresponding reversing levers. For systems provided with electric motors blades are adjusted by mechanical components driven by a centralised electric motor (e.g. by head spindles, toothed gearing). For more recent designs each rotor blade is provided with its own drive to adjust every blade optimally towards wind direction.
- Energy supply. The energy supply, also applied for blade adjustment, is located inside the nacelle. To ensure shut-down of the wind energy converter also in case of malfunction, the nacelle is provided with corresponding energy accumulators (e.g. pressure accumulators in case of hydraulic systems and batteries for electric-mechanical drives) (see Fig. 7.11).
- Emergency adjustment system. In case of a power failure or other kind of malfunction the emergency adjustment system safely locks the rotor. Rotors are, for instance, moved into feathered pitch, thus avoiding "racing" of the no-load rotor. Since such safety systems must by all means be redundant, usually additionally a mechanical brake is installed.

Gearbox. To convert the kinetic energy of the rotor into electrical energy, for conventional converters equipped with common four or six-pole synchronous or asynchronous generators, generally revolutions of 1,000 or 1,500 r/min are required when adhering as much as possible to grid specifications (50 Hz). Current rotor revolutions of 10 to 50 r/min with wind energy converters of installed capacities ranging from several 100 kW up to the multi-megawatt range thus require a transmission gear if no specific generators are applied (Fig. 7.11).

In this case the gearbox usually is part of the power train, which joins wind turbine shaft and generator shaft. It divides the drive system into the "slow" and the "fast" generator shaft. The gearbox is located inside the nacelle of the wind energy converter and often also serves as the main rotor bearing.

Currently, single or multiple-stage spur and planetary gears are applied. For blade-controlled converters, spur gears offer the advantage of reaching the rotor hub through the main shaft by means of e.g. feed lines. Disadvantages are larger building mass and the wider nacelle that is usually required. Planetary gears are comparatively more compact and lighter; however, the respective blade adjustment design is much more expensive.

Efficiency amounts to about 98 % per gear level. Energy losses occur due to the inevitable gearwheel friction causing heat transmission and sound emissions. The latter may constitute a limiting factor with regard to wind energy exploitation in view of the acceptance by the population. However, sound emissions may be reduced by appropriate design measures. Particularly transmission of sound waves from the gearbox to the wind energy converter body (i.e. nacelle, tower) has to be avoided in order to prevent system components from acting as resonance bodies. As for the automotive industry, transmission can, for instance, be prevented by a corresponding rubber buffer of the gearbox fastening.

Gearless wind energy converters are also increasingly used. A multi-pole ring generator is applied which is operated at variable revolutions using a direct current intermediate circuit. Since revolution transmission is no longer necessary under the described conditions, these kinds of wind energy converters do not require any gearbox.

Generator. The generator converts the mechanical rotation energy of the power train into electrical energy (Fig. 7.11). For this purpose slightly adapted commercially available generators are used for conventional converters while especially designed three-phase alternators are applied for gearless converters. The main commonly applied generator types are synchronous and asynchronous generators /7-6/.

Synchronous generator. Synchronous generators are equipped with a fixed stator at the outside and a rotor at the inside located on top of a pivoting shaft. In most cases direct current is transmitted to the rotor by slip rings. Direct current creates a magnetic field inside the rotor winding (excitation). When driving the shaft, a certain voltage is created by the rotating magnetic field inside the stator whose frequency matches exactly the rotational speed of the rotating field of the rotor. To prevent expensive maintenance, slip rings are often avoided by the application of so-called brushless synchronous generators, whose pivoting shafts are provided with small rotating exciters.

If a synchronous generator is connected to a stable grid, as for the European grid which is operated at a frequency of 50 Hz, it can only be operated at the number of revolutions pre-determined by the grid (Fig. 7.12). This is not desirable for

for wind energy converter operation as it creates high strain inside the power train, especially during gusty wind. By means of a direct current intermediate circuit or by isolated operation, synchronous generators may be operated at variable numbers of revolutions and frequencies.

Synchronous generators also provide idle power which is required for the operation of various consumers (e.g. motors). When compared to asynchronous generators, synchronous generators are characterised by slightly higher efficiencies.

Asynchronous generator. Asynchronous generators are also provided with a fixed stator and a pivoting rotor. However, excitation (creation of a magnetic field) of the rotor is performed differently. Rotors of asynchronous generators are provided with windings that have direct or shunt short-circuits. When an idle asynchronous generator is connected to an alternating current grid, voltage is induced into the rotor winding, similar to a transformer. The applied frequency is equal to the frequency of the applied voltage. As this winding is short-circuited, there is heavy current flow, so that a magnetic field is created inside the rotor. Since the rotor magnetic field tends to follow the stator magnetic field the rotor is accelerated. The faster the rotor turns, the lower is the resulting relative speed of the rotor winding and the rotating field and thus the voltage induced into its winding. During motor operation, the synchronous number of revolutions will be approached until the weakening rotor magnetic field is still sufficient to compensate for the friction losses of the rotor in idle mode. However, the synchronous number of revolutions cannot be reached as there would be no current induced into the rotor windings, no magnetic field and thus no torque. The specific difference between both numbers of revolutions of the rotor and the rotating field in relation to the rotating field is referred to as slip. Machine operation is thus asynchronous. The more weight is put on an asynchronous generator, the higher is the resulting slip, as higher capacities require stronger magnetic fields. More slip is associated with more induced voltage, more current and a stronger magnetic field. During motor operation, the operating speed is always below, and during generator operation always above the synchronous number of revolutions. Due to these excitation conditions, voltage and current are not in phase, so that reactive power is required. Depending on the respective power, appropriate condensers need to be connected or disconnected. This disadvantage is even more severe for isolated systems. In countries such as Germany or the Netherlands, the respective reactive power required for public grid operation may be supplied by the available power stations equipped with synchronous generators.

This "flexible" operational behaviour (Fig. 7.12) is desirable for asynchronous generators operated in conjunction with inflexible grids (e.g. European 50 Hz grid), to reduce the strain of the wind energy converter, and particularly the power train, during gusty wind. However, without respective adaptations, only very small asynchronous generators present a slip of up to approximately 10 %. Slip decreases with an increasing machine size. Common generators with capacities varying between several 100 kW and the multi-megawatt range only present slip from 0.5 to 1 % /7-6/ and are thus almost as inflexible as synchronous generators.

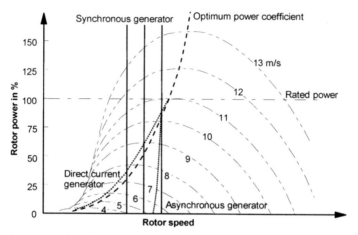

Fig. 7.12 Power rotational speed mapping of a typical wind energy converter including generator characteristic curves (see /7-3/, /7-4/, /7-6/)

However, by incorporating appropriate resistors into the rotor circuit, the slip is intentionally enhanced, but heat losses are increased and efficiency is reduced. Direct incorporation of the resistors into the rotor circuit triggers airflow and thus provides cooling. Since the air sucked in is saline, particularly at coastal sites, this design is prone to corrosion of the winding insulation. Currently, research is being conducted on outer rotor resistors which allow for a closed design of the actual generator.

A further possibility to influence the slip of asynchronous generators is the so-called double loaded asynchronous generator. Slip capacity is fed into or obtained from the grid by a frequency converter, whereas the stator is directly connected to the grid and the rotor is connected via the frequency converter. A modern insulated gate bipolar transistor allows for dynamic slip control and thus enables variable numbers of revolutions and idle power generation. A hybrid solution is an asynchronous generator in the form of an oversynchronous static Kraemer system (cascade conversion system). Slip power is unidirectional and can only be fed into the grid (Table 7.1) /7-6/.

Asynchronous generators also serve for motor start-up of wind energy converters. They are usually less expensive, more robust and require less maintenance than synchronous generators.

Wind direction yaw mechanism. This system component serves for adjusting the machine nacelle, and thus the rotor, as exactly as possible to the respective wind direction. The wind direction yaw mechanism joins the machine house (nacelle) and the tower head, as its components are incorporated into both system components (Fig. 7.11).

The nacelle is usually adjusted to the respective wind direction by a gear wheel mounted on top of the tower and operated by mechanical, hydraulic or electro-mechanical adjustment mechanisms. Small wind energy converters, rarely built nowadays, are provided with mechanical yaw mechanisms driven by wind vanes, servomotors or small size windmills. Bigger converters are usually provided with hydraulic, electromotive or electro-mechanical servo drives and are characterised by lower costs, smaller size and bigger torque at comparable construction costs.

All converters are additionally equipped with a stopping brake to lock the respective rotating mechanism. The brake compensates for low fluctuations in wind direction that may exert strain on the rotating mechanism and thus reduce its technical service life. It also permits to lock the nacelle during prolonged downtimes (e.g. during maintenance).

For bigger converters, the azimuth or tower head bearing is designed as anti-friction bearing, whereas small converters are provided with friction bearings with sliding (e.g. plastic) elements. The entire wind direction yaw mechanism is controlled by a special control system that receives all relevant data from a wind direction measuring device mounted at the nacelle shell.

Tower. The main function of the tower of a horizontal axis converter is to enable wind energy utilisation at sufficient heights above ground, to absorb and securely discharge static and dynamic stress exerted on the rotor, the power train and the nacelle into the ground (Fig. 7.11). Another key factor regarding tower dimensions and design is the natural vibration of the tower-nacelle-rotor overall system in view of the prevention of dangerous resonance, particularly during rotor start-up. Further influencing factors are dimensions and weight regarding transport requirements and thus available roads, erection methods, cranes and accessibility of the nacelle as well as long-term properties such as weathering resistance and material fatigue.

Most towers are made of steel and/or concrete. As far as steel constructions are concerned, besides the lattice towers usually observed for dated converters, there are also anchored and self-supporting tubular steel towers in closed, commonly conic design; the latter being the most common tower type applied nowadays.

The minimum tower height is determined by the rotor radius. Any additional tower height is a(n economic) compromise between the increased costs at enhanced heights and the increased mean wind speeds and thus increased power yield. Hence, the optimum between maximum energy yield and acceptable tower costs needs to be determined. This is why currently tower heights vary considerably with regard to site conditions; common tower heights vary between 40 and 80 m. On the mainland, due to generally lower wind speed increase at enhanced heights, when compared to coastal sites, usually higher towers (e.g. of heights of 90 or even 100 m or above) are built.

At equal power yield offshore installation of wind energy converters allows for a reduction of hub heights by approximately 25 %, when compared to onshore installation, due to different wind conditions at increased heights above ground (i.e. the mean wind speed at sea increases faster at enhanced heights above ground than on the mainland). Decreased hub heights correspondingly reduce tower costs.

Foundations. The type of foundation used to anchor towers, and thus wind energy converters, into the ground depends on the plant size, meteorological and operational stress and local soil conditions. On principle, support structures are subdivided into shallow and deep foundations. Both are state-of-the-art technologies but differ considerably with regard to costs. The optimum foundations design is determined by appropriate soil investigations.

Anchoring wind energy converters on the coastline is much more costly (see also /7-7/, /7-8/). There are various technologies that ensure stability. Foundation technologies include the support structures (i.e. foundation structure plus tower) and the technology required for anchoring the converters on the ocean floor. Installation is generally aimed at low manufacturing costs (e.g. by series production, material selection), low assembly costs (with regard to logistics, fast installation) and a long service-life (considering factors such as corrosion and fatigue).

Currently, floor-mounted support structures are preferred for water depths below 50 m. Floating support structures are also technically feasible, but will most probably be applied in areas with water depths exceeding by far 50 m. The exact dimensions of such floor-mounted support structures depend, among other factors, on the expected wind, wave and ice charges and geographical site conditions (e.g. water depth, soil conditions).

Floor-mounted support structures (Fig. 7.13) are subdivided further into gravity, monopole, and tripod foundations, outlined as follows. A forth type, currently of less importance, consists of a four-pile trelliswork construction, similar to the lattice towers applied for onshore wind energy converters.

Gravity foundation (Fig. 7.13, left). The gravity foundation principle is based on gravity force utilisation. Gravity foundations consist of concrete or steel frames which are filled with ballast on site. Foundations are put on the seabed on a levelled surface provided with a compensating layer to prevent transmittance of tensile forces onto the seabed and to make the system sensitive for extreme hydrodynamic loads. Since the maximum wave height, which partly influences the forces effective on the foundation, depends on the water depth, among other factors, larger foundations are required with increasing depths. According to current economic knowledge, this technology should only be applied up to a sea depth of 10 m. From a physical point of view, it should only be applied up to 20 m. Gravity foundations have, for instance, been applied for the offshore wind parks of Vindeby and Middelgrunden (both located in Denmark); the mass of such a foundation amounts to approximately 1,500 t at a water depth of about 5 m and an installed wind energy converter capacity of approximately 1.5 MW.

Monopile foundation (Fig. 7.13, centre). This foundations type consists of a single foundation pile (monopole), which quasi extends the wind energy converter tower up to the seabed. Depending on the plant size and support structure design its diameter varies between 3 and 4.5 m and its mass between 100 to 400 t. Installation on the seabed is performed by pile-driving, vibrations, and drilling, whereby the penetration depth varies between 18 and 25 m. The actual tower of the wind energy converter and the foundation pile are subsequently connected by a joining element to compensate for a possibly oblique position of the foundation pile. From a current viewpoint, the maximum water depth, up to which the installation of monopile foundation is sensible, is reported to be 25 m. Monopile foundation technology seams comparatively economic for water depths up to 20 m and favourable ocean floor conditions and has, for instance, been applied for the offshore wind parks of Bockstigen and Utgrunden (both located in Sweden).

Tripod foundation (Fig. 7.13, right). Tripod foundations consist of a central column element connected to the tower of the wind energy converter and a three-legged three-dimensional steel truss, which transmits the forces and the torque effective onto the construction corners. The diameter of the foundation piles anchored into the seabed by pile-driving, drilling and vibrations, amounts to about 0.9 m. Depending on the respective soil conditions, the desired seabed penetration depth varies between 10 and 20 m. Tripod foundations are most suitable for water depths exceeding 20 m, as landing water vehicles may collide with the steel truss submerged into the water at water depths below 7 m. To date, the described tripod foundations have not yet been applied for offshore wind energy converters. However, their application at a water depth of about 30 m is planned for the first German offshore wind park "Borkum West", currently being planned for the North Sea. The mass of such a foundation designed for the planned multi-megawatt converters amounts to approximately 800 t.

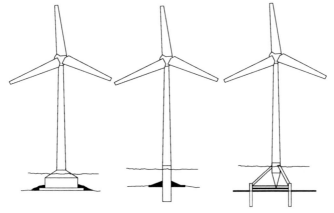

Fig. 7.13 Floor-mounted support structure of offshore wind energy converters (gravity foundation (left), monopile (centre), tripod (right); see /7-7/, /7-8/)

Grid connection. With regard to the connection of wind energy converters to the public power grid or any isolated power grid, direct and indirect grid connections are distinguished /7-6/; for both types asynchronous and synchronous generators are suitable (Table 7.1).

- For direct connection to an invariable frequency power grid, as it is the case for public European supply, synchronous generators turn at a constant number of revolutions and asynchronous generators at an almost constant number of revolutions according to the grid frequency (Fig. 7.12). Due to the inevitable "hard" connection, particularly in case of synchronous generators, high dynamic stress may be exerted onto the power train (hub, shaft, gearbox and generator rotor). This is why in most cases asynchronous generators are applied for direct grid connection.

Table 7.1 Comparison of direct and indirect grid connection with regard to the applied generator type (see /7-6/)

	Synchronous generator	Asynchronous generator
Direct grid connec-tion	$n_G = f$ constant number of revolutions; hard grid connection	$n_G = (1 - s) f; \ 0 \geq s \geq -0.01$ slightly declining number of revolutions that decreases with increasing converter size; simple grid synchronisation; reactive power consumer; relatively hard grid connection
Indirect grid connec-tion	$0.5 f \leq n_G \leq 1.2 f$ variable number of revolutions; grid connection via a rectifier with connected inverted rectifier (i.e. intermediate direct current circuit or direct converter); soft grid connection	$0.8 f \leq n_G \leq 1.2 f$ variable number of revolutions; squirrel cage induction machines via direct current intermediate circuit or direct converter (reactive power consumer); slip ring induction machine via dynamic slip control, oversynchronous static Kraemer system (both reactive power consumers) or double loaded asynchronous generator with direct stator and indirect rotor connections (e.g. via direct current intermediate circuit) (reactive power generation); soft grid connection

n_G generator number of revolutions; s slip (deviation from nominal number of revolutions); f grid frequency.

- Indirect grid connection converters may be connected via a direct current intermediate circuit, which allows for the operation of wind energy converters at a variable number of revolutions and generates alternate current at variable frequencies. Current is first converted to direct current by a rectifier and subsequently reconverted into alternate current by an inverted rectifier to match the voltage and frequency specifications of the power grid. This allows for optimum aerodynamic operation of the rotor within a revolution range from 50 to 120 % of the nominal number of revolutions (Fig. 7.12). The variable number of revolutions reduces the dynamic stress exerted on the converter. However, the direct current intermediate circuit incurs in additional costs and increases electrical losses. Grid connection via direct current intermediate circuits is common practice for medium to large converters, preferably in conjunction with synchronous generators.

For older converters equipped with direct current intermediate circuits, frequently inverted rectifiers have been used, which in some cases produced considerable overtones. They may interfere with the operation of other equipments in weak power grids. However, recent investigations in terms of power semiconductors have led to the use of inverted rectifiers, which supply current with relatively little distortion and are also partly suitable for idle power provision (e.g. IGBT inverted rectifiers with pulse-width modulation (PWM)).

Provided that the required network specifications are strictly adhered to, wind power converters may also be indirectly connected to the grid via direct converters.

Wind power converters can either be connected to the grid as individual converters or in the form of a wind park. The connection grid interference caused by the converter or the wind park has to be determined at the respective grid connection point. Short-term power fluctuations, sensitively perceived as flickering lights by the human eye, but also continuous voltage changes and possible overtones need to be considered. Fluctuations may be measured by the ratio of converter power to grid short-circuit power at the grid connection point. If certain values are exceeded, connection is only possible at a connection point of superior grid short-circuit power (e.g. at the bus bar of a substation), in order not to interfere with other consumers connected to the grid.

The main components of a grid connection are the electric coupling of the wind power converter or the wind park to the transformer, a transformer (if required) including a distributing substation equipped with medium voltage substation as well as a medium voltage connecting line up to the grid connection point.

The design of every wind power converter including control and protection equipment has to rule out all potential damage to the grid due to a failure of the wind power converter (such as power failure or short-circuit). Complete disconnection also has to be ensured for the safe performance of operation and maintenance.

Losses occur when power generated by wind energy converters is fed into the grid. They are mainly caused inside the transformer when power is converted into heat. However, they are comparatively low and at the most amount to a few percent.

System aspects of offshore installation. The available technologies need to be adapted to suit the changed framework conditions of offshore operation, when compared to onshore installation. Technologies may also be optimised accordingly.

Since onshore wind energy converters are in most cases optimised with regard to minimum noise creation, and as blade tip speed is the deciding parameter regarding noise emissions, the maximum number of rotor revolutions is subject to certain restrictions with increasing rotor diameters. If noise characteristics are less important, as it is likely for offshore wind energy converters, a higher numbers of rotor revolutions should be admissible. These would result in decreased tower head mass (i.e. power train mass) due to the reduced driving torque and would also help cut costs. However, considerably enhanced blade tip speeds may damage rotor blades due to erosion at high concentrations of particulate matter in the air. A high content of water drops and salt particles in conjunction with a relatively high air humidity and throw of spray, as it can be expected for offshore installations, additionally require effective protection of all offshore wind power converter components against corrosion and detrimental deposits.

Electric and electronic system components (such as operating control, sensors, generator, and transformer) require special protection against spray and deposits. For this purpose hermetic protection against ambient air or an installation under slight excess pressure, ensuring appropriate air-conditioning, appear suitable. However, the climate has to be controlled with regard to temperature and humidity to prevent overheating and moisture condensation.

The design of offshore wind energy converters includes the installation of hoisting equipment on board to remove and insert all major components, such as generator, gearbox, if applicable, etc. without expensive external hoisting equipment, from and into the nacelle. Furthermore, each offshore wind energy converter needs to be provided with a landing platform to enable personnel landing and material provision.

To ensure plant safety and a safe start-up in the case of power failure, an emergency power generation unit must be provided. Available options include batteries or power stand-by units for electro-mechanical systems, hydraulic accumulators for hydro-mechanical systems and spring brakes for exclusively mechanical systems. However, such short-term accumulator units only secure shut-off of wind energy converters. Special stand-by units are required for prolonged power failures.

To achieve the same plant availabilities as for onshore converters, in spite of the meteorologically more difficult plant accessibility, particularly during winter months, quality and robustness of all plant components need to be ensured over long periods under hard environmental conditions. For this purpose a highly sophisticated operating control system is required. The system should allow for an early detection of damages, start itself after possible grid damages and revert the wind energy converter to normal operation. Remote complete reprogramming and re-initialisation should also be possible from the onshore control panel. For this purpose, effective tele-monitoring and reliable communication technologies are required. Furthermore, the converter should be adequately equipped with all major sensors (such as vibration monitors and temperature sensors) and all indispensable system components.

7.2.3 Energy conversion chain, losses and characteristic power curve

Energy conversion chain. Wind energy utilisation converts energy extracted from moving air masses into electrical energy. Conversion usually involves several steps as illustrated in Fig. 7.14.

As shown in the illustration, kinetic energy of moving air masses is first converted into rotation (of the rotor) and thus into mechanical energy of the power train. Power trains of conventional wind energy converters include an inserted mechanical gearbox to increase the number of revolutions for the synchronous or asynchronous generator which generally require a higher number of revolutions than the rotor. Yet, there are also wind energy converters whose generator is adapted to the feasible number of revolutions and which do not require any gearbox (Fig. 7.14). Subsequently, the mechanical energy of the power train is converted into electrical power by a mechanical-electric converter (generator). Since the specifications of the generator outlet do not necessarily match the grid specifications, in most cases a second electric/electric converter is needed. The simplest version consists of a transformer; however, an indirect grid connection by either a direct current intermediate circuit or a direct converter is suitable.

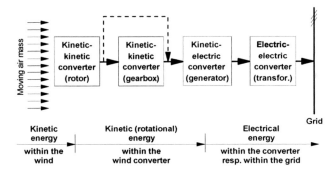

Fig. 7.14 Energy conversion chain of a wind energy converter (transfor. Transformator, resp. respectively; see /7-1/)

Losses. The different conversion steps are illustrated in Fig. 7.14. According to this they are subject to several loss mechanisms which may reduce the overall efficiency significantly in comparison to the Betz power coefficient of 59.3 %. Commercially available wind energy converters only transform 30 to 45 % of the energy contained in undisturbed wind into electric power. The difference between the maximum physical efficiency and maximum values that can be achieved at present is caused by a series of different and unavoidable losses which are inherent to the commercially available wind energy converters and any other type of energy conversion plant (Fig. 7.15).

The electric power output available at the generator outlet of a wind energy converter is determined by the power quantity contained in the wind minus aerodynamic, mechanical and electric losses. Net energy yield may additionally be re-

duced by the energy consumption of accessories such as the wind direction yaw
mechanism and the blade adjustment mechanism. The different loss mechanisms
are explained in the following.

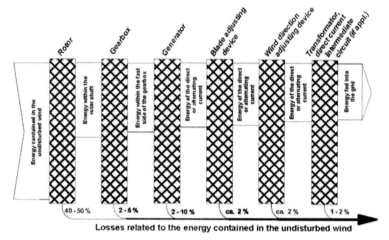

Fig. 7.15 Energy flow through a wind energy converter (see /7-1/)

– Aerodynamic losses are attributable to the blade shape that is never optimal
  with regard to the total surface covered by the rotor; aerodynamic losses are
  considered by the real power coefficient (i.e. the power portion contained in the
  airflow which can be extracted from the wind with regard to the ideal power
  coefficient and the given losses of a determined wind energy converter). Power
  coefficients are first and foremost determined by the number and the shape of
  rotor blades (and thus by the blade tip speed/wind speed ratio) and thus differ
  tremendously with regard to rotor designs. Fig. 7.16 shows the power coefficient c_p in relation to the ratio of blade tip speed at the outer rotor edge to the
  current wind speed λ. This ratio produces the characteristic curves $c_p(\lambda)$, typical
  for wind rotors of different designs. The most important parameters are
  summarised as follows:
  • number of rotor blades,
  • aerodynamic profile properties and
  • twisting of rotor blades /7-3/, /7-4/, /7-5/.
  Fig. 7.16 shows the tremendous variations of power coefficients of the illustrated rotor designs. The clear difference between the maximum power coefficient according to Betz and the maximum power coefficient of an ideal wind
  turbine is due to twisting losses. Fig. 7.16 also illustrates the advantages of
  high-speed propeller-type converters (i.e. wind energy converters of a large
  number of revolutions and only a few rotor blades, i.e. one, two or three-blade
  rotors) compared to low-speed converters whose twisting losses are considerably higher (i.e. converters with a low number of revolutions and many blades,
  i.e. Dutch windmills, American Western rotors). The respective maximum

power coefficient that may amount to almost 50 % at the turbine design point has a favourable effect on high-speed wind energy converters. Thanks to the flatter curve – in comparison to slow wind energy converters – of the characteristic lines $c_p(\lambda)$, the relatively high power coefficient is furthermore maintained within a relatively wide range of the tip speed ratio. For converters equipped with only a few rotor blades, deviations from the maximum tip speed ratio only slightly reduce the power coefficient. All in all, according to state-of-the-art technology, modern two or three blade rotors yield the highest power coefficients of all rotors built to date.

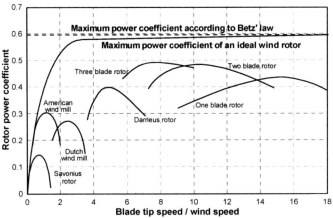

Fig. 7.16 Power coefficient-blade tip speed/wind speed ($c_p(\lambda)$) curve of different wind energy converter designs (see /7-3/)

- Mechanical losses are mainly due to friction losses and the resulting heat creation inside the rotor shaft bearings and, if a gearbox is provided, the rotating speed conversion which is prone to losses.
- Electrical losses comprise conversion losses inside the generator, power losses of the grid and, as the case may be, losses incurred during power conversion by the direct current intermediate circuit (losses attributable to semi-conductors, throttle valves, etc.) or the direct converter. Depending on converter dimensions, additional losses may result from electrical energy conversion by the transformer.

Characteristic power curve. The power yield of wind energy converters, for instance on the basis of the average power generated within 10 minutes, can be determined by means of a converter-specific characteristic power curve. It reveals the dependency of the average electrical power from the respective average wind velocity and thus shows the operational characteristics of the converter. With regard to operational characteristics, four different phases (Fig. 7.17) can be distinguished.

– Phase I. If the respective wind speed is lower than the converter-specific minimum speed, the converter will not start. This applies in particular to high-speed converters that only have a small blown surface when the converter is idle. The power of the useful speed difference is insufficient to surmount the converter's friction and inertia forces and to enable converter operation. Thus, there is no electrical power generated at the generator outlet.

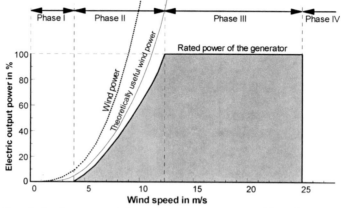

Fig. 7.17 Correlation between wind speed and generator capacities of typical marketable horizontal axis converters (see /7-1/, /7-3/, /7-4/, /7-5/)

– Phase II. If the airflow speed exceeds the wind speed required for start-up, the converter will start and generate electrical energy. The theoretical useful wind power increase is proportional to the wind speed in its third power. Yet, the useful electrical energy at the generator outlet is not exactly proportional to the theoretic useful energy as losses which are not linear to speed (e.g. aerodynamic friction losses) occur within this range of the characteristic power curve. Power control also needs to be considered up to the nominal wind speed of the wind energy converter and thus up to the nominal capacity of the installed generator. For this operation electric original power is equal to the product of the aerodynamic, mechanical and electric efficiency and the total power contained in the wind. For commercially available converters this phase covers a range from a cut in wind speed of approximately 3 to 4 m/s up to a nominal wind speed of 12 to 14 m/s.
– Phase III. Due to the limited generator capacity, matching the respective converter dimensions, the power absorbed by the rotor of a given wind energy converter must not exceed the installed nominal generator power over an extended period. For wind energy quantities that exceed the nominal wind speed and that are below the cut out wind speed of the converter, which theoretically enable energy absorption beyond the installed capacity, an appropriate control must ensure that the rotor axle transmits at most the installed generator capacity to the generator (Section 7.2.4). Within this wind speed range the yielded elec-

tric power is thus almost equal to the installed generator capacity. The cut-out
wind speed which protects the operation phase from higher wind speeds ranges
from 24 to 26 m/s.
- Phase IV. If the wind speed exceeds a certain speed limit determined by the
converter design and type the wind energy converter must be shut down to prevent mechanical deterioration. Under the described meteorological conditions
no electric power will be yielded.
On the basis of such a characteristic power curve (Fig 7.17) the electric power
generated by means of a wind energy converter within a given period of time (Fig.
7.18) can be determined if the corresponding frequency distribution of the wind
speed (Chapter 2.3.2; Fig. 2.35) is a known quantity. The frequency distribution
refers to the probability of the occurrence of a certain wind speed or a defined
wind speed interval within a given period.
    Electric power output of a wind energy converter E_{WEC} can thus be calculated
by Equation (7.22), whereas h_i represents the wind occurrence probability within
a certain speed interval i within the course of period t. $P_{el,i}$ indicates the electric
power corresponding to the defined wind speed interval i according to the characteristic power curve. The entire energy output is thus equal to the total of the respective interval-specific product of the available wind quantity and corresponding power within a given period over all observed wind speed intervals.

$$E_{WEC} = \sum_{i=1}^{n} h_i P_{el,i} t \qquad\qquad (7.22)$$

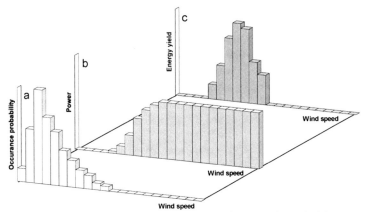

Fig. 7.18 Determination of the energy yield (c) within a certain period by means of the
wind speed frequency distribution curve (a; see Fig. 2.35) and the characteristic power
curve (b; see Fig. 7.17) (see /7-4/)

7.2.4 Power control

Wind energy converters require appropriate control mechanisms to limit power extraction at higher wind speeds (Fig. 7.17). Power controls prevent mechanical deterioration of the rotor and are also required due to the capacity (thermal) limitation of the generator (i.e. matching the installed electric capacity).

On principle, power and speed controls need to be distinguished /7-3/. If the number of revolutions must be kept constant, or almost constant, power has to be controlled accordingly. It must not exceed the installed generator capacity, as the latter would become thermally overloaded and deteriorated in the end. If, by contrast, the number of revolutions is variable within certain limits (Table 7.1) the maximum number of revolutions must not be exceeded to prevent mechanical deterioration of the rotor and other movable components. Moreover, the capacity must be monitored.

Currently, two different control methods are applied for commercially available wind energy converters to limit extractable wind power. They are referred to as stall and pitch controls. Both methods are suitable to limit the power absorbed by the rotor.

Stall control. Power absorption from wind can be limited by the so-called stall effect (intentional flow separation) (Section 7.1). For this purpose wind energy converters must be connected to a sufficiently strong grid and have to be operated at a constant number of rotor revolutions – regardless of the actual wind speed. Due to the described operational changes, inflow conditions at the rotor – at a constant number of rotor revolutions and its individual blades – change to an extent that airflow breaks down at certain (elevated) wind speeds (Fig. 7.19); because of the resulting eddy currents, the rotor slows itself down or maintains the effective torque at a constant level.

At wind velocities above the cut-in wind speed and below the nominal wind speed, the lifting force at the rotor blade required for rotor drive is obtained by the airflow effective on the profile. The aerodynamic angle of attack α between effective approach (inflow) velocity vI and the rotor profile chord increases with increasing wind speeds and a constant, or almost constant, number of rotor revolutions.

When the nominal wind speed range is reached, the angle of attack becomes so high that, due to the strong deviation, the airflow can no longer follow the surface contour. The flow stalls from the upper profile side (suction side; refer to Fig. 7.19). Because of the airflow stall the rotor lift is reduced; thus power extraction from the wind can ideally be kept at a constant level.

Flow stall at a rotor blade does not always occur at the angle of attack measured at stationary airflow for a profile (so-called static stall). The phenomenon of flow stall rather depends on the progress of the angle of attack (e.g. during gusts) and on the three-dimensional flow incident on the rotor blade (radial flow induced by centrifugal force) (so-called dynamic stall). Both may delay the stall, i.e. air-flow only breaks down at higher angles of attack. Thus, the maximal absorbable air forces or the nominal capacity may be exceeded, overloading the wind energy converter and particularly the generator. Another effect is the dynamic stall that occurs for a short while over a certain angle range of the rotor rotation. The result is a periodically dynamic stall due to a cyclical alternation of the angle of attack (e.g. while performing one turn in the bottom boundary layer). Periodic air forces may stir up the wind energy converter to such an extent that the rotor may burst. Thus, rotor blades need to be dimensioned accordingly to prevent mechanical deterioration of the converter, and of the rotor blades in particular, within the course of its technical service life (even intentional stall creating eddy currents to limit capacities may exert considerable strain on the rotor blades). Due to the discussed, not fully deterministic stall effect, generator capacities can only be maintained within certain ranges.

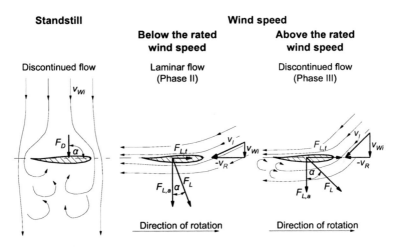

Fig. 7.19 Flow conditions present at a rotor blade profile of a stall-controlled wind energy converter (for an explanation of symbols see text; phases refer to Fig. 7.17; see /7-5/)

The characteristic power curve of a stall-controlled wind energy converter is shown in Fig. 7.20. For wind speeds a little above the nominal wind speed its power yield considerably exceeds the installed nominal generator capacity (i.e. planned generator overload up to approximately 110 %). At wind speeds within the cut out wind speed range, electricity output, by contrast, is a little below the nominal generator capacity.

Besides stall control, a so-called active stall control has established itself in the marketplace which allows transferring the rotor into stall position by means of a blade angle adjusting device (i.e. adjusting the blade angle β towards smaller angles of attack). Active stall controls ensure a smoother shape of the characteristic power curve similar to that of a pitch-controlled converter.

Pitch control. Due to the outlined disadvantages of stall controls, especially for large wind energy converters, pitch controls are applied to control wind power extraction. By twisting the rotor blades flow conditions and thus the air forces incident on the rotor blade are influenced to keep the wind power absorption by the rotor at speeds above the nominal wind speed almost at a constant level (Fig. 7.20).

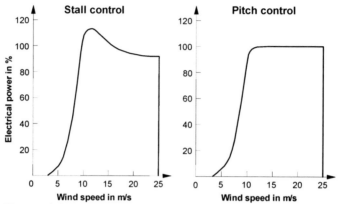

Fig. 7.20 Characteristic power curve of stall-controlled (left) and pitch-controlled (right) wind energy converter

Blade adjustment enables to continuously adjust the angle of attack of the rotor blade to the airflow, so that rotor power absorption can be controlled. The blade angle, at which the profile chord corresponds to rotor level at a blade radius of 70 % is referred to 0° setting angle. At a setting angle of approximately 90° the rotor blade reaches its so-called feathered pitch.

Within the normal operation range of a wind energy converter, airflow is always incident on the rotor blade; thus, pitch control avoids stall effects. This is especially true if the blade angle is adjusted towards smaller angles of attack for power control. In most cases, the adjustment area covers the range from 90 to 100° to reach the so-called feathered pitch from the operating angle (almost 0°). In feathered pitch position the blade resembles a wind vane and produces no or very little rotor rotation.

The blade adjustment device triggers rotor rotation when the setting angle is transferred to starting position (for instance at 45°). At an increasing number of revolutions the blade angle needs to be adjusted continuously to the optimum operating angle in order to maintain a positive, lifting angle of attack in spite of the increased circumferential speed.

Once the optimum operating angle has been reached, blades are usually no longer adjusted until they reach nominal speed, although better setting angles might exist. If wind speed increases further, this installed nominal generator capacity is extracted by continuous twisting of the rotor blades. This measure maintains the installed generator capacity comparatively well (Fig. 7.21).

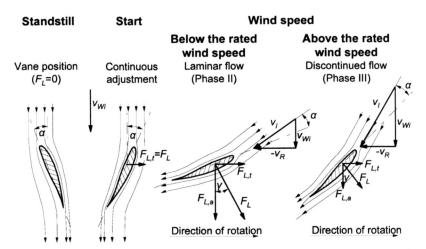

Fig. 7.21 Flow conditions at the rotor of a pitch-controlled wind energy converter (for an explanation of symbols see text; phases refer to Fig. 7.17; see /7-5/)

For isolated grids which, unlike grid-connected operation, are not necessarily aimed at achieving a maximum energy output, blade adjustment offers the additional ability of controlling power output by blade adjustment and of adjusting the power yield to the current energy demand (i.e. on principle, demand-side converter operation is possible).

Due to demand or grid-related reasons (for instance, in case of isolated grids with high wind energy converter capacities) the energy fed into the grid by the generator may be inferior to the theoretical capacity to be achieved according to the characteristic power curve. At decreasing flow speed the capacity may be maintained within certain limits by readjustment measures. Thus, like other types of power-controlled power plants, wind energy converters can also be controlled to a certain extent.

In comparison to stall control, pitch control allows for intentional and relatively smooth shut-down of the converter upon exceeding cut out wind speed (transition from phase III to phase IV; see Fig. 7.17). Pitch control avoids abrupt transition from the installed capacity to zero and thus prevents the resulting high mechanical strain exerted on the converter and on the power grid or conventional reserve capacity power stations.

7.2.5 Wind parks

Wind park design. Wind energy converters may be installed individually in exposed positions, for instance at the hilltops of low mountain ranges with free airflow, in rows (e.g. converters positioned along a dike) or in groups (e.g. positioned in lines, one behind the other). For the latter two variants certain minimum distances, depending on the respective site conditions, need to be observed, to

prevent reciprocal shut-down of the converters and to ensure relatively undisturbed wind conditions for every converter. Shadowing refers to the subsumed
effects of closely arranged wind energy converters that take away the wind from
each other and to the enhanced dynamic strain placed on downwind installed wind
energy converters due to the increased turbulence within the downwind flow of
the converters placed in front.

On principle, with regard to minimised shadowing there are two different arrangements of wind energy converters within a wind park of limited space. Besides optimised converter arrangement towards a preferred wind direction there is
also the possibility of an optimised converter installation without a preferred airflow direction (Fig. 7.22), maintaining a certain distance between the individual
converters to even out the imbalance between the reduced airflow speed due to
energy extraction by the rotor and the undisturbed airflow. For the following converter almost undisturbed wind conditions may be assumed. The required distance
between the individual converters depends on meteorological, topographical and
other site-specific conditions (such as legal or administrative restrictions) and may
vary widely.

The minimum distance between two converters standing next to each other is
referred to by the distance factor k_A which is defined as the ratio of converter distance to rotor diameter. The required distance between the individual converters is
thus a multiple of the rotor diameter.

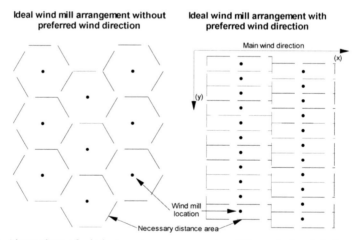

Fig. 7.22 Alternatives of wind converter arrangements within a wind park (schematic representation; see /7-1/)

If a determined site has a preferred wind direction, and provided that its topographic conditions are favourable to the installation of wind energy converters (e.g.
plains near the coast), converters may be arranged in several rows behind each
other (Fig. 7.22, right). Since under the described conditions wind mainly blows
from one direction, shadowing effects only need to be minimised for the main

wind direction. Depending on the specific site conditions, the respective distance factor varies between 8 and 10, with regard to the main wind direction $k_{A,x}$, and between 4 and 5 crosswise to the main wind direction $k_{A,y}$. Due to space restrictions as well as for economic purposes also lower distance factors $k_{A,x}$ of about 4 are possible. The minimum space required around a wind energy converter A_{WEC} is thus calculated by Equation (7.23), whereas d_{Rot} refers to the rotor diameter of the converter.

$$A_{WEC} = k_{A,x} k_{A,y} d_{Rot}^2 \qquad (7.23)$$

If there is no preferred wind direction, as this may be the case for the inland, and if there are no topographic restrictions impeding optimised converter arrangement, shadowing must be optimised for all cardinal directions. Around any wind energy converter an approximately circular area should thus be reserved, which, for the sake of simplicity, is represented by an isosceles hexagon. The distance factor k_A to be observed for this kind of converter arrangement varies between a range similar to that of converter arrangement with preferred wind direction (i.e. according to site and wind conditions between 6 and 15). The area to be reserved to meet the above requirements is calculated by Equation (7.24).

$$A_{WEC} = \sqrt{\frac{3}{4}} \left(k_A d_{Rot}\right)^2 \qquad (7.24)$$

If the site-specific optimum distances between the individual converters are observed, shadowing is minimised while space consumption is optimised. The losses incurred when compared to undisturbed operation are indicated by the wind park efficiency; depending on the specific site conditions efficiencies vary between 90 and 98 %. In spite of the losses caused by shadowing, regardless whether optimum distances are observed, installing converters in the form of wind parks is usually more effective, since within the scope of an economic overall analysis the described losses are usually overcompensated for by grid connection, access roads and the generally low mean costs for maintenance, repair and monitoring of losses.

With regard to offshore wind parks, located far off the shoreline, optimum wind park sizes exceed by far the size of onshore wind parks. This is due to the high grid connection costs. Costs are rather determined by the length of the required underwater cables than by the connected capacity /7-9/. For the actual dimensioning of offshore wind parks the generally higher wind speeds and lower turbulence must be considered. However, experience has shown that the reduced mean wind speed and the increased turbulence within the downwind flow of offshore wind energy converters are much more severe when compared to onshore converters. When comparing offshore and onshore wind parks of identical dimensions, offshore wind parks are characterised by lower efficiencies and more fatigue of the wind energy converter in the downwind flow of the main wind direction. Therefore, on average larger distance factors are assumed for offshore wind park designs than for onshore wind parks. Usually, they do not have the same space restriction requirements either.

Grid connection. There are two options for connecting the converters of a wind park to the power grid:
– connection via a direct current bus or
– direct grid connection by a common alternate current bus.

For the first alternative, synchronous generators operated at a variable speed in conjunction with a direct current intermediate circuit as well as line-commutated inverters are applied. In comparison to other concepts, major circuit feedback and stability problems are disadvantageous. Furthermore, direct current lacks zero-crossing which may generate stable electric arcs in case of malfunctions. Because of these disadvantages this type of grid connection is rarely applied.

Synchronous or asynchronous generators with direct grid connection may also use the same three-phase electrical grid. According to the respective nominal capacity, current is fed into the medium voltage grid via one or several transformers. If the low voltage cables are sufficiently dimensioned, a common low voltage bus may also be applied. Otherwise, the generators must be connected via separate transformers. This has become common practice. Currently, a wind converter is normally provided with its own transformer, for instance, feeding into a 20 kV cable that represents the actual grid connection.

Due to the long distances to be overcome, grid connection of offshore wind parks is more difficult. The main deciding parameters are the distance from the shoreline and the installed capacity. For instance, cables laid on the seabed must be designed and laid to sufficiently withstand deterioration, for example by draw nets. This is why cables are either buried or washed into the seabed.

The individual converters of an offshore wind park may either be connected with each other by medium voltage alternate current connections, as for onshore wind parks, or by medium voltage direct current connections. In the latter case every wind energy converter needs to be provided with its own controlled rectifier system. To transmit electrical energy to the on-shore grid connection, besides conventional alternate current transmission also high voltage alternate current or high voltage direct current transmissions are possible; additionally, the respective substations are required. According to the respective wind park capacity and the transmission voltage level, one or several underwater cables must be laid. High voltage direct current transmission is advantageous as it allows the transmission of high capacities over long distances without any idle power compensation installations as required for alternate current transmission. However, high voltage direct current transmission requires additional facilities for rectifying and inverting current and voltage. There are two different rectifying options: conventional Thyristor technology and Insulated Gate Bipolar Transistor (IGBT) technology with pulse width modulation (PWM) which also provides its own alternate current grid for isolated operation facilitating the control of current, voltage and reactive power. From the current viewpoint, high voltage direct current transmission results cost-efficient from a distance of approximately 60 km onwards.

7.2.6 Grid-independent applications

The majority of all individual wind turbines and wind parks are connected to the power grid. However, there is also a multitude of grid-independent and wind-diesel systems designed for small power grids which are outlined throughout the following sections.

Wind-battery systems. In most cases, grid-independent wind battery systems are small capacity wind turbines which, coupled with a battery and a battery charge regulator, supply one or several power consumers. The nominal capacities of the majority of these systems stretch from a few 100 W up to several 100 kW. Particularly in developing countries, where sites often offer both sufficient solar radiation and above-average wind conditions, the described wind-battery systems may additionally be combined with one or several photovoltaic modules (wind-battery-photovoltaic systems).

Wind-battery systems are generally provided with smaller wind energy converters than those outlined in the preceding sections. Their nominal capacities usually amount to several kilowatt, wind turbines of rotor diameters up to 6 m with nominal capacities ranging from 10 to 20 kW are referred to as small wind energy converters /7-10/; however, this term is also applied for converters with rotor diameters of approximately 16 m and nominal capacities of up to 100 kW /7-5/. These small converters encompass fast converters provided with 2 to 3 rotor blades as well as slow converters of numerous rotor blades. Within this range there are also more converters equipped with vertical axes (Savonius type or H rotor).

The following framework conditions should be ensured for technically and economically sensible application /7-11/.

- The wind speed should exceed 6 m/s at least on 200 days per year.
- Maximum wind lull period must not exceed 48 h.
- If possible, the lower edge of the rotor should be positioned at least 10 m above the highest obstacle in the surrounding area.
- Secure starting torque at about 4.5 m/s (i.e. smooth running gearbox or no gearbox, special electronic starting control).
- Generator capacity must not exceed 1.5 times the maximum supplied capacity.
- Noise emissions must not exceed 60 dB at a distance of 20 m from residential areas and 45 dB for installation on vessels.
- There should be no considerable vibration resonance from idle mode up to the operating point of the converter, tower and guying, if provided.

Common tower heights of small wind energy converters stretch from several metres up to approximately 30 m. Towers are often of simple design, e.g. lattice work or guyed tube towers.

The described energy supply systems are applied in the windy areas of developing countries (particularly in rural areas lacking power connection) as well as in industrial countries (in case retrofitting of grid connection is too expensive). Typical applications include lighting of remote sites, energy supply for communication facilities or signals on freeways, power supply to remote huts and houses. Further areas of application are battery-charging systems of vessels, power supply for electronic fences or cathodic corrosion protection.

In the meantime also larger wind-battery systems have been developed that may be connected to the power grid to ensure supply of a determined minimum capacity within periods of low wind availability and to compensate for fluctuating wind flow.

Wind pumps. Wind pump systems are sub-divided into systems equipped with electric and mechanical pumps.

Wind pump systems equipped with electric pumps first convert wind energy into electric energy and subsequently reconvert it into mechanical energy. They are usually applied for high capacities. Standardised wind turbines may be used. Thanks to the isolation of the mechanical energy of both wind rotor and pump, the pump location is entirely independent from the wind energy converter location, which is advantageous in many cases, as irrigation systems are often located in valleys and areas where there is little wind.

Mechanical wind pumps directly convert the kinetic energy contained in the air into mechanical pump energy without energy being first converted into electric energy. Fig. 7.23 shows such a system. Typical fields of application include potable and livestock water supply as well as irrigation and drainage.

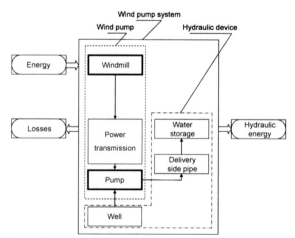

Fig. 7.23 Irrigation system using wind energy

To convert wind power into hydraulic energy, hydraulic capacity is of particular importance. It is the product of water density, local gravity, water flow and delivery head. Hence, a large volume flow at low flow height or a low volume flow at high flow height can be obtained. Whereas the former usually applies to irrigation and drainage, the latter is true for potable water supply.

For the described systems predominantly piston pumps as well as single or multi-stage centrifugal pumps are used. However, also diaphragm pumps, eccentric worm pumps and chain pumps may be applied.

For system dimensioning, the ideal combination of pump and turbine results indispensable. For this purpose torque characteristics of wind turbine and pump need to be compared. By adapting the respective pump and turbine design on the one hand, the operating speed of the wind turbine and the pump may be adapted by appropriate transmission reduction. On the other hand, adequate torque adaptation ensures a favourable start of the wind pump.

In order to understand wind pump operation, the respective characteristic flow curve of the wind pump system and the interdependence of wind speed and volume flow are important. Fig. 7.24 illustrates the characteristic flow curve for combinations with piston and centrifugal pumps. Extraction only starts at a certain minimum wind speed. With the piston pump volume flow increases rapidly upon slight wind speed alternations, whereas, at a higher level, volume flow is only slightly increased by enhanced wind speed. For the centrifugal pump, by contrast, the characteristic curve is almost linear over a large part of the wind speed curve.

While wind pumps have been applied for irrigation and drainage in Europe in the past centuries, they are currently predominantly used in developing and newly industrialised countries.

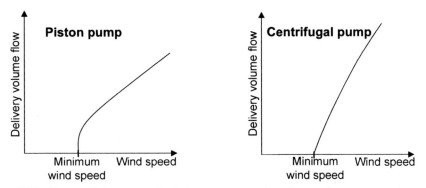

Fig. 7.24 Characteristic curves of wind pump systems based on a piston pump and a centrifugal pump

Wind-diesel systems. Wind-diesel systems are used for very small systems of capacities below 100 kW and even for isolated grids designed for cities or whole islands and reach maximum capacities of 10 MW. The major problem encountered with this system, when compared to the wind parks of large grids, is its inappropriateness as a buffer. Under the described conditions, grid penetration with wind energy, or the ratio of electric power generated by wind energy to the total energy demand within a given period, are either very low or suitable measures need to be taken to compensate for wind energy fluctuations or to store wind power. Furthermore, the existing power grids are often very weak when combined with such wind-diesel systems, particularly if the wind energy converter or wind park is to be connected to the power grid. Besides power generation regulation difficulties (adaptation of diesel and wind turbine operation), further problems are encountered with regard to adequate power distribution. Thus, increased wind portions in wind-diesel systems often require grid extension or reinforcement measures.

The described wind-diesel systems are mainly applied in developing or emerging countries. However, they can also be sensibly applied on the islands of industrialised countries.

Wind-sea water desalination. Distillation and reverse osmosis are the two different processes available for sea water desalination.

For distillation, sea water is first evaporated by heat and subsequently recondensed in order to obtain a desalinated distillate. The remaining condensed seawater is discarded. Most processes use the available condensation heat to preheat seawater or for evaporation at a low pressure level. Thermal energy consumption per m^3 of distillate amounts to between 60 and 80 kWh. Additionally, about 3 kWh/m^3 are required for pumping.

For reverse osmosis seawater is pressed through a semi-permeable membrane at high pressure (between 50 and 80 bar). Most of the ions dissolved in water stay back so that a low-salt permeate of potable water quality is obtained. Also for this process the remaining and highly condensed seawater is discarded. The energy of the highly pressurised concentrate may be recovered by pelton turbines or pressure exchangers. The energy consumption of reverse osmosis amounts to between 6 and 8 kWh/m^3 excluding energy recuperation, and to between 3 and 4 kWh/m^3 including energy recuperation. Since reverse osmosis plants have considerably lower energy consumption when compared to distillation plants they are more suitable for wind energy converters.

Fig. 7.25 shows the main system components of a seawater desalination plant based on the reverse osmosis principle. First, particulate matter is removed from sea water to protect the sensitive membranes. Subsequently, the required pressure is built up by a high-pressure pump. Part of the seawater is pressed through the reverse osmosis membrane (approximately 30 to 50 %), while the remainder of the water stays back in the form of concentrate. The pressure energy contained in the concentrate flow may subsequently be recuperated by turbines or pressure exchangers. Prior to distribution to customers permeate is usually chlorinated if necessary pH value and water hardness are manipulated.

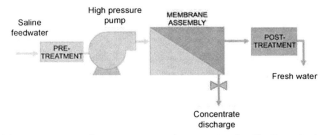

Saline feedwater — PRE-TREATMENT — High pressure pump — MEMBRANE ASSEMBLY — POST-TREATMENT — Fresh water — Concentrate discharge

Fig. 7.25 Main components of a reverse osmosis sea water desalination plant

Combined systems of seawater desalination and wind power stations are sensible, as contrary to power, storage of water is fairly simple. Hence, in periods of low electric load, excessive power may be used for desalination, whereas the desalination plant may be disconnected from the grid at times of high power demand and consumer water requirements may be covered by stored water.

Especially for islands, combinations of wind seawater desalination plants are advantageous to adjust the usual imbalance between freshwater scarcity and high water demand (e.g. by agriculture, tourism). There is thus a demand for desalination plants whose energy demand can ideally be covered by wind power stations, since wind conditions are usually very favourable on islands. Exemplary plants have been built on the Canary Islands as well as in the Aegean Sea. Desalination plants are also often combined with wind-diesel systems whose supply is usually more reliable.

7.3 Economic and environmental analysis

This section deals with the economic and environmental aspects of power generation by means of wind energy converters.

7.3.1 Economic analysis

Three grid-connected wind energy converters operated on the mainland (i.e. on-shore installation) of different capacities (1.5, 2.5 and 5.0 MW) are considered as representative. The converters are installed within a medium size wind park (consisting of about 10 converters).

The rotors are provided with three blades made of glass fibre-reinforced plastic, whereas towers are made of steel. For converter manufacturing, series production is assumed. Converters are mounted on regular good load supporting soil; shallow foundations are sufficient to ensure safe operation. Table 7.2 contains further important reference parameters of reference wind energy converters and reveals the key figures of the most common converter models out of the vast variety of the currently available designs. The referenced converters only refer to the currently most common technologies. The following comparisons may vary tremendously for different converter types made of different materials or manufactured by different technologies.

Table 7.2 Characteristic parameters of reference wind power stations

		1.5 MW class	2.5 MW class	5.0 MW class
Nominal capacity	in kW	1,500	2,500	5,000
Rotor diameter	in m	70	80	126
Tower height	in m	85	100	135
Reliability	in %	98	98	98
Park efficiency	in %	92	92	92
Technical life time	in a	20	20	20
Full-load hours	in h/a			
Site 1		1,800	1,800	1,800
Site 2		2,500	2,500	2,500
Site 3		4,500	4,500	4,500

Since wind power generation depends essentially on the available wind quantities, additionally three sites are assumed. Site 1 represents an average location on a foothill. Site 2 is a favourable place at the coast. And site 3 represents a very promising location on a windy island.

The potential power generation for any converter type based at one of the reference sites is calculated by the characteristic power curve of the converter and the wind distribution (see Section 2.3) on site. With regard to wind distribution, Weilbull distribution of form factor 2 has been assumed. The resulting full-load hours (i.e. the ratio of approximate annual power generation to nominal converter capacity) of various wind converters of the same capacity are illustrated in Table 7.2. Furthermore, generation availability time of 98 % (i.e. reliability) and a wind park efficiency of 92 % have been assumed.

To assess the costs incurred by wind power utilisation first variable and fix costs of the reference plants are discussed. Subsequently, power generation costs are calculated in relation to the respective available wind quantity.

Investments. Investment costs encompass the production costs, transportation and assembly, foundation, grid connection as well as miscellaneous costs (including design and infrastructure costs), whereby converter size and site conditions have a considerable influence on the cost structure /7-24/.

Overall costs of current converters reveal the magnitudes indicated in Table 7.3. They have been calculated on the basis of a representative market average and in relation to the tower height assumed for Table 7.2. The total cost (i.e. converter costs including grid connection, installation and miscellaneous costs) for a converter of an installed capacity of 1.5 MW add up to approximately 1.6 Mio. € (1,090 €/kW). Three quarters of the costs account for the converter. For the 2.5 MW converter total investment costs amount to about 2.6 Mio. € (1,040 €/kW). And the costs of the 5 MW wind mill are in the order of magnitude of roughly 4.9 Mio. € (980 €/kW). If the systems would be installed offshore the overall investment costs are significantly higher. According to current knowledge for the 2.5 and 5.0 MW wind mill the costs would roughly double.

An analysis of the specific costs of the actual wind converter per m² of rotor surface reveals that smaller converters yield better values than large converters. The approximate values of 1.5 MW converters, for instance, amount to 350 €/m², whereas the maximum specific costs of larger converters within the higher MW range are increased to about 430 €/m².

The above-mentioned specific investment costs are mainly list prices per individual converter. For wind farms, however, usually many converters are sold under the same contract. Under such framework conditions specific investment costs are generally much lower due to discounts granted by manufacturers within the scope of major projects. Experience shows that price reductions amount to 25 % of the list prices.

Table 7.3 Mean investment and operation costs as well as power generation costs for the reference wind turbines defined in Table 7.2

Capacity	in kW	1,500	2,500	5,000
Investments				
Wind energy converter	in k€	1,250	2,180	4,150
Grid connection	in k€	185	200	380
Miscellaneous	in k€	205	219	418
Total	in k€	1,640	2,599	4,948
Yearly costs[a]	in k€/a	95	151	295
Power generation costs				
1,800 h/a	in €/kWh	0.082	0.078	0.075
2,500 h/a	in €/kWh	0.059	0.056	0.054
4,500 h/a	in €/kWh	0.033	0.031	0.030

[a] operation, maintenance, insurance, miscellaneous

Additional investment expenses include grid connection, the foundation as well as development and design costs (i.e. miscellaneous costs) and vary tremendously with regard to project type (number and size of converters) and site conditions. Usually, grid connection costs account for the major portion. They reach from 5 to 30 % of the converter costs, whereas foundation expenses amount to approximately 3 to 9 %, development fees to roughly 1 to 5 %, and engineering costs to about 1.5 to 3 % of the converter costs. Miscellaneous expenses account for 5 to 8 % of the investment costs of a wind energy converter. Overall additional investment costs for onshore sites are between 16 to 52 % of the overall converter costs. Wind parks of low capacity converters usually have considerably reduced costs when compared to wind parks of much higher capacities. For very large wind parks provided with numerous wind turbines (e.g. construction of a substation and coupling a higher voltage level) may incur in even higher specific costs due to grid connection expenses.

With regard to offshore wind parks it is expected that additional investment costs will considerably exceed those of onshore wind parks due to the increased foundation and grid connection costs. Current estimates assume additional investment costs of 70 to 105 % of the actual converter investment. The costs depend tremendously on the water depth and the distance from the shoreline.

Operation costs. Operation costs include land lease expenses, insurance, maintenance and repair as well as technical operation. As data are based on solely a few years of experience, for the most common commercially available converters, cost estimates vary tremendously. However, the average annual costs are estimated at 5 to 8 % of the total investment costs (Table 7.3), whereas maintenance, service, repair, insurance and land lease costs account for the major share. However, administrative and managerial costs are also relevant. For offshore wind parks, by contrast, significantly higher expenses for maintenance, repair and insurance are expected.

Electricity generation costs. Specific power generation costs may be determined by annuity calculation derived from the total investments, the annual costs as well as the expectable energy yields. As usually a real discount rate of 4.5% and an amortisation period corresponding to the technical converter service-life of 20 years are assumed. Furthermore, constant operation costs are assumed for the entire service-life.

Table 7.3 shows the power generation costs pertaining to three converter types defined in Table 7.2 located at the three reference sites. According to the figure, power generation costs decrease tremendously with an increasing yearly full load hour and thus an increasing mean wind velocity.

Sites with annual mean wind speeds that allow for 4,500 h/a are very rare and can only be found very close to the coast especially on windy islands or in very exposed locations on shore. The power generation costs corresponding to these conditions range from 0.030 to 0.033 €/kWh for the conditions analysed here. Compared to that, sites with 2,500 h/a are more frequently available at good locations close to the coast. Under these conditions average real specific power generation costs between 0.054 and 0.059 €/kWh could be achieved. If the full load hours are even lower (e.g. 1,800 h/a) the electricity provision costs increase significantly (Table 7.3). Comparisons of different converter technologies with different installed capacities do not reveal any significant difference in terms of power generation costs for the same wind speeds respectively for the same full load hours.

Independent of the above observations, especially for the capacity range from 50 to 1,000 kW, which play no significant role in Europe any more, specific power generation costs decrease with increasing turbine capacity. This is mainly due to the tendency towards reduced investment costs and increased tower heights for enhanced capacities which permits to achieve higher wind speeds at hub height at the same site as the mean airflow speed increases with increasing height above ground. The smallest examined converter (converter of 1,500 kW, Table 7.2) thus results much more cost-effective than wind energy converters of lower installed capacities; they have thus lost importance within the course of the last decade.

Power generation costs are influenced by a multitude of parameters. To estimate and assess their impact Fig. 7.26 shows variations of the main influencing variables and related effects on resulting power generation costs. For this purpose a site with 2,500 h/a and a 2.5 MW converter is analysed.

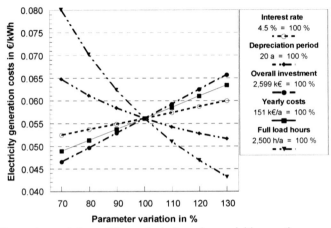

Fig. 7.26 Parameter variation of the main influencing variables on the power generation costs

Fig. 7.26 shows that the full-load hours – and thus the site-specific annual mean wind speed – have the biggest impact on power generation costs. Due to the increase in the mean wind speed at increased heights above ground there is a tendency towards ever higher towers to allow for optimum exploitation of the available wind quantities at a potential site. Besides annual full-load hours, investment costs represent the second deciding factor with regard to the specific power generation costs. Operation costs, interest rate and amortisation period, by contrast, only have a slight impact on the power generation costs.

7.3.2 Environmental analysis

There are also environmental effects caused by wind energy converters. Such effects are discussed below with regard to manufacturing, normal operation, possible malfunctions and the end of operation.

Construction. Conventional industrial branches, such as classic mechanical and electrical engineering, are involved in the manufacturing of wind energy converters. Hence, wind energy converter manufacturing has the same environmental effects on soil, water and air that these sectors are known for. Due to the strict environmental regulations, the corresponding environmental effects are relatively low. Also, the propensity toward malfunctions during manufacturing is relatively low – with the exception of e.g. iron and steel manufacturing.

Normal operation. Wind power stations do not directly release any toxic substances. Nevertheless, the operation of wind energy converters has certain effects on the natural environment. The main effects are described as follows.

Audible sound. With their rotating rotors wind power stations represent acoustic sources. Sound is primarily caused by aerodynamic noise at the rotor blades and by sound radiation by gearbox and generator.

The latter acoustic source was considerably reduced by converter casings and by gearless converters built over the last few years. Furthermore, the gearbox and generator on the one hand and the nacelle on the other hand are separated in almost every modern converter design. Thus, nacelles no longer act as resonance bodies and sound impact has been significantly reduced compared to converters built in the early nineties.

Aerodynamic sound emissions of wind energy converters are due to the airflow around the rotor blades and the rotor blade coming through the tower shadow; they mainly occur at medium and high blade tip speeds. For the environment solely the noise created at low and medium wind speeds is of importance, since natural wind sound is predominant at high airflow speeds. In the past, aerodynamic sound creation has been reduced by an optimised shape of rotor blade and blade tip. On the whole, this and further measures have reduced sound emissions considerably (by about 5 to 10 dB(A)) over the past decade /7-12/. This is why wind energy converters can only be clearly distinguished from background noise at comparatively low wind speeds.

Sound emissions created by wind power stations represent an essential parameter for converter design. Favourable arrangement of wind energy converters and a corresponding selection of converter models may allow to reduce the sound level by approximately 10 dB(A) at constant power generation. Also the legislation for noise reduction may prescribe certain sound levels for residential, industrial and mixed areas which must not be exceeded. In order to obtain a building permit for a wind energy converter or a wind park, observation of certain sound limits must be proven by corresponding survey reports to prevent unacceptable noise.

Infrasonic sounds. Wind energy converters mainly emit aerodynamic infrasonic sounds within a frequency range from 0.6 to 1.5 Hz. At these frequencies the human detection limit of 120 to 130 dB(A) is very high. At a distance of 120 m from e.g. a 500 kW converter infrasonic sounds of 75 to 85 dB(A) were measured which were reduced to 67 to 77 dB(A) at a distance of 300 m. Since certain distances have to be observed with regard to the building permit and in order to comply with the legislation there is no impact on the population by infrasonic sounds. According to current knowledge, there is also very little impact on animals.

Disco effect. In times of high direct solar radiation, luminous reflectance may occur at rotor blades, if solar radiation is reflected by the mirror-like surface of the rotor blades. However, luminous reflectance occurs accidentally and is only perceived temporarily at certain sun positions. Due to the concave rotor surfaces disco effects are very low. A constant impact over several consecutive hours can be almost ruled out. The described effects may be further reduced by low reflecting rotor blade design, which has become common practice for rotor blade manufacturing.

Shadow impact. The term shadow impact refers to the moving shadow cast by the rotor blades at times of sunshine /7-13/. Shadow impact depends among other factors on the weather and the solar altitude as well as on the converter size and the form of operation. For e.g. a 1.5 MW converter maximum shadow impact is perceivable up to a distance of about 1,000 m. Shadow impact is measured by the theoretical maximum impact duration (for any case sunshine, rotating rotors and unfavourable wind direction, impacting the rotor position towards the sun, are assumed) and real impact duration (the shadow is calculated for the actual weather conditions). Actual shadow impact time amounts to approximately 20 % of the theoretical absolute shadow impact duration.

Nevertheless, shadows may impact the people living in the neighbourhood of wind energy converters or wind farms. For this reason a survey has been conducted on the impact of periodic shadows in terms of perceiving and managing the problem and behavioural aspects. The test persons exposed to a mean shadow duration of 5 to 10 h/a showed individual effects with regard to the above three criteria; for instance for over 15 h/a of weighted shadow duration, strong effects have been observed for all three areas of investigation. Further investigations have been conducted to determine whether periodic shadow impacts with duration of 30 minutes and single occurrence may cause stress. Although none of the test persons experienced considerable strain, the proven increased strain on psychological and physical resources are indicative that cumulative long-term effects may fulfil the criteria of being an added stress factor /7-14/.

However, adequate planning in the preliminary stages of wind converter projects and appropriate micro-siting allows to considerably reduce the duration of shadow impact /7-15/, /7-25/ and thus minimises all related problems for residents.

Ice throw. Under certain meteorological conditions icing may occur at the blades of a wind turbine that may become loose and fall off during start-up. The risk of icing mainly depends on the respective meteorological framework conditions and thus on the particular site (e.g. low or high mountain ranges). However, the mortal risk of ice throw at a distance of approximately 200 m is comparatively low and almost equal to that of a lightning stroke /7-16/.

Natural scenery. Wind energy converters are technical buildings that inevitably alter the natural scenery. As both plant size and pole height have increased over the last 10 to 15 years, impacts on the natural scenery have also become increasingly important.

With regard to wind turbines two different impacts on the scenery are observed; on the one hand they alter the scenery's dimensions and on the other hand they have significant distant effects /7-17/.

This applies far and foremost to plains and exposed sites based in low mountain ranges, as wind turbines are still visible from far distances. Also the number of wind turbines and the tower height play a major role. However, using appropriate colouring, tower designs and rotor blade number and speed can reduce subjective irritation. For instance, in most cases solid towers fit better into the scenery than lattice masts, and, due to their smooth running, rotors provided with three rotor blades are generally more easily accepted by spectators than rotors equipped with one or two rotor blades.

Optical assessment of wind turbines is not subject to any objectively defined parameters but mainly depends on the associations and the personal taste of the respective spectator. The surroundings of a wind turbine are also a deciding factor. To date looking at wind turbines is not considered disturbing in most cases. Furthermore, computerised applications permit to assess and minimise the effect on the scenery still prior to wind turbine erection.

Preservation of bird-life. With regard to wind power utilisation, in terms of environmental effects, interference with feeding and resting birds, impacts on flying/migrating birds and hitting of birds have been reported.

Resting birds tend to avoid any kind of wind turbine and usually keep at a distance of several hundreds of metres. In this respect, certain bird types (e.g. curlews, ruffs) are much more sensitive than seagulls, for instance /7-18/.

For many bird species usually no conspicuous behavioural changes have been observed with regard to wind turbines /7-18/. Yet, also contrary survey results have been reported; for instance, white storks react most sensitive in the vicinity of breeding grounds /7-19/. On the whole, different bird types react very differently to wind turbines. While there is no interference with lapwings or oyster catchers and wind turbines, impacts on redshanks and blacktailed godwits cannot be ruled out /7-20/.

With regard to wind turbines, birds hitting has only been observed occasionally, probably due to rotor sounds, and is only of little importance when compared to the effect caused by other buildings or road traffic. Nine wind turbines sites have been observed over two years with regard to birds killed by hitting. There were 32 victims of 15 different bird types at 7 sites /7-18/.

On the whole, negative impacts of wind power utilisation on bird-life are reduced if determined areas are entirely excluded from wind power utilisation (e.g. natural reserves, flora fauna habitats (FFH)) as it has already become common practice for the design of wind power applications.

Further effects on fauna. Within the scope of site investigations mass, flight of migratory insects has not been observed to date /7-18/. Any serious interference with wildlife (such as hares, roe deer, red foxes, grey partridges, carrion crows) have been reported either /7-21/. Thus, potential effects on the fauna are low.

Space consumption. Space consumption for wind power utilisation is generally low. Direct space consumption is attributable to foundations, access roads and administration buildings, if required; the space required for the foundation of a 1.5 MW wind turbine amounts to roughly 200 m^2. The distances required in between the turbines of a wind park may almost entirely be used for agriculture. Compared with other power generation techniques using renewable and fossil energy sources, space consumption of wind power generation is relatively low.

Offshore wind power utilisation. Due to the long distance to housing settlements, for offshore installations audible sound, infrasonic sound, luminous reflectance and shadow impact are of no importance. However, the following potential effects on the natural environment are currently being discussed /7-22/:
– Sound radiation of wind turbines, also into the water, and thus possible noise irritation of animals.
– Effects on seabed symbioses and fish fauna.
– Risk of vessel crashes and ensuing environmental effects.
Future developments will show whether offshore wind power utilisation has a significant effect on the environment. From the current viewpoint environmental effects are expected to be lower in comparison to onshore wind power generation.

Acceptance. The degree of social acceptance of wind turbines is indicative of their effect on humans and the environment. Against this background the analysis of the correlation of wind power utilisation and tourism has repeatedly been subject to scientific investigations. The investigations revealed that vacation areas with wind turbines are esteemed less attractive than others without converters or which have other difficulties. However, wind power converters have never been a reason to change the holiday location. Furthermore, wind turbines are often positively associated with environmental-friendly and sustainable energy generation, which more than compensates for the loss of attractiveness.

Malfunction. From the current knowledge there are no detrimental wind turbine-specific effects to be expected in case of malfunction; in the worst case limited local effects may ensue. In order to minimise even those limited effects, converters with lubricated gearboxes have been provided with oil-collecting pans.

Regardless of the above effects, fires of electrical components (including cables) may release limited amounts of substances to the environment. However, substance release is not specific for wind power converters and can also be observed for other kinds of power plants. Furthermore, such occurrences can easily be avoided if the relevant guidelines are adhered to.

Also mechanical failure (such as rotor fracture) may damage vegetation (e.g. turf). However, if the prescribed safety distances to residential areas are observed, the risk of human injury in case of mechanical failure is very low, as rotor fractures tend to occur during storms when personal damages are unlikely.

End of operation. Wind turbines mainly consist of metallic materials which are disposed of in compliance with established procedures. However, disposal of the rotor blades consisting of glass reinforced plastic (GRP) is still unsettled. Material-thermal disposal seems most appropriate. However, it is assumed that many turbine components may be recycled. Recycling involves all related environmental effects or avoids the described environmental effects, if manufacturing of new material is avoided.

8 Hydroelectric Power Generation

8.1 Principles

Hydropower plants harness the potential energy within falling water and use classical mechanics to convert that energy into electricity. The theoretical water power $P_{Wa,th}$ between two specific points on a river can be calculated according to Equation (8.1) (see Chapter 2.4.1).

$$P_{Wa,th} = \rho_{Wa} g \dot{q}_{Wa} (h_{HW} - h_{TW})$$

(8.1)

ρ_{Wa} is the water density, g the gravitational constant, and $\dot{q}_{Wa}$ the volumetric flow rate through the hydroelectric power station. h_{HW} und h_{TW} describe the geodetic level of head and tailwater.

Due to the physically unavoidable transfer losses within a hydroelectric power station, only part of the power according to Equation (8.1) can be utilised. In order to show this, the Bernoulli equation (see Chapter 2.4.1) can be converted accordingly. Thus all terms adopt the unit of a geometrical length and can be represented graphically (Fig. 8.1).

If the energy balance between two reference points – up- and downstream of a hydroelectric power station – is set forth, the Bernoulli equation can be written according to Equation (8.2).

$$\frac{p_1}{\rho_{Wa,1} g} + h_1 + \frac{v_{Wa,1}^2}{2g} = \frac{p_2}{\rho_{Wa,2} g} + h_2 + \frac{v_{Wa,2}^2}{2g} + \xi \frac{v_{Wa,2}^2}{2g} = const.$$

(8.2)

In the following, the different terms of Equation (8.2) are defined as pressure energy $p/(\rho_{Wa} g)$, potential energy h, kinetic energy $v_{Wa,2}/(2\,g)$ and as lost energy $\xi\, v_{Wa,2}/(2\,g)$. ξ is the loss coefficient, and p_i and $v_{Wa,i}$ represent the pressure and the flow velocity at the corresponding reference points respectively. The lost energy is thus the part of rated power that is converted into ambient heat by friction and therefore cannot be used technically.

System setup. A hydroelectric power station, depending on scale, normally consists of a dam or weir, and the system components intake works, penstock, in some cases a headrace, plus the powerhouse and tailrace (Fig. 8.1; see also Chapter 8.2 and Fig. 8.2). The flow is led to the turbine via the intake structure, the headrace and penstock. Afterwards it streams through the draft tube into the tailrace.

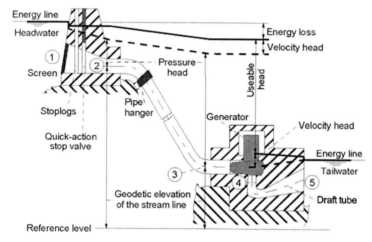

Fig. 8.1 Physical correlations in a hydroelectric power station

In Fig. 8.1 the lines enable the graphic representation of the Bernoulli equation. The dotted line represents the geodetic level of the water flowing through the hydroelectric power station. The so-called energy line is at the top left corner of the diagram. It shows the locations and respective energy losses. The distance to the broken line below the energy line corresponds with the kinetic energy of the water. This becomes apparent at the intake structure, where the water flow increases due to the narrowing of the cross-section and the kinetic energy therefore increases at the same time. The difference between the geodetic level and the broken line is the pressure energy level.

Intake. The intake structure is the connection between headwater and penstock or turbine. In the example (Fig. 8.1) there is a screen at the entrance of the intake structure keeping floating debris out of the plant. Furthermore the intake structure has stoplogs and a quick-action stop valve. The stoplogs enable the hydroelectric power station to be drained during maintenance work. The quick-action stop valve stops the flow into the hydroelectric power station in the case of an accident.

Within the intake structure, a partial conversion of potential energy into kinetic energy takes place (reference point 1 to reference point 2; see Fig. 8.1). Because of local energy losses in the intake and the flow resistance at the screen, part of the energy is lost before it can be utilised in the turbine. These losses are combined in Equation (8.3) as the loss correction value, ξ_{IS}, for the intake structure. As the headwater flow velocity usually can be neglected, the respective term on the left side of the Equation (8.3) is dropped. The water density ρ_{Wa} can be regarded as constant, up and downstream of the intake structure. Therefore the losses within the intake structure manifest as a decrease in the pressure level. They are represented in Fig. 8.1 by the drop in the energy line at reference point 1.

$$\frac{p_1}{\rho_{Wa}\,g} + h_1 = \frac{p_2}{\rho_{Wa}\,g} + h_2 + (1+\xi_{IS})\,\frac{v_{Wa,2}^2}{2g} \tag{8.3}$$

Penstock. The penstock bridges the distance between the headwater or intake structure on one side with the turbine on the other (balance point 2 to balance point 3; see Fig. 8.1). A further conversion of potential energy into pressure energy takes places. Because of the friction losses in the pipes, some of the energy is lost. The Bernoulli Equation for the penstock can be written according to Equation (8.4).

$$\frac{p_2}{\rho_{Wa}\,g} + h_2 + \frac{v_{Wa,2}^2}{2g} = \frac{p_3}{\rho_{Wa}\,g} + h_3 + (1+\xi_{PS})\,\frac{v_{Wa,3}^2}{2g} \tag{8.4}$$

The loss coefficient, ξ_{PS}, of the penstock is a result of the friction factor and the diameter of the penstock, and it increases proportionally to the length of the conduit. The friction factor is again dependent on the diameter, the flow velocity and the surface roughness of the penstock; for practical utilisation it can be obtained from relevant diagrams (e.g. /8-1/).

While the length of the penstock is dependent on the plant specifications, the diameter can be varied. A diameter increase reduces the friction losses and the turbine power increases. However, the penstock-related costs increase at the same time. Therefore the aim is always to achieve a technical and economic optimum. Run-of-river power stations with low heads do not have a penstock; the water flows directly from the intake structure into the turbine.

Turbine. In the turbine, pressure energy is converted into mechanical energy (reference point 3 to reference point 4; see Fig. 8.1). The conversion losses are described by the turbine efficiency $\eta_{Turbine}$ (Chapter 8.2.3). Equation (8.5) describes that part of usable water power that can be converted into mechanical energy at the turbine shaft $P_{Turbine}$.

$$P_{Turbine} = \eta_{Turbine}\,\rho_{Wa}\,g\,\dot{q}_{Wa}\,h_{util} \tag{8.5}$$

h_{util} is the usable head at the turbine, and the term $(\rho_{Wa}\,g\,\dot{q}_{Wa}\,h_{util})$ represents the actual usable water power $P_{Wa,act}$.

Losses within the turbine are differentiated as volumetric losses, losses due to turbulence, and friction losses. As a result of these losses the power at the turbine shaft $P_{Turbine}$ in Equation (8.5) is less than the usable water power $P_{Wa,act}$.

Outlet. Reaction turbines (e.g. Kaplan turbines, Francis turbines) enable a better utilisation of the head using a draft tube.

How this system operates can be demonstrated by following the stream line of current from the tailwater to the turbine outlet. The tailwater energy line is determined by its geodetic level and the ambient pressure (Fig. 8.1). When entering the tailwater, the water looses its remaining kinetic energy through turbulence. This is shown in Fig. 8.1 by the drop of the energy line at reference point 5. If – as shown in Fig. 8.1 – the turbine outflow and the draft tube outlet (reference points 4 and 5) are at the same geodetic level, the Bernoulli equation between these two points can be drawn up in a simplified way according to Equation (8.6).

$$\frac{p_4}{\rho_{Wa}\,g}+\frac{v_{Wa.4}^2}{2g}=\frac{p_5}{\rho_{Wa}\,g}+\frac{v_{Wa.5}^2}{2g} \tag{8.6}$$

Since the cross-section of the flow at the end of the draft tube is bigger than that directly behind the turbine, $v_{Wa.5}$ has to be less than $v_{Wa.4}$. The draft tube causes a reduction of the flow velocity before entering the tailwater. Thus the pressure p_4 at the turbine outlet has to be lower than the pressure determined at the tailwater, at the end of the draft tube p_5. Ultimately, this leads to a decrease in losses due to turbulences and thus a better head utilisation.

Overall system. In a hydroelectric power station hydraulic losses mainly occur in the intake structure, the penstock and possibly in the outflow (balance point 1 to balance point 5; Fig. 8.1). The actual usable water power $P_{Wa,act}$ is calculated by deducting the various loss terms from the theoretical water power (i.e. losses in the intake structure, the penstock, the outflow); this can be described with Equation (8.7).

$$P_{Wa.act}=\rho_{Wa}\,g\,q_{Wa}\left[\left(h_{HW}-h_{TW}\right)-\xi_{IS}\frac{v_{Wa.2}^2}{2g}-\xi_{PS}\frac{v_{Wa.3}^2}{2g}-\frac{v_{Wa.5}^2}{2g}\right] \tag{8.7}$$

The losses are thus dependent on the flow velocity and can consequently be minimised with an optimised plant design and layout. The power that can finally be obtained at the turbine shaft is a result of the actual available water power and the turbine efficiency.

8.2 Technical description

Based on the discussed physical correlations for the use of hydroelectric power, the technical requirements of hydroelectric power generation are described below.

The current state of technological development is the basis for this description (i.e. /8-2/, /8-3/, /8-4/).

8.2.1 Schematic layout

The components described in Fig. 8.2 are required for the technical conversion of energy in flowing water to electricity in a run-of-river power station. These components are the water intake at the headwater, the weir, the inlet and outlet of water to and from the turbine, the outlet at the tailwater, and the powerhouse with machinery and electrical equipment. These system elements are usually combined in the dam that enables the use of the head and the powerhouse.

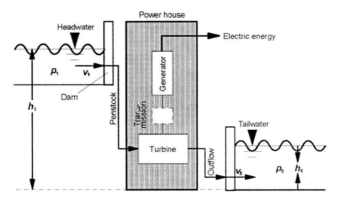

Fig. 8.2 Schematic layout of a hydroelectric power station (p pressure, h height, v velocity, indices according Fig. 8.1; see /8-5/)

Two system components are the main elements of the actual energy conversion within a typical hydroelectric power station. Together with the turbine that draws the energy from the water and converts it into mechanical energy, the second component is the generator for further conversion into electrical energy, and thus into the final product. Depending on the plant configuration, a transmission is additionally required if turbine and generator rotational speeds are different or if both components are not on the same axis. In smaller plants the transmission is often replaced by a simple belt drive.

8.2.2 Categorisation and construction types

Hydroelectric power stations can be divided into low, medium and high head power stations (Fig. 8.3); additionally run-of-river power stations and hydroelectric power stations with reservoirs can be distinguished. The differences between the various types are not clear-cut; in practice, there are a number of combinations and mixed types. In the following the significant aspects of the different types are discussed briefly.

In addition, the term "small hydropower" is used quite frequently, generally without defining it clearly. In Germany, for example, 1 MW is often used as the limit between large and small hydroelectric power; in Russia, however, hydroelectric power stations below 10 MW and in Switzerland below 300 kW are called small hydroelectric power stations. Another distinction is made between "traditional" and "non-traditional" small hydroelectric power stations.

The so-called "traditional" small hydroelectric power stations are run-of-river power stations and power stations with reservoirs, with corresponding low electrical output, situated next to natural or artificial waters. In these power stations, the main emphasis is on electrical or mechanical energy production (i.e. they are not run as auxiliary plants within another higher-level technical plant). Examples are mills with one or several millwheels for power generation or small hydroelectric power stations with generally one or at most two turbines – mainly as run-of-river power stations or as diversion-type power stations. Particularly very small plants are often used to drive machinery directly.

In addition to these "traditional" small hydroelectric power stations, that are not very different from large power stations, there are the so-called auxiliary plants (i.e. power generation is a by-product of the entire plant). This occurs for example if water has to be reduced to a lower energy level for a particular reason. Instead of a throttle valve a turbine is used for converting the flow energy into electrical energy. Examples for such plants are drinking water supply grids, sewage and domestic water systems, turbines for the delivery of in-stream flows at dams or existing power stations, and the water supply for the "attracting flow" at fish ladders.

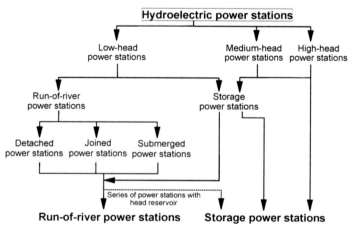

Fig. 8.3 Systematic of hydroelectric power utilisation

Low-head plants. Low-head plants harness the flow of a river, in most cases, without any storage; they are typical run-of-river power stations. They are characterised by a generally large flow and relatively low heads up to approximately 20 m. For example most of the run-of-river power stations in Germany are low-head plants /8-6/.

Depending on the plant layout, diversion-type and run-of-river power stations can be differentiated.

Diversion-type plants. In the case of diversion-type power plants, the actual powerhouse is outside the riverbed, somewhere along the course of a canal, into which the water is diverted. This water is taken from the river at a dam, is then diverted through the headrace or a pipeline to the power station, and transferred back to the riverbed at the end of the tailrace (Fig. 8.4). The so-called instream flow remains in the original riverbed. It is set according to ecological and economic criteria.

Additionally, side channel types and meander cutoff types can be distinguished in diversion-type power stations. For the former, an artificial waterway with a very low gradient concentrates the drop of longer river stretches together in one place, where the power station is built (so-called canal power stations; Fig. 8.4). Meander cutoff type run-of-river power stations have less damaging effects on the landscape than side channel-type power stations.

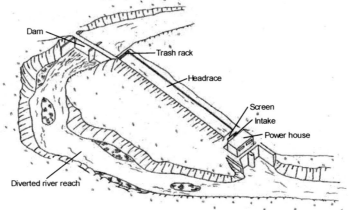

Fig. 8.4 Example of a typical diversion-type power station

Run-of-river plants. The term "run-of-river" power station is also used for power stations that are built into the actual riverbed. Depending on the setup of the dam and the turbine house the different designs described above can be used. Often, run-of-river power stations have additional tasks in the context of flood management, navigation (i.e. shiplock operation), and groundwater stabilisation. To give one example, Fig. 8.5 shows a possible design typical for a stream that has not been converted into a navigable waterway.

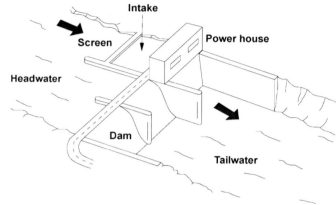

Fig. 8.5 Example of a typical run-of-river power station (see /8-5/)

For run-of-river power stations different arrangements of the powerhouse and the dam within the riverbed are possible (Fig. 8.6). In principle, block, twin block or multiple block designs can be distinguished; this differentiation is related to the respective locations of the dam and the powerhouse (Fig. 8.6). The third setup, the submergible design, is independent of the other two. Here, the powerhouse is within the dam (Fig. 8.6). These different designs can be subdivided further into the designs described below.

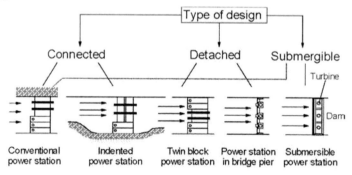

Fig. 8.6 Systematisation of run-of-river power station design

- Conventional block design. In a block layout, the longitudinal axis of the power house and the dam are perpendicular to the course of the river. This design is only possible if the highest flood can still be conveyed without problems through the retaining dam sections, without upstream flooding. The power station is normally located in bends because of the lower amount of bed load at the outer riverbank (Fig. 8.5, Fig. 8.6).
- Indented power station. The powerhouse is set up outside the actual riverbed in an artificially created bay (i.e. connected design; Fig. 8.6). This setup is re-

quired in very narrow streams for the dam to use the entire width of the river as a floodway.

- Twin block power station. Here, the powerhouses are situated on both sides of the river (i.e. detached design; Fig. 8.6). Such plants are normally built on border rivers, as it enables both countries to generate energy separately.

- Power station in pier. In a pier-type power station the mechanical installations and thus the powerhouse are identical with the piers that support the gates of the barrage (i.e. detached design; Fig. 8.6). This type of power station is characterised by favourable flow conveyance characteristics and a space-saving design.

- Submersible power station. Power station and dam are built in one block (Fig. 8.6). Thus the space required for machine groups and the required dam are reduced to a minimum. Therefore the plant fits in well with the surrounding landscape. There are hardly any plant parts visible above the headwater level.

If several run-of-river power stations are set up in a river directly one after the other, they form a power station chain. In extreme cases the backwater curve of the impoundment of a power station reaches upstream into the tailwater area of the next upstream power station; there are hardly any free flowing reaches. In some cases such power station chains are equipped with a large storage reservoir at the beginning or compensation reservoirs in between; they have quite large volumes with the capacity to store water for a certain amount of time to optimise performance.

Medium-head plants. The medium-head plants exclusively built as barrages mainly consist of a dam and a powerhouse at their base. Thus these plants use the head created by the dam, which can be between 20 and approximately 100 m high. The average discharges used by the turbines are partly obtained by appropriate reservoir management. If medium-head plants are built as diversion-type power plants, they sometimes use water from one or several streams, which is led through channels, free-flow or low-pressure tunnels to a compensation and storage reservoir and from there through pressure tunnels or shafts to the powerhouse.

High-head plants. High-head plants have a head of between 100 and 2,000 m (max.). They can be found in low and high mountain ranges and are normally equipped with a reservoir to store the inflowing water. The flow rates are relatively low. In contrast to the low-pressure plants, where the available power is a result of large flows, power is a result of high heads in this case. As the available water often comes from very small catchment areas, the effort it takes to capture water in the reservoirs is sometimes quite significant. Often, small streams are diverted from parallel valleys into the valley with the reservoir.

High-pressure plants can be planned as diversion-type or dam power stations. Diversion-type high-head power stations divert the water from the reservoir through tunnels or low pressure pipes via a so-called surge tank (reduction of water hammer); from there it flows to the turbine through penstocks or high pressure tunnels. The entire plants can be built into the neighbouring rocks (cavern power station). In the case of dam power plants, the power station is at the foot of the dam. The large reservoirs of the European Alps are all part of diversion type hydropower schemes. The power stations themselves are often far away from the reservoir in the lower valley of the main river.

Water from the reservoirs is sent to the power station for power generation according to demand. A distinction is made between daily, weekly, monthly and annual reservoirs plus interannual reservoirs. Annual reservoirs, for example, store the water from snowmelt in spring and summer in order to produce electricity in the following winter to cover the peak demand. The higher the available head, the smaller the reservoir may be, still producing at the same amount of energy.

Pumped storage power stations can have the same tasks as power stations with a reservoir. In addition, water can be pumped into the upper reservoir and thus store water from base-load power stations for a certain period of time; with turbines this stored energy can be converted into electrical energy again at times of peak electricity demand. Pumped storage power stations thus serve as electricity refiners (i.e. "conversion" of base-load into peak-load electricity). In addition, they can be used for frequency control. Pumped storage power stations with and without natural inflow have to be differentiated here; both types can be found.

Auxiliary plants. Nowadays, auxiliary plants can be increasingly found within drinking water supply systems. Pressure pipes take the water from high-level reservoirs to the consumer, and turbines or pumps running backwards, can be implemented in order to regain surplus energy. Thus only part of the energy used to pump the water into the high-level reservoirs can be regained (i.e. no utilisation of renewable energy). Renewable energy can only be obtained from wells or springs at a naturally higher elevation. The advantage of such plants is that only a turbine in the form of a reverse pump or a special axial turbine linked to a submersible generator is added. Furthermore reverse-running spiral pumps are used in some cases in sewage plants where the water level of the reservoirs is significantly above the outlet level. Thus the input energy can partly be regained.

A similar situation occurs in plants where dams were built for flood protection, to increase low flows or as drinking water or irrigation water reservoirs. The water that needs to be distributed can also be released via a turbine. The main purpose of such a reservoir is in those cases no longer energy production; thus the plant efficiency of such power stations is relatively low compared to the size of the dam or reservoir.

Another type of auxiliary plant is a turbine built into the dam of large diversion-type power stations to release certain instream flows into the river bed as required for ecological reasons (see Fig. 8.4). Only the direct height of the dam can be used as the head in this case and not the available overall head at the power station itself; however, part of the energy can be regained that would normally be lost due to the release of in-stream flows. The discharge required for a (not shown in Fig. 8.4) fish ladder cannot be used in this turbine. But the outflow from the turbine ought to be combined in a useful way with the attracting flow for the fish ladder. The advantage of constant or seasonally varying in-stream flows is that comparatively simply built – and therefore relatively cheap – turbines can be used because of the constant discharge.

8.2.3 System components

The turbine, the generator, and the transmission that might be required in some cases (Fig. 8.2), are all located in the powerhouse. Furthermore, structural and mechanical components and other system elements are required. In the following they are going to be described and discussed according to current state-of-the-art of technology. The technical information generally applies to medium and larger-size power stations, but it also applies to small hydroelectric power stations in most cases as well. The latter, however, might have a series of special components. Those will be mentioned separately – if relevant.

Dam, weir or barrage. The dam has to bank up the water in order to enable a controlled supply of water from the impounded water body to the power station. Thus the natural head of a stream is concentrated in one place. Upstream of the barrage or the dam is the reservoir with a certain storage volume. The dam and spillway need to be able to convey floods and maintain the water level at the required elevation in times of low water.

Dams can be built in many ways; they can take the form of fixed weirs, barrages, earth or rockfill embankments, masonry or concrete dams. Weirs can be fixed or movable, whereas barrages always have movable gates.

For smaller hydroelectric power stations the selection of retaining works depends on whether the headwater needs to be kept at a constant level. For very low heads and for plants in run-of-river design, the former is usually the case. Barrages or weirs with moveable gates have to be used in that case. These maintain a constant water level upstream of the dam and the turbine is regulated such that this water level is maintained. If the inflow to the impoundments exceeds the design discharge of the turbines, the gates have to be opened partly to release the excess inflow directly into the riverbed without further increasing the headwater level.

Mainly fixed weirs, fitted with flap gates and inflatable rubber dams are used in this case (Fig. 8.7). Flaps, often called fish-belly flap gates because of their profile, are driven hydraulically or by racks or chains; during a flood they are lowered on to the dam. Inflatable dams consist of a very firm, multi-layered rubber membrane and are filled with either air or water. A connected pump regulates the level of the dam crest using the internal pressure.

In the recent past inflatable dams have been increasingly accepted and – as long as there was no risk of vandalism – also proved to be reliable; however, the lack of aesthetics of unsubmerged inflatable dams may have to be considered. Both variations can be run without problems and can handle very high discharges. Ice, driftwood and bed load can be conveyed without causing any damage.

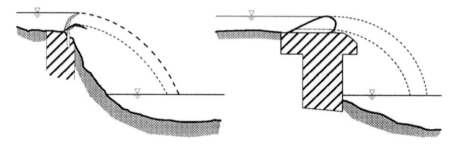

Fig. 8.7 Cross-section through a dam installation with fitted fish-belly flap gate (left) and inflatable dam (right) (see /8-7/)

For plants where the headwater level does not have to be kept constant, dams without moveable parts are also applicable. This is mainly the case for diversion-type power stations with a higher head. With higher heads the slightly varying headwater level has only little influence on the usable head and can thus be accepted in this case. Using such a fixed dam designed hydraulically to convey the design flood while maintaining a certain upstream water level, reduces the investment costs significantly. For fixed dams and weirs, very natural designs should be used (e.g. rough ramps), that might be used as fish passes at the same time.

For steeper streams and higher heads Tyrolean weirs are used for the diversion of water via the bottom without impoundment. A grid or a perforated plate built into the bottom of the riverbed conveys the water into a channel below. Tyrolean weirs have no moveable parts and can be overloaded very well. Bed load exceeding the diameter of the space between screen bars is conveyed downstream. There are ecological benefits if Tyrolean weirs do not stretch across the entire width of the stream and therefore part of the water remains in the river.

Reservoir. Mountainous regions (e.g. the European Alps) create the natural conditions for water storage. Such natural or artificial lakes – as daily, weekly, monthly or annual reservoirs or in some cases as interannual reservoirs – can create a balance between the fluctuating natural water supply and the equally fluctuating demand for electrical energy over time. Pumped storage power stations can also temporarily store a surplus supply of base-load power from thermal power stations or run-of-river power stations for its later use as peak-load power.

The delayed outflow from the reservoirs during low water periods in winter also contributes to an increased flow in the lower rivers and thus to an increase in power generation in the power stations installed there.

Intake. The intake establishes the connection between the headwater and the turbine. At the entrance of the intake, there is normally a trash rack keeping floating debris away from the plant. Gates or stoplogs are also integrated into the intake structure in order to seal the hydroelectric power station during maintenance works or to interrupt the flow into the power station in case of an accident. For very small plants these security installations are not necessary or are installed as simple sliding gates.

Headrace/Penstock. Water from the reservoir either flows directly through an intake or initially via a headrace channel, tunnel or pipe, and a penstock to the turbine. It is important to keep hydraulic losses at a minimum, which usually is achieved by large enough cross-sections and hydraulically suitable cross-section geometry.

If required, the so-called surge tank is located in front of the intake into the penstock. It reduces water hammer and pressure fluctuations that occur when starting and shutting the plant down, but also during each load alternation because of the water's inherent inertia.

The hydraulic connection between the headwater or the intake structure, and the turbine is established through the tunnel and the penstock and the distance between the plant components is overcome. Because of losses along the upstream waterway (i.e. intake structure, tunnel, penstock) a small part of the water's potential energy cannot be used for energy production.

Depending on the topography, but also the ecological and economic framework conditions, various combinations of upstream waterways are possible. As one example, the inflow from the headwater can be realised via an open upstream channel or as a pressure-free low gradient tunnel. In run-of-river power stations with low head, the water can flow directly from the intake structure to the turbine. In that case the tunnel, surge tank and penstock are not required.

Penstocks are normally built from individual pipe segments of welded steel pipes. Tunnels can be built for open channel flow, low-pressure or high-pressure flow. Depending on the hydraulic requirements, they either have a reinforced concrete lining or, in particular with high-pressure plants, a steel armour. In small power plants other materials are used for the upstream inflow (e.g. PVC pipes or more recently, wooden pipes with straining rings).

Powerhouse. In the powerhouse the main parts of a hydroelectric power station are situated. These are turbines, sometimes transmissions, generators, control systems, in some cases a transformer with an electric power substation and sometimes shut-off valves for the pipelines.

Turbines. The hydraulic machine that converts the upstream energy into a rotation is called turbine. Water wheels were the forerunners, but they are only rarely used nowadays.

Because of the different heads and flow rates and the resulting varying water pressure and speed conditions, various turbine types are built. They can be categorised, according to the energetic conversion, into impulse turbines and reaction turbines.

- Reaction turbines. Reaction turbines convert the potential water energy mainly into pressure energy that is transferred to the turbine blades, where it is converted into rotation. Reaction turbines are i.e. Francis, propeller, Kaplan and Straflo turbines. Currently, the maximum power produced by Kaplan turbines is 500 MW per unit, and for Francis turbines, approximately 1,000 MW per unit.

- Impulse turbines. In impulse turbines the potential and the pressure energy of water is converted completely into velocity energy. This energy is then transferred to the turbine which converts it into mechanical energy. The pressure before and after the turbine is the same; it is roughly the same as the atmospheric pressure. Impulse turbines are Pelton turbines and cross-flow turbines. Currently, the maximum power output for Pelton turbines is approximately 500 MW per unit.

Nowadays, heads from 1 to almost 2,000 m are exploited /8-8/. The common areas of use for medium-size and larger capacities are (Fig. 8.8):

- Pelton turbines   approximately 600 to 2,000 m
- Cross-flow turbines   approximately 1 to 200 m
- Francis turbines   approximately 30 to 700 m
- Kaplan turbines, vertical axis   approximately 10 to 60 m
- Kaplan turbines, horizontal axis   approximately 2 to 20 m

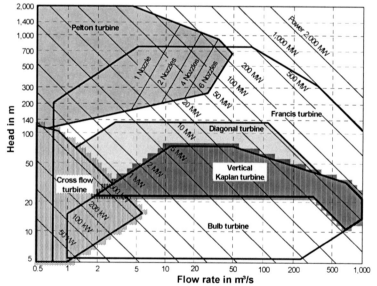

Fig. 8.8 Application of different turbine types (see /8-6/)

For smaller capacities, Pelton turbines are sometimes used from 50 m, and Francis turbines from 6 m, in old small power stations even for heads as low as 2 m.

Depending on the turbine type and size, efficiencies between approximately 85 and 93 % are achieved at the design discharge and head. The turbine efficiency is defined as the power at the turbine shaft and the available hydraulic power between the turbine intake and the draft tube outlet, including turbine outlet losses. As the turbine is only designed for a specific discharge, the efficiency is dependent on the respective availability of water. Fig. 8.9 shows the sometimes very different efficiency curves for the main turbine types. Pelton turbines, for example, have a very good efficiency even when working at only 20 % of their maximum discharge. Propeller turbines, however, should not be operated at less than 70 to 80 % of their design flow, as high losses are generated otherwise.

Layout and function of different turbine types are discussed in more detail below.

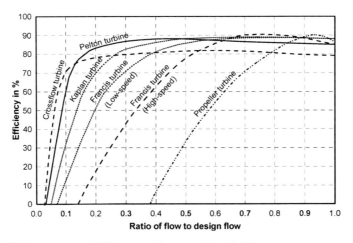

Fig. 8.9 Efficiency curve of different turbine types (see /8-6/)

Kaplan, propeller, bulb, bevel gear, S and Straflo-turbines. The Kaplan turbine and the designs derived from it in principle operate like a reverse operating propeller. With the exception of vertical shaft Kaplan turbines with radial approach flow, the flow runs through all these turbines axially; in the case of vertical Kaplan turbines the water is directed via the adjustable guide vanes to the rotating runner. Vertical, horizontal and slanting axis positions can occur. Additionally, Kaplan turbines and their derivatives have adjustable runner blades (so-called double regulation turbines). They allow a better adjustment to different flow rates and thus an efficiency improvement for various operating conditions.

Propeller turbines have rigid runner blades. With these blades they are not very adjustable to changing flows but show high efficiency at design discharge. They are mainly used if several machines within a hydroelectric power station are utilised in combination; thus it can be ensured that the individual machines are mainly used at design discharge.

Bulb turbines are Kaplan turbines with an almost horizontal axis. Their axial flow does not require any change of direction of the flow. This principle reduces hydraulic losses. Here, the generator is located directly in front of the turbine within a steel bulb. The generator can be accessed via a shaft. Their design is similar to the Straflo turbine described in Fig. 8.10. However, the bulb is bigger, as it accommodates the bearings of the turbine and the generator.

Bevel-gear turbines are built in a similar way. The submerged housing only contains a bevel-gear pair that converts the rotation of the turbine shaft to the generator axis that is vertical to the turbine axis; the generator is attached to the turbine housing and freely accessible.

The draft tube of S-turbines is bent down significantly immediately behind the almost horizontal located turbine. The turbine shaft is led outside to the generator; thus this is freely accessible, i.e. for maintenance works.

Straflo-turbines – Straflo is the abbreviation for straight flow – have a similar system design as Kaplan turbines and operate with external rim generators; i.e. the rotor sits on a ring attached to the runner blades of the turbine. This design is also shown in Fig. 8.10. In the steel bulb behind the runner is only the bearing of the turbine. The disadvantage of this design is the costly sealing between the runner and the generator.

Because of their design a flat efficiency curve at an overall high level can be realised for Straflo-turbines (Fig. 8.9). Due to their double regulation, all Kaplan turbines and derived designs can be operated over a broad partial-load range (30 to 100 % of the design power) with comparably high efficiency levels.

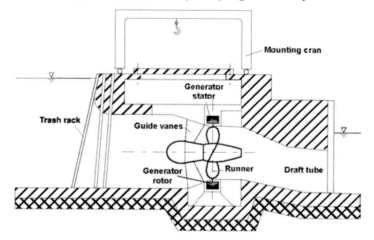

Fig. 8.10 Run-of-river power station with Straflo turbine (see /8-9/)

Francis turbines. In Francis turbines – they are typical reaction turbines – the water flows radially from the guide vanes into the scroll case, over the runner blades, and then flows out again axially, which causes a redirection of the flow. In contrast to the Kaplan turbine the Francis turbine (Fig. 8.11) runner blades are not movable. The flow is only regulated via the guide vanes; slow and fast rotating turbine wheels are differentiated here (so-called low-speed and high-speed runners). In larger machines with higher head, the water flows through a spiral to the guide vanes. In smaller machines the guide vanes are situated in a simple turbine pit.

In general, high rotational speed of the runner has to be aimed for, as this results in low torques at the turbine axis and thus smaller machine dimensions; thus the turbine costs, and in some cases the costs of other power station components, can be reduced. Francis turbines can normally be run from 40 % of their maximum power (Fig. 8.9). In the case of high-speed runners, high efficiencies only occur at about 60 % of the maximum flow.

Fig. 8.11 shows a run-of-river power station with a Francis-shaft turbine as an example. The adjustment of the guide vanes is done via a rotating shaft behind the turbine shaft that enables opening and closing the vanes.

Older small hydroelectric power stations often have Francis shaft turbines at heads as low as 2 m (see Fig. 8.8). If these plants were newly built nowadays, double regulation Kaplan tubular turbines or S-turbines would be used. With their very good efficiency curve over a broad range of discharges, they guarantee an optimum exploitation of energy. If old plants are reactivated, the entire in and outflow areas would have to be adjusted. The work involved in this is often so expensive that Francis machines are put in again, although they have a slightly less favourable efficiency curve.

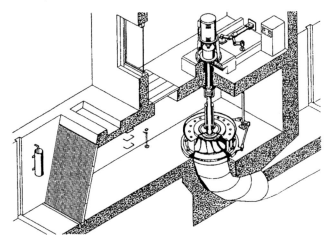

Fig. 8.11 Power station with a vertical Francis turbine (see /8-10/)

Pelton turbines. The Pelton turbine – an impulse turbine – has a runner (Pelton wheel) with fixed buckets. It is regulated by one or several nozzles with spear-type valves that control the flow and direct the water jet tangentially to the wheel into the buckets (Fig. 8.12). During this process the entire pressure energy of the water is converted into kinetic energy when leaving the nozzle. This energy is converted into mechanical energy by the Pelton wheel; the water then drops more or less without energy into the reservoir underneath the runner.

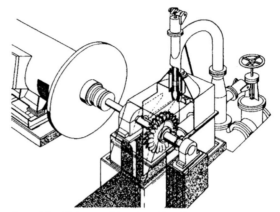

Fig. 8.12 Power station with a Pelton turbine (see /8-11/)

The Pelton turbine is characterised by a relatively flat efficiency curve (Fig. 8.9). It is therefore suitable for highly variable flows between approximately 10 and 100 % of the design discharge.

Pelton turbines can also be used in small hydroelectric power stations – but usually for lower heads than in large hydroelectric power stations (Fig. 8.8). Heads from 30 m can be enough to drive small Pelton turbines. For larger discharges, models with several nozzles (up to 6) are also used.

Cross-flow turbines. In cross-flow turbines the water reaches the runner after a simple guide vane and flows through the runner from the outside to the inside and then to the outside again after having passed through the interior of the wheel (Fig. 8.13). The runner is built like a roll and separated into two chambers in the direction of the axis at a ratio of two to one, thus allowing for better efficiencies at partial flows.

In a way similar to the Pelton turbine, the cross-flow turbine is particularly suitable for highly varying flows. The roll-shaped runner can be driven – depending on the available discharge – by opening only the small, the large, or both chambers under partial load. Thus the turbine efficiency curve is very flat (see Fig. 8.9). However, the absolute efficiency is not as high as for other turbine designs. Additionally, very low heads cannot be used entirely. Because of its good adjustment possibilities to varying flows and the simple, robust – and therefore inexpensive – design, the cross-flow turbines are often built into small hydroelectric power stations. For medium and low heads a draft tube is necessary to improve head utilisation.

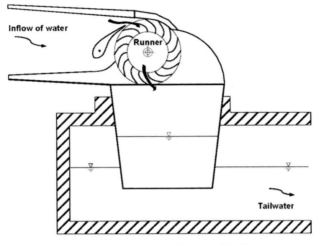

Fig. 8.13 Operation principle of a cross-flow turbine (see /8-6/)

Water wheels. As turbines are the most expensive parts of a small hydroelectric power station, water wheels can still be used if discharges are very small. They are suitable for heads of a maximum of 10 m; for lower heads there are hardly any technical limits. Discharges up to approximately 2 m³/s can be used with a single conventional wheel. Water wheels can reach efficiencies of 70 or 80 %. Under-shot, middle-shot and overshot water wheels are used – according to the available head (Fig. 8.14).

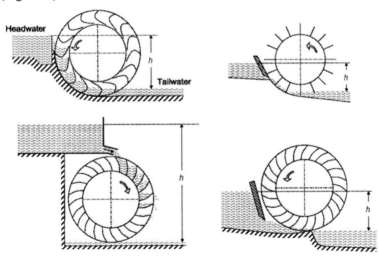

Fig. 8.14 Types of water wheels (see /8-6/)

Water wheels run with a lower rotational speed of approximately 5 to 8 rotations per minute. Therefore transmissions or belt drives are required to achieve the higher rotational speed necessary to drive the generator.

Initially, water wheels were mainly made of wood. Soon axles, wheel rims and spokes made of steel gained acceptance. Up to the present day the blades themselves are still made of wood in many cases. They are normally run on bearings on both sides and should, if there is a risk of heavy icing, be stored in a closed wheel chamber. Water wheels that run on bearings on one side only, coupled with a standard planet transmission and a generator unit are a recent development. The complete wheel-generator-transmission unit can be prefabricated and delivered to the place of installation. The customer only has to build the water intake, the flume for the wheel itself and the supporting foundation. The complete unit is put in and can virtually start operating immediately afterwards.

Outflow and tailrace. The water leaves the turbine into the so-called tailrace. For reaction turbines, a draft tube is used to make better use of the available head and reach all the way to the tailwater. The cross-section of the draft tube is increasing and shaped like a diffuser at the exit to the tailwater. Therefore the speed in the draft tube is reduced before entering the tailwater; thus part of the kinetic energy of the out flowing water can be used energetically within the turbines.

Shaft coupling and transmission. The turbine and the generator can be either directly coupled or indirectly, via a transmission. In large plants, turbine and generator are installed on the same axis. In this case only a simple coupler is used. In very small plants, however, transmissions are used in most cases because the increase of the rotational speed through the transmission allows the use of standard, high-speed generators. Toothed gearing for smaller units and toothed gearing or belt drives for very small machine units (up to 100 kW) are used in most cases. Transmission efficiencies, defined as the ratio of the power at the transmission output shaft and the capacity at the turbine shaft, reach approximately 95 to 98 %.

If the head is low, turbines in small hydroelectric power stations run with relatively low speed at e.g. 75 rpm. For cost reasons, generators are kept relatively small; they have to run at a much higher speed. Therefore in smaller power stations a transmission or a belt drive has to be inserted between the turbine and the generator.

Belt drives have proved to be efficient for up to approximately 50 kW electrical power and are used for significantly higher power levels nowadays. They are slightly cheaper than transmissions. But their life spans are generally shorter and they need more maintenance. As nowadays the belts can be repaired or exchanged in a very short time by mobile maintenance units, they are often the most cost-efficient alternative.

Generator. In the generator mechanical energy from the turbine or the transmission shaft is converted into electrical energy. Synchronous or asynchronous generators can be used here (for a functional description see Chapter 6.2.2). Synchronous generators are used if the plant is run in isolated operation or if the power generated by the hydroelectric power station is the main supplier of the grid. They can regulate the mains voltage and deliver reactive current. Asynchronous generators can only be run as part of an interlinked power system that supplies them with energising current. Their design is simpler than the design of synchronous generators and their efficiency is slightly lower. Overspeed can occur due to load rejections in the electricity grid, and the generators have to be able to withstand that.

The efficiencies of standard generators used in small hydroelectric power stations at rated capacity are between approximately 90 and 95 %. In large plants generator efficiencies between 95 and 99 % can be achieved. Generator efficiency is defined as the capacity at the generator clamp in relation to the capacity at the transmission shaft.

Transformer. A transformer that converts electrical energy from one voltage level to another is used if the output voltage of the hydroelectric power station does not correspond with the voltage of the grid it is supposed to be fed into. Such a component is characterised by efficiencies of up to 99 %.

Regulation. Depending on the mode of operation of the hydroelectric power station, different forms of regulation can be considered. Isolated operation always requires frequency regulation. A controlling device maintains the grid frequency at varying load. A simultaneous regulation of the flow or the headwater is not possible in that case.

For operation within an electricity grid, water level or flow regulations are common. The turbine opening, i.e. the guide vane and/or the runner blades, are opened enough to maintain the desired headwater level or to maintain the desired flow through the turbines. The turbines and generators always rotate at constant rotational speed, in line with the grid frequency.

8.2.4 Isolated and grid operation

Hydroelectric power stations are often operated in connection with the public services grid. A few exceptions are plants that drive machines directly (i.e. supply mechanical energy) or supply smaller industry and/or craft businesses (e.g. sawmills) with electricity.

Isolated operation requires an exact match between the demand for and the supply of electrical energy. This can only be achieved by a hydroelectric power station on its own if the plant's design flow and capacity is so low that there is always enough water available to generate the maximum amount of required power (i.e. the power in demand cannot be more than the power generated). Accordingly, water always has to be bypassed by the turbines when the achievable maximum power output is not required. This is generally uneconomic and can hardly be found in industrialised countries these days.

Like large hydroelectric power stations, small hydroelectric power stations can be operated in isolated or grid operation. They also require the corresponding synchronous or asynchronous generators that feed the grid with generated alternating current at low-voltage.

Very small hydroelectric power stations are also often used to supply remote settlements. Nowadays, synchronous generators can be bought for isolated operation, complete with voltage regulator, and suitable for reactive power generation. The frequency depends on the turbine speed and is mainly kept at a constant level by electro-magnetic or electronic frequency regulators.

8.2.5 Energy conversion chain, losses and power curve

Energy conversion chain. A hydroelectric power station consists of the components already described (Chapter 8.2.2). An energy transformation chain is caused by the interaction of these system elements, which is represented schematically in Fig. 8.15.

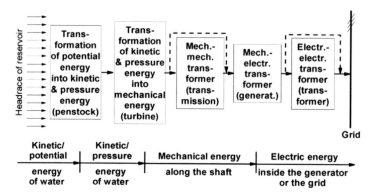

Fig. 8.15 Energy transformation chain for the use of hydropower (mech. mechanical, electr. electrical, generat. generator; see /8-5/)

Therefore in the intake structure and/or the penstock, the potential and the kinetic energy of the water in front of the dam is converted into pressure and kinetic energy before reaching the turbine. The water then flows through the turbine; here the water energy is transformed into rotary movement, and thus into mechanical energy of the runner and the turbine shaft respectively. This kinetic energy is sometimes transformed to another rotational speed by a transmission and then fed into the generator where the mechanical rotation is converted into electrical energy. Afterwards, an additional electrical-electrical transformation can be required within a transformer that enables feeding the electrical energy into the public services grid at its specific voltage level.

Losses. Within the energy transformation chain considerable technically unavoidable losses occur; the result is a lower level of energy at the plant exit than the available energy in the water between the head and tailwater. Therefore Fig. 8.16 shows the respective losses and their average range within the individual energy transformation steps and components.

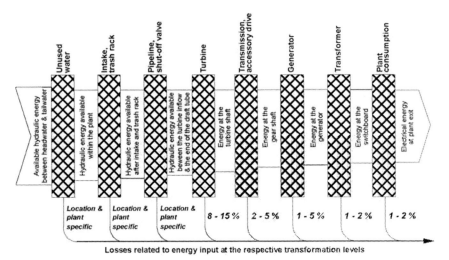

Fig. 8.16 Energy flow inside a run-of-river hydroelectric power station (see /8-5/)

The losses within the plant mainly occur at the water intake, at the screen, within the channels, pipes, penstocks, and the shut-off valves (if applicable), within one or more turbines, possibly within one or more transmissions and one or more generators. In larger plants there may also be corresponding conversion losses within the transformer. An additional loss is the potential energy of the water released at the dam (e.g. at times of flooding). The energy losses in the hydraulic part in particular are very location and plant-specific. A generally applicable and transferable indication of the magnitude is therefore not possible; at the very best there are only losses of a few percent. Together with the losses inside the other system components, total efficiencies clearly over 80 % can be achieved at full load; sometimes 90 % and above can be achieved. The total efficiency is the ratio of the electrical power at the plant exit and the power difference between head and tailwater minus the water released over the spillway. As hydroelectric power stations are often run at partial load, the annual average utilisation factors are generally lower; in modern and properly designed hydroelectric power stations they range between 70 and a maximum of 90 %; in older plants, especially within the low power range, they can be significantly lower, at 50 to 70 %. In connection with the total working capacity of the water the utilisation factors are significantly lower, as one part of the incoming water (i.e. floodways) is diverted, unutilised, across the dam since a portion of the annual water volume usually has to be released over the spillway during floods.

Operation behaviour and power curve. The systems performance and thus the operating performance of a run-of-river power station for the supply of electrical energy during the course of the year largely depend on the available flow and the current head. Fig. 8.17 shows the interaction of these components, including the corresponding turbine flow over the course of one year.

In the example the head rises during the summer months as the tailwater level drops with a decrease in flow. Because of the conditions at the dam, the headwater is kept at a constant level. The head correspondingly decreases during the winter and during spring. The outflow increases and thus the tailwater rises slightly.

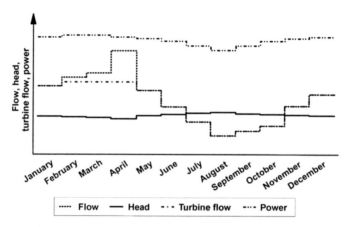

Fig. 8.17 Operation scheme of a run-of-river power station (see /8-5/)

The turbine flow is linked to the discharge in the river; it therefore decreases during the summer in line with the lower discharge. As the turbine is only designed for a given maximum flow (so-called design flow), only the design flow can be utilised, even if the discharge in the river is higher. The additional discharge has to be spilt unutilised. In the example represented in Fig. 8.17, this is the case in February, March and particularly in April.

The power output of the hydroelectric power station is almost proportional to the flow through the turbine (Equation (8.5)). Thus in the example given in Fig. 8.17, it decreases during the summer months, which are characterised by decreasing flows. Furthermore, power output is also dependent on the available head (Equation 8.5)); as the head normally does not change as much as the flow through the turbine, its influence is generally less significant. The dependency of the power generation of the hydroelectric power station on the head, however, is responsible for the slight power output decrease in April – in spite of appropriately high flows.

These correlations can also be described by the power duration curve schematically represented in Fig. 8.18 for a typical run-of-river power station. According to this plan, a run-of-river power station with a defined design is characterised by a specific flow at a corresponding head; together this results in a certain power output over time. Considering these design conditions, the values change with increasing and decreasing flows.

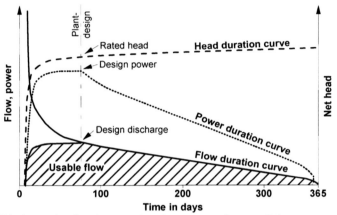

Fig. 8.18 Discharge, head and power duration curves of a run-of-river power station (see /8-2/)

- Power generation decreases with decreasing flows. The corresponding head rises as the tailwater drops slightly due to the smaller flow. If the discharge is very low, the plant has to be switched off at a certain point, as turbines cannot operate flows below a certain minimum. Under these conditions no power can be generated if the turbines are dimensioned correctly; this is only – if at all – the case on very few days of the year (Fig. 8.18).
- Power generation also decreases with increasing discharge. The turbine cannot process discharges exceeding the design capacity. The power generation is reduced as the difference between the head and tailwater level decreases with an increase in the discharge. The additional flow has to be spilt and can therefore not be used energetically. In the worst case, power generation is no longer possible, as the difference in height between head and tailwater is too small. That is normally the case if the gates of the dam or barrage are opened (i.e. during floods).

8.3 Economic and environmental analysis

Hydroelectric power stations have contributed to meet the energy demand for over 100 years. The costs related to this and the evaluation of the relevant environmental impacts is analysed in the following paragraphs. We will, however, define selected plants that will be subject to the following economic analyses. It has to be considered that run-of-river power stations – as opposed e.g. to wind energy conversion systems or photovoltaic plants – are power stations that are largely determined by the prevailing local conditions. Thus material input – and with it, the costs involved – can sometimes vary significantly for different run-of-river power stations. Generally applicable cost estimates can therefore only be made to a very limited extent.

8.3.1 Economic analysis

Hydroelectric power stations are characterised by a number of possible design types. Therefore only selected reference plants can be analysed here, that ought to be seen as examples for possible plant configurations defined by the specific local conditions. This is particularly true for the structural components (i.e. water intake, head and tailraces or penstock, power house), which sometimes vary a lot in different hydroelectric power stations.

Within the following analysis we will take a closer look at four hydroelectric power stations, as their reference technology is typical for extensive territories throughout the world (Table 8.1). According to this it is distinguished between two small hydroelectric power plants with an installed electrical power of 32 and 300 kW respectively, and two larger plants with a power of 2.2 and 28.8 MW respectively.

- 32 kW low-pressure (Plant I). The water is diverted at a small dam and then processed with a head of 8.2 m via a 110 m long fiberglass pipe with a Kaplan turbine. The electrical energy generated by a directly coupled, synchronous induction generator is fed into low voltage grid.
- 300 kW low-pressure (Plant II). The power station utilises the head of 4.6 m, created by a concrete dam followed by bypass reach of 200 m. From the intake structure, the headwater channel leads directly to the power station. The electrical energy generated by a Kaplan turbine connected by a transmission to the synchronous generator is fed into the medium voltage grid near the power station via a transformer.
- 2,200 kW low-pressure (Plant III). This run-of-river power station is designed as a river power station. The head of around 6 m is created with a dam. The backwater reach of almost 2,000 m is secured by longitudinal dykes. The hydraulic energy is converted into mechanical energy by a Kaplan turbine and then further converted into electrical energy by a synchronous generator. This electrical energy is fed into the medium voltage grid next to the power station via a transformer.
- 28,800 kW low-pressure (Plant IV). This run-of-river power station is also designed as a river power station; the powerhouse is located in the riverbed. With a catchment reach of approximately 8.8 km, it achieves a head of 8.3 m. The two Kaplan bulb turbines have a combined capacity of 425 m^3/s and the elec-

tricity from the synchronous generators is fed into the high voltage grid via two transformers and a metal-clad SF_6-(sulphur hexafluoride)-switchboard plant. The hours at peak load (theoretical) of the analysed plants are in line with the dimensions typical for such plants installed in Europe. The technical lifetime of the structural components is assumed to be 70 years and the machinery plant components to be 40 years. For the plants' own use, 1 % of the produced electrical energy has been assumed.

Table 8.1 Technical parameters of the analysed reference systems

Reference plant		I	II	III	IV
Nominal power	in MW	0.032	0.3	2.2	28.8
Type of power station		low-pressure	low-pressure	low-pressure	low-pressure
Turbine type		Kaplan	Kaplan	Kaplan	Kaplan
Head	in m	8.2	4.6	5.9	8.0
Design flow	in m^3/s	0.5	8	40	425
Full load hours	in h/a	4,000	5,000	5,000	6,000
Annual work (gross) in GWh/a		0.128	1.5	11	173

In order to be able to estimate the costs for these hydroelectric power stations the investment costs and the operation costs are analysed initially. The specific electricity generation costs are derived from these costs.

Investments. The plant costs are mainly the expenditure for the structural components (i.e. power house, dam, water intake, gates, screen and trash rack cleaner), for the mechanical components (i.e. check valves, turbines), for the electrical engineering components (i.e. generator, transformer, energy output) and the other incidental expenses (i.e. acquisition of land, planning, authorisation).

These costs are very location-dependent; overall and generally valid guidelines can therefore not be established. In many cases the costs for structural works are 40 to 50 % of the overall costs. Mechanical component costs (i.e. turbines, transmission, and regulator) are approximately 20 to 25 % for larger plants and up to 30 % of the overall costs for small hydroelectric power stations. Roughly 5 to 10 % have to be set aside for electrical installations. What remains are miscellaneous costs (i.e. planning costs, additional construction expenses, overheads, building interest rates (i.e. interest payments on invested capital during the building period)). Independently, costs for today's increasingly comprehensive ecological compensation measures can account for 10 to 20 % of the overall plant costs. Especially for run-of-river power stations with a large catchment reach, these expenditures (e.g. appropriate design and structuring of the reservoir fish ladders) can lead to a significant increase in the overall costs.

An increase in plant size, however, generally leads to significantly lower investments. To give an example, the specific investments for new plants with electrical power output of below 100 kW, the specific investments are between 7,700 and 12,800 €/kW. For new plants with an electrical power output of 1 to 10 MW, they are as low as 4,100 to 4,600 €/kW (Fig. 8.19).

In comparison with the additional construction of new plants, the costs of re-conditioning old plants or for their modernisation are significantly lower. For the reconditioning of plants between 1 and 10 MW, costs of approximately 1,500 €/kW and for the modernisation of approximately 1,000 €/kW are quoted. Reconditioning costs are largely dependent on the availability of plant components that can still be used.

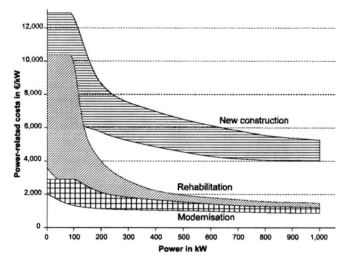

Fig. 8.19 Specific investments of hydroelectric power stations relative to plant capacity (see /8-12/)

Apart from the rated capacity, the costs of run-of-river power stations are also dependent on the head at the potential location. Therefore in most cases, plants with the same rated capacity are characterised by lower specific investments with increasing head between headwater and tailwater level.

Operation costs. In optimally designed and low-maintenance hydroelectric power stations the operation costs are very low. Variable costs occur i.e. for staff, maintenance, administration, provisions for plant renewals, disposal of screenings, and insurance. The individual cost allocations can vary a lot from plant to plant, depending on the local conditions. The annual operation costs are approximately around 1 to 4 % of the overall investment. For small and very small hydroelectric power stations, they are generally higher than for large plants.

Electricity generation costs. The annual average real production costs that remain constant over the amortisation period of a hydroelectric power plant are derived from the overall investments. For smaller plants with an output below 1 MW, and larger plants with an output above 1 MW (Table 8.2) the technical lifetime of the structural components is assumed to be 60 and 80 years. By the same classification, the technical lifetime for the machines and the electrical parts are 30 and 40 years respectively. In line with the methodology used so far, an interest rate of 4.5 % is assumed.

Starting from these framework conditions, the resulting electricity generation costs can be calculated for hydroelectric power stations being built today, considering the predicted annual energy production. For the 32 kW hydroelectric power station previously considered as an example (Plant I, Table 8.1), it comes to around 0.065 €/kWh. Due to the high investments lower generation costs can only be achieved at full load hours that exceed the full load hours assumed within this example, and at lower operation costs (e.g. very low costs for plant operation in the case of private ownership). In contrast, the analysed small hydroelectric power station with 300 kW rated capacity (Plant II) generates electricity at costs of around 0.073 €/kWh; this hydroelectric power station is therefore – because of the higher location-specific investment costs compared to Plant I – characterised by higher electricity generation costs. Thus this plant does not necessarily reflect the normally recognisable trend where electricity generation costs normally decrease with increasing plant capacity. As a contrast, the first large plant we looked at as an example shows this trend with an installed electrical output of 2.2 MW (Plant III); it is characterised by specific electricity generation costs of 0.049 €/kWh and thus has the lowest specific costs for the generated kilowatt hour of electrical energy of all the plants analysed here. On the other hand the analysed 28.8 MW plant generates electricity at higher costs (Plant IV); this is due to the higher costs in spite of the higher full load hours. This is mainly caused by high environmental protection regulations (Table 8.2).

Table 8.2 Investment and operation costs plus electricity generation costs of the analysed hydroelectric power stations

Reference plant		I	II	III	IV
Nominal power	in MW	0.032	0.3	2.2	28.8
Annual work (gross)	in GWh/a	0.128	1.5	11	173
Investments					
Struct. components	in %	63	57	60	52
Electric. components	in %	37	43	40	48
Total	in Mio. €	0.138	1.67	9.1	167
	in €/kW	4,310	5,570	4,140	5,800
Operation costs	in Mio. €/a	0.001	0.02	0.09	1.7
Electricity gen. costs	in €/kWh	0.065	0.073	0.049	0.058

The specific electricity generation costs are normally lower if already existing plants can be reactivated or thoroughly modernised. In spite of the high level of dependency on the location, under such conditions the costs would be between 0.03 and 0.08 €/kWh; the bottom margin of this range is again determined by plants with larger power output and the top margin by small and very small hydroelectric power stations. If only the hydroelectric mechanical components have to be renewed as part of a general overhaul, even lower electricity generation costs are possible; depending on the individual local conditions, they can be between about 0.025 €/kWh for plants in the MW-range and approximately 0.05 to 0.08 €/kWh for small and very small hydroelectric power stations in the range between less than 10 to some 100 kW.

The electricity generation costs are influenced by a number of different parameters. In order to show their influence, the main parameters have been varied in Fig. 8.20, using the example of a newly constructed run-of-river power station with an installed power of 300 kW (Plant II; Tables 8.1 and 8.2). It shows that the investments and the full load hours have the most significant influence on the specific electricity generation costs. If the overall investments increase for example by 20 %, the specific electricity generation costs increase from 0.073 to 0.086 €/kWh and thus by 18 %. On the other hand, the operating costs have hardly any influence on the electricity generation costs. The amortisation period equally has comparatively little influence on the specific costs of the electrical energy provided.

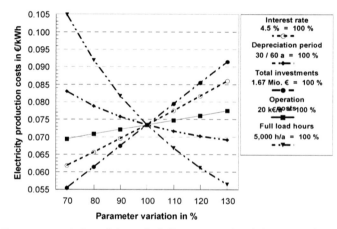

Fig. 8.20 Parameter variation of the main influences on electricity generation costs

8.3.2 Environmental analysis

When building, during normal operation, in the case of an accident and at the end of the operation of a hydroelectric power station, the environmental effects discussed below might occur.

Construction. Hydroelectric power stations are – similar to wind power stations – partly products of "traditional" mechanical engineering and electrical engineering. Therefore, for example, during the production of turbines or generators, all kinds of environmental effects on soil, water and air typical for these sectors can occur. Because of the current far-reaching environmental regulations, the corresponding environmental effects are generally kept at a comparatively low level. The potential for accidental damage during the production process is generally – with some specific exceptions (e.g. steel smelting) – relatively low.

Additionally, there are environmental effects linked to the construction of hydroelectric power stations – and hydroelectric power stations with reservoirs in particular – at the plant location. To give some examples, the following possible environmental effects in the context of the new construction, reactivation or modernisation of hydroelectric power stations during the building phase, are to be mentioned here /8-13/.

- Water pollution caused by construction material or fine soils getting into the stream, release of fine particles caused by excavation, improper cleaning of building machinery, etc.
- Oil losses due to improper handling during construction and maintenance works.
- Oil seepage, commonly from hydraulic systems, e.g. during demolition works.

These environmental pollutants can be avoided, or at least minimised by introducing the appropriate operational procedures and observing the existing safety and environmental regulations. Additionally, there is a corresponding accident potential – also with supra-regional environmental effects – especially when building hydroelectric power stations with large dams and reservoirs or larger run-of-river power stations (e.g. in the case of flooding). However, if the appropriate regulations are observed, no significant environmental effects have to be expected.

Normal operation. During the operation of hydroelectric power stations, no toxic substances are released directly, with the exception of potential lubricant losses. The resulting environmental effects can be kept low or ruled out by using biodegradable lubricants, or, in the case of small hydroelectric power stations, by using lubricant-free machines.

In addition, hydropower plants often serve multiple purposes (especially in the case of power stations with reservoirs) such as water sports, fishing, irrigation, flood protection, storage of drinking water. This may in some cases also lead to ecological advantages, for example a positive influence on the groundwater table, the creation of aquatic biotopes, the enrichment of water with oxygen through turbine operation.

However, the use of hydroelectric power can influence the environment in other ways. The main three problem areas are described below (see also /8-13/, /8-14/, /8-15/, /8-16/, /8-17/).

Impoundments. River and diversion type power stations generate backwater effects or reservoirs which influence the ecological conditions in the affected river sections and the adjacent riverline habitats. In the impoundments, the flow velocities are significantly reduced and thus the shear stress at the river bottom decreases. This leads to an increase in sedimentation of small-grained bedload material which is usually in suspension (e.g. fine sand, silt and clay). Coarse structures of the riverbed that had existed so far (such as riffles, runs or pools) and hence the habitats for fish and other small biota are covered by this fine matter. The usual habitats are taken away from the organisms that live in the hyporheic zone and at the river bottom; but that is exactly where the majority of biological processes occur in running waters in medium and high altitudes. This leads to a loss in the diversity of local habitat conditions.

Further consequences of the damming of rivers frequently are an increase in water temperature and thus a reduction in the oxygen content in deep reservoirs. Furthermore, every reservoir interrupts the continuous flow, i.e. it is a barrier that can only be overcome with great difficulty, or not at all, by the many organisms in running waters that migrate or move around for various reasons. In more or less stagnant reservoirs they cannot find the living conditions they need. Therefore, the natural running waters species composition changes; due to the different food available for predators, this can, for example, lead to a total shift in the composition of species that also includes mammals, birds and amphibians. In the worst case methane can occur because of the anaerobic decomposition of organic sediments (i.e. biogas).

Furthermore, the agglomeration of pollutants (especially heavy metals) in these fine-grained sediments can lead to an increase in the pollutant concentration in the reservoir. Apart from the eco-toxic dangers, the disposal of dredging material from the reservoir causes further problems.

Through flushing of the reservoirs, within very short periods large amounts of mainly small-grained sediments might be released. In the past, this has often led to a strong impairment of the ecosystems in the lower course of the rivers. Reservoir flushing during natural floods with a slowly increasing discharge, a high enough oxygen concentration in the water, an acceptable maximum concentration of suspended matter and pollutants, plus the adjustment of the flushing time to the development stage of the fish fauna, can help minimise these negative effects.

Apart from fine sediments, the coarser bed load is also held back in the reservoir. Due to the lack of sediment downstream of the dam, erosion, and thus an incision of the riverbed, can occur. This leads to a lowering of the groundwater table linked to the stream, which can lead to the draining of wetlands and to a change in the riparian vegetation.

The effects are most serious when linking up several power stations in series (i.e. power station chains). A longer river section then loses its flow characteristics and the end of backwater of one dam often stretches to the next power station upstream.

If reservoirs are designed naturally, valuable new habitats for flora and fauna can be created. They will not be the typical river biotopes; they are rather living communities similar to the ones in still waters /8-14/.

Barrier effect of dam and power house. A large number of aquatic animals migrate up or downriver; e.g. spawning migrations, migrations of invertebrates or fish into sections that are scarcely populated after flooding, or migrations looking for better feeding grounds. Reservoirs, dams and powerhouses are barriers for all fauna and sometimes also flora migrations and dispersion. The passive drift with the flow is no longer possible in reservoirs and only occurs during floods when organisms are washed across the dam or the spillway.

Such installations therefore interrupt the connectivity of the running waters. This leads to a split and a reduction in the size of the water habitat and impairs or even prevents spawning, feeding, expansion and compensation migrations. It can particularly lead to the following consequences of interrupted fish migration (see /8-13/).

– The potential for reproduction of certain types of fish is limited.
– The species diversity upstream of the dam is being reduced.
– The isolation of various fish populations increases.
– The recolonisation in regions impoverished by a catastrophic event such as flood or pollution is slower.

For migratory fish that move to small tributaries for spawning, dams are the most massive barriers; for small dams natural bypass channels or fish ladders might be a possible countermeasure. To enable the full functionality of fish ladders, they have to be allocated and dimensioned properly. As they have often not worked properly in the past, natural bypass channels are increasingly built that can offer habitats similar to those of the main waters in spite of lower flows.

If the running waters lose interconnectivity because of a dam or reservoir, this also has consequences for the downstream migration of fish. For example, fish that live in fast-running waters can only orient themselves with difficulty in the reservoirs with their lower flow velocities. Fish can also be injured if the water after an overflow weir crest is not deep enough. Furthermore, there is a danger for fish when moving through the turbine because of the prevailing pressure and flow conditions; additionally, the machines can cause mechanical injuries on the fish bodies. This can be partly prevented by tight gaps in the screening before the turbine inflow or correspondingly allocated devices that keep fish from entering into the turbines.

Diverted reaches. In diversion type power stations the water is diverted from the original bed into a headrace canal or pipe. This can lead to the following environmental effects /8-14/, /8-18/:

– reduction of flow in the riverbed,
– loss of natural annual and diurnal periodic flow variations,
– extension of the low flow periods,
– change in the water supply for riparian areas,
– change in the temperature regime in the diverted river reach,
– surge and sink effects in times of flood,
– increase in sedimentation,
– lower water quality,
– drop of numbers of species in flora and fauna,
– decrease in spawning sites,
– increase in algae growth.

For organisms living in running waters, the flow conditions are especially critical; many species need specific flow conditions as their habitats. A decrease in the flow velocity e.g. means a decline in the oxygen and food supply and the sedimentation of fine materials; this often leads to a change in the composition of species. At the same time, an appropriate flow for the waters is required for the transport of suspension matter and fine sediments – their sedimentation in the diverted river reach has similar effects here as in the reservoir.

A low water level in the diverted reaches also carries the risk of critical water temperatures in summer and winter. If the water is warmed up by strong sun radiation, the oxygen content decreases. The production of too much algae can lead to an oversaturation with oxygen during the day; once the algae die, the result is massive oxygen depletion.

Because of the reduced flow, a more or less significant part of the original riverbed is dry. That can lead to a decline in the aggregation of available cover for fish or spawning sites. If the aquatic habitat is reduced too much in its size, it normally leads to a quantitative reduction in the number of fish as well as a general reduction in the diversity, or to the development of a composition of species untypical for the habitat.

The reduction in the wetted area can, however, also lead to the development of new, valuable secondary biotopes. Free sand and gravel banks, for example, become habitats for specialist species such as bottom nesting birds, special beetles, locusts or spiders – for these species such extreme biotopes are often the last retreats.

The negative effects of diversion-type power stations can be limited if instream flows are fixed in advance to observe ecological requirements. Besides the minimum flow, there are a number of individual possibilities to achieve a composition of species in a diverted reach, in spite of the relatively low water level, similar to the river sections that have not been changed. The morphology of the riverbed plays a key role here.

More recent research has come to the overall conclusion that diversion type power stations are less damaging for the habitat "running waters" than river power stations with large reservoirs if they are designed appropriately and with an ecologically justified minimum water regulation.

Malfunction. If an accident occurs in an operation, lubricants might be released. If biodegradable lubricants (e.g. lubricant based on vegetable oil) are used, appropriate protection mechanisms (e.g. oil separators) are installed, and the lubricants are stored outside of the area that might be flooded, the risks of the potential environmental damage can be minimised. Furthermore, electrical parts of the plant might burn (e.g. cable) and release a limited amount of toxic pollutants into the environment, which are, however, not specific for hydroelectric power stations. Mechanical failures within the machine components in general are not at all harmful for people and the environment – or the danger is limited to a certain small area. If dams or weirs fail, this can have a large impact on the population and the flora and fauna. This creates a large accident potential, which is, however, limited in probability by the current – very far-reaching – regulations.

End of operation. The actual hydroelectric power components are mainly built from metallic material with recognised and generally environmentally friendly disposal methods. The demolishing of the structural components on site (e.g. dams, barrages) is more problematic. However, it has to be assumed that the plant location can still be used for the power industry after the technical lifetime of a hydroelectric power station has been exceeded. Considering this, such issues have not been significant so far. For the demolishing of dams, environmentally friendly recycling options could be developed. Thus an environmentally friendly disposal of hydroelectric power stations appears to be possible.

9 Utilisation of Ambient Air and Shallow Geothermal Energy

A typical characteristic of ambient air and near-surface heat it is the very low temperature level. The heat is mainly generated by solar radiation (Chapter 2). Only a small part of the energy in the soil (i.e. shallow geothermal energy) is not produced by the sun, but by the geothermal energy flow caused by the heat potential available within the earth (i.e. deep geothermal energy). The share of geothermal energy normally increases proportionally with an increasing depth underneath the surface of the earth (Chapter 2). Nevertheless, this type of energy is defined as geothermal energy – independent of the origin of energy in the near-surface ground (i.e. whether or not it results from solar radiation and/or geothermal energy stored in the deep underground, Chapter 2) (Fig. 9.1).

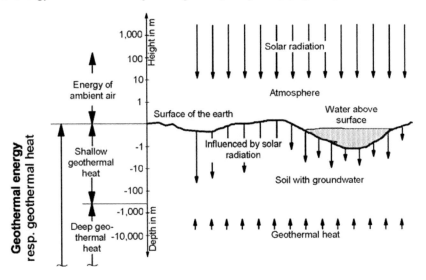

Fig. 9.1 Terminology definitions

In line with this terminology it has been agreed that utilisation of geothermal energy in general, and shallow geothermal energy in particular, starts at the surface of the earth (Fig. 9.1). The difference between the use of shallow geothermal energy and of geothermal energy from deeper layers (Chapter 10) is arbitrary and goes back to an administrative regulation originally set up in Switzerland. According to that regulation, systems for (deep) geothermal energy utilisation below 400 m were subsidised by carrying the risk of the drilling process. As there had not been any plants in operation at depths between around 200 and approximately 500 m for a long time, this was seen as the boundary. A value of 400 m as the approximate lower limit of shallow geothermal energy utilisation in the meantime has been adopted by other normative guidelines as well (e.g. German VDI guideline 4640). However, to set such an exact limit for the transition between shallow heat utilisation and heat utilisation from the deep underground is problematic, as the continuous technical development enables e.g. geothermal energy probes to reach increasingly deeper. Thus the limit between shallow and deep geothermics is becoming increasingly fluid.

Ambient air and shallow geothermal energy can be harnessed by a number of different technologies, methods and concepts. As the utilisable energy is normally generated at a low temperature level (mainly below 20 °C), a device to increase the temperature is generally required in order to enable the technical utilisation of the heat (e.g. to heat a residential building). This means that a heat pump needs to be built into the system. Alternatively, the subsoil temperature level can be increased by storing additional heat (e. g. from solar energy using solar collectors or excess heat from industrial processes). This option has hardly been put into practice so far. In order to harness ambient air and shallow geothermal energy, additional externally supplied energy is always required (e. g. electricity from the public grid, natural gas or biogas, fuels).

Hence a system to supply useful or final energy through ambient air and shallow geothermal energy utilisation generally consists of three system elements:
- Heat source system to enable the withdrawal of energy from ambient air and near-surface ground.
- Heat pump or another technical system essential to increase the temperature level and
- Heat sink; the system to feed or utilise the heat at a higher temperature level. This increase is obtained by using a heat pump.

The principles and their technical realisation, on which the first two main system elements are based, will be described in the following. Heat sink systems, however, are standard systems for heating and are thus not dealt with separately. Additionally – after a description of the principles of the heat pump as the basis for the utilisation of low-temperature heat – the different technical concepts of heat source systems will be discussed initially. Afterwards, the technical principles of the heat pump together with the resulting overall systems will be presented. This leads to analysing these systems from an economic and environmental point of view, also showing their potential and their forms of utilisation.

9.1 Principles

During isenthalpic throttling of a real gas, the temperature normally decreases if no external heat is added. In reversal, during isenthalpic compression of a real gas or a fluid, the temperature increases. This process, known as the Joule-Thomson-Effect is based on inter-molecular interaction between the gas molecules. During expansion work is required to counteract the molecular attraction. This work reduces the internal energy, which leads to a temperature decrease. Hence the Joule-Thomson effect is a measurement for the deviation from real to ideal gases.

Without auxiliary devices, heat flow is only possible from a higher to a lower temperature. In order to make ambient air heat or shallow geothermal energy utilisable, the direction of the flow has to be reversed. Heat is absorbed at lower temperatures (i.e. from the environment) and then released again at higher temperatures (e.g. to a radiator, for domestic water heating). To enable such a "heat pumping" process from lower to higher temperatures, the appropriate equipment and additionally high-quality energy (e.g. electricity) is required.

A refrigerating agent (refrigerant) is circulated to keep the temperature almost constant during heat absorption and release, mostly using evaporation and condensation (cold vapour process). On the low-pressure side heat is absorbed at a constant low temperature by the evaporating refrigerant (i.e. energy from the ambient air and/or the near-surface ground). Heat flows from this cold heat source to the refrigerant, which is even colder. Afterwards pressure is added to the evaporated refrigerant by a compressor. This leads to an increase in temperature caused by the Joule-Thomson effect and the almost isentropic and non-isenthalpic compression. At this higher pressure and temperature level, heat can be discharged in a condenser to the heat source that needs to be heated up. Again heat flows from a high to a slightly lower temperature level. During the real heat pump process the refrigerant is condensed in the condenser and sometimes supercooled slightly. Afterwards it is lead via a throttle where it is expanded isenthalpically and thereby cooled by the Joule-Thomson effect. Then the evaporation process starts again.

Fig. 9.2 shows two forms of presenting the thermodynamic cycle for a compression heat pump. The lg p-h diagram (pressure-enthalpy-diagram) on the right is clearer, as both pressure levels are clearly visible here. The internal energies in the individual system parts do not change assuming a stationary circular process. Thus the heat flow $\dot{Q}$ of each part of the circular process results from the mass flow $\dot{m}$ and the enthalpy difference Δh of that part of the circular process. Thus the heat flows of the individual partial processes can be checked from the lg p-h-diagram as sections on the x-axis (enthalpy) (Equation (9.1)).

$$\dot{Q} = \dot{m}\,\Delta h = \dot{m}\,T\,\Delta s \qquad (9.1)$$

In the T-s-diagram (temperature-entropy-diagram) on the left of Fig. 9.2, it is apparent that there is a significantly higher temperature at the compressor outlet than during the condensation that follows afterwards. One part of the heat sink can be at a considerably higher temperature level than the condensation temperature. By exploiting condensation and evaporation, a large part of the heat can be charged or discharged at a constant temperature, which is an advantage for most heat sources and heat sinks. The heat capacities have to be read from the *T-s*-diagram (temperature-entropy-diagram) as areas beneath the curve ($T \Delta s$) (Equation (9.1)).

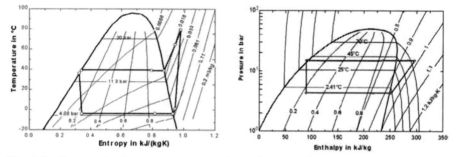

Fig. 9.2 Compression heat pump process shown in a *T-s*-diagram (temperature-entropy diagram) and an lg *p-h*-diagram (pressure-enthalpy diagram)

Heat pump principle. The heat pump is a "device which absorbs heat at a certain specific temperature (cold side) and releases it again at a higher temperature level (warm side) after adding drive work" /9-1/. Hence a heat pump can withdraw thermal energy from a heat source at a low temperature level (e.g. ambient air). The absorbed thermal energy including the drive work converted into heat can then be supplied for utilisation as thermal energy at a higher temperature level.

Depending on the functionality of the heat pump, the heat pump can be charged with the necessary drive energy in the form of mechanical energy or heat. Correspondingly, according to the resulting drive principle compression and sorption heat pumps are differentiated. In addition, sorption heat pumps are divided into absorption and adsorption systems; so far the latter have hardly been significant for the applications analysed here. The respective basic functionality principles – with the exception of the adsorption system – will be presented in more detail in the following. For compression heat pumps cold vapour processes will be assumed.

Compression heat pumps. In compression heat pumps a steam cycle takes place in a closed circuit that mainly consists of the four steps evaporation, compression, condensation and expansion. These systems thus consist of
– an evaporator,
– a compressor with drive plus
– a liquefier (condenser) and
– an expansion valve (Fig. 9.3).

In addition to the control components essential for operation, further system components and auxiliary devices such as valves, a manometer, security devices and other control instruments are required.

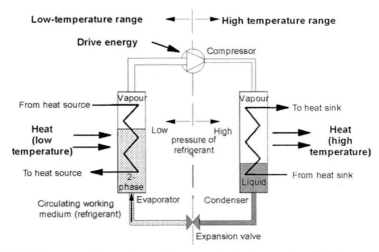

Fig. 9.3 Basic heat pump flow scheme of compression heat pump (see /9-1/)

The compressor is driven mechanically by an electric or combustion engine. Combustion motor drives can couple the heat generated by cooling the motor with the heating process.

The working medium circulating in the heat pump circuit is evaporated in the evaporator at low pressure and low temperature (even below 0 °C) by adding heat. The heat is made available by using ambient air or shallow geothermal energy via a heat carrier intermediate circuit or directly in the case of direct evaporation. The working medium, now gaseous after the energy withdrawal from the heat source, is suctioned and compressed by a compressor. During that process its temperature is increased to a level higher than the flow of the heat utilisation system (i.e. low-temperature heating system for a residential building). Still under high pressure, the working medium is then liquefied in the condenser, discharging heat to the heat utilisation system. Afterwards it flows to the low-temperature section via the expansion valve. Now the circuit starts again. Evaporator and condenser as the heat exchanger are the interfaces of the heat pump with the rest of the system.

Sorption heat pumps. Absorption heat pumps as important representatives of sorption heat pumps consist of an evaporator, an absorber a desorber and the condenser. Two expansion valves and a solvent pump are required for operation. Whereas a mechanical compressor is used for the compression heat pump, there is a "thermal compressor" used in the absorption heat pump. Drive energy for this "thermal compressor" is mainly required thermally (desorber); this thermal drive energy can be supplied e.g. by gas or oil combustion or by using (industrial) waste heat. Recently there have been attempts to supply this heat – at least partially – by thermal solar collectors.

A two-component mix (so-called working pairs) circulates within the solvent circuit of the absorption heat pump. One of the components (working medium) is highly soluble in the second component (solvent). Classic combinations of working pairs are water/lithiumbromide and ammonia/water. The first substance is always the working medium and the second the solvent.

The process within the condenser, expansion valve and the evaporator of the absorption heat pump is identical with the compression heat pump. In contrast, the compression process originates from two overlapping circuits with a different level of pressure (Fig. 9.4).

The solvent pump enables the connection between the two pressure levels. It needs considerably less drive energy than a compression heat pump, as a liquid medium can be pushed to a high pressure level requiring less energy than a gaseous medium.

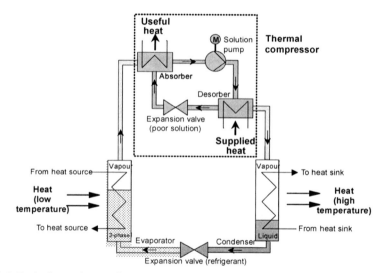

Fig. 9.4 Basic flow scheme of an absorption heat pump (see /9-1/)

In the absorber, the gaseous working medium from the evaporator (water(water(H_2O)/lithiumbromide(LiBr)) and ammonia (ammonia(NH_3)/water (H_2O))) is absorbed by the concentrated solvent. Heat is released in the process. The diluted solution is then pumped by increasing the pressure through the solvent pump into the desorber, where the working medium is driven out of the solvent again by adding heat (drive energy) and then reaches the condenser (liquefier). In a condensation process, it releases heat there. The working medium now undergoes the same steps with the expansion valve and the evaporator as in compression heat pumps. It again reaches the absorber in its gaseous form, while the reduced solvent is transported directly from the desorber to the absorber again by means of a throttling device to reabsorb the working medium. Utilisable heat is thus generated in the absorber and the condenser.

The purity of a working medium after entering the heat pump circuit is important for an efficient heat pump operation. It depends on the difference between the boiling temperatures of the working pair. If a salt and a fluid (e.g. water/lithium-bromide) are used, there is a big gap and the working medium water is available in very pure condition. If ammonia and water are used, ammonia takes the role of the working medium as it has a lower boiling point. Further components are integrated here in order to ensure a high level of purity of the working medium in spite of the small difference in the boiling points.

All in all, the absorption heat pump also absorbs heat at a low temperature level (e.g. ambient air or shallow geothermal energy) in the evaporator. Drive energy has to be used in the desorber and the solvent pump. The main input of drive energy in the absorber is done in the form of heat (i.e. "thermal compressor"). The energy input to drive the solvent pump for pumping and the pressure increase of the enriched liquid solvent is comparatively low.

Parameters. According to the first law of thermodynamics, the energy balance of a compression heat pump is drawn in Equation (9.2). $\dot{Q}_{Evap.}$ describes the heat flow to the evaporator, P_{Drive} the compressor drive power and $\dot{Q}_{Cond.}$ the heat flow delivered by the condenser.

$$\dot{Q}_{Evap.} + P_{Drive} = \dot{Q}_{Cond.} \tag{9.2}$$

The efficiency of a heat pump can be quantified by a parameter similar to the efficiency or the utilisation coefficient of other appliances. The efficiency or the utilisation coefficients are generally defined as the ratio of "output" to "input". Thus it is always below one.

This definition raises the issue of the level of "input" for heating the evaporator of the heat pump. This is carried out from near-surface ground or ambient air in this case. Amounts of heat that would largely be unutilised otherwise and are now used by the heat pump are the "input". Therefore they are not considered when calculating the energy parameter – like it is normally done for a system that exclusively uses fossil fuels.

Thus the resulting parameters can be above "one" – as not the entire energy utilised by the heat pump is balanced – (comparable to the "efficiency" or the "utilisation coefficient"). For that reason special parameters are defined to describe the efficiency or the utilisation rate of a heat pump (i.e. the coefficient of performance (COP), the (seasonal) performance factor (SPF) and the heating rate). Additionally, as the reciprocal value of the COP and the performance factor respectively, the input rate or the annual input rate (Table 9.1) are analysed. As some of the terms are only used in German speaking countries, the German terms are given in brackets. In the following only the two parameters COP and SPF used in the English speaking area as defined in Table 9.1 are used. The main parameters are explained in more detail in the following.

Coefficient of performance (COP). The COP for electrically driven heat pumps is defined as the ratio of the discharged utilisable heat flow in the condenser to the electric drive power of the compressor, for specific heat source and heat sink temperatures. Thus it can be compared with the efficiency of conventional heating systems. It is dependent on the operational conditions of the systems. In this case only the amount of energy is considered as "energy input" that is used to drive the heat pump (e.g. the electrical energy used to drive the electric heat pump) (Equation (9.3)). ε describes the efficiency rate, $\dot{Q}_{Evap.}$ the heat flow to the evaporator, $P_{Drive.}$ the compressor drive power and $\dot{Q}_{Cond.}$ the condenser heat flow.

Table 9.1 Parameters of heat pumps (German phrases in brackets for the definitions of German speaking countries)

	Symbol	Calculation	Comments
Efficiency rate (Leistungszahl)	ε	Heating capacity/ electrical drive power	Only for certain operational conditions characterises electrically driven compression heat pump
Work rate (Arbeitszahl)	β	Heating work/ / electrical drive work	Also annual work rate (β_a), characterises electrically driven compression heat pump
Heat rate (Heizzahl)	ζ	Heating capacity / Energy content of the end energy carrier	Only for certain operational conditions for absorption and combustion motor heat pumps
Annual heat rate (Jahresheizzahl)	ζ_a	Heating work / Energy content of the end energy carrier input	For absorption and combustion motor heat pumps
Input rate (Aufwandszahl)		Drive power/ Heating capacity	To replace efficiency rate (e.g. VDI 4650)
Annual input rate (Jahresaufwandszahl)		Drive work/ heating work	To replace work rate (e.g. VDI 4650)
Coefficient of Performance	COP	Heating capacity/ power input	English-speaking area, combines efficiency rate ε and heating rate ζ
Seasonal Performance Factor	SPF	Heating work / work input	English-speaking area, combines annual work rate β_a and annual heating rate ζ_a

$$\varepsilon = \frac{\dot{Q}_{Cond.}}{P_{Drive}} = \frac{\dot{Q}_{Evap.} + P_{Drive}}{P_{Drive}} = 1 + \frac{\dot{Q}_{Evap.}}{P_{Drive}} \tag{9.3}$$

The temperature difference between the heat source and the heating system (i.e. heat utilisation system) has a considerable influence on the efficiency rate. Additionally, the refrigerant and the design of the heat pump play a role. With an increasing temperature difference between the heat source and the heat utilisation system, the efficiency rate of the heat pump decreases. This can also be derived from the lg p-h-diagram (pressure-enthalpy-diagram) in Fig. 9.2. If the temperature difference between the evaporator and the condenser increases, the pressure difference between those two points increases at the same time. Thus the compressor has to overcome a higher pressure ratio and therefore has to contribute a higher enthalpy difference or specific heat input. The enthalpy difference in the condenser, however, remains almost the same if the pressure is increased. In order to achieve a high COP, the temperature of the heat source ought to be as high as possible and the flow rate in the heat utilisation system as low as possible.

Work rate. The efficiency of electrical heat pumps over a longer period of time is described with the work rate. Here the discharged useful heat is compared to the input of drive work. In addition to the drive work of the compressor, the energy consumption of auxiliary components belonging to the heat pump (e.g. pumps) plus losses through unsteady operation are considered. This enables the description of the system efficiency for a defined period of time (e.g. with the annual work rate or seasonal performance factor (SPF) over the course of one year). Whereas the COP is determined under given operational conditions (temperatures), these conditions are defined by the practical operation within the heating system. The work rate (mostly the annual work rate or seasonal performance factor) is thus more meaningful to describe the efficiency of heat pump systems.

Heat rate. For absorption heat pumps and heat pumps driven by combustion motors using natural gas, propane or diesel as their drive energy, the heat rate is used instead of the efficiency rate. Instead of the annual work rate the annual heat rate or SPF is provided. For the latter the utilisable energy is compared to the energy content of the fossil energy carriers over a defined period of time (mostly one year).

Considering the primary energy efficiency of electricity generation and distribution, the annual work rate of electrically driven heat pumps can be compared with the annual heat rate. For the English speaking countries heat rate and annual heat rate are defined as COP and SPF.

9.2 Technical description

Systems using ambient air or shallow geothermal energy as heat source – not taking into account the heat distribution system (i.e. heat sink) in the building that is similar for many common energy supply systems and will be not analysed further – consist of the two main components: heat source system and heat pump. These individual system elements are described and discussed in the following – together with resulting overall systems.

9.2.1 Heat source systems for ambient air utilisation

Air is generally available nearly everywhere as a heat source. It can supply a wide range of required heat at very varying temperatures. In order to achieve the optimum design, the seasonal and the daily course of ambient temperature and, if possible, also humidity that can supply latent heat through condensation, are required. The utilisation of the heat source "ambient air", however, causes some specific problems /9-2/.

– The low specific density (the water density to air density ratio is 1,000; i.e. water density is 1,000 times higher than air density) and the specific heat capacity, which is smaller by a factor of 4, requires large volumetric displacements and thus large machines. If dimensioned too small, acoustic problems could occur. Additionally, a suitable auxiliary system is required for the fans.
– Strongly fluctuating ambient temperatures during the heating season – very low and very high temperatures occur very rarely and the ambient temperature is likely to be in the average range of –3 to +11 °C in most countries e.g. in Europe – require a correspondingly high equipment input.
– In the case of heating buildings as the main form of utilisation, the substantial difference between the heating capacity of the heat pump influenced by the ambient temperature and the heating requirement of a building plays an additional role. The lower the ambient temperature, the higher the heating requirement of the house. Simultaneously, the temperature difference between the heat source and the heat sink (high inlet temperature to the heat utilisation system due to a high level of heating requirement of the house) increases. A higher temperature difference leads to a lower heating capacity and a lower COP of the heat pump (Fig. 9.5).

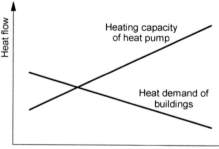

Fig. 9.5 Divergence between the heating capacity of heat pumps designed for ambient air usage and the heating requirements of a building (see /9-2/)

Heat can be withdrawn from ambient air in various ways. The most common way is that ambient air flows directly around the evaporator of the heat pump that withdraws heat. Generally, the evaporator is built as a finned tube heat exchanger with bundles of tubes with fins on the air side running in parallel, through which a refrigerant flows. The air flow rate of heat pump evaporators ought to be between 300 and 500 m³/kW of heat source capacity (evaporator capacity). The flow velocity of the air through the heat exchanger should be below 2 m/s to prevent the production of too much noise and an excessive use of electricity for the fan, an integral part of the heat pump that has a negative effect on its COP /9-3/. If the air at the heat exchanger's surfaces is cooled down to below 0 °C, the humidity condenses and settles as frost on the walls of the evaporator. This can occur even at air entry temperatures below 6 °C. In order to prevent the evaporator from being "blocked", it has to be defrosted occasionally under such operational conditions. The related standstill time leads to losses in heating capacity, which result in a decline in the SPF.

If no fan is used, the evaporators are called "silent evaporators". The ambient air is then only moved through free convection, this leads to a lower heat transfer coefficient. Because of the lower level of air-side heat transfer, such silent evaporators thus need a larger surface. The advantage, however, is that they operate in complete silence. In order to implement this form of evaporator, acceptance problems would probably have to be fought due to the large construction volume. Defrosting a silent evaporator is also problematic.

Three types of utilisation for ambient air as a heat source can be differentiated in general.

– Outdoor installation. Fig. 9.6 on the left shows this form of ambient air utilisation with heat pumps. Accordingly, the heat pump is entirely installed in the open. The heat is transferred into the house by well-insulated tubes. A general advantage of installing the heat pump in the open is the noise minimisation inside the house. Additionally, this form of installation only requires a small area within the building. However, it has to be ensured that the heating tubes do not cool down below 0 °C in order to avoid freezing.

– Split installation. One possible way of preventing the heating tubes from freezing is the split installation. Here the evaporator of the heat pump is installed outside the house (advantage: noise minimisation in the building) and the remaining heat pump is installed inside the house (Fig. 9.6 on the right). Both parts of the heat pump are linked via the refrigerant tubes. Compared to the outdoor installation, this form of installation requires more space inside the building. The part of the heat pump installed on the inside, however, can be hung on the wall, thus saving space. The split installation can also be used in old buildings, as there is generally no major installation work required. All it requires is breaking through some walls to lay the refrigerant tubes. The condensate of the air humidity at the evaporator has to be discharged in a way that prevents icing below the evaporator.

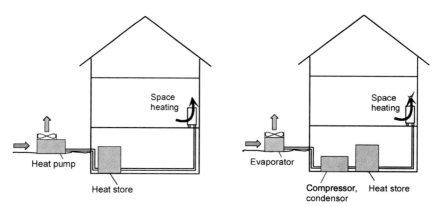

Fig. 9.6 Heat pump using ambient air as heat source (left: outdoor installation; right: split installation)

- Indoor installation. Another form of installation is the complete integration of the heat pump into the building (Fig. 9.7). In that case the ambient air has to be transported to the heat pump through air ducts that are insulated well against heat and noise. The cooled down air then has to be discharged again to the environment. Intake and outlet opening have to be built in a way that avoids a "shortcut" between the cooled down exhaust air and the intake air.

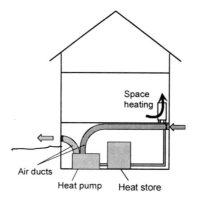

Fig. 9.7 Heat pump using ambient air as the heat source (indoor installation)

Flat-plate absorbers with integrated evaporator tubes combine the utilisation of ambient air and solar radiation. Often a brine circuit is installed between the absorber and the evaporator, as the oil recirculation into the refrigerant circuit is not ensured in the case of flat-plate absorbers. As flat-plate absorbers use diffuse and direct solar radiation, the location and the direction are important. Additionally, there has to be a way of releasing the condensate created by the absorber.

A special form of ambient heat exchanger, which belongs to the group of flat-plate absorbers are, the solid absorbers. The heat exchanger tubes of this absorber type are embedded in solid concrete components and thus utilise the ambient heat indirectly absorbed by the external concrete surfaces. A large amount of this ambient heat is solar radiation energy. Because of the large component masses, the solid absorbers can store large amounts of heat and thus largely compensate for the fluctuations in ambient air and solar radiation. Normally solid absorbers also have some kind of functionality as a building component of the house. To give some examples, estate walls, noise protection walls, external walls of buildings or concrete garages can be built as solid absorbers.

9.2.2 Heat source systems for shallow geothermal energy utilisation

Heat sources using the near-surface ground generally utilise the heat stored in the ground (i.e. in the soil or the rocks and their pore filling (mostly groundwater)). They differ mainly in the form of heat withdrawal from the subsoil or heat release into the subsoil. Two basic variations can be distinguished (Table 9.2).

– Closed systems. One or more heat exchangers are – either horizontally or vertically – installed in the ground. A heat transfer medium (or heat carrier) (e.g. water (with antifreeze compound in most cases, or refrigerant) flows through it in a closed circuit. This process withdraws heat from the subsoil (i.e. from the soil or rock matrix and the pore filling) (or charges it with heat for space cooling in summer). Heat transfer between the heat carrier and the subsoil takes place by heat conduction. The heat carrier is not in direct contact with the soil or rock matrix and the pore filling. Therefore systems of that type can theoretically be used almost everywhere.

– Open systems. When utilising groundwater, the water is pumped directly through wells from the layers with groundwater (aquifers). Thus the groundwater itself is a heat carrier. It is cooled down afterwards (or heated up for space cooling in the summer) and transferred into the same aquifer via an injection well. In the subsoil, a heat transfer takes place between groundwater and the soil or rock matrix. Groundwater as the heat carrier is not circulated in a defined circuit and that, furthermore, is in direct contact with the aquifer; therefore these systems are called open systems (Table 9.2). The prerequisite for such systems is the existence of appropriate layers with groundwater in the subsoil.

– Other systems. In addition, there are other variations that do not fit exactly into the categories mentioned so far. These are systems that are not entirely sealed off from the groundwater; systems utilising water from artificial hollow underground spaces and a system for air preheating, where the heat carrier is sealed off against the underground, but does not circulate, as new air is continuously sucked in.

These different heat source systems for the utilisation of near-surface ground energy are described in the following.

Table 9.2 Variations of shallow energy utilisation

	Depth	Heat carrier	Remarks
Closed systems			
Ground coupled collectors (horizontal)	1.2 – 2.0 m	Brine[a]	Influence of the climate, large surface
Direct evaporation (horizontal)	1.2 – 2.0 m	Heat pump working medium	Material copper, sometimes galvanised
Ground probes – pile-driven (vertically or diagonally)	5 – 30 m	Brine[a]	Material steel, sometimes synthetic material, only in lose rock
Drilled (vertically)	25 – 250 m	Brine[a], possibly water	Material HDPE[b], ideal in solid rock
Heat transfer poles ("Energy poles"; horizontal or vertical)	5 – 30 m	Water, possibly brine[a]	static function most important, if possible no frost temperature
Open systems			
Groundwater wells (Doublet)	4 – 100 m	Water	Minimum of 2 wells (production & injection well), groundwater pump
Other systems			
Coaxial wells (vertical)	120 – 250 m	Water	High bore costs, overload not possible
Pit-/tunnel water		Water	Possibilities limited to certain areas
Air preheating/cooling (horizontal)	1.2 – 2.0 m	Air	Tubes in the ground sucking air in

The depth values are typical mean values; [a] water-antifreeze mix (in the past salts, nowadays rather types of alcohol or glycol); [b] high density polyethylene

Closed systems. The ground-coupled heat exchanger for closed systems are differentiated as horizontally and vertically installed heat exchanger. Additionally, there are special forms that cannot be allocated unambiguously (i.e. soil-contact components) and are generally not primarily used for energy generation (i.e. double utilisation).

Horizontally installed ground-coupled heat exchangers. Two forms of installation of horizontal ground-coupled heat exchangers common in Europe in closed systems (also called ground-coupled heat collectors) in the form of tube registers are shown in Fig. 9.8. The (galvanised) metal tubes in direct evaporation systems and the mainly plastic tubes in systems with a brine intermediate circuit are sunk into the ground at a depth of approximately 0.5 m below the frost level (normally between 1.0 and 1.5 m below the earth's surface). The distance between the individual tubes ought to be approximately between 0.5 and 1.0 m. In order to avoid damage, they are embedded in a layer of sand.

According to current heat insulation regulations, the utilised surface for buildings should be approximately 1.5 to 2.0 times the space to be heated in order to be able to withdraw enough heat from the ground even during longer cold periods.

For houses with a low energy standard, the space can be smaller. Depending on the quality of the ground, the withdrawn heat capacities vary between 10 and 40 W/m^2 (Table 9.3). This enables a heat generation of approximately 360 MJ per square metre ground during the heating period.

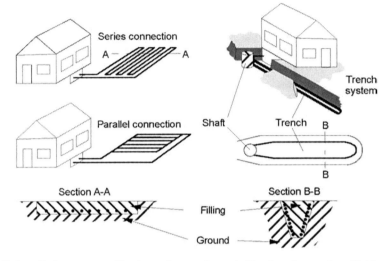

Fig. 9.8 Installation pattern of horizontal ground-coupled heat exchanger (see /9-4/)

A significant reduction in the space required can also be achieved by the installation pattern of a trench collector also shown in Fig. 9.8. According to this concept, the heat transfer tubes are installed at the side walls of a trench with a depth of approximately 2.5 m and a breadth 3.0 m. The required length of the trench depends on the quality of the soil and the heating capacity of the heat pump. A specific trench length of 2 m per kW heating capacity can be taken as a guide value /9-5/.

Table 9.3 Mean withdrawn heat capacities from the soil (see VDI 4640, sheet 2)

Type of soil	Withdrawn heat capacity
Dry, sandy soil	10 – 15 W/m^2
Humid, sandy soil	15 – 20 W/m^2
Dry loamy soil	20 – 25 W/m^2
Humid loamy soil	25 – 30 W/m^2
Water saturated sand/gravel	30 – 40 W/m^2

A further attempt to reduce the required space is to install the tubes spirally. There are two main possible designs of spiral collectors, which are mainly built in North America. The fundamental disadvantage of such collectors is that ventilation problems can occur.
– For the slinky or cunette collector /9-6/ a roll of commercially available plastic tube is laid out on the floor of a broad trench and stretched to the sides (verti-

cally to the wound axis) in a way that makes the windings overlap. Afterwards the trench is filled up again. Such a collector can also be sunk vertically into a small, slot-die trench.

- For the Svec-collector /9-7/ a plastic tube is wound up on a roll during production. When sunk into a prepared trench, the tube can then be stretched and fixed like a coil (vertically to the wound axis). Afterwards the trench is filled up again.

With all these more compact ground-coupled collectors there is a danger that the necessary heat recovery does not take place during the summer if they are only used for heating purposes. The reason is that the boundary marking area of the surrounding soil and the earth's surface is relatively small compared to the accessible volume. Therefore such a set-up is more appropriate for energy storage, thus compact ground-coupled collectors are most appropriate for heating and cooling systems. For heat pumps exclusively used for heating purposes, plane ground-coupled collectors are more suitable.

There are two ways of withdrawing heat from the ground and transferring heat from the heat source to the heat pump.

- Heat withdrawal and transport can be achieved with an intermediate circuit using a heat carrier ("brine") that absorbs heat from the ground and discharges it to the heat pump evaporator. In Germany, a mix of monopropyleneglycol and water (partly also monoethyleneglycol) has been most effective, which is frost proof at 25 % glycol up to a temperature of approximately -10 °C and at 38 % of glycol up to a temperature of approximately -20 °C. Plastic tubes with an exterior diameter of up to 40 mm are used for the ground-coupled heat exchanger. These materials have a sufficient ageing and corrosion resistance and are elastic and chemically stable at the given temperatures. The individual tube lines are welded or screwed together.

- Heat withdrawal and transport can also be realised through the so-called "direct evaporation". In that case the working medium of the heat pump circulates directly in the tubes of the ground-coupled heat exchanger. It evaporates there and thus withdraws heat from the earth. The evaporator of the heat pump is sunk into the ground. In general, copper tubes with a plastic coating to protect them from corrosion are used. The advantage of this direct evaporation is in the lower level of machinery input and the higher achievable SPF of the heat pump. However, it requires a system design that is exactly adjusted in terms of refrigeration. Furthermore, the filling volumes of the working medium are much larger than for systems with an intermediate circuit.

Vertically installed ground-coupled heat exchangers. Vertical ground-coupled heat exchanger for closed systems (so-called ground probes) require significantly less space compared to the horizontal heat transfer media. They are normally used in tight spaces or for the retrofit of heating systems, because only little garden space is affected during the installation.

Ground probes are vertically sunken into, boreholes up to, and over, 250 m deep. Their principal layout variations are shown in Fig. 9.9. A good heat transfer between the soil and the probe has to be ensured. This can either be done by injecting a bentonite-cement suspension or by additional filling with quartz sand.

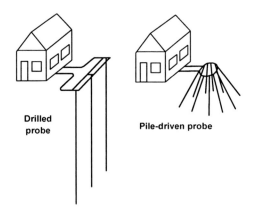

Drilled
probe

Pile-driven probe

Fig. 9.9 Different layouts of vertically arranged drilled and pile-driven ground-coupled heat exchanger (see /9-4/)

By using pile-driven steel probes and drillings with small machinery (up to a depth of approximately 30 m) the layout presented in Fig. 9.9 on the right can be achieved. The pile-drive and the drill are installed in one spot enabling them to rotate. They can sink the ground probes without further relocation. For pile-driving, mostly metal-coaxial probes are used. If stainless steel is not used, a cathodian protection from corrosion has to be applied. Other processes have been developed to sink U-type slings of plastic tubes directly into soft soil, using the appropriate auxiliary mechanisms.

The layout of the most common ground probes is shown in cross-section in Fig. 9.10. Single or double-U tube probes consist of two or four tubes that are linked at the bottom end to enable the heat carrier to flow down in one tube and up in another. In the coaxial basic form the withdrawal of heat from the soil only takes place on one flow section (depending on the system either upwards or downwards).

The material mainly used for ground probes is High-Density-Polyethylene (HDPE) (e.g. PE 80 or PE 100 according to DIN 8074 or DIN 8075). Typical tube dimensioning is 25 x 2.3 mm at a probe design length of 60 m and 32 x 2.9 mm at 100 m. For coaxial ground probes, plastic-coated high-grade steel or copper tubes can also be used – at high costs. In general, the danger of leakage due to corrosion of the ground probe has to be kept as low as possible by choosing the right material.

Like for horizontally ground-coupled heat exchangers, there is also a danger for ground probes that the soil is cooled down too much due to under-dimensioning and a corresponding excessive heat withdrawal. This leads to lower temperatures of the heat carrier and thus a reduction of the COP of the heat pump. In contrast to the horizontal heat exchanger installed at a depth of 1.0 to 1.5 m, the deeper layers cannot recover entirely during the summer. Artificial heating would have to be provided for (e.g. by solar collectors or from industrial waste heat).

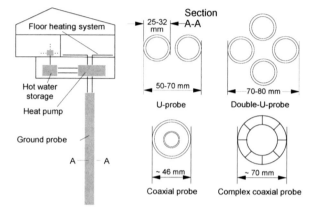

Fig. 9.10 Types of common ground probes (see /9-4/)

Table 9.4 Specific heat withdrawal capacities of ground probes in smaller systems at different utilisation hours at full load (analogous to VDI 4640, sheet 2)

	1,800 h/a	2,400 h/a
General guidelines		
Bad subsoil (dry lose rocks)	25 W/m	20 W/m
Solid rock subsoil, water-saturated lose rock	60 W/m	50 W/m
Solid rock with high heat conductivity	84 W/m	70 W/m
Individual soils		
Gravel, sand, dry	< 25 W/m	< 20 W/m
Gravel, sand, carrying water	65 – 80 W/m	55 – 65 W/m
Gravel, sand, strong groundwater flow, for small systems.	80 – 100 W/m	80 – 100 W/m
Clay, loam, moist	35 – 50 W/m	30 – 40 W/m
Limestone (solid)	55 – 70 W/m	45 – 60 W/m
Sandstone	65 – 80 W/m	55 – 65 W/m
Acidic magmatites (e.g. granite)	65 – 85 W/m	55 – 70 W/m
Alkaline magmatites (e.g. basalt)	40 – 65 W/m	35 – 55 W/m
Gneiss	70 – 85 W/m	60 – 70 W/m

The requirement for using the table: only heat withdrawal (heating incl. hot water) takes place; length of the individual ground probes between 40 and 100 m; smallest space between two ground probes would be a minimum of 5 m for ground probe lengths of 40 to 50 m or at least 6 m for ground probes with lengths of over 50 to 100 m. Suitable ground probes are double-U probes with an individual tube diameter of 25 or 32 mm or coaxial probes with a diameter of at least 60 mm. The values given above can fluctuate considerably, depending on rock formations such as crevasses, foliation and weathering.

Table 9.4 shows guidelines of a possible heat withdrawal for smaller systems and different types of soil. In order to keep a long-term balanced situation, an annual amount of withdrawn heat between 180 and 650 MJ/(m a) must not be exceeded – with exclusive recovery by solar energy penetrating the surface of the earth and geothermal energy flowing up – depending on the individual subsoil conditions /9-4/.

The values shown in only give very rough guidelines. A more exact definition of the specific heat withdrawal capacities can be calculated if the thermal subsoil conditions are known (Fig. 9.11). For larger ground probe systems only calculations are possible for the system design in order to determine the required amount and length of ground probes. These calculations can be done by existing computer programs. For difficult cases a simulation with numeric models should be carried out, especially if the influence of flowing groundwater has to be considered. In order to obtain reliable input parameters for such calculations, the Thermal Response Test was developed. It allows the determination of thermal subsoil parameters on site.

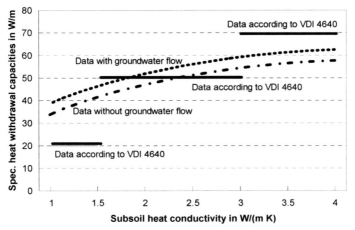

Fig. 9.11 Exemplary specific heat withdrawal capacities for small ground probe installations (calculated for a detached family house requiring 10 kW for heating, on the basis of 2 ground probes and 1,800 hours at full load per year, excluding hot water; see /9-2/)

The bore method to sink the ground probes is dependent on the expected ground layers and the available space /9-4/.
– In lose rock boreholes can be sunk with a hollow drill screw. The bore material is transported and/or displaced.
– When drilling with jetting methods, that can also be applied in lose rock; the drilling fluid continuously carries the bore material out through the ring section from the deepest point of the borehole. Additionally, the drilling fluid pumped into the borehole also stabilises the borehole wall and maintains it in line with the diameter of the bore. It also cools and greases the bore tools and seals off

the groundwater against the borehole with the filter cake caused by excess pressure in the borehole, opposed to the surrounding soil. For jetting purposes, water is normally used, sometimes with certain added substances (e.g. bentonite; see also DVGW-Guidelines W 115 and W 116); this supports the jetting in adopting the described characteristics and fulfilling its tasks better.

– The borehole hammer method has been largely accepted. Air driving the bore hammer and transporting the drillings above ground at a rate of ascent of 15 to 20 m/s is used for jetting. Depending on the borehole hammer, air pressures of over 10 bar and air volumes of over 10 m3/min are required /9-4/. If necessary, a foaming agent can be added to the air to enable an improved transport of drillings and to avoid detritus in the borehole.

After sinking the ground probe into the borehole, it has to be filled up again in order to guarantee a good heat transfer between soil and probe. The filling can be a bentonite-cement suspension.

There have been attempts to also apply the method of direct evaporation for ground probes instead of using a heat carrier circuit. Around 1990, several of these systems were built in Austria and the USA. Various problems occurred e.g. with the return of the compressor oil and the large amount of filling, with then still largely utilised working media that were damaging for the ozone layer. These problems stopped this development. Recently, direct evaporation has been discussed again. Nowadays, ammonia is used as a working medium. A pilot plant was built in Coswig near Dresden, Germany.

A more promising new development in the field of ground probes is their design as a heat pipe and the use of CO_2. Thus using water-damaging antifreeze compound can be avoided. Energy in the circulation pump is saved due to the heat pipe functionality and the disadvantages of direct evaporation are avoided by separating the heat pipe and the refrigerant circuit. Ground-coupled heat pumps with such heat pipes are already run successfully in Upper Austria.

Components with earth contact (energy piles, slot-die walls). A further variation of vertical ground-coupled heat exchanger are the heat transfer piles, so-called "energy piles" /9-8/, /9-9/. They are foundation piles, used for difficult subsoil conditions for laying the foundation of buildings. These piles are equipped with heat transfer tubes and allow the installation of ground-coupled heat exchanger at low additional costs in locations where foundation poles have to be used anyway.

Energy piles can basically be combined with all well-known geotechnical structure-on-pile foundation methods. So far, cast-in-situ piles (Bore piles) and ready-made piles (ramming piles) from reinforced concrete with a full cross-section and hollow piles as well as steel piles have been used. Every pile type has specific advantages and disadvantages. Cast-in-situ concrete piles are very flexible, but from a technical and an economic point of view should only be used starting at a minimum diameter of approximately 600 mm. Their production is quite cost-intensive and requires a lot of care. Ramming piles are easy to produce in a factory; however, during the ramming process adequate protection for the tube connections has to be provided. Heat transfer tubes can only be added to the pre-fabricated pile length. Hollow piles, where heat transfer tubes can be added to the hollow in the pile at a later stage, allow the utilisation of the entire pile length. However, they reduce the available diameter for the tubes.

Apart from foundation piles, other concrete components can also be used as heat exchanger in the earth (e.g. claddings of foundation trenches made of slot-die or pile walls), as these fixtures are generally no longer required for static purposes after the building is finished. Supporting walls, cellar walls or foundation plates can also be used as heat exchangers. In these cases, a good insulation against the interior is equally necessary, as with the manifold of energy pile systems which are laid under the floor plate; this enables the actual withdrawal of heat from the subsoil and prevents for example, the cellar from becoming cold and damp.

Open systems. Open systems for near-surface ground energy utilisation are groundwater wells. They are discussed in the following.

Because of its relatively constant temperature between 9 and 10 °C, groundwater is very suitable as a heat source for heat pumps. Limitations are the lack of availability of the heat source. Rich enough and not too deep groundwater layers (aquifers) with a suitable water quality are not available everywhere. Further limitations might be caused by regional water legislation.

The heat source system for groundwater utilisation consists of a production well that provides the groundwater, and an injection well that is used to recharge the groundwater layers with the cooled down water (doublet). The extraction and the injection well have to be at a reasonable distance in order to avoid a thermo-hydraulic shortcut. The extraction well should also not be within the cold zone of the injection well as this reduces the efficiency of the heat pump system.

The well capacity needs to ensure a continuous extraction for the nominal flow of the connected heat pump, this corresponds to approximately 0.2 up to 0.3 m^3/h for each kW of evaporator capacity. The capacity of the well depends on the local geological conditions. The temperature change of the groundwater that is recharged to the injection well or wells should also not exceed ± 6 K. The extracted amount and the minimum recharging temperature should be in line with the respective regulations.

Fig. 9.12 shows the typical construction of a heat pump system for groundwater utilisation. Common well depths are 4 to 10 m (e.g. /9-1/, /9-3/), which can be deeper in larger systems (the transition to hydrothermal energy utilisation is fluid in this case, see Chapter 10). The clay barrage above the gravel fill holds air and gravitational water back. The gravel filling between the well borehole and the filter tube should be between 50 to 70 mm in thickness. The suction tube, the intake of the tailwater pump in the extraction well and the downspout in the injection well always have to end below the water surface in each operating state.

Before designing the well, hydro-geological analyses should clarify the chemical structure of the groundwater, the aquiferous and non-water permeable layers, plus the groundwater level and the permeability of the aquiferous layers. For these purposes a pilot drilling has to be carried out that can potentially be used as a well later.

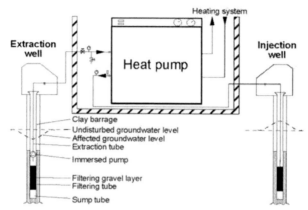

Fig. 9.12 Elementary diagram of a groundwater-heat pump installation (see /9-10/)

One particular problem is the sedimentation of iron ochre in the injection wells. It occurs very often in oxygen-free groundwater with a low redox potential. Such groundwater should not get in contact with ambient air. Therefore the entire system needs to be closed and be kept under excess pressure all the time, otherwise water treatment by deferrisation and de-manganesing would be required. Lime precipitation, however, does not play a role at temperature fluctuations of a maximum of ± 6 K.

Under certain conditions groundwater heat pumps are possible that exclusively consist of one or several production wells. Such concepts rule potential problems with injection wells out. Technically, this requires that the aquifer has enough newly produced groundwater and that the water can be channelled appropriately or sunk again. In Germany, such systems are generally not authorised.

Other systems. Other systems are the utilisation of groundwater with a coaxial well, the utilisation of pit and/or tunnel water and air preheating or cooling in near-surface soil.

Coaxial wells. Coaxial wells ("Standing Column Wells") /9-11/ are positioned between ground probes and groundwater wells. An ascending tube with a filter at the bottom end and surrounded by a stack of gravel is built into a borehole. Towards the rock, the stack of gravel can be separated with a plastic liner. Water is pumped from the ascending tube with a submersible pump - in a similar way as in a groundwater well. It is then cooled down in a heat pump (or heated up) and then seeps out again through the stack of gravel in the ring section. During the sinking process the water absorbs heat from the surrounding subsoil or discharges heat into the subsoil.

Due to the lack of separation from the natural subsoil (a plastic liner does not seal off completely), antifreeze cannot be used in coaxial wells. The heat pump has thus been run in a way that prevents freezing – in the same way as using groundwater. A maximum annual number of operation hours is generally fixed in advance for that reason. Furthermore, long seeping paths, large amounts of water in the borehole ring section and an increased temperature at the bottom of the borehole have been perceived as useful. Therefore coaxial wells are normally between 100 and 250 m deep.

Measured specific heat withdrawal capacities of coaxial wells under normal operation are between 36 and 44 W/m and under short-term operation at full load at around 90 W/m /9-11/. Hence they have similar dimensions as those of the ground probes. The average heat source temperatures are, however, a little higher compared to ground probes. This achieves a better COP of the heat pump.

Cavity and tunnel water. Artificial hollows in the subsoil can serve as collectors of groundwater or groundwater reservoirs. They are mainly mines (no longer or still operational) or tunnels, where the hollows had not primarily been built for a thermal utilisation. This special creation of hollow spaces is normally ruled out due to high costs (with the exclusion of thermal subsoil storage). At times we move away from the field of shallow geothermal energy when dealing with pits and tunnels. To give an example, the water for thermal use from a coal mine in the Eastern Ruhr area in Germany would be obtained from depths significantly below 1,000 m and from the interior of an Alpine tunnel e.g. in Switzerland sometimes at a depth of over 2,000 m.

Water from mines can be obtained e.g. through drillings from above ground. Above all, the depth of the water level in the pit determines the heat withdrawal method. It may lead to high pumping heads and correspondingly high energy input to operate the pumps. In general, after cooling down, the water has to be transferred back through another borehole into the pit. The flow between the withdrawal and the intake borehole should be as long as possible (achieved e.g. by drilling at different levels). Mines in the low mountain range areas that ascend via drifts from valleys, water flowing naturally from these drifts can also be used as a heat source.

Water from large tunnel constructions normally flows to the portals and can be utilised as a heat source there. In some Alpine tunnels this water has temperatures which are significantly above the annual mean temperature.

Preheating/precooling of air. Utilisation of air preheating in the subsoil (without heat pump) already existed in the eighties in the farming sector. Intake air for the pig pens was sucked in through tubes in the ground. Winter and summer temperature peaks were broken. As a further development, in order to extend the operation time of heat pumps utilising the heat source air during the winter, some systems were operated that transferred air through tubes in the ground, pre-heated it there and then transported it to the heat pump evaporator /9-1/, /9-3/ (Table 9.5).

Such heat sources are called concrete collectors, air wells or air registers. As air has a very low heat capacity, comparatively large amounts of air have to be moved. Lately, preheating and pre-cooling of intake air in tubes in the ground (without heat pumps) have gained significance for the ventilation of buildings with a low-energy and passive-energy standard.

Table 9.5 Designs and configurations of tubes for preheating of air in the ground

Types of design
Concrete tubes (can absorb humidity), PVC-tubes (low pressure decrease)
Tubes free in the ground, tubes insulated at the top, tubes flat underneath the foundation
Single tubes or tube registers
Types of operation
Fresh air is always transported through the tubes
Fresh air is only transported through the tubes if the outflow temperature is above the ambient air
Fresh air is transported through the tubes every time the outlet temperature is below the ambient temperature, for evaporators heat sources with a higher temperature are always used (additional charging of the ground)

9.2.3 Heat pump

A heat pump – like any other technical system – consists of various system elements. In the following, they are primarily explained further for electrical compression heat pumps, as they have the highest market share (e.g. /9-1/, /9-2/, /9-3/, /9-12/). It can be differentiated by the heat exchanger utilised in the evaporator and the condenser, the compressor, the expansion valve, the lubricant as well as the working medium (refrigerant).

Heat exchangers. Heat exchangers are devices that transfer heat following the temperature gradient between two or more substances. At the same time they enable a change of thermodynamic state of these substances (cooling, heating up, evaporating, and condensing). For heat pumps, they are mainly used for the heat transfer between the heat source and the heat pump (i.e. the evaporator) and between the heat pump and the heat sink (i.e. the condenser).

The size of the heat exchanger and thus the heat transfer surface is primarily determined by the driving temperature difference (i.e. the gradient) between the cooled heat source and the evaporation temperature at the evaporator or the condensation temperature and the temperature of the heated heat transfer medium in the case of a condenser.

For a given capacity of the heat exchanger a small temperature difference (gradient) requires a large heat transfer surface. In the opposite case, a large temperature difference needs a small heat transfer surface. In order to achieve a high heat pump COP, the mean temperature gradient in the evaporator and the condenser should be as small as possible. Thus the temperature difference between the heat carrier on the condenser side (e.g. heating water) and the heat carrier of the evaporator (e.g. brine) is not increased unnecessarily by temperature gradients that are too big. Values of around 5 K have been accepted as a good compromise.

Heat exchangers can be differentiated according to the flow direction of the substances involved: there are parallel, cross-current, or countercurrent flow heat exchangers. Mixed types exist. If the heat carriers are brine or water, shell and tube, plate or coaxial heat exchanger can be used.

- Shell and tube heat exchangers consist of a bundle of tubes usually connected to plenums (sometimes called water boxes) through holes in tubesheets. They are fitted into a jacketed tube (shell). The two media that are involved in the process flow in the tubes, and around the tubes within the shell.
- Plate heat exchangers consist of plates that are welded, soldered or screwed together. The two media alternate in their flow between the plates. In comparison with bundled shell and tube heat exchanger of the same capacity they require less space.
- Coaxial heat exchangers consist of one internal tube and an external tube fitted around it. One of the two media is flowing through the internal tube, whereas the other – mainly in countercurrent flow – flows in the space between the internal and the external tube.
- Finned tube heat exchangers are air-to-liquid heat exchanger and consist of several parallel tubes that are redirected a number of times in order to lengthen the path. The entire stack of tubes is joined together by fins. The air mainly flows in cross or countercurrent between the fins around the tubes, the fluid flows within the tubes.

In the case of heat pumps, heat exchangers of these types are mainly used to transfer heat between the heat source and the heat pump (i.e. the evaporator) or between the heat pump and the heat sink (i.e. the condenser). The respective characteristic features are discussed in the following.

- The evaporator is the connecting element between the heat source and the heat pump. The temperature difference between the heat source and the evaporation temperature of the refrigerant determines its size. Dry evaporation, flooded evaporation and evaporation during pumping can be differentiated.
  - During dry evaporation as much refrigerant is injected by the expansion valve into the evaporation tubes as can still be evaporated entirely. It is simultaneously overheated slightly (superheating means heating the working medium up to a temperature above the evaporation temperature here). Superheating is a measurement for the injection of the refrigerant in this case.
  - Within the flooded evaporator one part of the evaporator is flooded with a liquid refrigerant. Evaporation takes place around the tubes. The saturated steam is released from the heat exchanger, superheating is therefore not possible. In a separator connected at the end, drops of fluid that have been carried along during the evaporation process have to be separated.
  - For evaporation during pumping the refrigerant is evaporated in the tube. A significant excess of fluid is pumped off into a secondary circuit and from

there into a so-called "pump container". Only there steam and fluid are finally separated. Thus even large heat transfer surfaces can be operated at constant load.

– The condenser is the interface between the heat pump and the heat sink. It discharges utilisable heat to the liquid or gaseous operating medium. Like the evaporator it is designed as a heat exchanger. Its mean temperature gradient reflects the temperature difference between condensation of the refrigerant and of the heat consumer (heat sink). Depending on the design, heaters of fluid (shell and tube, coaxial or plate heat exchanger) and air heaters (mainly finned design for air heaters) can be differentiated.

Compressors. Within the heat pump compressor the gaseous refrigerant moving in a closed circuit between evaporator and condenser is compressed. Fully hermetic, partly hermetic and open compressors can be differentiated here.

– For fully hermetic compressors the compressor and the drive motor are installed together in a gas tight, soldered/welded capsule (i.e. encapsulated housing). The drive capacity can be up to several kW.

– For semi-hermetic compressors the motor is flanged to the condenser. Like the fully hermetic compressors, they have a common shaft. The drive capacities are between 4 to 150 kW in this case.

– In open compressors the drive motor is outside of the actual condenser. Motor and compressor are linked by a shaft and a coupling. Open compressors are mainly used in larger systems. The drive can be driven electrically or through a combustion engine.

Important compressor designs are piston, scroll, screw-type, and turbo compressor.

– In piston compressors the pressure increase is achieved by reducing the size of sealed off compressor spaces (i.e. by using displacement machines). They are built as fully hermetic compressors with drive capacities of up to around 25 kW, as partly hermetic up to around 90 kW and as open machines for even higher capacity levels. The suction volume flows can be up to 1,600 m³/h. The machines are built for smaller and also larger capacities with 1 to 16 cylinders.

– In scroll compressors a disc with spiral fin moves eccentrically above a fixed disc with the corresponding counter-fin. The spaces separated by the fins are getting smaller in size during the operation of the compressor. Thus the encircled gas is compressed and discharged again through openings before the gap is enlarged again (i.e. the displacement machine). The production of the fin requires high precision in order to maintain the compression spaces as gas-tight as possible. The advantages of this design are the circular movements and the few moving components and the good performance at partial load.

– Screw-type compressors can be differentiated as oil-free systems on the one hand and systems with oil-injected cooling. Oil works as a cooling agent and a lubricant and is supposed to seal off the gap between the rotor blades themselves and towards the frame. The part load efficiency of such compressors is generally slightly lower than that of piston compressors. Heat pumps with a screw-type compressor additionally require an oil separator on the pressure side connected to the end of the compressor. After separation from the working medium and the cooling in the oil cooler at the end, the oil is again available for a renewed injection into the screw-type compressor. Screw-type compressors have a comparatively long operational life as they have only a few movable parts (i.e. no working valves).

– Turbo compressors are dynamical type compressors structured with one or several stages of compression. One compression stage consists of a running wheel with a fixed mounting of blades and guide blades to convert kinetic into potential energy. Within one turbo compressor up to 8 of these bladed wheels can be installed. Thus a pressure between 8 and 11 bar can be achieved. Radial and axial machines are used. However, mainly radial turbo compressors are utilised as they achieve a higher pressure ratio per compression stage and the specific production costs are lower than for axial machines. The power adjustment is achieved through the size of the housing, the amount and the width of the bladed wheels. For a steady adjustment of power, the rotational speed can be changed and/or the guide blades in the air intake can be adjusted. As the supply with lubricating oil is totally separated from the working medium in the turbo compressor (lubricating oil free compression of the working medium), the solvent capacity of the working media is not relevant for the lubricating oil. Advantages of the turbo compressors are: a low level of wear and tear due to their simple construction, the steady power regulation from between 10 to 100 %. They also require comparatively little space – even for higher capacities. Such turbo compressors are only offered for large capacities.

If required, compressors can be coupled in two different ways. For the compression in several stages several compressors are connected in series if the pressure difference between the evaporation and the condensation pressure cannot be managed any longer by a single compressor. However, in the case of the heat pump cascade, each compressor has its own condenser and evaporator. The ideal refrigerant can thus be used at the respective temperature. However, this type of connection causes higher system costs and heat losses (temperature gradient) due to the higher number of required heat exchangers.

For lower capacities that dominate ambient air and shallow geothermal energy utilisation and near-surface ground energy, piston and scroll compressors are normally used. They are generally built into a hermetically sealed off capsule together with an electric motor (i.e. fully hermetic compressors). In contrast, screw-type, turbo and similar compressors with semi-hermetic or open design are generally reserved for larger capacities.

Expansion valves. In the throttle or expansion valve the pressure of the liquid refrigerant is released from condenser pressure to evaporator pressure. Additionally, the mass flow of the working medium circulating in the heat pump circuit is controlled. Selecting the expansion valve depends on the refrigerant, the size of the compressor and the capacity of the heat pump respectively. Possible designs are thermostatic or electronic expansion valves or capillary tubes.

– Thermostatic expansion valves are used for dry evaporation. They are controlled by the evaporation pressure and the superheating of the refrigerant discharged from the evaporator. As superheating takes place in the evaporator and requires a corresponding heat transfer surface, superheating has to be kept as low as possible with regard to high evaporator efficiency.

– Electronic expansion valves are regulated by using mathematical algorithms containing the thermophysical properties of the refrigerant and the relevant parameters of the heat pump. These are the transit time of the refrigerant through the evaporator, control characteristics, compressor data, etc. They enable a controllable evaporation temperature and superheating.

– Capillary tubes are used in combination with fully hermetic compressors. They are very thin (mostly between 1 and 2 mm internal diameter). They are up to 1 or 2 m long in order to guarantee the required throttling effect. However, capillary tubes can only guarantee supercooling in the condenser, but not superheating in the evaporator. Therefore they need a low-pressure accumulator to ensure that the compressor does not take any liquid refrigerant in. Capillaries are also used for coarse regulation in combination with thermostatic or electronic expansion valves. Refrigerators are, for example, almost exclusively built with capillaries.

– Newer developments, where a small turbine is installed instead of the valve and thus the compression energy is regained in parts ("Expander"), have only been known from larger heat pumps that are hardly ever used for shallow geothermal energy utilisation.

Lubricants. The use of the lubricant should minimise the wear and tear of the compressor. Depending on the compressor design, lubricants (oil) and refrigerants have more or less contact.

– In turbo compressors the separation of refrigerant and oil is easily achieved. Oils that do not blend with the refrigerant can be used.

– Screw-type compressors require large amounts of oil for sealing off as they are directly in contact with the refrigerant. Oil separators are used here to avoid oil losses.

– Piston compressor surfaces are continuously moistened with oil by the piston movement. Lubricant and refrigerant are in contact here, too.

– Scroll compressors are also lubricated with oil. Lubricant and refrigerant are in contact.

With the exception of turbo compressors, an optimum adjustment of the lubricant characteristics to those of the refrigerant is of particular importance. As a small amount of oil is always discharged into the refrigerant circuit in the case of oil-lubricated condensers, it has to be ensured that it is transported through the entire refrigerating circuit.

Working media (refrigerants). In the past, mainly fully or semi-halogenated chlorofluorocarbons CFCs and HCFCs have been used as heat pump working media in compression heat pumps (Table 9.6). As chlorofluorocarbons contribute to the depletion of the stratospheric ozone layer to a large extent, nowadays only refrigerants can be used that are not harmful for the ozone layer. They should also have a fairly low global warming potential.

Table 9.6 Characteristics of refrigerants with their environmental relevance (see /9-13/)

R-Number	Name	Formula	Boiling temperature[a]	WDC[b]	ODP[c]	GWP[d]
CFCs[e] and CFC mixes (no longer legal)						
R12	Dichlorine-Difluoro-Methane	CCl_2F_2	-30 °C	2	0.9	8,500
R502	R22/R115 at a ratio of 48.8 to 51.2 % (R155 – Monochlorine-Pentafluoro-Ethane, C_2ClF_5)		-46 °C	2	0.23	5,590
HCFCs[f]						
R22	Monochlorine-Difluoro-Methane	$CHClF_2$	-41 °C	2	0.05	1,700
HFCs[g]- and HFC-mixes						
R134a	Tetrafluoro-Ethane	$C_2H_2F_4$	-26 °C	1 – 2	0	1,300
R407C	R32/R125/R134a at a ratio of 23 to 25 to 52 %		-44 °C	21	0	1,610
R410A	R32/R125 at a ratio of 50 to 50 % (R32 – Difluoro-Methane, CH_2F_2; R125 - Pentafluoroethane, C_2HF_5)		-51 °C	2	0	1,890
Halogenated and chlorine-free working media (Propane and propylene can burn)						
R290	Propane	C_3H_8	-42 °C	0	0	3
R1270	Propylene	C_3H_6	-48 °C	0	0	3
R717	Ammonia	NH_3	-33 °C	2	0	0
R744	Carbon dioxide	CO_2	-57 °C	0	0	1

[a] Boiling temperature; [b] Water damage category; [c] stratospheric ozone depletion potential (relative, R 11 is 1.0); [d] global warming potential (relative, time frame 100 years, CO_2 is 1.0); [e] fully halogenated chlorofluorocarbons, [f] semi-halogenated chlorofluorocarbons, [g] Fluorocarbons.

The refrigerants used here are often named with their abbreviations. In the past, mainly the chlorofluorocarbons R12, R22 and R502 were used in compression heat pumps. The semi-halogenated chlorofluorocarbon R22 is sometimes still used. This nomenclature according to the German DIN 8962 refers to the chemical composition of the substances. The figures or letters added to the letter "R", the abbreviation for refrigerant, reflect the atomic composition of the refrigerant. The first figure refers to the number of carbon(C)-atoms minus one. The second figure names the number of hydrogen(H)-atoms plus one. The third figure contains the number of fluorine(F)-Atoms. The remaining free carbon valences have to be formulated as chlorine(Cl)-atoms. In the case of fluoro-methane compounds (one carbon(C)-atom) the first figure is not used. Added small letters stand for isomers. To give an example, tetrafluoro-ethane ($C_2H_2F_4$) is thus called R134a and difluoro-dichlorine-methane (CF_2Cl_2) is called R12. This method can also be applied to chlorine and fluorine-free hydrocarbons (e.g. propane (C_3H_8) is called R290). Working media from fundamentally different substance groups are allocated numbers starting with seven (e.g. water (R718) or air (R729)).

According to the regulations of the CFC-Halon-Ban Regulation since 1995 no CFCs (i.e. fully halogenated chlorofluorocarbons) can be used as a refrigerant in new systems e.g. in Germany. R22, a semi-halogenated chlorofluorocarbon, has been banned for new systems since January 1st, 2000. From January 1st, 2015 semi-halogenated chlorofluorocarbons are no longer allowed in existing systems either (EU guideline).

Due to the many requirements towards refrigerants it was costly to find suitable replacement substances for the media that have been used so far. If they contain a lot of hydrogen atoms, they are normally inflammable. If the share in chlorine or fluorine is high, mean life in the atmosphere can be expected to be high. The stratospheric ozone depletion potential is high in that case. The categorisation of refrigerants according to the ones used currently and in the past according to these criteria is shown in Fig. 9.13. Thus from a current point of view mainly the following two HCFC mixes are considered together with halon and chlorine-free working media (e.g. propane, propylene, ammonia).

– R407C as a short-term substitute for R22.
– R410A as a substitute for R22, that can only be used in heat pumps after the construction has been changed accordingly.

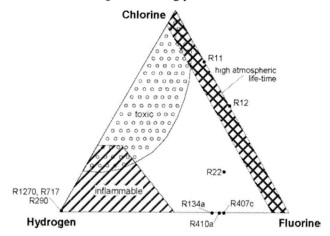

Fig. 9.13 Characteristics of refrigerants (see /9-14/)

In new heat pump systems, e.g. in Germany, the refrigerants propane (R290) and propylene (R1270) are used sometimes. These substances have no stratospheric ozone depletion or global warming potential and agree well with the materials and lubricants used so far. Additionally, the filling amount can be reduced significantly in comparison with the semi-halogenated chlorofluorocarbon R22. The required amounts for small systems up to 10 kW are at only around 1 kg. Due to the combustibility of the refrigerants R290 and R1270, special security measures have to be taken in line with the filling amount. They can generally be realised without problems in practice. For larger systems mainly R134a is currently used as the substitute refrigerant.

9.2.4 Overall systems

Heat source systems (Chapter 9.2.1 and 9.2.2) and heat pumps (Chapter 9.2.3) are integrated into overall systems that enable the utilisation of ambient air or shallow geothermal heat as end and useful energy respectively. Therefore firstly typical system configurations for characteristic applications are described. Afterwards system aspects of such overall systems will be discussed.

System configurations. In the following, a heating system with exhaust air heat recovery and exhaust-air to inlet-air heat pump, with a ground-coupled heat pump and a heat pump system for heating and cooling purposes will be introduced as typical overall system configurations.

Heating systems with exhaust-air to inlet-air heat pump. Over the last few years, exhaust-air to inlet-air heat pumps have been especially developed for houses with a very low heating energy demand and controlled ventilation systems. Not only do they cover the entire demand for space heating energy by heating up the inlet air, but can also cover the domestic hot water demand to a large extent. Fig. 9.14 shows exemplary such a heat pump unit. The inlet air is heated further by the heat pump condenser after exhaust air heat recovery. The evaporator is allocated in the exhaust air duct after the exhaust air heat recovery heat exchanger. In order to reduce icing of the evaporator on the exhaust air side, one option is to position a ground-coupled heat exchanger between ambient air and the exhaust air heat recovery heat exchanger. Thus the exhaust air in the heat exchanger cannot cool down so much. The way the cooling fans are allocated enables their exhaust heat to contribute to the heating process (i.e. in the exhaust air duct before the heat exchanger of the exhaust air heat recovery and after the condenser in the intake air duct). If enough heating energy is available for the house, the heat pump switches to the condenser of the domestic hot water generation system. Furthermore, an additional thermal solar system can be used to generate domestic hot water. During cold and cloudy periods in winter, an electric immersed heater is available as a back-up for the domestic hot water supply. Such heat pump systems can achieve SPFs of up to 3.5 /9-15/.

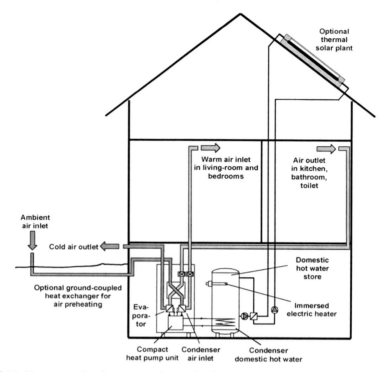

Fig. 9.14 Heat pump heating system for passive houses equipped with exhaust-air to inlet-air heat pump for air space heating systems and generation of domestic hot water plus ground-coupled heat exchanger for air preheating (see /9-15/)

Heating systems with ground-coupled heat pump. As an example, Fig. 9.15 shows a heat pump heating system with a horizontal ground-coupled heat exchanger. The heat pump primarily supplies space heating water in this case. Domestic hot water generation that would require a higher temperature level, in principle could also be realised using a heat pump, but also using a separate heater. The heat pump directly feeds a low-temperature floor heating system. Due to the storage capacity of floor heating systems, a buffer storage might not be required. Only for an increased demand towards indoor temperature quality (e.g. balancing of temperature within a building, for example with appropriate solar radiation) or if throttling of individual heating circuits is required, a buffer storage may be needed. An appropriate control device steers the heat pump that is normally run with On/Off operation – depending on the required space heating inlet temperature and ambient temperature. Due to the high storage mass of the heated building surfaces (e.g. stone floor, concrete floor) the On/Off operation of the heat pump does not lead to a loss in comfortable room temperatures.

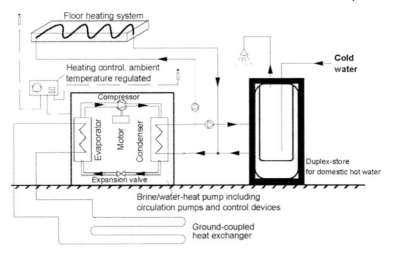

Fig. 9.15 Heat pump heating system with ground-coupled heat exchanger (see /9-1/)

Heat pump systems for heating and cooling purposes. Ground-coupled heat pumps cannot only be used for heating, but also for space cooling purposes, as the operation of heat pumps can generally be reversed. Heat is then transferred from the building into the ground. Through the double use of this cost-intensive system component in the ground, such systems for heating and cooling operate at comparatively reasonable costs. In the case of vertical probes, hollow spaces can occur around the tube due to water vapour diffusion away from the tube because of the heat transfer into the ground. This has to be avoided by adequate backfilling of the tube.

Under Central European climatic conditions it is also possible to achieve space cooling without operating the heat pump as a refrigerating aggregate (i.e. by using cooling ceilings or convectors and the cooled brine from the ground), if the cooling demand is low. Fig. 9.16 therefore shows three possible types of operation of ground-coupled heat pumps with direct cooling.

– In winter, the heat pump is in charge of heating. This generates low temperatures (or a deficit in heat) at the heat pump evaporator. The subsoil cools down (heat mode in Fig. 9.16).

– During transition time or – for suitably designed systems with a cooling capacity that is not too high – during the entire cooling period, heat from the building can be transferred to the ground in line with the natural temperature difference between ground and cooling device. Thus the building is cooled (direct cooling; refrigeration mode 1 in Fig. 9.16). The temperature of the heat carrier can exceed the original temperature of the ground. This is possible as long as it is low enough to still ensure the required cooling of the building. In most cases, a dehumidification of the intake air is not possible for direct cooling (refrigeration mode 1), as only at the beginning of the cooling period the temperature is guaranteed to sink below the defrosting temperature between 14

guaranteed to sink below the defrosting temperature between 14 and 16 °C in an air register.

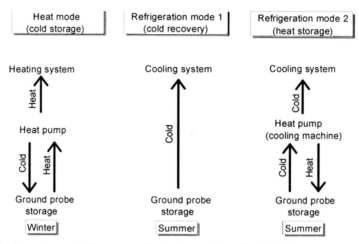

Fig. 9.16 Ground-coupled heat pump for heating and cooling /9-16/

- In refrigeration mode 2 (Fig. 9.16) the heat pump operates as a refrigerating machine. Space air is cooled by the heat pump evaporator and the generated heat is transferred into the subsoil. In this operational mode each operational condition (including air dehumidification) can be achieved – like in a conventional refrigerating system. The saving of drive energy is an advantage compared to conventional refrigerating machines releasing condenser heat to the ambient air.

Split system air conditioners for space heating and cooling. Split system air-conditioners are air-to-air heat pumps with an indoor and an outdoor unit connected by two tubes of the refrigeration cycle. They cover 70 % of the total market for room air-conditioners (RACs), which constitute a growing electrical end-use in the European Union. RACs may be used in households and can be bought or ordered directly by their occupants and as such they can be classified as domestic appliances. However, the same appliances are also commonly used in offices, hotels and small shops. Quite often, split system equipment is also used for manufacturing and process cooling applications. The widespread use of split systems (about one million plants were sold in Europe already in 1996, and the market has constantly grown since then) is an affirmation of their advantages /9-17/.

   In the cooling mode the indoor unit comprises the evaporator and its fan and a water drain. The outdoor unit consists of compressor, condenser (including fan), expansion device (often only a capillary tube) and the filter dryer unit. In most modern commercial applications, the compressor and the condenser are combined into a single piece of equipment called condensing unit (for the cooling application).

Some split system air conditioners can reverse the refrigeration flow by using a four-way valve near the compressor (reversible systems). Such units can also serve as space heating devices with air as heat source and air as heat sink. The indoor unit becomes the condenser and the outdoor unit the evaporator (including compressor). For the heating mode another expansion device is used. This means that such machines have also two magnetic valves around the expansion unit that allow the use of different expansion devices for heating and cooling. For the heating mode a water drain has to be placed on the outdoor unit that is led into the room in order to prevent ice formation in the drain in winter. Additionally a defrosting mode for the outdoor unit is necessary.

Modern systems have remote control with programmable timer, variable speed compressor and fans with inverter technology, adjustable louvers and are therefore very easy to use for the consumer. A typical split air-conditioning system for cooling only application is shown in Fig. 9.17. The main advantages of such systems are outlined below.

– Low investment cost and moderate operating costs.
– Cheap, flexible and easy mounting (only two holes in the wall for the cooling mode and an additional hole for the drainage of the outdoor unit for the heating mode).
– Easy to use for the consumer.
– High reliability.
– Decentralised application.
– SPF of about 2.5 for the cooling mode /9-17/.
– Low space demand.

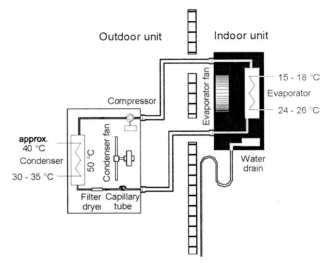

Fig. 9.17 Typical split system air conditioner (cooling-only mode)

The disadvantages of split systems may be as discussed below.

- One of the main disadvantages of cooling buildings in general, and in particular with such systems, is the increasing electricity demand in hot summer climates during the day (e.g. Southern Europe). This results in different price levels in summer during day and nighttime in some European countries.
- Fan(s) used in the condensing unit can be a relatively loud noise source that may require special consideration depending on the application.
- Air handling equipment near the centre of the building requires special provision to admit outdoor air. Units with economiser cycles must usually be located near an outside wall.
- An element of care is necessary to design refrigerant piping, especially for long piping runs. Improperly designed piping can cause the system to lose capacity. An improperly designed piping can even cause compressor failure.
- Each installation has to be equipped with drainage for the water from dehumidification of indoor air (cooling mode) or outdoor air (heating mode).

Systems aspects. In the following selective systems aspects of heat pump systems are discussed.

Types of operation. With regard to operating a heat pump system, the following variations can be differentiated.
- Monovalent operation. Only the heat pump supplies the required domestic heat. Furthermore, the following operational modes can be distinguished
  - without interruption (i.e. the heat pump always supplies the required heat on its own),
  - with interruption (i.e. operation of the heat pump can be stopped temporarily by the Utility Company supplying the end energy (i.e. electrical energy) to run the heat pump. If the heat distribution system does not have the required heat storage capacity to bridge these operational interruptions, a buffer storage has to be connected to the heat pump) and
  - with additional electrical resistance heating for electric heat pumps to cover demand peaks. This so-called monoenergetic operation is not used for ground-coupled heat pumps in Central Europe.
- Bivalent operation. The heat pump supplies the required heat together with other systems. Bivalent-alternative and bivalent-parallel operation are distinguished. In addition, there are bivalent mixed types. Bivalent operation is not very significant for the utilisation of ground as the heat source. Only for larger systems it might be of a certain interest.
  - For bivalent-alternative operation the heat pump covers the entire heat demand up to a certain switch-over point (i.e. a certain ambient temperature). Afterwards, an alternative additional heater takes over the entire heating supply (e.g. a gas-fired boiler). The heat pump system is only designed for certain percentage of the maximum heating demand. The additional heater, however, has to be able to cover 100 % of the entire heating demand.

- For bivalent-parallel operation the heating demand is simultaneously covered by the heat pump and an additional heating system from a certain temperature onwards.

In principle, if the return-flow temperatures are acknowledged and the heat storage is suitably large, various further heat sources can be integrated into a heat pump system, for example solar collectors or a fireplace.

The operation of ground-coupled heat pumps is normally monovalent. This is possible as the ground as a heat source only shows low seasonal temperature fluctuations and is thus available throughout the entire course of the year. Bivalent-alternative operation is only useful for systems with a non-adjusted heating system (high temperature). For bivalent-parallel operation, the flow of the heat pump is fed into the return-flow of the heater, where the water of the heater is heated further. Especially for larger systems with marked peaks in demand this mode of operation can make sense from an economic point of view. If ambient air is utilised, a bivalent mode of operation can also ensure a corresponding security of supply.

Areas of application. Areas of application for heat pump systems are mainly in space heating and domestic hot water supply. The generation of process heat for commercial and industrial use – also in the low-temperature segment – has played a minor role so far /9-18/.

- Space heating. For space heating almost exclusively electrically driven heat pumps are used. In comparison, compression heat pumps with a combustion engine drive and absorption heat pumps are not been very widely spread so far. Ambient air as well as ground (including groundwater) and possibly surface water can be used. In order to achieve high SPFs, a temperature difference between heat source and heat sink (space heating system inlet temperature) that is as small as possible has to be aimed for. The dependency of the COP on the inlet temperature of the heating system and the heat source temperature shown in Fig. 9.18 clarifies that heat pump systems are preferable for low-temperature heating systems. The use of heat pump systems in older heating systems with flow temperatures of up to 70 °C is therefore uneconomic compared to low-temperature systems due to their resulting low annual work rates. As a substitute or an alternative for single heaters (e.g. storage or coal heaters) single-space heat pumps using ambient air as the heat source can be utilised. These compact appliances (depth around 20 cm) are fixed directly to the wall of the space to be heated.

- Domestic hot water. Heat pumps for domestic hot water supply are offered as compact appliances and generally use ambient air as their heat source. They are normally installed in basement rooms, which act sometimes also the source of the used air. However, it is better to also use ground or ambient air as heat sources for domestic hot water supply pumps. At air temperatures below 7 °C, the heat pump becomes non-operational and the electric heating that has to be part of the domestic hot water storage takes over and heats the water up. Thus enough domestic hot water can be supplied even at times of peak demand. Additionally, higher water temperatures can be generated short-term to avoid legionella.

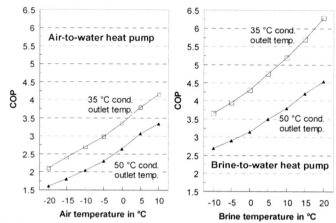

Fig. 9.18 Influence of the heat source temperature level and the space heating inlet temperature on the heat pump Coefficient of Performance (COP) (brine with 25 % antifreeze compound; temp. temperature, cond. condenser see /9-18/)

- Space heating and domestic hot water. A joint generation of domestic hot water and space heat using the heat pump with a heat exchanger does generally not make sense due to the normally higher temperature level of the domestic hot water and the resulting lower SPFs for the entire system. It is therefore useful to separate the generation of domestic hot water and space heating. For that purpose, domestic hot water can be generated with a separate heat pump or an electrically heated domestic hot water store respectively. A cheaper alternative option is the supply of heat to domestic hot water and space heating using the same heat pump. Together with the heat exchanger for the heating system a second heat exchanger for the domestic hot water supply is installed in the condenser part of the heat pump. If the generated heat is discharged into the heating circuit, a lower condenser temperature is required than if it is discharged to the domestic hot water circuit. The heat pump can thus be run under the individual optimum conditions with a COP that is as high as possible. A further possibility to combine the generation of space heating and domestic hot water is the utilisation of the heat that is released at high temperatures after the compressor up to the condensation point when desuperheating the refrigerant, down to the point of condensation in a separate heat exchanger for domestic hot water generation. The condensation itself is used to warm up the space heating water.
- Further uses. Apart from the supply of space and domestic hot water heating, heat pump systems can also be used for space cooling, as the operation of the heat pump is generally reversible. In those countries where space cooling is a

standard feature in residential buildings (e.g. North America, Japan), such systems are widely spread. The "heat pump" often operates primarily as a refrigerating machine for space cooling and only secondly as the proper heat pump. Given the climatic conditions in Central Europe with its comparatively low need for refrigerating, space cooling can be achieved without running the heat pump as a refrigerating aggregate. The heat exchanger that consists of ground probes can deliver sufficient temperatures of 8 to 16 °C with a simple cooling with cooling ceiling or blast convectors and between 9 and 10 °C for the groundwater.

COP characteristics. The quality of heat pump systems can e.g. be measured with the COP, the SPF and the heating rate (Chapter 9.1). In the following dimensions that can be achieved nowadays will be discussed.
- Coefficient of performance (COP). The COPs given by the producer always refer to certain operational conditions (heat source and sink temperature). Given ideal conditions, between 40 and 65 % of the COPs generated by a loss-free Carnot Process can be achieved /9-12/. However, this gives reason to assume a certain level of energetic development potential for the current state of technology. Fig. 9.19 shows an example for the use of the heat source ground (brine circuit, no direct evaporation) and groundwater. Mean values of currently achievable COPs and those that can be expected in the future are given as an example using a maximum flow temperature of 35 °C in this case.

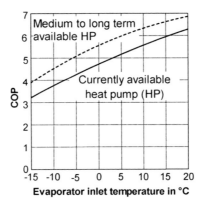

Fig. 9.19 Coefficients of Performance (COPs) of ground-coupled electric heat pumps (flow temperature of heating system 35 °C; evaporator inlet temperature approximately –10 to +10 °C for brine and approximately +5 to +15 °C for water; see /9-19/)

- Seasonal performance factor (SPF). For groundwater heat pumps the SPFs of new systems are between around 4.0 to potentially slightly above 4.5. When utilising ground as the heat source, currently SPFs of approximately 3.8 to 4.3 can be achieved. For direct evaporation, the SPF of the system is mostly around 10 to 15 % higher. The decisive factor for high SPFs of the heat pump systems

are a sufficient dimensioning and a flow temperature of the heat utilisation systems that is as low as possible (e.g. 35 °C for floor heating).
- Heating rate. Depending on the electricity generation technology it is based on and the SPF of the heat pump system, currently achievable heating rates are between 1.1 and 1.8.

The dependence of the COP on the temperature difference between the heat source and the heat sink is thus a main characteristic of heat pump systems. Fig. 9.20 shows the characteristic curves that are currently achievable in operation when utilising heat pumps withdrawing the energy from the ambient air or the near surface ground.

Accordingly, e.g. an air-coupled heat pump at an annual mean temperature of the heat source at currently 0 °C has a COP of approximately 3. In comparison, a ground-coupled heat pump achieves a COP of around 4.5 under the same circumstances. This is one of the reasons for the increasing installation of heat pumps for shallow geothermal energy use. The importance of air-coupled systems is gradually declining in Central Europe.

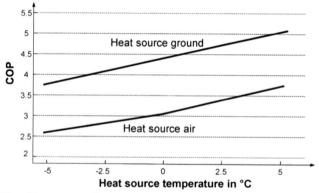

Fig. 9.20 COPs of heat pumps for low-temperature heating systems withdrawing heat from ambient air and near-surface ground

Heat regime in near-surface ground. Artificial withdrawal and/or discharging heat into the near-surface ground leads to a disturbance of the heat regime in the ground. Therefore the heat deficit or the excess heat have to be balanced by heat transfer. Meanwhile this is mainly ensured by the system itself for systems with an approximately balanced energy balance in the subsoil (e.g. heat pumps for heating and cooling purposes). For ground-coupled heat pump systems with an exclusive heat withdrawal from the near-surface ground this does not apply. The heat deficit has to be balanced by the natural ambient heat flow (i.e. mainly consists of solar energy and the geothermal energy of the deep subsoil).

Through measurements from a system and by extrapolating using numeric simulation it becomes obvious that ground-coupled heat pumps can be run constantly, even if only heat is withdrawn (i.e. system exclusively for heating purposes). The withdrawn heat is balanced with the ambient heat flow (Fig. 9.21)).

Overall, this shows that a lasting heat supply can be provided by ground probes if the design is correct (e.g. in line with the German VDI 4640). Especially given a large number of systems in limited spaces, a suitable probe length and – in an extreme case – artificial heat supply during the summer have to be used.

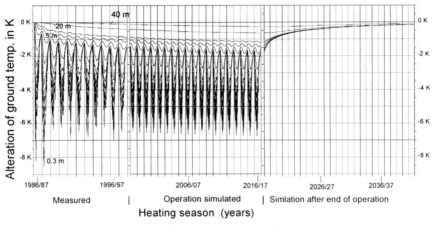

Fig. 9.21 Measured and simulated variations of subsoil temperature (cooling down compared to temperatures not manipulated) at a depth of 50 m and at different distances from the ground probe during 30 years of operation and 25 years of rest after terminating operation /9-19/

9.3 Economic and environmental analysis

In the following, the costs and selected environmental effects for selected heat pump systems with technical parameters reflecting the current market spectrum will be analysed.

9.3.1 Economic analysis

In Central Europe, heat pumps for space heating and domestic hot water generation are mainly built as monovalent compression heat pumps driven by an electric motor. The following analyses will deal with systems that can cover various supply tasks (three single family houses with different types of insulation (SFH) and one multi-family house (MFH); see Chapter 1.3) – in each case with a different heat demand (Table 9.7). For that reason, system configurations of heat pump systems with direct evaporation, a brine circuit with a horizontal or a vertical heat exchanger and heat pump systems connected to the groundwater are defined (Table 9.7). These heat pump systems generate space heat and domestic hot water, with a priority for the generation of domestic hot water. The working medium used in all systems is R407C. The heat pump is always installed in the basement of the building to be supplied. The achievable SPFs are determined by the heat pump technology and the characteristics of the heat source plus the share of domestic hot water generation as part of the entire demand for heat. Due to the higher temperature level, the domestic hot water generation is characterised by lower COPs than space heating. When measuring the SPF auxiliary energy consumption for e.g. control purposes the brine or the groundwater pump has to be considered as well.

Table 9.7 Reference-configurations of the analysed heat pump systems

System		SFH-I[a]	SFH-II[b]	SFH-III[c]	MFH[d]
Space heating demand	in GJ/a	22	45	108	432
Domestic hot water demand	in GJ/a	10.7	10.7	10.7	64.1
Nominal heating requirement	in kW	5	8	18	60
Heat source					
Ambient air (without preheating of air) (AWO)		X	X		
Ambient air (with preheating of air) (AW)		X	X		
Ground-coupled horizontal collector with brine circuit (GB)		X	X	X	
Ground-coupled horizontal collector with direct evaporation (GD)		X	X	X	
Ground-coupled probe with brine circuit (GP)			X	X	X
Groundwater (GW)			X	X	X

[a] Single family house (SFH-I) with low-energy design; [b] single family house (SFH-II) according to current heat insulation standards, [c] single family house (SFH-III) as an old building with an average heat insulation, [d] multi family house (MFH); for the definition of SFH-I, SFH-II, SFH-III and MFH see Table 1.1 and Chapter 1.3 respectively.

The SPFs assumed in the following are higher for those single family houses that are not so well insulated. The relative share of domestic hot water generation compared to heating decreases in that case. Domestic hot water generation requires a higher temperature level than heating, which leads to a lower COP of the heat pump.

– Ambient air with/without preheating (AW/AWO). For systems without preheating, the air is transported to and from the heat pump via insulated galvanised sheet steel ducts. Opposed to that, in the system AW, the ambient air is preheated with a so-called air well. This is a concrete duct of approximately 60 m length and a diameter of 25 cm. It is sunk into the ground at a depth of 1.5 m. The SPFs of the analysed reference systems are assumed to be 2.17 (SFH-I) and 2.37 (SFH-II) for systems without air preheating (AWO) and 2.40 (SFH-I) and 2.65 (SFH-II) for systems with preheating (AW).
– Ground-coupled heat pumps with brine circuit (GB). HDPE tubes are sunk 1.2 m deep as collectors. The heat carrier – like in all analysed media with a

brine circuit (ground-coupled heat pump and vertical probe) – consists of 30 % propylene-glycol and 70 % water. Due to their relatively high surface requirement, ground collectors are only used for comparatively low heat capacity levels (in general smaller than 20 kW). Therefore only the systems SFH-I, SFH-II and SFH-III can be operated as heat source systems with ground collectors. For the compound system of domestic hot water generation and space heating SPFs of 3.43 (SFH-I), 3.65 (SFH-II) and 3.85 (SFH-III) can be achieved.

- Ground-coupled heat pumps with direct evaporation (GD). For the assumed systems with direct evaporation, copper tubes with a plastic coating are sunk at a depth of 1.2 m on a layer of sand. Due to the similarly large surfaces required, heat pump systems for the supply tasks SFH-I, SFH-II and SFH-III are analysed. The refrigerant R407a serves as the heat carrier from the collector to the heat pump. The annual work rates of these systems are at 3.76 (SFH-I), 4.00 (SFH-II) and 4.20 (SFH-III).
- Vertical ground probe with brine circuit (GP). At an assumed heat withdrawal capacity of 50 W per m ground probe, ground probe lengths of 2 x 60 m (SFH-II), 3 x 90 m (SFH-III) and 12 x 75 m (MFH) can be derived for the systems under review. The HDPE probes are designed as double-U-tubes and installed in boreholes that are afterwards filled with a suspension of bentonite, cement and water. The SPFs for domestic hot water generation and space heating are at 3.59 (SFH-II), 3.77 (SFH-III) and 3.73 (MFH).
- Ground water wells (GW). For the systems SFH-II, SFH-III and MFH, production and injection wells that are 20 m deep each are excavated. The lining and walling of the boreholes is done correspondingly. The extracted groundwater that serves as a heat carrier is discharged via injection well into the ground again after heat withdrawal by the heat pump. The SPFs are at 3.95 (SFH-II), 4.20 (SFH-III) and 4.15 (MFH).

In order to be able to give an estimate of the costs involved in supplying low-temperature heat with the heat pump systems defined above, investment and operation costs plus the specific heat generation costs for the reference systems defined in Table 9.7 will be presented below. Due to the location-specific geological conditions (e.g. condition of the ground, heat conductivity of the subsoil, distance of the groundwater conductor from the top edge of the terrain) significant differences in the design of the heat source system and thus in the cost structure of the compound system can occur. Additionally, the costs for electrical energy and the connection of the heat pump to the public electricity grid are widely dispersed depending on the respective conditions of the local utility. The costs discussed in the following can therefore only show a certain scale and average reference values. In individual cases and depending on the local framework conditions, lower but also higher heat generation costs can be possible.

Investments. The amount of specific investments into heat pump systems is largely determined by the applied technology and the size of the system. In general, the specific costs decrease with an increase in system size. This is mainly true for the heat pump aggregate including domestic hot water generation. In contrast to that, the heat source systems show a slight decrease of costs, with the exception of groundwater utilisation as a heat source. Thus specific investment costs of the analysed brine/water and water/water heat pumps are between 220 and 1,000 €/kW. The costs for heat pumps of direct evaporation systems are slightly lower. For heat source installations, costs for systems with vertical ground probes are between 540 and 600 €/kW, for systems with groundwater utilisation between 240 and 600 €/kW and with horizontal ground collectors using brine or direct evaporation between 240 and 300 €/kW. If rammed wells are used instead of the assumed boring wells, the costs can be reduced significantly, especially for smaller systems. For rammed wells, the investment costs for the entire heat source system of the analysed systems are at around 3,000 € for an 8 kW installation, at approximately 4,000 € for an 18 kW installation or at 13,000 € for a 60 kW installation (Table 9.8 and 9.9).

Table 9.8 Investment and operation costs plus heat generation costs of heat pump systems for the generation of domestic hot water and space heating for the reference configurations SFH-I and SFH-II (Table 9.7)

System	SFH-I[m]				SFH-II[n]					
Heat source	AWO[a]	AW[b]	GB[c]	GD[d]	AWO[a]	AW[b]	GB[c]	GD[d]	GP[e]	GW[f]
Seasonal performance factor (SPF)[g]	2.17	2.40	3.43	3.76	2.37	2.65	3.65	4.00	3.59	3.95
Investments										
Heat source in €	0	2,725	1,514	1,514	0	3,028	2,267	2,267	4,845	4,784
Heat pump in €	6,662	6,056	4,966	4,542	8,660	7,500	6,056	5,450	6,056	5,753
Hot water[h] in €	1,671	1,671	1,671	1,671	1,671	1,671	1,671	1,671	1,671	1,671
Other[i] in €	1,514	1,514	1,575	1,696	1,514	1,514	1,514	1,635	2,120	3,149
Total in €	9,847	11,966	9,726	9,423	11,845	13,713	11,508	11,023	14,692	15,357
O&M costs[j] in €/a	197	212	172	166	237	243	196	187	221	307
Electr. costs[k] in €/a	628	568	397	362	979	876	636	580	646	588
Heat gener. costs[l]										
in €/GJ	53.0	55.3	43.9	41.8	41.4	41.1	33.3	31.3	38.5	40.6
in €/kWh	0.191	0.199	0.158	0.150	0.149	0.148	0.120	0.113	0.139	0.146

[a] Ambient air heat pump without air preheating; [b] ambient air heat pump with air preheating; [c] heat pump with horizontal ground collector brine; [d] ground coupled heat pump with direct evaporation; [e] heat pump with ground probe; [f] heat pump with groundwater; [g] for domestic hot water generation and space heating; [h] domestic hot water storage and connection to the heat pump; [i] e.g. costs for the heater room, hydrological licensing according to the local water legislation plus mounting and installation; [j] operating and maintenance costs without electrical energy costs for e.g. drive of the heat pump compressor, control, brine circulation pump, etc.; [k] costs of electrical energy e.g. for driving the heat pump compressor; [l] annuity calculation at an interest rate of 4.5 % and an amortisation period over the technical system life (heat source systems 20 years, heat pumps, domestic hot water generation and heat storage 15 years and building components 50 years); [m] single family house (SFH-I) in low-energy design; [n] single family house (SFH-II) according to current heat insulation standard; for the definition of SFH-I and SFH-II see Table 1.1 and Chapter 1.3 respectively.

With 1,500 to 2,400 €/kW specific investment costs for ambient air heat pump combined systems are higher than the overall costs of comparable ground-coupled systems.

Apart from the investment costs for the heat source system and the heat pump, there are also costs to be covered for the generation of domestic hot water and the heat storage (Table 9.8 and 9.9). Furthermore, costs for mounting and installation need to be added as well as the partial costs for the installation space in the basement of the buildings to be supplies. Costs for the hydrological application and reporting that the heat pump system is using the groundwater and the ground, to the authority in charge also has to be included. With an increased system size, the main share of the costs shifts from the heat pump to the heat source system. While e.g. between 51 and 68 % of the total costs have to be allocated to the heat pump for the analysed reference systems of system SFH-I, their share is between 26 and 29 % for the system MFH.

Table 9.9 Investment and operation costs plus heat generation costs of heat pump systems for the generation of domestic hot water and space heating for the reference configurations SFH-III and MFH (Table 9.7)

System		SFH-III[k]			MFH[l]	
Heat source	GB[a]	GD[b]	GP[c]	GW[d]	GP[c]	GW[d]
Seasonal performance factor (SPF)[e]	3.85	4.20	3.77	4.20	3.73	4.15
Investments						
Heat source in €	4,239	4,239	10,295	5,027	35,126	14,535
Heat pump in €	9,266	9,266	9,266	9,266	17,442	17,442
Hot water[f] in €	1,671	1,671	1,671	1,671	3,634	3,634
Other[g] in €	2,846	2,846	3,452	4,179	7,328	7,631
Total €	18,022	18,022	24,684	20,143	63,530	43,242
O&M costs[h] in €/a	297	297	339	403	744	865
Electricity costs[i] in €/a	1,285	1,178	1,312	1,178	5,542	4,981
Heat generation costs[j]						
in €/GJ	26.3	25.4	31.6	28.3	23.4	19.4
in €/kWh	0.095	0.092	0.114	0.102	0.084	0.070

[a] Heat pump with horizontal ground collector brine; [b] ground-coupled heat pump with direct evaporation; [c] heat pump with vertical ground probe; [d] heat pump with groundwater; [e] for the generation of domestic hot water and space heating; [f] domestic hot water storage and connection to the heat pump; [g] e.g. costs for the heater room, hydrological licensing according to the local water legislation plus mounting and installation; [h] operating and maintenance costs without electrical energy costs for e.g. driving the heat pump compressor, control, brine circulation pump, etc.; [i] electrical energy costs e.g. for driving the heat pump compressor; [j] annuity calculation at an interest rate of 4.5 % and an amortisation period over the technical system life (heat source systems 20 years, heat pumps, domestic hot water generation and heat storage 15 years and building components 50 years); [k] single family house (SFH-III) as an old building with an average heat insulation; [l] multi family house (MFH); for a definition of SFH-III and MFH see Table 1.1 and Chapter 1.3 respectively.

Operation costs. The operation costs consist of the maintenance costs for the heat pump system (e.g. exchanging the refrigerant or the heat carrier (brine); exchange of sealings). Table 9.8 and 9.9 show a list of these costs for the reference systems defined in Table 9.7. Depending on the size of the system, the operation costs are therefore between approximately 166 and 865 €/a. Electrical energy costs to drive the heat pump compressors and costs for e.g. the brine circulation pump, groundwater extraction, ambient air ventilator or control are not included. These costs are listed separately in Table 9.8 and 9.9. The electricity price for heat pumps is estimated at 0.15 €/kWh compared to 0.19 €/kWh being the normal electricity price.

Ground-coupled heat pump systems have the lowest variable costs. The ambient air heat pump systems have higher energy or electricity costs due to their lower SPF. Groundwater-coupled heat pump systems carry significantly higher operation costs (excluding electricity) than ground-coupled heat pumps. These higher costs also result from the heat source system.

Heat generation costs. With an interest rate of 4.5 % and an amortisation period over the technical life, the heat generation costs that are also presented in Table 9.8 and 9.9 can be calculated – on an annuity basis (see Chapter 1) – for the reference systems defined in Table 9.7. For the heat source systems 20 years are assumed, for the heat pump, the domestic hot water generation and the heat storage 15 years and for the building components 50 years are assumed as the technical system life.

Depending on the size of the system and the annual working rate (Table 9.8 and 9.9) the heat generation costs – in line with the installed power – are between 19.4 and 55.3 €/GJ. Systems using ambient air show the highest heat generation costs, systems with vertical ground probes an average amount and systems with horizontal ground collectors and direct evaporation are characterized by the lowest heat generation costs. This becomes also obvious in Fig. 9.22. It shows a comparison of heat generation costs of the analysed variations according to Table 9.8 and 9.9.

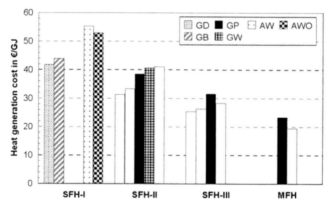

Fig. 9.22 Comparison between the heat generation costs (see Table 9.8 and 9.9; AWO ambient air heat pump without air preheating; AW ambient air with air preheating; GB ground-coupled heat pump with horizontal ground collector brine; GD ground-coupled heat pump with direct evaporation; GP heat pump with vertical ground probe; GW heat pump with groundwater; SFH-I single family house I; SFH-II single family house II, SFH-III single family house III; MFH multi family house; see Table 9.7)

According to Fig. 9.22, the heat generation costs decrease significantly with an increased demand in heat. Additionally, in all analysed cases air-coupled systems always show comparatively higher heat generation costs than systems with a ground collector.

In order to give a better estimation and evaluation of the influence of the different parameters on the generation costs, Fig. 9.23 shows a variation of the main cost-sensitive parameters of an 8 kW heat pump system with ground collector and direct evaporation (GD). According to this, investments and the required heat amounts (which correspond to the full load utilisation hours) have the main influence on the heat generation costs. It is assumed that heat capacity and thus the costs of the heat source system do not change under the influence of heat discharge variations. The duration of the amortisation period also has a significant influence on the heat generation costs – like the SPF achieved by the system. Electricity and operation costs and the assumed interest rate have less influence on the heat generation costs.

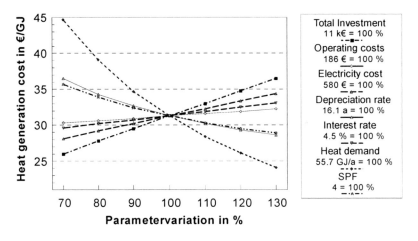

Fig. 9.23 Parameter variation of the main influencing variables on the specific heat generation costs using the example of an 8 kW heat pump system equipped with a horizontal ground-coupled heat exchanger and direct evaporation (GD) (reference system SFH-II; amortisation period of 16.1 years corresponds to the weighted average of all system components; see Table 9.8 and 9.9)

9.3.2 Environmental analysis

For the systems utilising ambient air and shallow geothermal energy analysed so far (Table 9.7) a number of selective environmental parameters are discussed below occurring during construction, normal operation, in the case of an accident and at the end of operation.

Construction. The environmental effects caused by installing a heat pump system for ambient heat utilisation are mainly due to the use of water as a heat source and, for ground probes, sinking the boreholes. Potential environmental effects caused by drilling are pollutant emissions into the subsoil by drilling equipment, drill tubes and auxiliary equipment. Chemical/biological changes due to drilling fluids can also cause environmental effects. Such pollutant emissions can be largely avoided by using preventive measures (DIN 4021 and DVGW W 116) to avoid contamination, bacteriological pollution and chemical-biological changes in the subsoil, etc., plus an adjusted drilling method /9-20/. Furthermore, noise effects can occur which are normally within the legal boundaries if the regulations of noise pollution are observed.

The installation of the heat pump system itself normally does not cause any other environmental problems than would occur during the installation of a conventional heating system. Dangerous moments that existed in the past due to the stratospheric ozone depletion potential of the refrigerant, e.g. due to losses during the filling process, ceased to exist after the ban of those refrigerants. If the refrigerant still has a certain global warming potential, it can potentially harm the climate. The industrial production of heat pumps also causes the environmental effects common for the mechanical engineering industry. Due to sometimes very far-reaching legislation, they are at a comparatively low level.

Normal operation. The discussion of environmental effects of geothermal energy utilisation during normal operation mainly covers the following areas: environmental effects of heat pump working media, thermal effects on the soil, groundwater and the atmosphere, hydraulic changes in the subsoil due to withdrawal of groundwater, noise effects and environmental effects caused by drillings. These aspects will be discussed in the following.

Environmental effects of heat pump working media. Refrigerants have an influence on the global and the local environment.

The global effects (e.g. damaging the stratospheric ozone layer, contribution to the anthropogenic greenhouse effect) depend on the use of the system and how well it is sealed off. They also depend on the type of system, the amount of refrigerant that is used for filling, the way the refrigerant is dealt with and the type of agent. Refrigerants based on chlorofluorocarbons are damaging for the stratospheric ozone layer described by the so-called ODP-value (Ozone Depletion Potential). The use of ozone-depleting refrigerants in new systems has been banned e.g. in Germany since January, 1st, 2000 /9-20/.

Certain types of refrigerants additionally have a direct effect on the climate. The respective contribution is generally stated in relation to CO_2, often for a time span of 100 years. It is called GWP-value (Global Warming Potential). Refrigerants used nowadays have a GWP of 1,300 (R134a), 1,610 (R407C), 3 (R290), 0 (R717) and 1 (R744) (Table 9.6). The trend that is becoming increasingly obvious is to use refrigerants with either no or a very low GWP.

If no leakage is assumed for normal operation – which should be easy to realise applying modern system technology – no such environmental effects will occur.

Thermal effects on the soil, the groundwater and the atmosphere. The utilisation of heat in the ground, the groundwater or the near-surface atmospheric layers by heat pumps leads to a corresponding cooling effect.

In vertical ground probe systems, for instance, temperature decreases of up 2 K occur within a distance of 2 m. If systems are dimensioned correctly, a thermal balance is established in the long-term. Additionally, the influence of the heat withdrawal remains at a local level. Furthermore, a moderate cooling of the ground has no known effect on its structure. Icing caused by an excessive heat withdrawal and the ensuing thawing process can change the structure in fine-grained soils (e.g. clay), which can cause the soil to sink around the built-in vertical ground probes /9-21/. As there are normally no living creatures or plant parts at the normally used depths, the cooling does not lead to any known ecological impairment. Additionally, the influence towards the surface of the earth is negligible and is offset by the solar radiation heat. A negative impact on the groundwater can be excluded as well.

If vertical ground probes are used, they have a certain influence on the ground fauna and vegetation. The range of influencing factors depends largely on the system design. If the collectors are too small in dimension, the activity level of the ground fauna (e.g. earthworms) is reduced – due to an excessive cooling of the ground – to a large extent. It also leads to e.g. a considerable delay in vegetation and a reduction in harvest amounts and flowers. If systems are designed appropriately, which should always be aimed for, these effects are comparatively small. For example no systematic change could be discovered in certain beetle populations which could have been caused by heat withdrawal by heat pumps with ground-coupled collectors /9-21/. During the summer, the same temperature level is achieved as without using collectors. A significant impact on the groundwater can also be excluded – due to fairly shallow excavations to lay the probes.

The situation is similar in the case of heat withdrawal from the atmosphere. The heat exchange – compared to heat withdrawal from the ground or the groundwater – between individual fractions of air in near-surface atmospheric layers is significantly more intensive. Thus a possible impact of cooling can be balanced immediately. Related environmental effects have therefore not been observed so far. Furthermore, the heat withdrawn from the ambient air for heating purposes is discharged to the ambient air again by the building.

Additionally, due to cultural influences, the temperature of the ground, the groundwater or the near-surface atmospheric layers has increased in many places. Therefore cooling down can be positive. Up to now, no significant negative aspects of the cooling of ground, groundwater or near-surface atmospheric layers by heat pump systems have become known.

Hydraulic changes in the subsoil caused by groundwater withdrawal. The withdrawal of groundwater and the ensuing discharge leads to a lowering of the groundwater level around the production well and a rise in the groundwater level around the injection well. This leads to an adjustment of the flow that is limited to a certain area.

Noise effects. Negative effects for the environment have often occurred due to high sound power levels of systems in the past. Compared to earlier system generations, significant reductions in sound radiation have been achieved for new systems in the market. Heat pumps with a heating capacity around 10 kW sometimes achieve a sound intensity of below 45 dB(A). Thus sound emissions are practically not a problem nowadays.

Effects caused by boreholes. Harmful situations for the groundwater occur if the borehole is insufficiently sealed off from the top edge of the terrain. This can lead to water-contaminating substances seeping in from the surface of the earth /9-22/. If sinking and completion of the boreholes is done professionally, such aspects practically do not occur.

The groundwater flow conditions can be influenced in a negative way if development drillings (e.g. for vertical ground probes) sink through two or several groundwater levels with different levels of pressure without control. The hydraulic contact of various groundwater layers is undesirable, particularly if one of the layers contains highly mineral or contaminated groundwater. By including a barrage, possible harmful effects can be largely avoided /9-22/.

Malfunction. Accidents in the context of heat pump utilisation can occur if the materials sunk into the subsoil corrode easily and cannot withstand the stress. The materials used for vertical ground probes, for example, have to be able to withstand the pressures during deep drillings and not tear easily. So far no tube fractures have occurred for PE-tubes, for instance, at depths up to around 150 m /9-22/.

The extent of the environmental effects caused by the heat pump, and the heat source part involved, that occur in the case of an accident depend on the types of refrigerant and antifreeze compound that are used. Antifreeze additives that are currently used most are ethylene-glycol and propylene-glycol – both in the hazard 1 class for water. The environmental effects that can, for instance, be caused by leakage, are generally small. The same is true for heat pump working media for direct evaporation, as the used refrigerants – with the exception of ammonia – are either not harmful for water or can only cause very little damage. Experiments with R290 as the refrigerant have shown that only relatively small, temporary hazardous effects occur for soil and groundwater. They are also very limited to a certain area /9-23/. New developments with CO_2 heat pipes are definitely not hazardous for the water.

Environmental hazards can occur in the case of a fire or an explosion of the heat pump due to the toxic character of the heating agent. According to EN 378-1 refrigerants are divided into three groups. The refrigerant that is used very often, R290, belongs to group A3 (higher combustibility, lower toxicity), R717 to group B2 (less combustible, more toxic) and R407c, R134a and R744 to group A1 (no spreading of flames, low toxicity). Carbon dioxide, which will probably be increasingly used as a refrigerant in the future, is known as the most environmentally friendly refrigerant (ODP = 0 (Ozone Depletion Potential), GWP = 1 (Global Warming Potential), non combustible and non toxic). Health risk potentials only occur due to bursting caused by a mechanical explosion and due to leakage through system components /9-24/, /9-25/.

If the existing security measures are observed (installation requirements depending on space volume and the filling amount of the refrigerant, ensuring ventilation, etc.) according to UVV VBG 20, EN 378 and DIN 7003 E, accidents can be avoided or at least their consequences can be minimised.

Additionally, environmental pollution of soil and groundwater can be caused by lubricants in the case of an accident. If synthetic oils are used, their low level of water damaging effects and their good biodegradability can minimise such environmental dangers. This is particularly important in the context of direct evaporation systems as larger amounts of oil are used there.

Altogether, the potential environmental effects in terms of the absolute damage are also limited in the case of an accident and only take effect on the system site itself.

End of operation. Potential environmental effects in the context of the end of operation can occur when groundwater and deep vertical ground probes are used, if the borehole is not sealed off properly. Furthermore, refrigerants can leak during the dismantling of the system. However, if current regulations are observed, this is unlikely to occur. As far as it is currently known, recycling the system components does not cause any particular environmental effects.

10 Utilisation of Geothermal Energy

10.1 Heat supply by hydro-geothermal systems

The expression 'hydro-geothermal energy utilisation' refers to the exploitation of the energetic potential of cold thermal (40 to 100 °C) and hot thermal (above 100 °C) water available within the deep underground.

Usually, geothermal fluid is brought up to the surface from underground via boreholes, where the heat is used by potential consumers. Except for cases in which geothermal fluid is consumed as such, they are re-transferred to the underground via a second borehole. This closed cycle helps to conserve the mass balance and thus avoids hydraulic problems. Moreover, for environmental reasons, especially highly mineralised geothermal fluids cannot be disposed of above the surface.

Within the following section the technical basics, as well as economic and environmental aspects for systems for the use of hydro-geothermal energy are described.

10.1.1 Technical description

When describing the system we distinguish between the downhole and the uphole part of the geothermal fluid circuit, the district heating system, and the incorporation of geothermal energy into supply systems /10-1/, /10-2/, /10-3/, /10-4/. Fig. 10.1 provides an overview of the basic concept of such an energy system, illustrating the two-hole system (consisting of one production well and one injection well) required to tap hydro-geothermal deposits inside the ground. But firstly the drilling of a geothermal well is described briefly.

Geothermal well drilling. Due to severe drilling conditions at high temperatures especially in geothermal fields, or for the drilling of hard abrasive rock types (like for Hot-Dry-Rock reservoirs; see Chapter 10.3), the drilling of geothermal wells is usually more challenging compared to oil and gas wells. Therefore within the following explanations the basics of deep well drilling is discussed with a special focus on unlocking geothermal reservoirs.

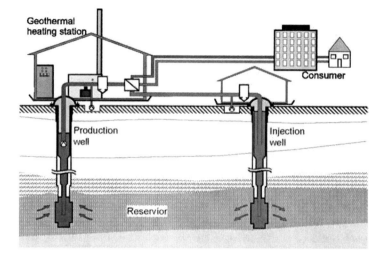

Fig. 10.1 Basic layout of a hydro-geothermal heating station designed to tap hydro-geothermal heat from the deep underground (see /10-2/, /10-5/)

Drilling technique. The technique for the drilling of geothermal wells is very similar to the drilling of oil and gas wells /10-6/. Almost exclusively, the rotary drilling technique is applied. The drilling tool is in most cases a tricone bit. The bit is rotated via the drill pipe, a steel pipe of much smaller diameter than the bit, by a rotary table in the floor of the drilling platform. The rock cuttings are transported to the surface by "drilling fluid", which is pumped down through the drill pipe and ascends in the annulus between drill pipe and borehole wall. Due to the relatively large cross sectional area of this annulus high flow rates of the drilling mud are required to achieve the flow velocities necessary for transporting the rock particles. During drilling the weight of the drill pipe is hanging on a hook in the tower of the drilling rig. A string of heavy thick walled large diameter drilling collars sitting on top of the drilling bit provides the load for the support of the drilling bit. The weight of the collars and their large diameter stabilise the drilling direction, and cause a vertical trend of the well. Further stabilisation of the drilling direction and smoothening of the borehole wall is provided by reamers installed between the drilling collars and equipped with hard metal cylinders rolling around the borehole wall /10-1/, /10-2/.

The material of the tricone bits depends on the rock properties, and are selected according to the experiences made during earlier drilling operations in the corresponding rock type. For more soft rock formations like mudstone, hard metal cones are sufficient. For hard rock the tricones are equipped with tungsten or even with diamond particles. The lifetime of the drilling bits range from several decametres to several hundred meters of drilling. The lifetime of the bit is one of the most important factors for the drilling costs. This is not due to the costs of the bits themselves but due to the costs for the round trip (pulling and lowering of the drill string) necessary for the replacement of the worn drill bit.

Traditional drilling rigs for depths deeper than 3,000 m are heavy constructions. The height of their tower is in the range of 40 to 60 m. The hook load often amounts to 600 t. Including tanks, pumps, and pipes for circulating and cleaning the drilling mud, the ramp and stock area for the drill pipes, containers for personnel, material, recording units the whole construction covers an area of up to one hectare. Heavy engines for the mud circulating pumps, for the rotary table, and the winch for the round trips are required (Fig. 10.2). Their total power amounts to several MW. The space for the drill site and the noise can become a severe problem in densely populated areas; for example electrical motors instead of diesel engines can improve the situation regarding noise.

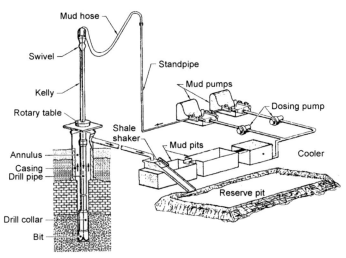

Fig. 10.2 The rotary rig and its components /10-2/

The drilling fluid, or mud, is an essential element of the drilling operation. Its function is not only the transport of the rocks cuttings. Of similar importance is its ability to stabilise the borehole wall and to prevent the invasion of formation fluid into the well. Both functions are provided by the pressure resulting from the weight of the fluid column and require sufficient density of the drilling mud. On the other hand losses of the drilling fluid in the rock formations have to be minimised. This can be realised by choosing a drilling fluid which builds a mud cake (thin layer of particles) on the borehole wall as well as by a proper choice of the density of the drilling fluid. The realised drilling fluid density is therefore a compromise and needs careful consideration. It must also consider the possibility of accidental hydro-fracturing of the rock, resulting in a total loss of the circulation. Total losses can also happen in highly permeable rock formations, in faults or in fracture zones. Last but not least the drilling fluid is also needed for cooling of the drill bit and for reducing the friction between borehole wall and drill pipe. Most of the drilling fluids are water based and contain bentonite or other tixotropic materials. Adjustment of fluid density is achieved by adding salts or barite. The stability of bentonite drilling fluids becomes a major problem at temperatures above 150 °C since they start to degenerate significantly /10-6/ and 190 °C seems to be the limit for water based drilling fluids. This does not mean that they cannot be used at these rock temperatures since the rock is cooled as long as the circulation of the fluid is maintained, but it can become a problem during longer breaks of circulation. In hard crystalline rock formations brine with a friction reducing agent proved to be very efficient at temperatures up to 200 °C.

For many of the exploitation schemes mentioned above directional drilling is essential. This technique was pushed forward for offshore drilling in oil and gas reservoirs where multiple wells are drilled from the bottom of a single well. Today, even the drilling of a several kilometre long horizontal well section from the bottom of a vertical well is possible. In most cases down-hole motors are used for rotating the drilling bit under these circumstances. These are either turbines or Moineau-motors driven by the drilling fluid pumped through them by strong injection pumps at the surface. Today down-hole motors with a driving power of more than 1,000 kW are available. Directional drilling is also possible with the conventional rotary technique.

Drilling direction is continuously monitored by using the so-called "Measuring While Drilling (MWD)" technique. A pressure pulse generator transmits the signals of directional sensors installed in the bottom part of the drill pipe via the drilling fluid to the surface. Reverse signal transmission and a hydraulically driven actuator allow adjustment of the drilling direction at any time and depth. The technique is successfully applied in soft rock formations at temperatures up to 150 °C.

There is only little experience with this technique in hard crystalline rock at high temperature and depth. Its application in the Hot-Dry-Rock (HDR) project Soultz (see Chapter 10.3) showed that the more intense vibrations of the drill string in this type of rock requires some improvement of this technique.

Well completion. To prevent a collapse of the borehole and protect freshwater aquifers at more shallow depths the geothermal well like any other deep well is cased down to the reservoir by inserting and cementing in steel pipes (Fig. 10.3). This is done in several stages. After the drilling of the first 15 m or so the large diameter conductor pipe is set and cemented. This pipe has a diameter of 24 to 30 in and lends structural support to the well and the well head. After drilling another 30 to 100 m the surface casing (13 3/8 in outer diameter) is set and cemented in. This casing provides a more sound foundation and protects the freshwater aquifers from contamination /10-6/. After reaching the top of the reservoir the technical casing is set and cemented in. This casing needs the most complex design considerations. The design has to take into account the stresses expected

by mechanical unstable rock formations, the fluid pressures in the various formations including the geothermal reservoir, the flow rates expected from the reservoir, the fluid properties as far as they give reason for corrosion and scaling, the
thermal expansion or shrinking induced during production or injection, and other
factors. Thermal expansion is especially important in geothermal wells since it can
lead to buckling or rupture of the intermediate casing especially in sections were
the borehole diameter is enlarged due to unstable rock conditions and incomplete
cementation. Depending on the geological conditions and the depths of the reservoir the technical casing may be installed in two stages.

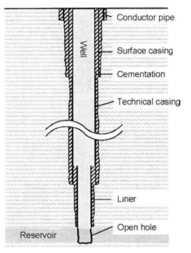

Fig. 10.3 Production well completion (see /10-2/)

Within the reservoir two different methods for the completion are in use: open
hole completion and cased hole completion.
– According to open hole completion the last casing ends above the reservoir,
  leaving the reservoir open (the reservoir itself is thus not protected with any
  casing). Thus the borehole section within the reservoir remains uncased when
  the rock is stable. Additionally a gravel pack may be installed in sandstone reservoirs to prevent the infiltration of small sand or clay particles mobilised in
  the formation. Open hole completion is the least expensive solution and gives
  maximum access to the reservoir resulting in higher production efficiency. But
  it always bears the risk that the open hole section may be blocked by loose rock
  particles or borehole breakouts during the production period.
– For unstable rock formations the borehole section within the reservoir has to be
  protected with a casing and the annulus between casing and reservoir is being
  filled with cement. This can be done by a casing (i.e. a pipe running up to the
  surface) or by a liner (i.e. a pipe running up only to the bottom of the technical
  casing). Access to the reservoir fluid after cementation is provided by perforating

Once the wells have been completed and the hydraulic conditions have been confirmed by a performance test, production and injection wells are provided with casings for the production and injection of the geothermal fluid. They are an integral part of the anti-corrosion system of the entire geothermal fluid circuit. Commonly inner-coated steel or plastic tubes are used.

Once a production pump has been installed below the water level (established during production) the well is provided with inert gas (like Nitrogen) to prevent corrosion.

Testing and modelling. The evaluation of the hydrodynamic properties of geothermal reservoirs is referred to as a "test". In order to evaluate the properties of geothermal reservoirs by specific tests, beginning with the drilling phase, the production rate, the permeability of the reservoir, the temperature and pressure of the geothermal fluid produced from the reservoir, the chemical properties and the gas content of the geothermal fluid as well as the rock stability need to be determined.

Experience with geothermal heating stations in Northeast Germany has shown that for flow rates between 50 and 100 m^3/h for production and injection wells, sandstone reservoirs (porous rock reservoirs) with a minimum useful porosity of 20 to 25 %, a minimum permeability of 0.5 to 1.0 μm^2 and an effective minimum depth of 20 m are required.

The possible injection index (i.e. the defined injection pressure for a related geothermal fluid volume flow) can be derived from geo-scientific investigations of the reservoir rock and the performed production tests. Injection tests which always bear the risk of damaging the reservoir can thus be avoided /10-7/.

The reservoir layers to be tested are chosen on the basis of well measurements and experiments carried out on reservoir rocks removed during drilling. From each of these layers a certain quantity of geothermal fluid is extracted. On the basis of the obtained results the most promising reservoir layer to be used for the production is chosen. Once the reservoir has been determined, it will be conclusively evaluated with regard to its possible production and injection volume.

The data and parameters obtained during the drilling phase as well as by regional geologic investigations allow for the elaboration of models describing the hydrothermal and thermodynamic behaviour of the geothermal reservoir during the production of geothermal heat by such a geothermal heating station. The most important information is the expected water level within the production well (to confirm the technical feasibility of the production volume flow) and the time scale for temperature depletion of the reservoir, with a view proving the required technical lifetime of the geothermal heating station.

Numerous numeric models are available to simulate the reservoir operation. The required modelling depth does not only depend on the dynamics investigated, but is rather determined by the specific knowledge of the reservoir, which is mostly quite limited, especially during the planning phase of such a geothermal heating station. Extensive knowledge on the reservoir can only be gained by opening up the reservoir and evaluating the discussed tests. This knowledge will also have a major impact on the question of which model should be chosen.

Geological faults, interactions with neighbouring reservoirs and stratifications within the reservoir form significant framework conditions and require detailed modelling. This can only be provided by a local discretisation of these variables on the basis of numerical simulation. Especially for the evaluation of long-term behaviour, complicated numerical models are required, which allow for consideration of the existing complex geological circumstances.

To date, the 3D models CFEST (Coupled Flow, Energy and Solute Transport) and TOUGH (Transport of Unsaturated Groundwater and Heat) are recognised as the most appropriate models to simulate geothermal-related problems. If there is only very little knowledge available on the reservoir during the planning phase, or in the case of a homogenous reservoir with an even parameter distribution, the 2D model CAGRA (Computer-Aided Geothermal Reservoir Assistant) is also suitable /10-8/.

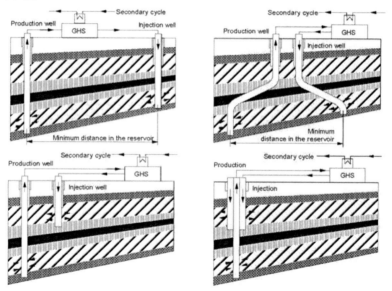

Fig. 10.4 Possibilities of tapping geothermal fluids (GHS – geothermal heating station)

Design of the downhole system. Since the geothermal fluid circuit is responsible for the major share of the investment for a geothermal heating station, investigations aim at finding a more cost-effective solution than the two-well system (Fig. 10.4). The aboveground pipeline connection can be kept to a minimum, if the production and injection well are drilled as divergent boreholes from a single location.

If several reservoirs are found at one site, a simultaneous exploitation is generally possible. Such a system layout can minimise the distance between the production and the injection well. Additionally, the damage of the production reservoir with contaminated geothermal fluid, for instance due to the use within public swimming pools, is impossible. Limitations of such so-called two layer systems are due to the nonexistent mass balance compensation, and thus to an expected change within the reservoir pressure. Experience has shown that for reduced production volumes of 30 to 60 m³/h, production and injection of geothermal fluid can be performed by a single well.

Uphole part. The aboveground geothermal fluid circuit, i.e. the actual geothermal heating station connected to the production and injection well, links the available geothermal energy to the demand for heating, warm water and process heat, which is subject to significant local and temporal variations. The aboveground geothermal fluid circuit has to meet the following requirements:
- production and transfer of geothermal fluid,
- heat transfer to a secondary thermal loop,
- geothermal fluid processing to ensure appropriate injection water quality,
- pressure increase prior to injection,
- injection of geothermal fluid and
- process safety.

The following discussions contain a description of the specific aspects, dimensions and the operation of the aboveground plant, as well as illustrations of the main components. Fig. 10.5 shows a diagram of the general system layout. However, we need to keep in mind that the system layout largely depends on the characteristics of the respective geothermal reservoir and thus on the characteristics of the extracted geothermal fluid.

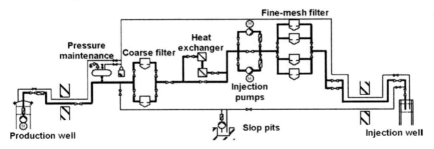

Fig. 10.5 Operation principle of the uphole part of the geothermal fluid circuit

Production of geothermal fluid. Geological conditions often require the use of a pump to produce geothermal fluid. The two basic solutions are either a gas lift, operating as a siphon, or mechanical pumps.

The use of gas lift according to the siphon principle based on ambient air is necessarily associated with a ventilation of the geothermal fluids. This might result in a change of chemical properties and maybe the precipitation of solids. Therefore gas lifts based on ambient air are generally limited to tests conducted within the drilling phase of a geothermal well. Due to its high cost, gas lifts based on inert gases (such as nitrogen) do not constitute a practicable solution for continuous operation either.

This is why, for the production of geothermal fluid, exclusively pumps installed below water level within the production wells are exclusively used so far.

If low installation depths are possible, borehole shaft pumps are used. For such units, the drive mechanism is located aboveground and linked to the pump via a shaft. For that reason almost the entire borehole cross section is available for extraction.

For high installation depths, borehole motor pumps are used. They are preferably powered by an engine driven by electrical energy. Also, turbine pumps are available powered by diverted volumes of the geothermal stream, compressed aboveground and pumped to the turbine/pumping unit below the surface. In this case annulus between the well casing and the turbine feed pipe is used to pump the geothermal fluid aboveground.

Such borehole motor pumps usually consist of an assembly of pump, protector, and engine. The pump is installed within the production well below the geothermal fluid level. The required energy is provided to the pump by means of a cable located within the annulus section, between the cemented casing and the production pipe.

The pump is usually a multi-stage centrifugal pump, for which the number of pumping stages is related to the height difference to be overcome, and the impeller design influencing the volume flow. In most cases, asynchronous three-phase-motors are used. The special grade oil ensures the isolation of the motor coils, lubrication of the motor bearing and heat transfer to the motor housing. The interior oil pressure varies as a result of volume changes to the motor oil filling at different temperatures (starting and stopping the motor). To prevent water from entering the motor, the pump is provided with a pressure compensation system allowing for compensation between the geothermal fluid inlet pressure and the internal oil pressure. This protection device usually contains cascade systems preventing water from entering the motor, taking advantage of the existing density difference between geothermal fluid and oil.

The required pump installation depth depends on the expected maximum reduction of the water level during pumping operation, and the necessary minimum immersion depth. During the operation of the pump (i.e. the production of geothermal fluid), the water level falls below the idle level, and is dependant on the delivered volume flow. The reason for that are the friction pressure losses between the borehole bottom and the pump inlet, and the pressure reduction between the non-influenced reservoir and the interior of the downhole filter. Additionally, water level changes occur due to temperature-related density changes of the liquid column within the well.

The required production level between the geothermal fluid level and the well-head determines pump dimensions. Additionally the pressure losses within the production pipe and the production level have to be taken into consideration, so that the required flow throughout the aboveground part of the system is ensured. If the produced geothermal fluid contains large gas quantities, the bubble point must not be below the pump installation depth. Depending on geological conditions, well completion and the pump type, the installation depth could vary between 100 and 400 m and even more.

In order to adapt the production of geothermal fluid to a given heat demand, the pump is controlled by varying the rounds per minute (rpm) of the electro motor. Even though geothermal heating plants should preferably be operated at base load, i.e. with a continuous volume flow, the described control technique is recommended

– to allow for a reaction to changing reservoir conditions,
– to ensure a smooth start-up and shutdown operation of the plant, and
– to consider the aboveground plant characteristic changes according to the increasing pressure losses due to an increasing filter load.

Quality assurance of re-injected water. Re-injection of cooled geothermal fluid is comparable to filtering (i.e. the geothermal fluid is pumped through the injection well, the reservoir rocks influenced by drilling operations and the undisturbed reservoir rocks). Hence, particulate matter must necessarily either be prevented from entering the reservoir, or its entering must at least be delayed.

Within an uninfluenced reservoir there is a chemical balance between geothermal fluid and reservoir rocks. However, within the geothermal fluid circuit realised within a geothermal heating plant, this balance could be influenced by pressure and temperature changes. This is true for the following aspects:

– pressure relief during extraction,
– temperature reduction and/or a possible oxygen entry into the aboveground geothermal fluid circuit,
– pressure increase during re-injection,
– mixture of the geothermal fluid from the production well with those of the injection well that may not necessarily be identical in terms of their chemical characteristics,
– temperature increase.

Due to pressure relief, additional degasification may occur during production, leading to changes of the pH-value and the redox potential. The redox potential is additionally increased by oxygenation.

The predefined quality of the re-injected geothermal fluid could be ensured by avoiding possible contamination within the geothermal fluid circuit through the overall geothermal heating plant and by filtration of re-injected water. Both options are briefly explained below.

– Avoiding contamination of re-injected geothermal fluid. Most geothermal fluids are highly complex, highly saline, undersaturated solutions with low contents of dissolved gases. In most cases, precipitations of soluble salts are unlikely. Geochemical models have shown that, provided that a mixture of the geothermal fluid with foreign waters is prevented, geothermal fluid used as an energy carrier will only be subject to iron precipitations. Sulphate, carbonic, and silicate mineral precipitations within the aboveground geothermal fluid circuit are not expected. But to achieve this, the overall geothermal fluid system must by all means be guarded against the penetration of oxygen; this helps additionally also to prevent corrosion. Therefore the overall system has to be maintained under permanent excess pressure during system operation, as well as in idle mode. For this purpose a highly sophisticated gas shield system has to be installed. Moreover, chemical changes due to microbiological activities within the geothermal fluid circuit have been observed which may also cause built-up of particulate matter (thus precipitations). In particular, bacterially induced formation of hydrogen sulphide (H_2S) will result in pH-changes, and thus in sulphide precipitations. Also, bacteria are organic material and will therefore enhance particulate matter contamination. Moreover, during start-up, traces of oils and greases (e.g. from the borehole motor pump) may enter the geothermal fluid circuit; which has to be prevented by appropriate measures.

– Filtration of re-injected geothermal fluid. In spite of all process control preventive actions, the re-injection reservoirs must nevertheless continuously be protected by a filtration of the injected geothermal fluid. Appropriate filters implemented within the aboveground part of the geothermal fluid circuit ensure good to excellent separation results within the predefined particulate matter categories, and thus a good purification.

Heat transfer. The heat contained within the produced geothermal fluid needs to be transferred to a heating system or a heating water circuit as efficiently and cost-effectively as possible. For this purpose mainly screwed plate heat exchangers are used which offer the following benefits:
– low temperature differences of up to 1 K between the two media,
– high heat transmission coefficients,
– low building volume and mass,
– sufficient pressure resistance for common geothermal pressures of up to 16 bar,
– low water contents to be processed or disposed of in case of damage or service,
– easy maintenance of the plate surface thanks to easy disassembly.
Mainly, heat exchangers made from titanium are used, as this material is highly resistant to the frequently highly corrosive geothermal fluid.

In order to avoid the possibility of geothermal fluid entering the heating water circuit (e.g. in the case of leakages), the heat exchanger is operated by overpressure referring to the heating system side. Whenever this is not possible or inefficient other measures need to be taken (e.g. permanent monitoring of heating water by measuring the electrical conductivity, installation of double-wall heat exchangers between the two media; however, this will considerably impair heat transmission properties).

Corrosion prevention and suitable materials. The equipment used within geo-thermal fluid circuits has to meet certain requirements in terms of operation safety, environmental protection and cost-efficiency.

– Wall openings due to corrosion and the resulting environmental contamination need to be diligently prevented.
– Corrosion products damage the reservoir as the geothermal fluid is re-injected and thus increases the filtration requirements.

Besides the described exclusion of oxygen, careful material selection is therefore of major importance.

Some geothermal fluids are poor in oxygen but contain high contents of aggressive carbonic acid. Furthermore, corrosion is mainly controlled by chloride ions. Due to these circumstances corrosion becomes uncontrollable once oxygen has entered the system. Experiences with unalloyed or basically alloyed steels have shown, for instance, that corrosion of 0.05 to 2.0 mm/a in relation to volume flow occurred due to a very small oxygen entry. This material removal is not evenly distributed over the whole surface. Little lamina are removed and leaving a scarred surface.

However, there is a wide range of suitable materials to be used for geothermal fluid circuit. The choice is mainly determined by the nature and temperature of the geothermal fluid, its pressure capabilities and required material manufacturing. For instance, plastics, composite materials (plastics/glass fibre), coated and rubberised metals as well as several combinations of high-alloy steels are suitable. Besides the requirements of the geothermal fluid circuit the demands of individual components such as sensors and their connections, fittings and sealing materials need to be carefully taken into consideration.

Leakage monitoring. It is important to protect the system against leakages and to contain geothermal fluid quickly, if leakage occurs. The following combination has to be provided

– anti-corrosive casing material,
– safe casing technology (stable casing connections, double casing systems etc.) and
– leakage monitoring device (they have to ensure permanent control and must quickly detect even small leakage's precisely and reliably).

Besides the observation of the geothermal fluid system related to the heat exchanger, the monitoring of the most deeply buried production and injection pipes, connecting the heating station with the production and injection well, requires special attention. For this purpose, control systems originally developed for district heating systems are applied, which, however, require double-casing installation to prevent moisture from entering the control section. Alternatively, wireless monitoring processes could be used.

Slop system. The slop system collects geothermal fluid that is outside of the geothermal fluid circuit. This system reprocesses these fluids and transfers them back into the circuit. Slop waters result from
− rinsing out the production well and the aboveground system after first start-up and prolonged standstills,
− filter replacements,
− repairs,
− evacuation of the casing system,
− at some sealing of e.g. pumps and filters, and
− leakages within the overall system.
The main slop container is located very close to the injection well. This container is designed to be able to absorb the volume of geothermal fluid to be held back necessarily to protect the reservoir rock within the injection well during the start-up operation. It is furthermore equipped with a sedimentation tank for particulate matter.

Additional slop containers are located at other operating locations of the overall geothermal heating station. They are needed for leakages collected by floor intakes.

Re-injection of geothermal fluid. During the re-injection of the geothermal fluid into the reservoir rocks, injection well pressure losses inside of the injection well casing need to be overcome. This is also true for the required overpressure between the injection well and the uninfluenced reservoir. For a given configuration, these two variables are dependent on the volume flow.

During standstill the geothermal fluid reaches a certain water level inside the injection well. This level depends on the temperature of the geothermal fluid inside the well and thus on its density. During plant operation, however, the fluid level inside the injection well rises according to the given pressure losses. Up to a certain volume flow there is thus the risk of negative pressure at the wellhead, which has to be avoided by appropriate measures (e.g. such as foot valves inside the well).

If the volume flow to be re-injected continues to rise, the required wellhead pressure can be built up to a certain level. This pressure level is pre-determined by the pressure to be built up by the borehole motor pump installed within the production well as well as the allowed pressure level of the overall aboveground system (in most cases up to 5 bar). Only if a higher pressure is needed, injection pumps are integrated into the geothermal fluid circuit close to the injection well. Under these circumstances the piping downstream of the pumps has to be designed for this pressure.

District heating systems. Except for the supply of industrial consumers, characterised by a high low-temperature heat demand, household customers in conjunction with district heating systems can use geothermal heat for the provision of space heating and domestic hot water. For this purpose, usually water-operated district heating systems, equipped with one, two, three, or four lines, are applied; however, two-line systems are currently predominant.

The technical structure of district heating systems depends foremost on the given urban conditions (such as the arrangement of houses, routes), the grid size and the number of (geothermal) heating stations that feed into the grid. Fig. 10.6 illustrates different types of typical district heating systems: radial, ringed, and meshed networks.

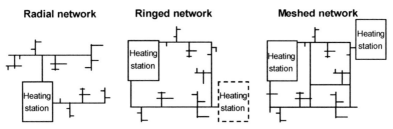

Fig. 10.6 Main distribution networks of district heating systems (see /10-2/)

Although ringed networks allow for incorporation of several heating stations, they are in general more expensive compared to other network types, as their pipeline route length and the nominal diameter of the ringed lines are relatively big. Nevertheless, this disadvantage is compensated by high supply reliability and a very good extensibility. However, due to the high investment costs needed to realise such a ringed network, they are only suitable for large to very large heat distribution networks. For small and medium-sized district heat systems radial networks – the typical network type for geothermal energy utilisation – are more appropriate since their investment costs are considerably lower due to a reduced pipeline length compared to ringed networks /10-2/.

There are also various systems available for sub-distribution and house connections. One option is to connect every consumer individually to the network. Secondly, routing from one house to the other is possible. The first procedure offers a high degree of flexibility and is preferred for partly developed areas. In case of densely built-up areas, routing from house to house is in general more economic. In the latter case, the houses are grouped together, whereas only one house is connected to the distribution pipeline. All remaining houses are connected to this one house, so that fewer connections from the main distribution pipeline are required (Fig. 10.7). Usually, a combination of both distribution types is realised to benefit from the advantages of both systems.

The main installation types are aboveground installed pipelines as well as pipelines passed within ducts or directly within the ground (so-called ductless lines). Aboveground pipelines are characterised by a visual impact and could restrict the use of the limited space in densely populated areas. Direct earth laid pipelines generally prevail within the range below 20 MW of thermal capacity (i.e. typically for geothermal systems) because of their lower costs due to little space consumption and short installation periods compared to systems installed in ducts.

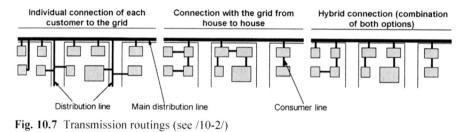

Distribution line   Main distribution line   Consumer line

Fig. 10.7 Transmission routings (see /10-2/)

Because of the moisture within the soil, corrosion resistance is the major requirement to be met by the distribution system. Also, moistening of the insulation needs to be prevented with regard to limiting heat losses. Primarily, plastic jacket pipes equipped with steel medium pipes are applied. Alternatively, plastic medium pipes can be used. For sub-lines and house connections flexible metal or plastic medium tubes are suitable, which are laid "from the coil". These kinds of tubes are laid faster and reduce the risk to damages.

Last but not least house substations are required to link the district heating system to the existing heating system within the house. Such substations are available in a standardised design and are supplied with all accessories required for the connection of the building to the district heating system. Direct and indirect substations are distinguished.

– In direct substations heating water from the district heating system flows through all components of the heat distribution system installed within the respective house. The temperature is controlled by simply adding cooler water. In most cases direct systems are more cost-effective compared to indirect systems /10-2/.

– Indirect systems are characterised by a heat exchanger located between the district heating system and the heat distribution system installed within the respective house. The main benefits of such an indirect system result from the independence of the house heat distribution system from the pressure conditions and water properties of the district heating system.

Overall system layout. Due to the high investment costs for tapping geothermal energy resources, in general, geothermal heating stations have necessarily to be connected to large district heating systems. For this reason, it is advisable to first check all opportunities of establishing or extending a district heating system at the planned site before considering geothermal heat supply. Supply systems of a thermal capacity of 5 MW constitute in most cases a lower limit for the supply of mainly household customers. This might not be true if there is a really favourable annual load duration curve of the given heat demand with a high volume of full load hours (e.g. low-temperature heat demand by a big company running on three-shift-operation) and/or the geothermal fluid can be used (partly) as a raw material (like for the production of mineral water or spa purposes).

Geothermal heating stations are operated in base load. To achieve an optimum performance of thermal capacity the goal is always to achieve a high amount of full load hours. Furthermore, to be energetically most efficient, geothermal heat should be directly transferred from the geothermal fluid to the heating water used within the district heating system. Due to temperatures of the geothermal fluid between 40 and 80 °C, in general, the return temperatures from the district heating system should be below 40 °C. Only in case of lower temperatures, heat pumps should be applied for additional cooling of geothermal fluid.

Fig. 10.8 shows some examples of geothermal heating stations that provide base and peak load in a cost-efficient way. According to this, the plants may consist of one or several heat exchangers for direct heat transfer. Optionally, heat pumps may be integrated within the overall system depending on the geothermal fluid temperatures. In this case, absorption heat pumps, electric heat pumps, and compression heat pumps equipped with gas motors (see Chapter 9.2) may be used. The technology to be applied is determined by the given site-specific circumstances, the prices for fossil fuel energy, as well as the site-specific relation of prices between electricity and fuel.

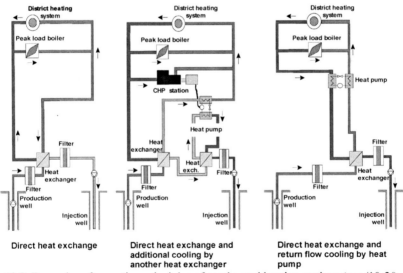

Fig. 10.8 Examples of operating principles of geothermal heating stations (see /10-2/)

Currently, boiler-driven absorption heat pumps are preferred as the boiler may be dimensioned cost-effectively also as a peak load boiler. The boiler can thus realise heat provision in case of a malfunction of the geothermal fluid circuit.

Many geothermal heating stations are equipped with a combined heat and power (CHP) station. Such systems cover the electricity demand for geothermal fluid production and injection of 80 to 200 kW with a high amount of full load hours. The additionally produced heat can be integrated within the overall system.

In general, a peak load and back-up boiler fired with fossil fuel energy (e.g. light heating oil, natural gas) completes the entire heating system. Therefore the boiler is designed according to the maximum heat load of the overall system.

Geothermal heating stations are designed according to site-specific requirements. Due to the multitude of existing impacting factors (such as the characteristics of the geological resources, the geothermal fluid characteristics, the demand and the consumer behaviour, the (existing) heating network parameters, regionally influenced prices of fossil or renewable energy carriers, the organisational structure of the plant operator) general statements can only serve as reference values.

10.1.2 Economic and environmental analysis

The following sections provide an economic and environmental analysis for a selection of hydro-geothermal heating plants.

Economic analysis. For all analysed geothermal heating stations geothermal fluid is produced through one well and, after cooling, is pressed into the reservoir through another well (Fig. 10.1). The water used within the district heating system is directly heated up with heat exchangers. The heat can be increased to the desired temperature by means of heat pumps, if required by the reservoir characteristics or to satisfy consumers. The required electricity can be provided by natural gas-fired or light fuel oil fired CHP stations. The heat produced together with electricity is additionally fed into the district heating system.

In order to compensate for demand peaks or setbacks, heat insulated water tanks acting as energy storage are frequently integrated into the overall system. To cover peak demands or in case of malfunctions, geothermal heating plants are commonly equipped with a peak-load heating station based on light fuel oil or natural gas.

Geothermal heating plants for the provision of space heating or domestic hot water can only cost-effectively be realised and operated with a thermal capacity of 5 MW and more. Out of the demand cases defined in Chapter 1, only the district heating systems II (DH-II) and III (DH-III) are appropriate for geothermal heat generation. However, the defined demand cases (i.e. three single family houses (SFH-I, -II, and -III) and the multiple families home (MFH)) have been considered for supply by the two district heating systems.

Heat distribution is ensured by a heating network of an average flow temperature of 70 °C and a return flow temperature of 50 °C. The district heating systems are designed as plastic-sheathed tubes with a steel medium tube and an indirect grid connection of the consumer. The latter are additionally equipped with an intermediate storage for domestic hot water (80 % utilisation ratio).

- Reference system DH-II. The geothermal heating plant is equipped with one production and one injection well of a total depth of 2,450 and 2,350 m, respectively, with a distance of 1,500 m between the two boreholes (Table 10.1). The temperature of the geothermal fluid used for energy provision is approximately 100 °C. Depending on the actual heat load 30 to 60 m³/h of geothermal fluid are produced by the production well and delivered to the heating plant. The production and injection well are connected to the heating station via an underground pipeline of glass fibre reinforced plastic. Within the geothermal heating station heat is supplied by two load-controlled titanium plate heat exchangers of a maximum thermal capacity of 1,700 kW each, which are incorporated into the overall system by means of an intermediate circuit. This permits an optimum temperature adaptation of the geothermal fluid parameters "temperature" and "mass flow" to the heat capacity required by the district heating system. About 85 % of the heat provided throughout the year is covered by geothermal fluid, whereas the remaining 15 % are provided by a light fuel oil-fired peak-load heating station. The geothermal heating station is designed for a thermal capacity of about 5 MW, of which 3.4 MW directly result from geothermal fluid. Generated heat is distributed according to the heating network defined in Table 10.1.
- Reference system DH-III. For this system geothermal fluid of a temperature of 62 °C is produced from a reservoir layer located at a depth of 1,300 m. The geothermal fluid is delivered by means of a borehole pump installed at a depth of 390 m within the production well (Table 10.1) at a volume flow of 72 m³/h. Subsequently, the geothermal fluid is pumped to the geothermal heating station, which, according to the specific operating conditions, withdraws thermal capacities of 450 to 1,400 kW by heat exchangers. Two heat pumps driven by electrical energy with a thermal capacity of 1,450 kW subsequently cool down the geothermal fluid to about 25 °C, providing a temperature of 70 °C, available to heat up the heat transfer medium used within the district heating system. Subsequently, the cooled geothermal fluid is re-injected to a reservoir at a depth of about 1,250 m. The electricity required for the operation of the heat pumps and other auxiliary units (such as pumps) are partly provided by two natural gas-fired CHP stations with a thermal capacity of 630 kW and an electric capacity of 450 kW. The heat generated beside the electric energy is additionally used within the district heating network. In order to cover peak demands, and to ensure a secure heat supply in case of malfunction of the geothermal part, two natural gas-fired boilers are integrated within the overall system. For such a geothermal heating plant with an installed thermal capacity of 10 MW about half (50.6 %) of the annual heat demand is covered by geothermal heat (direct heat transfer and geothermal share of the heat provided by the heat pumps). Together with the heat from the CHP station and the overall heat provided by the heat pumps this system accounts for almost 85 % of the total annual heat demand. The remaining 15 % are provided by the peak-load heat-

ing station fired with natural gas. The heat is distributed by two independently operated district heating systems as shown in Table 10.1.

In the following the investment and operation costs as well as the heat production costs for these geothermal heating stations are discussed in detail.

Table 10.1 Technical data of geothermal reference plants

District heating system		DH-II	DH-III
Heat demand	in GJ/a[a]	26,000	52,000
Heat at heating station[g]	in GJ/a[b]	32,200	64,400
Geothermal share	in %	85	50,6[c]
Base-load heating station			
geothermal capacity	in MW	3.4	2.1[c]
geothermal heat free heating station	in GJ/a	27,400	32,600[c]
production temperature	in °C	100	62
maximum flow rate	in m³/h	60	72
production/injection well depths	in m	2,450 / 2,350	1,300 / 1,250
auxiliary units			CHP station / heat pump[d]
Peak-load heating station			
fuel		light fuel oil	natural gas
burner technology		Low-NO$_x$	Low-NO$_x$
heating capacity	in MW	5	2 x 5
Boiler efficiency	in %	92	92
District heating network			
length	in m	6,000	2 x 6,000
feeding/return temperature[c]	in °C	70 / 50	70 / 50
network efficiency	in %	85	85
house substations efficiency[f]	in %	95	95

[a] heating demand of the consumers connected to the district heating system (see Chapter 1.3); [b] including losses of the heat distribution network and house substations / hot water intermediate storage; [c] direct heating transfer and geothermal share of heat provided by heat pumps; [d] two natural gas-fired CHP stations each (630 kW thermal capacity, 450 kW electric capacity and a thermal/electric efficiency of 52 respectively 38 %) and heat pumps driven by electric motor (coefficient of performance of 4 at a thermal capacity of 1,450 kW); [e] annual average; [f] average efficiency (80 % domestic hot water systems and 98 % for space heating systems; [g] including the losses of the heat distribution network and the building substations).

Site-specific impact factors and local conditions (such as temperature, salinity, production rates of geothermal fluid, local consumer structure and demand behaviour) result in considerably different designs for such geothermal heating systems. And such differences are responsible for significantly different costs of geothermal heating plants. Due to these uncertainties the costs outlined below can therefore only describe general magnitudes. This is the reason why, depending on the individual case, more or less favourable heat production costs may be achieved.

Investments. Besides the applied technology, investment costs of geothermal heating plants (Table 10.2) are mainly influenced by the system size. Specific costs generally decrease with increasing plant size. Basically, the costs for the production and injection well dominate the overall costs for a geothermal heating plant.

Besides the production and injection well, the main components of geothermal heating stations are production and injection pumps, heat exchanger, filters, and slop systems. If the geothermal fluid temperature is too low with regard to the temperature level of the district heating network, heat pumps are additionally required.

With a share of approximately 51 respectively 30 % (DH-II respectively DH-III) in both case studies, the two wells account for the major portion of the investment costs. Depending on the depth of the well, the drilling costs for the production and the injection well vary widely; for average geological conditions they might be in the order of magnitude of about 1,000 €/m.

Depending on pipe dimensions, site conditions (e.g. rocky or sandy ground) as well as settlement structure the costs of the district heating network may vary widely. However, an average price for a district heating network suitable for geothermal heating stations is in the order of magnitude of 385 €/m. The investment costs for house substations according to Table 10.2 include the costs for the connection between the heating grid and the building as well as the expenses within the building (e.g. domestic warm water storage).

The remaining expenses such as filter and slop system of the aboveground geothermal fluid circuit as well as the planning of the heating plant influence the overall investment costs only to a minor extent. This is also true for the heat exchangers as well as the construction costs (building including ground).

The indicated costs for heat supply from geothermal and, to cover peak loads, fossil fuel energy include furthermore the costs for the peak load heating plant (including boiler, burner, fuel storage or natural gas connection) as well as the expenses for the CHP stations and the heat pumps (Table 10.2). However, with regard to total costs they account only for a small portion of scarcely 4 and 11 %, respectively.

Operation costs. The running expenses (see Table 10.2) include the cost for repair and maintenance, personnel, insurance and electricity (needed for example to circulate geothermal fluid and to drive heat pumps) if the electrical energy is not produced within the geothermal heating plant. However, when compared to investment costs they are relatively low (see Table 10.2). Fuel costs of the geothermal/fossil systems are indicated separately within Table 10.2.

Heat generation costs. On the basis of an interest rate of 4.5% and a depreciation period according to the technical lifetime, the heat production costs are calculated for the reference plants described in Table 10.1 and shown in Table 10.2.

With regard to the technical lifetime the following assumptions have been taken: 30 years for all geothermal plant components, 50 years for buildings and the district heating network, 20 years for house substations, heat pumps and the peak load boiler and 15 years for the CHP station.

Following these assumptions, depending on the share of geothermal energy within the overall heat provision, the plant size and the capacity of the house substations the heat production costs amount to between 36 and 50 €/GJ (Table 10.2).

Table 10.2 Investment and operation costs as well as heat production costs of the geothermal plant components, heat distribution within the district heating network within a geothermal/fossil system providing domestic warm water and space heating

System		DH-II				DH-III			
geoth. capacity	in MW	3.3				1.45			
geoth. heat supply	in GJ/a[a]	22,100				26,300			
total heat supply	in GJ/a	26,000				52,000			
geothermal share	in %	85				50.6			
Supply case		SFH-I[b]	SFH-II[c]	SFH-III[d]	MFH[e]	SFH-I[b]	SFH-II[c]	SFH-III[d]	MFH[e]
Heat demand	in GJ/a	32.7	55.7	118.7	496.1	32.7	55.7	118.7	496.1
Geothermal heating station									
prod. well	in Mio. €	2.5				1.7			
injection well	in Mio. €	2.5				1.7			
heat exchanger	in Mio. €	0.034				0.014			
heat pump	in Mio. €					0.22			
CHP station	in Mio. €					0.32			
peak-load heat	in Mio. €	0.45				0.72			
buildings	in Mio. €	0.20				0.33			
miscellaneous	in Mio. €	1.74				1.66			
district heat. system	in Mio. €	2.31				4.62			
Total	in Mio. €	9.73				11.28			
Operation costs	in Mio. €/a	0.13				0.16			
Fuel costs	in Mio. €/a	0.06				0.75			
Building connection									
investment costs	in €	5,520	5,810	7,120	13,440	5,520	5,810	7,120	13,440
operation costs	in €/a	102	107	129	237	102	107	129	237
Heat production costs	in €/GJ	49.9	44.1	40.0	35.8	48.7	42.8	38.8	35.8
	in €/kWh	0.18	0.16	0.14	0.13	0.18	0.15	0.14	0.13

[a] geothermal share of the total heat supply without heat losses due to the district heating network, the house substations as well as the warm water intermediate storage systems; [b] low-energy detached family house (SFH-I); [c] detached family house equipped with state-of-the-art heat insulation (SFH-II); [d] detached old building family house equipped with average heat insulation (SFH-III); [e] multiple-families house (MFH); for the definition of SFH-I, SFH-II, SFH-III, and MFH see Table 1.1 or Chapter 1.3.

These heat provision costs relate to the total heat production costs of a combined heat supply based on geothermal and fossil resources. This is true because, according to the plant design, the base load is covered by geothermal energy whereas the peak load is provided based on fossil fuel energy (DH-II und DH-III). For the system DH-III additionally part of the base and peak load heat is provided by the CHP station and the heat pumps (DH-III). The heat production costs of a geothermal heating system without the described use of fossil fuel energy would be considerably higher, as the geothermal plant components would also additionally have to cover peak demand, leading to a reduced number of full load hours of the geothermal part and thus also of the heat provided by geothermal resources.

In general, heat production costs decrease with increasing plant size. Due to the relatively high investment costs of tapping geothermal heat resources heating plants with relatively small thermal capacities are characterised by relatively high investment costs. Also, the heat production costs of small heating stations equipped with deep wells and direct heat transfer are higher compared to geothermal heating systems characterised by relatively shallow wells supplemented with heat pumps. The use of heat pumps is thus more cost-effective than deeper wells according to the assumptions made here. This might change with strongly increasing costs for the electricity to be purchased from the grid and/or with increasing costs for fossil fuel energy.

Heat production costs are influenced by many factors which, depending on the site-specific conditions, have to be optimised with regard to minimising the overall system costs. Heat costs can be further reduced by decreased costs for drilling a well, optimised district heating networks as well as by an optimised heating plant configuration.

To better estimate and evaluate the impact of the various factors influencing the described heat production costs, Fig. 10.9 illustrates a variation of the predominant sensitive parameters of geothermal heat supply systems with peak load coverage (DH-II) for a house connection with a heat demand of 8 kW (SFH-II). For the house substation constant costs are assumed. Also, the geothermal efficiency of 85 % remains unchanged.

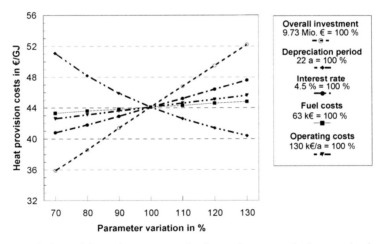

Fig. 10.9 Variations of the main parameters having an impact on the heat production costs of geothermal/fossil-fuel fired heat supply systems (DH-II) according to the supply case SFH-II with a capacity of 8 kW (the depreciation period of 22 years corresponds to the average depreciation period of all plant components, house substations not varied; see Table 10.2)

Hence, investment costs and full load hours are the factors mostly influencing the heat provision costs. The full load hours – and thus the useable heat – might change under otherwise unchanged framework conditions (such as installed thermal capacity) with customers showing a demand behaviour different from the household customers assumed here. Heat supply to industrial enterprises with a heat consumption constant throughout the overall year (e.g. a dairy) permits for example to increase the provided heat dramatically (i.e. the number of full load hours) and reduce the heat production costs accordingly (see Fig. 10.9). Interest rate and depreciation period also have a major impact on the production costs of geothermal heat supply systems. In contrast, operation and fuel costs are of minor importance.

Environmental analysis. As with any other form of geotechnical utilisation, also hydrothermal energy generation constitutes an intervention into the natural balance of the earth's upper crust. The intervention causes energetic and material changes, mechanical fracture events and, to a lower extent, also mass displacements /10-3/. However, since there are no cavities created by hydro-geothermal exploitation, the created environmental effects are considerably lower when compared, for instance, to the exploitation of an oil field or even an open top mine.

On this basis, for hydrothermal energy utilisation the following environment effects are identified /10-5/. For this purpose, it is distinguished between the construction of a geothermal heating station, its normal operation and possible accidents (malfunction) and the end of operation.

Construction. For well drilling the same processes as for mineral oil and natural gas exploration or exploitation and water procurement are applied. However, when constructing a hydrothermal plant this process is the one that is connected with the greatest environmental dangers. Due to well depths up of to 3,000 m and above, there is the danger of a hydraulic short circuit of various layers and aquifers, respectively. However, by observing the corresponding regulations developed for the oil and gas industry, the environmental effects resulting from such a short circuit can be minimised. The moderately perceptible noise during well drilling only occurs during this time-limited process and not during the operation of the geothermal heating plant. The additional environmental impacts due to the use of appliances, intermediate storage of the drilling mud, space consumption for the drilling rig etc. are restricted to a short period of time and to the site. After drilling and completion of the well the original state of the area around the well head is restored.

Normal Operation. No substances or particles are released during normal operation of a hydro-geothermal heating plant. All emissions possibly released during the operation of the heating station at the site of the geothermal plant are attributable to fossil fuel-fired additional plant components or peak-load boilers.

During the operation of geothermal heating stations the water balance within the used aquifer can theoretically be affected by improper re-injection. This may lead to a modification of the pore pressure and – as a consequence – microseisms which might be recognised by a minor earthquake. However, these effects have only rarely occurred in the past. Also, cooling of the aquifer may lead to changes of the reservoir chemistry with all resulting effects, such as precipitation of minerals. However, since the aquifer is located far below the surface and is generally not connected to the biosphere, environmental effects on flora and fauna have not been reported so far.

Furthermore, gases such as CH_4, NH_3, H_2S, CO_2, etc. may be emitted, and minerals may be released by subterranean fluid circulation. However, since geothermal fluid is transported within a closed circuit in the geothermal heating station, no material releases are observed during normal operation. This could be different when the geothermal fluid is used for balneology purposes. However, the described effects can be and are avoided by appropriate technical measures. Solid precipitations separated by filters integrated within the geothermal fluid circuit must be disposed of according to the applicable regulations.

Within normal operation, environmental effects are thus mainly restricted to thermal effects on the upper and lower layer of the exploited reservoir (i.e. the rock bed above and underneath), the rocks surrounding the production and injection wells, and geo-mechanical impacts during long-term operation.

– Thermal impacts on the aquifer. By re-injection of the cooled geothermal fluid into the aquifer during the operation of a hydrothermal plant, the initial aquifer temperature is continuously reduced, which leads to a temporarily varying temperature gradient between the storage and its surrounding layers /10-5/. The result is a conductive heat flow from the overburden into the reservoir, partly reheating the injected waters and simultaneously cooling the overburden. The calculation of the temperature changes within the surrounding rocks has shown that in a worst case scenario, after 30 years of operation only a depth of 160 m has been thermally affected while temperature reductions exceeding 10 K have only been observed in depths up to 70 m around the well /10-5/. Moreover, no direct environmental impacts on the biosphere have been observed that are attributable to a cool-down of the underground and there has not been reported any organic life underneath the reservoir that could be affected.

– Thermal impact on the rocks surrounding the well. The well transmits heat or cold to the environment and influences it thermally. An analysis of the well surrounding of a typical well used at the geothermal heating station of Neustadt-Glewe/Germany shows that the maximum heat transmission to the direct environment amounts to 230 kW at the beginning of operation and to 180 kW after 30 years of operation. After this time period and the assumption of an uninterrupted heating the thermal impact, characterised by the ratio of the fluid temperature to the undisturbed rock temperature, would reach a radius of 60 m around the well. The thermal impact would be reduced to 56 % at a distance of 10 m and to only 34 % at a distance of 20 m. Hence, there is no far-reaching thermal impact of the underground due to geothermal wells. Moreover, only the circular area around the well could have a slight impact on flora and fauna. However, up to now no such effects have been observed.

– Geo-mechanical impacts. Around the injection well a cold water area spreads out within the aquifer as injection progresses. This could lead to contractions of the layer within the deeper underground, which theoretically may reduce the layer thickness and cause the earth's surface to sink. However, simulations have shown that if such lowering occurs, it only has a slight impact and only occurs over a very long period of time. According to these investigations such contractions amount to approximately 1 to 3 mm/100 m aquifer thickness. Compared to the lowering which is common for deep mining of hard coal and ore or for mineral oil or natural gas exploitation, or that occurs with regard to the settlement of building grounds, the effects observed during the geothermal heat production are almost negligible. Also geo-mechanical consequences during the post-operation period are not expected /10-5/. Impacts on the earth's surface, leading, for instance, to a deterioration of the building infrastructure are thus very unlikely.

All in all, the environmental effects occurring during normal operation of hydrothermal heating plants are thus either clearly lower or comparable to those of fossil fuel-fired plants (i.e. influence on the landscape, covered surface).

Malfunction. In case of malfunctions hot geothermal fluid may penetrate to the earth surface. Due to the high salinity that is observed in some reservoirs the flora and fauna may be deteriorated if geothermal fluid is discharged into the surface water. However, by appropriate planning and monitoring (such as leakage monitoring systems, pressure balance, slop systems, etc.) the described risks are considerably reduced. Theoretically, further environmental impacts may occur if, due to the high mineral content of geothermal fluid, harmful chemicals are applied to eliminate precipitations in, and obstructions of, pipelines. However, since the chemicals are subsequently transferred back into the underground together with the cooled geothermal fluid, the corresponding environmental impacts are limited according to current knowledge. Furthermore, harmful substances may be released to the environment due to fires at electrical components (such as cables). However, these fires are not specific for geothermal plants and can be avoided by observing the relevant fire protection guidelines.

End of operation. To avoid detrimental effects on the environment with regard to the end of operation, appropriate well sealing is of major importance. The borehole seal must exclude any penetration of harmful substances from the surface into the hole as well as any hydraulic short-circuit of different layers. The disposal of the applied plant components does not raise extraordinary environmental problems, as it is comparable to the disposal of conventional machine parts having a relatively slight impact due to extensive legal guidelines.

10.2 Heat supply by deep wells

If geothermal fluid cannot be tapped by deep wells, the borehole can still be utilised by means of a geothermal deep well; whose functionality is similar to that of
geothermal probes applied near the surface (see Chapter 9.2). Thus the following
paragraphs contain a brief description of geothermal deep well technology. Subsequently, economic and environmental aspects are discussed.

10.2.1 Technical description

To exploit heat stored in the underground by means of a deep well a cased borehole is required (see Chapter 10.1). Additionally, the hole must be equipped with
a double coaxial casing. To access the heat of the deep underground a heat transfer medium is pumped into the underground from aboveground through the annulus section between the production tube and the well casing. From the bottom of
the well the heated heat transfer medium is pumped to aboveground within the
production well (Fig. 10.10). The well casing seals the borehole fully against the
bedrock (i.e. closed sy

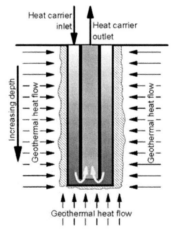

Fig. 10.10 Operation principle of a geothermal deep well

As the rock temperature rises with increasing depths in proportion to the geothermal gradient, the heat transfer medium heats up on its way to the bottom of
the borehole, and thus withdraws energy from the underground making the geothermal heat technically available. However, only as much heat can be withdrawn
from the underground as is provided by the natural geothermal heat flow (on average 65 mW/m^2). As the heat transfer medium does not penetrate the tube casing
there is no mass transfer between the deep well and the underground (i.e. closed
system) as it is the case for hydro-geothermal heat utilisation (i.e. open system).

In most cases water is used as heat transfer medium. However, the water is generally treated with inhibitors to minimise corrosion of the downhole components. For this purpose the experience gained within existing district heating systems are applied.

Subsequently, the ascending pipeline respectively the production tube re-transfers the heat transfer medium – "charged" with geothermal heat – upward from the bottom of the borehole to aboveground. To ensure high temperatures at the well exit and in order to minimise heat losses, the entire production tube is heat-insulated.

The heat carrier which passes through the well decreases the temperature at the surface of the casing cemented into the underground. In spite of the relatively low heat conductivity of the rocks available within the underground, due to the temperature difference between the heat transfer medium and the surrounding rocks, heat is transferred into the heat carrier, which may amount to 200 W/m /10-2/. As the surrounding rocks do not keep their initial temperature, the heat transfer medium can only reach temperatures considerably lower than those of the uninfluenced rocks.

The thermal capacity of such a closed system (i.e. a geothermal deep well) is influenced mainly by
- geological parameters, such as the local geothermal gradient and the existing thermal physical properties of the rocks available at the respective depth around the well,
- the technical configuration of the well (i.e. diameter and materials, insulation properties of the applied tubes, heat transfer between primary rocks, cement and casing) and above all
- the operating principle of the entire system.

For common well depths within the range of 1,000 to 4,000 m and average geological conditions, geothermal capacities between 50 and 400 kW are expected.

The heat transfer medium is circulated within the deep well by means of a pump which acts as the major uphole component of such a system. The required pump capacity is lower than that of a circulation pump used for hydro-geothermal utilisation, since there are no pressure losses in the actual heat exchanger and, contrary to geothermal fluid extraction, the medium flows through a closed pipeline.

Since the temperature at the well exit is generally lower than 40 °C, a heat pump is indispensable. Due to the relatively low thermal capacity of several 100 kW for such a deep well, an electric or gas motor-driven compression heat pump is usually applied. Such a system configuration allows for an extensive cooling of the heat carrier circulated within the deep well. The produced heat is at the same time lifted to a temperature level appropriate to be used in small district heating networks. To obtain a favourable coefficient of performance (COP) moderate temperatures at the heat pump exit are advantageous (i.e. inlet and outlet temperatures of the district heating network). If the outlet temperature of the heating network is considerably lower than the temperature at the well head, a direct heat exchanger can optionally be installed in front of the heat pump. Fig. 10.11 illustrates the example of deep well incorporation into a heat supply system.

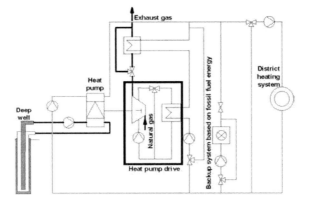

Fig. 10.11 Incorporation of a geothermal deep well into a heat supply system

The influence of the operating principle on the geothermal capacity mainly depends on the temperature of the heat transfer medium fed into the deep well. The capacity can therefore be influenced by the aboveground system components integrated within the overall energy provision system as well as by the operating mode. In times of high heat demand, this allows for an operating modus ensuring a high share of geothermal energy within the overall heat provided to the customer. But very low temperatures of the heat carrier result in an increased temperature gradient within the casing wall and thus within the rocks surrounding the well in general. Continuous operation under such operating conditions leads to a considerable reduction in temperature of the surrounding rocks, which may result in permanently decreased system capacities. Operating mode should thus strive for a maximum output by keeping the impacts on the underground at a tolerable level in order to ensure maximum service life (i.e. sustainable operation of the well).

Since geothermal deep wells are very capital-intensive within heat supply systems, they should only be designed to cover the base load. In wintertime, peak load is provided by a conventional fossil fuel-fired boiler permitting also to further increase the heating temperature (Fig. 10.11). From an economic viewpoint, geothermal deep wells should be used at locations where a sufficient heat demand is given (e.g. small district heating network, large scale individual consumers, commercial enterprises, local authorities) /10-9/, /10-10/.

10.2.2 Economic and environmental analysis

Compared to hydro-geothermal heat generation systems, deep well heat supply systems are almost location-independent as such deep well systems are characterised by a closed circuit system (i.e. closed system). Such systems consist of a well and the pertaining completion, a heat exchanger which usually includes a heat pump and a fossil fuel-fired peak load plant. Depending on the consumer structure, an additional heat distribution network may be required. The following paragraphs contain an economic and environmental assessment of such systems.

Economic analysis. Within the following explanations we assume a 2,800 m geothermal deep well with a total thermal capacity (including the fossil fuel-fired peak load boiler) of 4 MW. Furthermore, a heat demand according to the small district heating system (DH-II) of approximately 26 TJ/a off consumer or 32.2 TJ/a off plant is assumed (see Chapter 1.3). The geothermal capacity amounts to approximately 500 kW. The system is equipped with a heat pump to provide the required inlet temperature to the heating network. The heat pump is characterised by a coefficient of performance (COP) of 4 and has an electricity demand of approximately 600 MWh/a. The share of heat delivered by the heat pump referring to the overall provided heat amounts to almost 44 % (i.e. approximately 10.5 TJ/a geothermal heat are integrated within the overall system). The fossil energy-fuelled peak load plant has a thermal capacity of 6 MW (to ensure supply security) and an overall efficiency of 92 %.

Heat is supplied to a new housing estate by a small district heating network (DH-II) (heating network temperature 70/50 °C, 15 % network losses, 1,800 full load hours; Chapter 1.3). Two options are analysed: the drilling of a new well (reference system I) and the use of an existing well (reference system II). However, one should keep in mind that it is not very likely to find an existing well useable to be extended as a deep well close to customers. The technical lifetime of such a plant amounts to approximately 22 years.

Within the following explanations the investment costs, the operation costs and the heat provision costs are discussed.

Investment costs. Table 10.3 illustrates the investment costs of the analysed systems. Besides drilling of a new well the costs are mainly dominated by the completion of the well (downhole part). Compared to these expenses, the costs for heat exchanger, heat pump and peak-load plant are relatively low. Additionally, high costs are caused by the heat distribution network (Table 10.3).

Operation costs. The operation costs are mainly due to the use of auxiliary energy (electrical energy for running the pump for the heat carrier circulation, fossil fuel required for operating the peak load plant (separately indicated in Table 10.3) and electrical energy required for the heat pump). Further expenses are given by maintenance, repairs, and miscellaneous costs. The total annual costs amount to approximately 0.54 Mio. €/a to cover the defined heat demand of the new building estate (Table 10.3).

Table 10.3   Investment, operation and heat production costs of heat supply systems equipped with deep wells and a heat distribution according to the small district heating network and house connection for the provision of domestic hot water and space heating

System		Existing well				New well			
Small district heating network		DH-II				DH-II			
Combination of systems		Light oil + heat pump				Light oil + heat pump			
Total heat demand	in GJ/a[a]	26,000				26,000			
Heat at heating stat.	in GJ/a [h]	32,200				32,200			
Geothermal share[i]	in %	43.6				43.6			
Supply case		SFH-I[d]	SFH-II[e]	SFH-III[f]	MFH[g]	SFH-I[d]	SFH-II[e]	SFH-III[f]	MFH[g]
Heat demand	in GJ/a[b]	32.7	55.7	118.7	496.1	32.7	55.7	118.7	496.1
Heating station and network									
Investments									
Production well	in Mio. €	1.7				3.0			
Heat pump	in Mio. €	0.22				0.22			
Peak-load plant	in Mio. €	0.63				0.63			
Buildings etc.	in Mio. €	0.8				0.8			
Heating network	in Mio. €	2.3				2.3			
Total	in Mio. €	5.65				6.95			
Operation costs	in Mio. €/a	0.16				0.16			
Fuel costs	in Mio. €/a	0.38				0.38			
House substation, house connect.									
Investments	in €	5,523	5,814	7,122	13,444	5,523	5,814	7,122	13,444
Operation costs	in €/a	103	108	130	237	103	108	130	237
Heat production costs	in €/GJ	52.1	46.2	42.1	39.1	55.7	49.8	45.7	42.7
	in €/kWh	0.19	0.17	0.15	0.14	0.20	0.18	0.16	0.15

[a] without heat distribution network losses and losses due to building substation/warm water intermediate storage; [b] without losses due to building substation/warm water intermediate storage; [c] at an interest rate of 4.5 % and an amortisation period covering the technical life time of the plant (geothermal plant components 30 years, buildings and heat distribution network 50 years, building substation, heat pump and peak load boiler 20 years); [d] low-energy detached family house (SFH-I); [e] detached family house (SFH-II) equipped with state-of-the-art heat insulation; [f] old building detached family house (SFH-III) equipped with average heat insulation; [g] multiple families house (MFH); for the definition of SFH-I, SFH-II, SFH-III and MFH see Table 1.1 or Chapter 1.3; [h] including the losses of the heat distribution network and the building substations; [i] delivered by the heat pump.

Heat production costs. According to the calculation method applied so far, heat production costs are calculated as annuities at an interest rate of 4.5 %; the amortisation period covers the entire technical lifetime.

Heat supply by geothermal deep wells is characterised by relatively high costs (Table 10.3); this is especially true if a new well needs to be drilled. As for hydrogeothermal utilisation, due to the high share of fixed costs, the specific heat cost decreases with the increase of the utilisation rate and the heat demand, respectively. Only the energy costs required for the circulation of the heat carrier and the heat pump operation as well as the operation of the fossil fuel-fired peak load system depends on the provided useful energy.

Fig. 10.12 illustrates variations of the essential variables determining the heat production costs, using the example of a new deep well designed for supply purpose SFH-II (Chapter 1.3). The figure shows that the full load hours and the amortisation period have a great impact on the heat production costs. But also operation and fuel costs as well as total investment costs greatly influence the heat production costs.

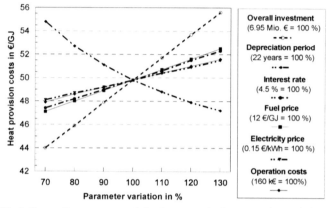

Fig. 10.12 Variations of parameters of the heat supply by a geothermal deep well (new well according to demand scenario SFH-II, house substations not varied; see Table 10.3)

Environmental analysis. Additionally, the analysed reference plants designed for geothermal energy utilisation by means of deep wells are assessed according to selected environmental aspects which are discussed in detail throughout the following paragraphs. These environmental effects are subdivided into effects caused during plant construction, operation, malfunctions, and at the end of the operation (i.e. demolition).

Construction. During the construction of deep well systems environmental effects are mainly caused during the drilling of the well due to e.g. uncontrolled drilling through two or several aquifers of different pressure levels, penetration of harmful substances into the underground caused by drilling operations, drill pipes and accessories or chemical-biological modifications due to drilling fluids. However, these effects can almost be entirely avoided if the corresponding legal guidelines are taken into consideration. During the construction phase noise emissions may occur which are restricted to this period only. This is also true for the usage of the area around the drilling location.

Operation. The site-specific environmental effects caused by geothermal deep well systems are comparable to those of every other kind of heat supply plant. However, since heat is partly provided by geothermal resources that are almost free of any emission released during operation, and which substitute fossil fuel energy, the release of harmful substances into the air as well as all other effects related to the utilisation of fossil fuel energy are reduced in proportion to the geothermal share. The additional energy required for circulation of the heat transfer medium, and for running the heat pump operation, also has to be considered. But in general the environmental impact is lower compared to the environmental effects connected with the direct use of fossil fuel energy.

Theoretically, during normal operation of the deep well environmental effects are only likely due to the operation of the well. Moreover, as deep wells are generally well sealed against the surrounding rocks (i.e. closed system), the different aquifer zones are hydraulically separated from one another; therefore environmental risks are particularly low.

Water, at the most enriched with anticorrosive agents, is commonly applied as a heat transfer medium within such deep wells. Environmental risks of polluting the underground by a glycol enhanced heat carrier, as could be the case when exploiting shallow geothermal resources, are therefore not given.

Additionally, there is neither the risk of any geomechanical impact, or modifications of stress as for hydro-geothermal utilisation, as there are no substances removed from the underground (i.e. closed system).

Further environmental effects at the plant site are related to the use of the heat pump, which are also particularly low during normal operation, provided that the corresponding guidelines are observed.

Hence, all in all the effects of useful energy provision by deep wells on humans and nature are particularly low; generally they are lower than those of fossil fuel-fired plants (also with regard to landscape and space consumption).

Malfunction. In case of malfunctions, environmental effects may arise from deterioration of pipelines and a subsequent hydraulic short-circuit of different aquifers or the penetration of harmful substances.

Further environmental risks may result from fires or explosions of the heat pump with regard to the toxicity of the heat transfer medium. However, if the relevant standards and safety regulations are observed, these kinds of malfunctions are very unlikely to occur.

Demolition. Regarding environmental effects at the end of operation, the same as for geothermal exploitation by means of wells near the surface applies. Potential environmental risks are mainly related to proper completion of the well and proper disposal of the heat pump.

10.3 Geothermal power generation

Geothermal power production has a relatively long history. Just 38 years after the invention of the electric power generator by Werner von Siemens and 22 years after the start of the first power station by Thomas A. Edison in New York in

1882, geothermal power production was invented by Prince P. G. Conti in Lardarello, Italy in 1904. Geothermal power production in Tuscany has continued since then and amounted to 128 MW of installed electrical power in 1942 and to about 790 MW in 2003. In 1958, a small geothermal power plant began operating in New Zealand, in 1959 another in Mexico, and in 1960 commercial production of geothermal power began in the USA within the Geysers Field in California. Today 25 countries use geothermal energy for power production, and the world-wide installed electrical capacity has increased to about 8,930 MW in the year 2004 with an average annual increase of about 17 % between the year 1995 and the year 2000 /10-6/, /10-11/, /10-12/. One of the main reasons for this success is the base load ability of geothermal power generation.

Today, geothermal power production is economic viable only when high temperatures are found at relatively shallow depth. In regions with a normal or a slightly above normal geothermal gradient of about 3 K per 100 m, one has to drill more than 5,000 m deep in order to achieve temperatures above 150 °C. Such deep wells are expensive (generally more than 5 Mio. €) and there is a high risk of failure. For this reason under economic considerations geothermal power production is mainly restricted to geothermal fields with extremely high temperature gradients and high heat flows. Such fields often show surface manifestations of geothermal activity like fumaroles or hot springs. A scheme of such a geothermal field is shown in Fig. 10.13. The heat source are hot magmatic bodies that have risen up in the earth crust from a greater depth (often several tens of kilometres) to more shallow regions. The high heat flow in the overlaying crust, resulting from these magmatic bodies, heats up water of meteoric or marine origin in porous or fractured rock formations, that are in most cases covered by a cap rock of low permeability. Due to buoyancy effects the water starts to converge within the host rock bringing high temperatures closer to the surface. Depending on temperature and pressure the fluid may start to boil at certain depths and vapour is produced. The amount of vapour characterises those systems as liquid or vapour dominated reservoirs. Typical fluid temperatures range from 150 to 300 °C at depths between a few hundred and 3,000 m. Even higher temperatures are encountered at greater depths.

Most of the geothermal fields used today are in zones with active volcanism. Generally, not the volcanoes, which cool down rapidly, but the magma chambers buried underneath the volcanoes are the heat source for geothermal manifestations over prolonged time periods and are the indirect sources for geothermal power production. Magma chambers contain silitic or basaltic magma. Their volume measures from about 1 to 10^5 km^3. Their heat content is tremendous (up to 10^{23} J) /10-6/.

Geothermal fields accompanied with volcanism /10-6/ are found along subduction zones at active continental margins where oceanic crust is pushed underneath the continental crust. Examples of these areas are aligned along the Pacific Ring of Fire, such as the Altiplano of the Andes, the Taupo Region in New Zealand, Kamchatka, and parts of Japan, Indonesia and the Philippines. Volcanic areas are also found above "hot spots" or mantle plumes outside the subduction zones. Examples are the Yellowstone volcanic fields or the Clear Lake Volcanic Field with the geothermal field of the Geysers. Continental rifts are also zones with recent and young volcanism. Rifting (spreading) of continental crust occurs in regions where the magma flow in the upper mantle is in direction of the earth surface causing an uplift of the mantle crust boundary and a thinning of the crust. A prominent example of a continental rift system with a high geothermal potential is the East African Rift or Graben. This Graben is extending from the Red Sea to Mozambique in the southern part of Africa. The largest mass of hot magma is ascending along the mid-ocean ridges. Most of the spreading centres are under water. There are, however, a few regions astride these ridges that are above water and accessible for extraction of geothermal heat. The best-known example is Iceland.

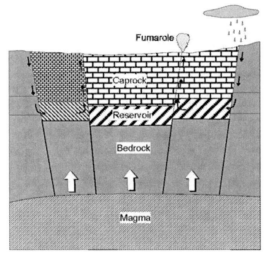

Fig. 10.13 Scheme of a geothermal system /10-11/

The potential for geothermal power production from geothermal fields accompanied with volcanism is tremendous. Recent estimates reach up to 22,000 TWh/a /10-11/. For comparison, the geothermal power production in the year 2004 amounted to about 57 TWh/a. This shows that geothermal power production from geothermal areas can be increased by at least two or three orders of magnitude compared to the current state. Many developing countries in South and Central America, in Asia, and Africa could and some do supply a major portion of their electricity consumption by geothermal power production from such geothermal fields associated with volcanism.

Such geothermal fields are generally characterised on the one hand by temperatures above 150 °C at relatively shallow depths (frequently under 1,000 m), and on the other hand by a high degree of hydraulic permeability. Though their poten-

utable to the ORC plant, the geothermal fluid circuit, miscellaneous and operation costs.

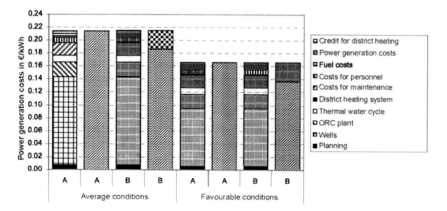

Fig. 10.22 Electricity generation costs of reference Plants A and B (the total costs indicated by the left bars refer to the costs related exclusively to the power generation (i.e. total power production costs of the entire plant without a possible income by heat sale through the district heating system to the household customers), whereas the right bars consider such a possible heat credit (and thus indicate the net power production costs))

Power production costs are considerably influenced by the temperature of the extracted geothermal fluid. Fig. 10.23 shows this effect using the example of exclusive geothermal power generation under the average site conditions. Under such circumstances, by means of the combined ORC/flash process and an approximate geothermal fluid temperature of 200 °C minimum power production costs amount to approximately 0.18 €/kWh. Compared to this, due to their lower efficiencies single or double flash processes only allow to yield lower power quantities at the same drilling costs. The specific electricity generation costs are thus considerably higher.

The illustration shows that the specific power generation costs increase with further increasing temperatures. This is mainly due to the influence of the well depth on the drilling costs because the temperature can only be increased with an increased well depth. The impact of increasing temperature on the efficiency of the respective cycle (and thus on the achievable power output) is more than compensated by the superproportional increase of drilling costs with increasing depths.

Electricity generation costs are considerably reduced if the geothermal power plant is located on a temperature anomaly (i.e. the increase of the temperature with increasing depth exceeds the average geothermal temperature gradient). If a well does not need to be drilled as deep to achieve a given temperature, production costs may decrease considerably, since drilling costs have a major impact on the specific energy production costs. In the present case, minimum production

stimulation methods, such as acid injection or hydraulic fracturing, may become very important when exploiting these kinds of reservoirs. In many cases the situation may not differ very much from the situation in the crystalline basement, where large artificial fracture surfaces will have to be created in order to achieve the desired flow rates. Experiments performed in Hot-Dry-Rock projects have shown that the waterfrac-technique is most suitable and possibly the only possibility for this purpose.

On the basis of the above mentioned framework conditions, the following approaches of geothermal power generation are distinguished.

– Power generation by open systems. Within open systems the heat carrier is circulated within an open circuit (i.e. the heat transfer medium is pumped into the underground, mixes up with potentially available geothermal fluid, and is produced again). In this respect, the following kinds of reservoirs are distinguished.

  • Hot water aquifers (i.e. fissured porous reservoirs). By means of a two or more well system, aquifers containing a hot geothermal fluid can be tapped in sedimentary basins. Assuming that the aquifer has a sufficiently high temperature and that a sufficient production rate is either naturally available or produced by stimulation, such a two or more well system can either serve for combined heat and power (CHP) generation, or exclusively for power generation.

  • Faults. Faults are potential flow paths of waters and are tapped e.g. by two wells similar to the hot water aquifers mentioned above, provided that sufficient permeability is available. As they generally reach deep into the underground, they allow achievement of very high temperatures.

  • Crystalline rocks (i.e. Hot-Dry-Rock (HDR) or Hot-Fractured-Rock (HFR)). By fracturing new or enlarging already existing small faults respectively the existing network of fissures within the basement rocks, the Hot-Dry-Rock (HDR) or Hot-Fractured-Rock (HFR) technologies artificially create a new heat exchanger in the underground. If this heat exchanger is connected with the surface by means of e.g. two wells, water can circulate and heat up. It is thus available for geothermal heat and/or power generation.

– Power generation by closed systems. Within closed systems the heat carrier is circulated within a closed circuit. Hence, the heat transfer medium that is pumped into the underground is entirely separated from any fluids possibly available in the ground. Two types of closed systems are distinguished.

  • One-way system. The underground is opened by means of a flow-through system, where a heat transfer medium is pumped one-way within a well into the underground and produced within the same well some kilometres away as a feed for heat and/or electricity generation.

  • Two-way system. Energy can also be withdrawn from the underground by means of a deep coaxial well, through which the heat transfer medium circulates as discussed in Chapter 10.2.

However, from a current standpoint only geothermal power generation concepts based on open systems seem to be promising from a techno-economic point of view. Thus within the following paragraphs only open systems are discussed.

10.3.1 Technical description

Systems for geothermal power generation consist of two major parts: The subsurface system for heat extraction (for drilling technique and well completion of a geothermal borehole see Chapter 10.1) and the power generating system at the surface (see /10-4/, /10-14/, /10-15/, /10-16/, /10-17/, /10-18/, /10-19/).

10.3.1.1 Subsurface system

The following explanations are aimed to discuss all aspects associated with the subsurface system of plants allowing for the production of electricity or heat and electricity from geothermal energy.

Exploitation schemes. To unlock the underground successfully different options are available which meet the needs of the respective geological conditions. Within the following explanations different appropriate exploitation schemes are described.

Geothermal fields. Most of the electricity produced from geothermal energy today, comes from vapour-dominated fields. Such geothermal fields are the most easy to exploit. A more or less dense array of vertical production wells is drilled in these areas and the vapour produced from these boreholes is, after removing the liquid phase, transported by insulated pipes above surface to a power station. The liquid phase, and the condensed geothermal steam from the power station are in many cases disposed of in surface waters. This caused environmental problems in some areas, and have led to a continuous decrease of the reservoir pressure, for instance in the geothermal fields of "The Geysers" and at Lardarello. For this reason the operators started to drill re-injection wells and to re-inject the geothermal fluids and/or make-up water (in case of Lardarello sea water) into the reservoir. This measure not only stopped the decrease of steam production observed before, but led to a continuous recovery of the production flow and geothermal power production.

The exploitation scheme in liquid-dominated fields is essentially the same. Similar to the vapour-dominated fields, the geothermal fluid is in most cases not actively produced by down-hole production pumps but by the buoyancy effect originating from the boiling of the geothermal fluid in the production wells. In liquid dominated fields, re-injection of the geothermal fluid is even more desired because of the higher mineral content of the produced fluids.

Hot water aquifers. Basically at least two wells are needed for tapping this kind of reservoir (i.e. one production and one re-injection well) since the often highly mineralised geothermal fluid must not be disposed of in surface waters, and re-injection may also be required to maintain the reservoir pressure. Production and injection wells must at least be located at a distance of approximately 1 km to avoid a fast arrival of the re-injected cold water, and to ensure a long technical lifetime of the system. For vertical wells this distance also separates the wellheads of these holes and a long insulated pipeline has to be installed at the surface. In densely populated regions this is not always an easy task and the appearance of an aboveground pipe may, in some cases, not be tolerable. Additionally, two drilling locations are required which have to be maintained throughout the technical life-time of the system. The additional costs spent aboveground associated with such a system layout are in many cases better invested for directional drilling. Directional drilling allows drilling both wells from one site. For wells reaching depths of 3,000 m or more, a separation distance of 1,000 m at depth is easily achieved with a moderate deviation from vertical. This is even true when only one well is devi-ated and the other is drilled vertically. And this can in many cases be the most economic solution. Drilling the first well vertically is often advantageous since it allows a better planning of the deviation of the second borehole and minimises the financial risk.

In most cases, pumping of geothermal fluid by using centrifugal pumps in the production well is mandatory for several reasons. Firstly, the water table in the borehole may be too deep to be counterbalanced by buoyancy effects so that a self-pumping mechanism is established. Secondly, the transmissivity of the aqui-fers is rarely high enough to enable economic production flow rates by buoyancy effects. Thirdly, boiling of highly mineralised fluids in the well may lead to severe scaling problems in the borehole and the surface pipes, for instance by precipita-tion of calcite. And last but not least, a certain overpressure has to be maintained in the geothermal loop to prevent infiltration of oxygen causing corrosion in the surface installations and in the casing of the re-injection well and to avoid plug-ging of the formation by iron oxide precipitation near the re-injection interval.

Fault zones. Major faults or fault zones have lengths of several tens or hundreds of kilometres and reach depths of 10 km and more. In contrary to the more or less horizontal hot water aquifers, fault zones are generally vertical or subvertical. The depth for tapping this resource can therefore be selected so that a sufficient tem-perature can always be achieved.

In most cases, also for this reservoir type at least one production and re-injection well need to be drilled. The distance between production and re-injection point has to be in the same order of magnitude as for hot water aquifers. However, the vertical extent of the faults, or fault zones, allows for realisation in the vertical direction, by drilling one deep (production) well and a less deep (re-injection) well. This can reduce the investment costs significantly. Drilling vertical wells may be sufficient in some cases. A better control of the intersection depths with the fault is achieved by directional drilling, especially when the orientation and dip of the fault zone at a certain depth is not precisely known. Directional drilling also enhances the chance for intersecting a greater number of individual fractures composing the fault zone.

In most cases, active pumping by using centrifugal pumps in the production well will be required to achieve the desired flow rates.

Crystalline bedrock. Although the crystalline bedrock at great depth is not completely impermeable due to the presence of open fissures, fractures, or faults its overall permeability is generally too low for geothermal power production. The basic concept of the Hot-Dry-Rock (HDR) technology thus consists of creating large fracture surfaces to connect at least two wells. During operation cold water is injected in one of the wells and heated up by the rock temperature, while circulating through the fracture system. It is then produced in the second well. To prevent boiling an overpressure is maintained in the geothermal loop. Steam for power generation is produced in the secondary loop using heat transferred from the primary loop via a heat exchanger. Depending on drilling depth (usually above 5,000 m) and temperature (usually above 150 °C) a doublet system of commercial size will operate at flow rates between 30 and 100 l/s with an installed electric capacity of 2 to 10 MW. To ensure a technical lifetime of at least 20 years a distance of approximately 1,000 m between the two wells at depth and a total fracture surface of 5 to 10 km^2 is required.

Due to the high flow velocities in the fractures especially near the injection and the production well the flow impedance of the fracture system is important for the performance of the system. For energetic and economic reasons the flow impedance of the system (difference between inlet and outlet pressure divided by the outlet flow rate) should not exceed 0.1 MPa s/l.

Fluid losses in the fracture system are another important factor for the operation of Hot-Dry-Rock (HDR) systems. Fluid losses above 10 % of the circulation flow are not tolerable over prolonged time periods. This is not only true for the very large volumes of freshwater lost in that way, but also for the additional pumping power required for the make-up water. The problem of water losses can be avoided or minimised by active pumping in the production well. In this way the fluid losses in the high-pressure zone on the injection side are compensated by a gain of fluid in the low-pressure zone on the production side.

In international projects performed during the last decades various types of Hot-Dry-Rock (HDR) systems have been proposed and tested (Fig. 10.14).

– The basic model ("Los Alamos concept", Fig. 10.14) consists of a doublet of deviated wells connected by planar parallel fractures. These fractures are created by injecting large quantities of water at high pressure into insolated intervals of the boreholes. This technique is known as hydraulic fracturing. Orientation and extent of the fractures is determined by monitoring the seismicity induced during the fracturing process and the second well is directionally drilled in order to penetrate the fracture system created in the first borehole.

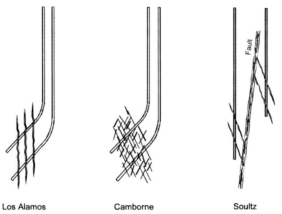

Los Alamos   Camborne   Soultz

Fig. 10.14 Hot-Dry-Rock (HDR) concepts

- The concept proposed by the Camborne School of Mines (Cornwall, England) ("Camborne concept", Fig. 10.14) is based on the existence of a network of natural fractures (joints). By massive water injection these fractures are opened up and widened so that sufficient flow is achieved between the wells. Since natural fractures at great depth are generally vertical to sub-vertical, inclined boreholes are advantageous.
- The concept applied in the European Hot-Dry-Rock Project Soultz (Fig. 10.14) is based on the existence of permeable fault zones. These large scale vertical or sub vertical discontinuities have a lateral extent of at least several square kilometres and a thickness of several meters to decametres. They are generally composed of highly fractured and altered rock material with a much higher permeability and porosity than the surrounding bedrock. Within Graben structures like the Upper Rhine Valley, a dense network of such discontinuities exists so that large hydraulic fractures created in a well will always connect to at least one or several of these faults or fracture zones. The purpose of hydraulic fracturing is therefore not to directly connect the boreholes via hydraulic fractures like in the Los Alamos concept, but to link the boreholes separately to these major natural faults. The permeability of the faults or fracture zones may be increased during the fracturing tests due to induced shearing. The Soultz concept has been successfully applied at Soultz in the Upper Rhine Valley. At a depth of 3,000 m a Hot-Dry-Rock (HDR) system of commercial size has been established and has successfully been tested in 1997. A bigger Hot-Dry-Rock (HDR) system with a central injection borehole and two inclined production wells has been established in 2004 at 5,000 m depth and a rock temperature of 200 °C. The system will be tested in the upcoming years.

Enhancing the productivity. Due to the relatively low energy content of steam or hot water compared to hydrocarbons geothermal wells require a very high mass flow in order to achieve a commercially viable power output. Commercial wells in vapour dominated reservoirs produce several 10 t/h of dry steam. In liquid dominated reservoirs a mass flow of one to several hundred t/h of hot water is required for economic reasons. This can only be achieved in highly porous, fractured, or karstic rock formations with a very high permeability. But the productivity of geothermal wells can be enhanced to some extent by acid treatment and by hydraulic fracturing. In case of the Hot-Dry-Rock (HDR) concept no permeability of the formation is required. The flow paths are artificially created by using the waterfrac technique. These methods are described in the following sections.

Acid treatment. Acid injection is a suitable means for productivity enhancement when flow impedance close to the well has to be removed or minimised. In clastic rocks, such as sandstone, the acid is dissolving the cement bridges between the quartz particles and is thus increasing the pore space. Due to the relatively high porosity (> 10 % of the rock volume) large volumes of the pore filling have to be removed in order to achieve a measurable effect. In practice, the acid affects only the first decimetres or metres around the borehole wall. Its effect on productivity enhancement is therefore very limited. Better results are obtained in fractured rocks. In this type of rock fluid flow is restricted to joints, natural flow pipes, or channels. This secondary porosity forms only a small fraction (often < 0.01 %) of the rock volume and can therefore be enhanced over much bigger distances around the well. The best results are obtained in karstic carbonate rock formations. In this type of rock formation the effect of acid injection often consists in establishing or improving a flow connection between the well and highly conductive flow pipes in the vicinity of the borehole.

Though quite often applied in oil or gas wells, acid treatment is still more an art than a technique. The choice of acid (hydrochloric acid, fluoride acid, acetic acid, formic acid and others), its concentration, the volumes and flow rates and of additives, e.g. corrosion inhibitors, is based more on experience and sense than on formulas. Quantitative predictions are hardly possible. Common acid volumes for an acid treatment are in the order of some to some tens of cubic meters. Since it requires specialised equipment for on-site mixing and injection, acid treatment is a relatively expensive technique /10-20/.

Hydraulic fracturing. The hydraulic fracturing technique for oil wells was invented in 1947 /10-21/ in order to improve their productivity. It is now a mature technique and has been applied since then in more than a million oil and gas wells. The basic process is to inject fluid into the borehole by using high-pressure pumps. When the fluid pressure reaches a certain critical level the rock bursts. Usually, a single axial fracture is initiated splitting the borehole in two halves over a length of several metres or decametres. The fracture extends into the rock on both sides of the borehole as the injection continues. It is assumed that the fracture plane is orienting vertically to the direction of the minimum principal compressive stress component in the rock. Since for most tectonic settings the direction of the minimum principal stress is horizontal, most fractures are vertical. For hydraulic fracturing operations in oil or gas wells, usually several hundreds of cubic metres of fluid are injected at flow rates between 10 and 100 l/s and at well head injection pressures of several tens of MPa. Fracture operations in oil and gas wells are designed such that the height of the fracture is limited to the thickness of the oil or gas-bearing layer. Typical fracture lengths in the horizontal direction are in the order of hundred metres. In order to keep the fractures open after pressure relief, sand or other fine grain substances (proppants) are mixed to the fracture fluid and pumped into the fractures. For this reason gel-type fluids are used, that are suitable to transfer the propping material. At the same time the high viscosity of these fluids minimises fluid losses into the surrounding rocks during the fracturing operation. The viscosity breaks down after the placement of the propping material thus enabling the transfer of oil or gas in the fracture.

Though hydraulic fracturing is a mature technique in oil and gas wells, there is only little experience with this technique in geothermal applications. It has to be stated however that the requirements in geothermal applications are considerably higher. The very high mass flows in geothermal applications (except for steam wells), for instance, require higher hydraulic fracture conductivity, higher temperature fracturing fluids, and for the more aggressive fluids a higher chemical resistance of the proppants. For these reasons the costs of fracturing operations are in the range of several hundred thousands to some Mio. €. Therefore, it is hard to predict, which role hydraulic fracturing will play in future geothermal applications.

Waterfrac technique. The Hot-Dry-Rock (HDR) and related concepts are based on creating artificial fractures or opening natural fractures with surface areas of several square kilometres. This is a great technical challenge and can never be achieved by the conventional hydraulic fracturing technique. The only promising technique to date is the so-called waterfrac technique. This technique is similar but simpler than the conventional frac technique. Instead of injecting small to moderate quantities of viscous fluids containing proppants, very large volumes of water are injected to fracture the formation.

The waterfrac technique has been widely applied for HDR projects /10-22/, and it was demonstrated that in crystalline rock the desired large fracture surfaces can be created by injecting water volumes of several ten thousands of cubic metres. It was also proved that these fractures are kept open after pressure relief due to a self-propping effect. This effect is not totally understood yet. The most likely explanation is that the two opposite fracture surfaces are laterally displaced against each other by irreversible shearing during the fracturing process and that due to the roughness and unevenness of the fracture surfaces they no longer match, and open spaces remain after releasing the fluid pressure. It has been demonstrated that the hydraulic conductivity of self-propped fractures in crystalline rock is considerably higher than the hydraulic conductivity of propped fractures created with the conventional technique in sedimentary rock. The success in these research projects give good hope that commercial HDR-systems may be installed in the near future with technical success. The most advanced project in this respect is the European HDR-project Soultz where a HDR-system of commercial size is in preparation at a depth of 5,000 m and rock temperatures of 200 °C.

Reservoir evaluation. Various parameters have to be determined to describe the characteristics of the geothermal reservoir. The most important are depth, thickness, temperature, formation fluid pressure, porosity, and permeability. In igneous rock types fracture distribution, orientation and characteristics are of similar importance. Many of these parameters can be determined directly or indirectly by geophysical borehole measurements. The given possibilities are discussed below. For Hot-Dry-Rock (HDR) systems it also necessary to determine the spatial extension the fracture system created in the first well in order to be able to link the second well into the periphery of this system. For this purpose, the method of seismic fracture mapping has been developed which is described below.

Borehole measurements. Geophysical borehole measurement (borehole logging) is a mature technique in the oil and gas industry, and many of the methods developed there can be applied for geothermal wells. The severe environmental conditions, especially the higher temperatures and more aggressive fluids, however, require some adjustments of the logging equipment for geothermal wells that is not always possible, and limits the number of available tools. High temperature versions of geophysical logging probes use either high temperature electronic parts, or the probes are shielded by Dewar housings, sometimes in combination with heat sinks in the inner of the probe. For temperatures up to 150 °C almost the whole suite of logging equipment from the oil and gas industry is available. About 180 °C is the limit for most of the high temperature electronic parts and the number of available tools for this temperature is already limited. Temperatures above 180 °C require specialised high temperature tools containing either selected high temperature electronic parts or using heat shields. Some tools, such as temperature or pressure probes, can operate without electronic parts and can therefore more easily be adjusted to higher temperatures.

Of special importance for geothermal wells are production wells. These measure temperature, pressure, and flow velocity during production or injection and are used for determining static and dynamic fluid pressure and for localising and quantifying the major inlets or outlets in the borehole.

For geothermal wells in igneous rocks, especially for Hot-Dry-Rock (HDR) systems image tools are very helpful. Today, two types of image probes are available.

– The acoustic borehole televiewer (BHTV) scans the borehole by means of an ultrasonic beam. While the ultrasonic beam is well reflected by the smooth well wall, it is more or less absorbed by fissures. Colour representations of the amplitude of the reflected signal generate optical illustrations of the wall surface and allow detection and characterisation of even fine fractures, including their orientation. The travel time of the signal is used to determine the radius of the borehole and allows the construction of a three-dimensional model of the borehole geometry that is of great interest in unstable well sections, where breakouts of the borehole wall are to be observed.
– The formation micro scanner uses the signals from electrical electrodes arrays placed on scratch patches along the borehole wall. Colour representations of electrical resistance between each pair of electrodes generate an oriented image of the borehole wall. Fluid filled fractures have a much higher electric conductivity than the rock and are easily detected. The resolution of the formation micro scanner is even higher than that of the acoustic borehole televiewer.

Seismic Fracture Mapping. For the completion of Hot-Dry-Rock (HDR) systems it is very important to determine the shape, the extent and the orientation of the fracture system created in the first borehole in order to design the trace of the second borehole. Various methods have been tested for this purpose. The best results have been achieved by using the seismicity induced during the fracturing process. Ten thousands of small seismic events are induced during massive water-frac-experiments in crystalline rock /10-23/. These signals can be recorded with geophones at the surface or in shallow boreholes around the Hot-Dry-Rock site. The detection times of the signals on the various stations allow localising the hypocentres of these signals. A three-dimensional representation of these hypocentres gives a detailed image of the induced or activated fracture system.

Also seismic tomography seems promising, which investigates the rock section between two boreholes by seismic waves generated by a probe lowered in one of the borehole. This method permits to localise fracture systems acting as seismic absorbers or reflectors.

In the vicinity of the borehole, fissure orientation can be determined by means of a Variable Acoustic Logging Tool. This method is based on the reflection of tube waves on fracture surfaces intersecting the borehole.

10.3.1.2 Aboveground system

The power plant technology applied for geothermal power generation is divided into three groups that are discussed as follows:
– open systems which directly use the geothermal fluid as a working fluid within the power plant,
– closed systems which transmit the heat of the geothermal fluid to another working fluid which is then used within the power plant and
– combined systems, a mixture of both open and closed systems.

In all these different options the steam pressure of the working fluid is reduced in work machines (such as steam turbines, screw or piston expansion machines). The generated mechanical energy of the rotating axle is then transformed into electric energy by means of an electric generator.

The related thermodynamic process is the Clausius Rankine process (see Chapter 5.1) that is state of the art within conventional power plant technology. Steam is heated at the same pressure (i.e. isobaric heat supply) and evaporated, relieved isentropically producing work, and subsequently condensed isobarically followed by an isentropic compression.

The different processes used for electricity generation from geothermal fluid based on this Clausius Rankine process are explained throughout the following sections. They differ with regard to the achievable efficiency and in terms of exploitation of the geologic resource. Fig. 10.15 thus illustrates average specific data.

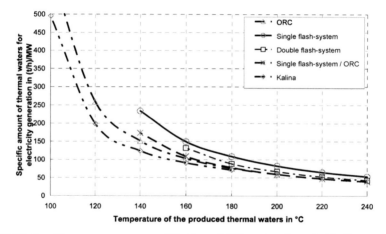

Fig. 10.15 Specific average resource utilisation of different cycles suitable for geothermal power production (ORC – Organic Rankine Cycle; Kalina – Kalina Cycle)

Open systems. Depending on the characteristics of the geothermal resource either direct steam utilisation systems or flash systems are possible. Flash systems are further divided into single flash systems without condensation, single flash systems with condensation and double flash systems. They will be explained as follows.

Direct steam utilisation. This process applies to geothermal resources where superheated steam is directly produced or where steam constitutes a high portion of the produced geothermal energy. After separation of particulate matter and water droplets, the steam pressure is directly transferred to a turbine where work it produced. This is the only difference compared to flash systems, because the flash vaporisation, and thus the energetically unfavourable pressure decrease, are not required where direct steam utilisation is possible.

Single flash process without condensation. The pressure of the produced hot water or water-steam mixture from the underground is fed into a flash container, where the pressure is slightly reduced. By this measure, the portion of the steam is increased. Afterwards the gaseous and liquid phases are separated from each other. While in most cases the separated liquid is re-injected into the underground, the produced steam is fed into the turbine to produce work. Parallel to this, the pressure is reduced to atmospheric conditions and the steam is subsequently released to the atmosphere via a diffuser.

Due to pressure reduction to atmospheric pressure, the energetic efficiency of such systems is very unfavourable in general. Nevertheless, such geothermal power plants are very cost-efficient since condensers and cooling towers are not required.

Single flash processes are thus sensible if high thermal fluid volume flows can be achieved or if thermal fluid temperatures are high (or if a high portion of non-condensable gases cannot be otherwise controlled).

Single flash process with condensation. Fig. 10.16 illustrates the principle of a single flash process with condensation.

The boiling liquid produced from the underground is thus evacuated into the flash vessel. Under a pressure level below the pressure level of the production well, a small quantity of dry saturated steam and a large quantity of boiling water is produced. Once the steam has been separated from the liquid, it is transported to the turbine where it performs work.

The cooling parameters of the condenser determine the minimum final pressure of the steam released, which in most cases is clearly within the vacuum area. Whether such a low pressure level can be utilised depends on the cost of evacuating non-condensable gases from the condenser.

At a temperature of approximately 160 °C the maximum resource exploitation amounts to 150 (t/h)/MW and to 50 (t/h)/MW at a geothermal fluid temperature of approximately 240 °C (Fig. 10.15).

Besides the technical and possibly environmental problem of releasing incondensable gases into the atmosphere, single flash processes are often characterised by the disadvantage of producing solid precipitations during flashing. Such precipitations remain as coatings on the equipment and have a detrimental effect on safety, or need to be removed and disposed.

A major energy portion of the geothermal fluid produced originally from the underground (i.e. the fluid that has been separated within the flash vessel and re-transferred to the underground) is not used for energetic purposes. This is the reason why the utilisation ratio of the entire system is relatively low.

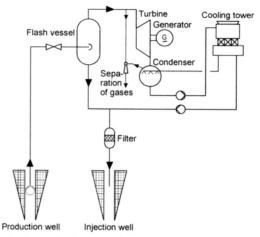

Fig. 10.16 Simplified schematic diagram of a geothermal power production facility operating according to single flash technology with condensation

Double flash process with condensation. The disadvantage of the relatively low utilisation ratio of the overall produced geothermal energy within single flash processes can be remedied by a simple appendage to the single flash system. The pressure of the boiling water drained from the first flash vessel (separator) is relieved a second time and the generated steam is again separated within another vessel. Subsequently, the separated steam is then used by another (low-pressure) turbine or within a complementary low-pressure part of the high-pressure turbine (Fig. 10.17).

The efficiency of the additional pressure reduction within the second flash vessel depends on the temperature level inside the condenser and the starting point of the pressure reduction above the atmospheric pressure.

At a temperature of approximately 160 °C the maximum resource exploitation varies between 130 (t/h)/MW and 40 (t/h)/MW at a geothermal fluid temperature of approximately 240 °C (Fig. 10.15). It is thus considerably higher than the utilisation of single flash plants.

Although flash systems at temperatures above 175 to 180 °C are no longer applied, double flash systems are especially suitable to control the difficulties related to low resource temperatures, such as a low flash pressure and a low specific steam volume. Hence, due their simple design, low operation and maintenance costs and their ability to produce their own cooling tower water, double flash systems are characterised by considerable advantages, especially when compared to binary systems. However, double flash systems need to be optimised economically, as they present higher investment costs when compared to alternative systems due to the needed additional turbine or turbine stages, additional vessels, pipelines, control systems, etc.

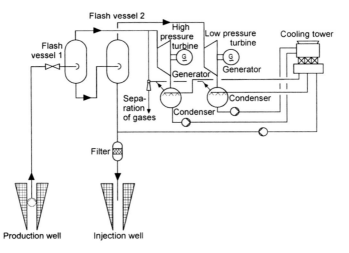

Fig. 10.17 Simplified schematic diagram of a geothermal power production facility operating according to the double flash technology

Closed systems. The term closed system refers to geothermal power plants that do not directly exploit the geothermal resources, and thus the extracted steam for power generation by a pressure reduction within a turbine, but rather apply a secondary medium. Within such closed systems the geothermal energy is transferred to a secondary medium via appropriate heat exchangers (evaporators). Due to the low temperature of the geothermal fluid or fluid-steam mixture in general, the secondary medium has to be characterised by a low boiling temperature.

Under the described circumstances the Rankine process, using organic working fluid, or the Kalina process are most suitable. They will be discussed within the following sections.

Such cycles are applied if the primary medium is not hot enough or if its pressure is too low to allow for the generation of the required pressure parameters for a thermodynamically wise pressure reduction. Furthermore, the application of a second working fluid is sensible if the hot water produced from the underground is characterised by unfavourable chemical properties (such as mineralisation, gas content, etc.), which can either not be directly controlled, or only at unreasonably high costs.

Organic Rankine Cycle. Except for the used working fluid, and thus the realised temperature and pressure parameters, the Organic Rankine Cycle (ORC) only differs slightly from the classic Rankine process based on steam realised in numerous conventional power plants.

As with common power plants cycles, the working fluid is preheated (in this case with the geothermal energy provided by the geothermal fluid), evaporated, and relieved by means of a turbine. It is then cooled by a recuperative heat exchanger (compared to relieved steam this medium is still superheated), condensed and lifted again to evaporator pressure by means of a pump. The corresponding schematic diagram is illustrated in Fig. 10.18.

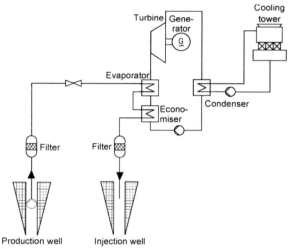

Fig. 10.18 Simplified schematic diagram of a geothermal power production facility operating according to the ORC (Organic Rankine Cycle)

Within the temperature range of fluids extracted from geothermal wells, usually hydrocarbons (such as n-pentane, isobutane) are applied as working fluids. In the past, also fluorocarbons (such as C_5F_{12}) have been used. Furthermore, the application of mixtures of hydrocarbons is under consideration, as they promise to enhance efficiencies due to their smooth evaporation temperatures.

In order to apply working fluids of organic origin the plant design needs to be adapted accordingly. The turbines differ from those used for water vapour due to the different molecular weight and the lower specific heat capacity. Furthermore, precautions have to be taken with regard to the higher corrosion of turbines and heat exchangers as well as for sealing the system with regard to the atmosphere.

Fig. 10.19 shows the power generation efficiencies of selected ORC-plants and, additionally, an average efficiency curve /10-24/; according to this illustration average efficiencies amount to between 5.5 % for a geothermal fluid temperature of approximately 80 °C and 12 % for a geothermal fluid temperature of approximately 180 °C. This corresponds to a resource exploitation exceeding 500 (t/h)/MW at 80 °C and 80 (t/h)/MW at 180 °C (Fig. 10.15).

Up to a temperature of the geothermal resource of about 135 °C net power generation efficiencies are below 10 %. At the upper limit of the temperature range under consideration (200 °C), efficiencies amount to between 13 to 14 %, provided that the heat content of the geothermal fluid is exploited exhaustively and that the desired cooling temperature is reached.

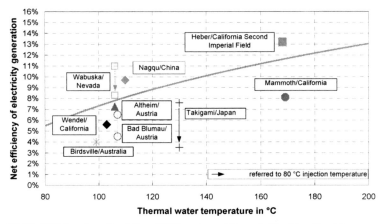

Fig. 10.19 Efficiencies of existing ORC systems (Organic Rankine Cycle)

Kalina Cycle. Similar to the ORC process, for the Kalina process a working fluid is used that circulates within a cycle isolated from the geothermal fluid. But here a mixture of ammonia and water serves as working fluid. Fig. 10.20 illustrates the simplest version of this cycle.

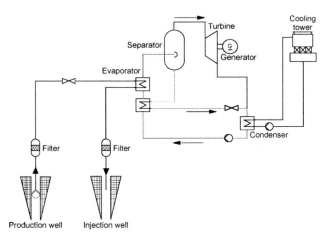

Fig. 10.20 Simplified schematic diagram of a geothermal power production facility operating according to Kalina cycle

The working fluid, composed of a mixture of two components, is preheated and evaporated by the geothermal fluid inside a heat exchanger acting as an evaporator. The different boiling points of the two compounds within the working fluid allow for a smooth transition to the evaporation temperature. An ammonia-rich vapour is produced and a fluid that is weak in ammonia remains. The vapour is then transported into a turbine and is released producing work. Afterwards, the remaining vapour and ammonia-weak fluid are again mixed and conveyed towards the condenser to liquefy the compound mixture. With the help of a pump, the liquid is then brought to evaporation pressure. To improve energetic efficiency, recuperators are installed within the cycle. One of them is shown in Fig. 10.20 between the hot ammonia-weak working fluid and the cold basic working fluid.

For a geothermal fluid temperature of approximately 80 °C the achievable efficiency amounts to about 8.5 %, and to 12 % for a geothermal fluid temperature of approximately 160 °C. Hence, resource exploitation rates between 500 (t/h)/MW and 70 (t/h)/MW, respectively (Fig. 10.15), are reached. However, these comparatively high efficiencies have only rarely been achieved by plant operation because the Kalina cycle is still under development, and there are very few demonstration plants under operation so far.

The big advantage of this cycle is that evaporation and condensation are not realised isothermally, as for pure substances (as used within ORC process). In fact, due to the mixture of two components a sliding temperature transition during evaporation and condensation is possible.

- The temperature curve of the geothermal fluid extracted from the underground (e.g. a temperature reduction from 150 to 85 °C), and the working fluid used within the Kalina cycle (a rise in temperature from 75 to 145 °C, provided that the water/ammonia solution is mixed to meet these characteristics) can thus be ideally adapted to one another. This helps to reduce the average temperature between both mass streams, and thus the heat transfer losses.
- Compared to the application of pure substances the average evaporation temperature is increased, whereas average condensation temperature is reduced. This is improving the Carnot efficiency of the cycle (i.e. theoretical maximum efficiency).

Besides energetic advantages this process also offers civil engineering benefits. Since ammonia and water have similar pressure release properties steam turbines can be applied. Additionally, ammonia/water mixtures have for quite some time been used on a large scale for other technical purposes (such as refrigeration). It should thus be possible to handle the ammonia/water solution within a large-scale cycle without any substantial technical problems. However, the low temperature differences within the heat exchangers and the poor heat transmission properties require much bigger equipment compared to conventional power plants based on a steam cycle.

Combined systems. Here, different processes are combined according to the site-specific circumstances; for instance, a single flash process is combined with a binary process. In general, various cycles are possible. For example, once the steam has passed the steam turbine and pressure is reduced, it can serve as a heat source for an ORC process. Another possibility is that the drained liquid from the first separation step is not reduced a second time within a second flash vessel, but heats the evaporator of an ORC plant (Fig. 10.21). Combined systems offer comparatively high efficiencies (Fig. 10.15).

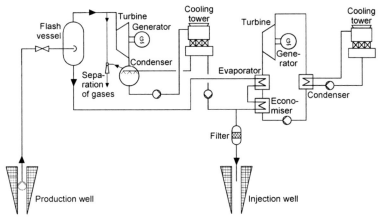

Fig. 10.21 Simplified schematic diagram of a combined facility of single flash process and ORC

10.3.2 Economic and environmental analysis

The following sections contain an analysis of economic and environmental aspects of geothermal power generation and combined heat and power generation.

Economic analysis. For the following considerations geothermal resources with a reservoir temperature of 150 °C and a production rate of 100 m^3/h are assumed. The geothermal reservoir is tapped by means of a two-well system consisting of a vertical and a diverted well. The exploited reservoir is located in areas of a normal geothermal gradient at about 4,600 m depth (average geological conditions) or in areas with much more favourable geothermal gradients at about 2,700 m depth (promising geological conditions). Successful stimulation ensures production rates of 100 m^3/h. Based on this concept, two plant configurations are examined.

– Plant A. The assumed exclusively geothermal power generation is provided by means of an ORC plant of an electrical efficiency of 11 % and an electrical capacity of 1,000 kW. Additionally, four further power plant cycles will be examined (i.e. Kalina, single flash, double flash and a combined single flash and ORC). The power generation plant is operated providing base load with 7,500 h/a. It covers its own electricity demand.
– Plant B. For Plant B combined heat and power (CHP) generation exclusively based on an ORC is assumed. As for Plant A the technical maximum electrical

power generation is realised. The low-temperature heat (of about 70 °C) produced after the electricity generation process is provided to consumers living in a housing estate (SFH-I) built according to the low-energy design (3,000 h/a) by means of a small district-heating grid (DH-II, return temperature of about 55 °C). Heat production costs are calculated by means of credits for the heat supplied by the cogeneration plant at a price of 0.032 €/kWh respectively 8.9 €/GJ (about 26,000 GJ/a).

Based on these frame conditions the investment costs for such plants are discussed first. Then the operation costs are analysed, and on this basis the energy provision costs are presented.

Investments. The costs for drilling the wells including the casings are subject to considerable variations due to changing geological conditions. On average, the costs for wells of depths between 4,000 and 5,000 m amount to approximately 1,500 €/m. However, costs do not increase linearly with increasing depths; up to a depth of 3,000 m costs amount to approximately 1,000 €/m.

The costs of such a well are predominantly determined by the costs for renting and operating a drilling rig (including personnel and costs for the Diesel fuel), which on average account for 36 % of the total cost of a well. About 4 % of the total costs are needed for site preparation and restoration of the drilling site, once the well is successfully produced, whereas approximately 15 % correspond to drilling bits and directional drilling services, 12 % correspond to the drilling fluid and cementation, 20 % to well casing including riser, and 12 % to well head completion. These shares may vary considerably depending on the local conditions. Additionally, approximately 235,000 € is needed for the required well pumps (Table 10.4).

Costs for stimulation are even more uncertain, as according to the geological conditions at a certain location different frac technologies, frac pressures, frac proppants, and the quantities to be pressed into the underground need to be applied. These frac methods are state-of-the-art processes as far as crude oil and natural gas production is concerned. However this is not true for geothermal power generation. This is why the costs for, for example, one frac of a fluid volume of 250 m^3 and approximately 60 t of proppants are conservatively estimated at 360,000 €. However, if a drilling rig needs to be transferred and installed at a specific site, costs may rise up to 550,000 €.

Costs for the basic aboveground geothermal fluid circuit including tubes, valves, and control equipment are strongly varying depending on the ground conditions (e.g. rocky underground densely covered with buildings or sandy agricultural underground). Here the costs are estimated at roughly 600 €/m. For additional system components directly linked to the thermal water cycle (such as slop system, filter, etc.) the costs, due to the very different and site specific arrangements, are globally assumed to correlate with the geothermal capacity as 25 €/kW.

The costs of an ORC plant of an electric capacity of 1,000 kW amount to approximately 1.9 Mio. €. Additionally, cost intensive heat exchangers are required. The price of the building, where all remaining system components are located, including the land property adds to that. For the calculations indicated below 500,000 € are assumed.

Moreover, planning and preparation costs (such as geological surveys, permission fees) need to be considered. For the case study analysed here, an average value of 4.8 % of the overall investments is assumed.

Table 10.4 Investment and operation costs as well as electricity generation costs

		Average conditions	Favourable conditions
Installed electric capacity	in kW	1,000[a]	1,000[a]
Investments			
doublette	in Mio. €	15.40[b]	9.4[c]
pump	in Mio. €	0.24	0.24
geothermal fluid circuit	in Mio. €	1.20	1.20
slop and filter systems	in Mio. €	0.22	0.22
power plant	in Mio. €	1.90	1.90
heat exchanger for DH	in Mio. €	0 (A) / 0.23 (B)	0 (A) / 0.23 (B)
building	in Mio. €	0.50	0.50
planning	in Mio. €	0.95	0.66
Total	in Mio. €	20.39 (A) / 20.64 (B)	14.11 (A) / 14.35 (B)
Operation, maintenance etc.	in Mio. €/a	0.34	0.34
Credit for heat	in Mio. €/a	0 (A) / 0.22 (B)	0 (A) / 0.22 (B)
Electricity provision costs	in €/kWh	0.22 (A) / 0.18 (B)	0.17 (A) / 0.14 (B)

DH district heat, (A) Plant A (see text), (B) Plant B (see text); [a] temperature of the geothermal fluid free power plant 140 °C, production rate 100 m³/h, two vertical wells, distance between the two wells 2,000 m; [b] well depth 4,600 m; [c] well depth 2,700 m

Operation costs. Annual operation costs basically consist of costs for personnel and maintenance. To minimise the costs, a plant operation without supervision is assumed (Table 10.4).

Maintenance costs are estimated related to individual plant components at 0.5 % of the investment costs for wells, and 4 % for pipes, heat exchangers and miscellaneous, respectively. For the ORC plant or any other type of conversion plant and the building, maintenance cost of 1 % of the investment costs are assumed.

Energy generation costs. Under the discussed frame conditions the electricity generation costs have been calculated for an ORC plant for a flow rate of the thermal fluid extracted from the underground of 100 m³/h and a well head temperature of 150 °C (Table 10.4). According to Fig. 10.22 they are thus within the range between 0.14 and 0.22 €/kWh.

For exclusive power generation (Plant A) the calculated electricity production costs vary between 0.17 and 0.22 €/kWh. Well costs, including drilling and stimulation as well as pumps account for approximately 70 %, whereas 30 % are attributable to the ORC plant, the geothermal fluid circuit, miscellaneous and operation costs.

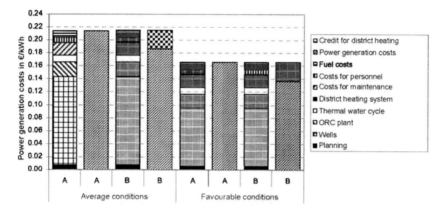

Fig. 10.22 Electricity generation costs of reference Plants A and B (the total costs indicated by the left bars refer to the costs related exclusively to the power generation (i.e. total power production costs of the entire plant without a possible income by heat sale through the district heating system to the household customers), whereas the right bars consider such a possible heat credit (and thus indicate the net power production costs))

Power production costs are considerably influenced by the temperature of the extracted geothermal fluid. Fig. 10.23 shows this effect using the example of exclusive geothermal power generation under the average site conditions. Under such circumstances, by means of the combined ORC/flash process and an approximate geothermal fluid temperature of 200 °C minimum power production costs amount to approximately 0.18 €/kWh. Compared to this, due to their lower efficiencies single or double flash processes only allow to yield lower power quantities at the same drilling costs. The specific electricity generation costs are thus considerably higher.

The illustration shows that the specific power generation costs increase with further increasing temperatures. This is mainly due to the influence of the well depth on the drilling costs because the temperature can only be increased with an increased well depth. The impact of increasing temperature on the efficiency of the respective cycle (and thus on the achievable power output) is more than compensated by the superproportional increase of drilling costs with increasing depths.

Electricity generation costs are considerably reduced if the geothermal power plant is located on a temperature anomaly (i.e. the increase of the temperature with increasing depth exceeds the average geothermal temperature gradient). If a well does not need to be drilled as deep to achieve a given temperature, production costs may decrease considerably, since drilling costs have a major impact on the specific energy production costs. In the present case, minimum production costs for exclusive power generation amount to about 0.17 €/kWh (for ORC with water cooling and Kalina cycle at 180 °C or for the combined Flash process/ORC between 200 and 230 °C).

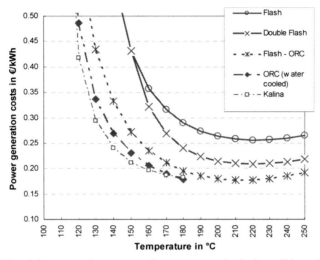

Fig. 10.23 Electricity generation costs under average geological conditions (Plant A, varying temperatures, 100 m³/h production volume)

Electricity generation costs are further reduced by increasing the volume flow per two-well system up to a rate of, for example, 200 m³/h. Under favourable site conditions minimum production costs for exclusive power generation amount to about 0.08 €/kWh (for ORC with water cooling and Kalina process at 180 °C or for the combined Flash process/ORC between 200 and 230 °C). But also pure flash processes are appropriate to generate power at a price of 0.09 or 0.10 €/kWh at a geothermal fluid temperature of approximately 220 °C.

If heat can be sold at the plant gate (Plant B), power production costs are decreased to 0.18 €/kWh, under average site conditions, and to approximately 0.14 €/kWh for sites with promising geothermal conditions. Although investment costs are slightly increased when compared to exclusive power generation, the additional heat credits allocated to the provided electricity tremendously reduce the corresponding electricity production costs. This is due to the assumed conditions of heat purchase which do not influence potential power supply. Hence, additional heat credits are obtained at very low extra costs which in turn tremendously reduce power production costs.

Electricity production costs are mainly influenced by investment costs and temperature (Fig. 10.24). If, for instance, investment costs are reduced by 30 %, power production costs are decreased from 0.17 to 0.13 €/kWh and thus by 0.04 €/kWh.

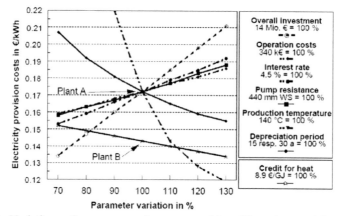

Fig. 10.24 Variations of parameters of power provision (Plant A, promising geological conditions; for reasons of comparison also Plant B is shown varying the credits for heat)

Environmental analysis. In the following section the geothermal power generation is assessed according to selected environmental effects. Again, assessment is performed with regard to the effects caused during construction of the plant, operation, malfunctions, and the end of operation.

Construction. Environmental effects can already arise during exploration and drilling of wells to unlock the geothermal energy. Effects are comparable to those of crude oil and natural gas exploration. However, besides noise effects which have to be within the limits defined by the existing legal framework, only very few environmental effects remain once drilling is completed, as it is possible to re-cultivate the drill site except for the well head. Furthermore, such drilling effects on the environment are limited to the short period of drilling only, which, depending on the respective depth, usually takes 3 to 6 months.

Normal operation. Due to the circulation of the geothermal fluid within the heat exchanger (i.e. within the underground), small concentrations of salts and minerals may be dissolved several kilometres below the earth's surface. Outside of volcanic areas, heavy metal or sulphurous compounds are either not available, or their concentrations are so low that they are hardly detectable. However, these substances do not have a detrimental effect on the environment, as the geothermal fluid is transferred within a closed circuit and dissolved substances are subsequently retransferred into the underground /10-1/.

If the underground is cooled off due to the re-injection of cooled geothermal fluid the mechanical stress field may change, which may generate local relief and thus lead to microseism or small earthquakes. However, during experiments performed at Soultz-sous-Forêts microseism has to date only be observed during extensive stimulations and thus during the production of the heat exchanger within the underground and not during ordinary circulation and thus production of the hot geothermal fluid. In seismic unstable areas, this microseism may produce small earthquakes before the natural event. However, this situation is very unlikely to occur as such sites would not be chosen due to their history or would be eliminated at least after the stress has been detected at site. For this reason, also the slightest acoustic signals are recorded, which occur when rocks burst during stimulation /10-1/.

Due to the relatively low temperatures, varying between 150 and 250 °C, the efficiency of geothermal power generation is relatively low when compared to conventional power plants. This is a result of the high quantities of waste heat to be released into the atmosphere and thus pollute the environment. However, if there is a certain demand for low-temperature heat in the locality of the power plant, the heat might be used to substitute other environmental harmful heat provision systems (e.g. coal oven). Furthermore, waste heat could also be returned into the underground.

Geothermal generation facilities are characterised – compared to other power plants based on renewables and conventional energy carrier – in general by little land demand depending on the reservoir and generation conditions only. Additional space is required for the pipelines that transfer the geothermal fluid from the well head to the power plant. In the case that the pipes are installed aboveground they also affect the landscape visually. Additional space is required for cooling towers and power stations. However, these space requirements do not differ essentially from those of fossil fuel-fired power plants.

For normal operation, altogether only very few environmental effects can be expected. From the current viewpoint no significant detrimental effect on humans and the natural environment is likely to occur.

Malfunction. Due to the salt and mineral contents, in case of a malfunction, minor environmental impacts may result from heat transfer medium leakage at the earth's surface (i.e. hot water or hot steam). However, the environmental impacts are low compared to those observed in the volcanic areas of the earth. Further environment impacts may result from a possible pressure relief of the heat transfer medium within the power plant (due to a malfunction (e.g. leakage) of the circuit). Small amounts of gases may be released; however, such a release can be prevented by appropriate safety measures which are well known from the conventional power plant technology /10-1/. As for any other technical plant, environmental impacts are thus by all means possible; however, they are always limited to a restricted area and do not have global effects. From our current knowledge they only have very little impact.

End of operation. To prevent unwelcome environmental impacts, at the end of operation the wells need to be properly sealed to permanently exclude penetration of harmful substances from and to the earth's surface and to prevent a hydraulic short-circuit of various subterranean layers. Disposal of plant components, in contrast, does not cause any environmental impacts as the components correspond to a large extent with conventional machine standards whose environmental impacts are very limited due to the given regulations.

Annex A: Harnessing Ocean Energy

A.1 Energy from wave motion

Wave energy is a form of energy that is mainly caused by wind energy. Such waves contain both potential and kinetic energy. For ideal deep water waves, which are not subject to any ground friction, the total capacity of a standard wave of 1 m width is directly proportional to the product of the square of the wave height and the wave period.

An assumed so-called standardised spectrum based on the above relation between wave height and wave period allows for the determination of the wave power or energy in relation to height or frequency. For instance, wave heights of 1.5 m at an average wave period of 6.2 s are typical of the German North Sea coast and lead to a significant weight height of 2.11 m and a total wave power of approximately 14 kW/m wave front. If it was possible to exploit the whole energy of a wave front of the length of the German North Sea coast (about 250 km), according to our model, theoretically an approximate power of 3.6 GW could be generated (see /A-1/).

Because of its considerable energy potential, for several decades, wave energy has been investigated with regard to power generation. However, the multitude of more or less unrealistic proposals that have been elaborated discredited this type of renewable energy. Thanks to the tireless commitment of several research teams over many years, this view now changes gradually. The use of wave energy became a more and more serious and important option for electricity generation on a small and medium scale.

Systems for electricity generation from wave energy can also be used for the protection of the coast because systems using wave energy convert the energy from the sea into electricity and remove this energy therefore from the ocean. Under these circumstances the wave energy is not only reflected or dissipated. A good combination of power generation and coastline protection could thus also enhance the economic attractiveness of wave energy exploitation.

The principle of converting the motion of sea waves into mechanically useful motion is trivial. By "inverting" the principle of a piston engine the motion of a body floating on waves (replacing the piston) can make a shaft rotate by means of a rod drive, whereas the rotating shaft in turn drives a generator.

The feasibility of this basic concept has already been demonstrated /A-3/, /A-4/. The aim of conversion plants using wave power is thus not to demonstrate the technical feasibility of transforming wave power into electricity, but to enhance the technical reliability of providing electric power at reasonable costs. In reality, this goal can rarely be fulfilled due to a series of requirements, which are summarised as follows:

- Hydraulic optimisation is absolutely necessary to achieve high electrical efficiencies. If, for instance, only the upward and downward wave motions are exploited, 50 % of the energy contained inside the wave is wasted. However, design principles have been elaborated which allow using the entire energy of the waves.
- The power plant design also needs to be designed in order to withstand the "wave of the century". If a wave energy converter is designed for harnessing waves of a height of 1 m, it also needs to withstand waves of ten times this size. In our case this would be a 10 m wave containing 10 times the above wave energy. The precautions to be taken into consideration necessarily result in considerable additional design costs.
- The power has to be designed very reliably even under unfavourable operation conditions. During the period when wave energy is most effective (e.g. autumn storms) maintenance or repairs cannot be performed for weeks. If a system fails during this period, efficiency is tremendously reduced due to long downtimes.

The following sections contain a discussion of different wave energy exploitation systems. Please note that for this purpose no differentiation is made between breakers energy and wave energy. The former is assumed to be a form of the latter.

A.1.1 TAPCHAN system

Within a TAPCHAN (tapered channel wave energy conversion device) system water advancing over the beach by breakers or swells is conducted into a raised reservoir via a converging inclined channel (Fig. A.1). This tapered channel concentrates waves of different frequencies, coming from different directions and simultaneously converts the kinetic wave energy into potential energy. Within this channel the wave height is increased due to the decreasing width. This has the consequence that the water level raises and the seawater eventually spills over the narrow end of the channel into the reservoir whose water level is located several meters above the average sea level. From this storage reservoir, the seawater, accumulated at a higher energetic level due to the difference in height, can flow back to sea via a turbine.

Due to the storage reservoir this system requires more space than most other wave energy conversion systems. Because of inflow losses (including shallow water effects) only a limited amount of the original wave energy (of deep waters) can be used. However, due to the levelled drainage of the storage reservoir and the applied low-pressure turbine, which is state-of-the-art technology on the markets for power plant equipment, operation of this system is much easier than of most other breaker or wave powered energy exploitation systems. An additional advantage is that the system components applied within this power plant are not subject to open sea conditions, and thus offer a longer technical lifetime. Also the maintenance can be easily conducted. Furthermore, power plant components permanently in motion do not touch the waves, and conversion from kinetic into potential energy is performed by solid reinforced concrete elements. The plant thus also withstands bad weather conditions ("wave of the century"). It is moreover beneficial that such plants are easily accessible from the shore. As fresh seawater is continuously conducted into the storage reservoir the latter is also suitable for fish farm operation. When compared to a straight overflow edge, parallel to the wave crest, a considerable benefit of such wave or swell-powered generators provided with a tapered channel is that basically all waves reach the required height at some point in order to fill the raised reservoir over the narrow end of the channel.

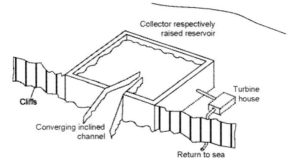

Fig. A.1 Operating principle of a TAPCHAN system (according to /A-1/)

The operating principle has already been proven by the TAPCHAN demonstration plant built in Norway at Toftestallen, near Bergen, in 1986.

A.1.2 OWC system

Another system to use wave or breaker energy is generally referred to as oscillating water column (abbreviated to OWC). From the current (at least short-term) viewpoint the OWC system is the most promising type of wave energy converters. Already applied in 1910, the system is probably the first wave-powered generator in history of human mankind. At that time air compressed inside the cave of a rocky shore was exploited. Current technologies, by contrast, aim at using the wave motion in artificially build chambers. For decades, the technology of the oscillating water column has also been applied to supply the energy for lighted buoys. In the following, first the operating principle of buoys and subsequently a large-scale power plant based on this principle are discussed.

OWC buoy. The OWC buoy is based on a vertical tube submerged so deep into the seawater that it is below the level where wave motion occurs. This buoy contains a water column that cannot directly follow the buoy or wave motion and is therefore caused to oscillate by motion. A water or air turbine installed inside the tube, located inside the upper part above sea level, rotates due to the upward and downward movements of the water column and drives a generator to generate power (Fig. A.2).

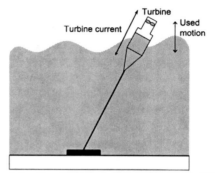

Fig. A.2 Operating principle of an OWC buoy (according to /A-2/)

A major problem of the OWC system is to transmit slow motion into faster motion that can be used for power generation. Since this transformation is technically only possible in conjunction with mechanical and hydraulic system elements that are furthermore very costly, this technology is not suitable for the small-scale systems, for fundamental reasons. For OWC systems, wave motion is thus transmitted by means of air; usually Wells turbines are applied.

Lighted buoys, operating according to the OWC principle, have been applied on sea for more than 20 years. The small air turbines installed inside such buoys have proven to be durable and cost-effective. Due to the fast movement of the turbine water does not penetrate into the turbine ball bearings; therefore corrosion is prevented.

OWC breaker-powered generator. To find solutions for large-scale wave or breaker energy exploitation, the current investigations focus on OWC wave and breaker-powered generators. Usually, such OWC systems are mounted at the bottom of the sea, preferably, near a steep coast. Through an opening below the average sea water level, the energy of moving water waves penetrates into a big chamber, thereby transmitting the wave frequency to the contained water column. As the water level is moving up and down the air volume above the water surface within the OWC plant is "breathed" in and out. Subsequently, the kinetic energy of the air that is sucked in or blown out is partly converted into electric power by means of an appropriate turbine. Optimum power generation is achieved if the frequency of the oscillation system consisting of inflow, water column, air quantity, turbine, and outflow corresponds to the frequency of the advancing waves.

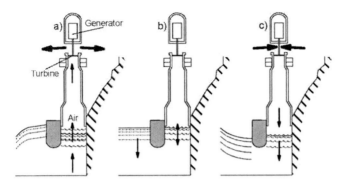

Fig. A.3 Operating principle of an OWC breaker-powered generator (a) advancing wave; b) wave crest; c) reflected wave; according to /A-1/)

Fig. A.3 shows an example of such a plant that has been built in Northern Scotland. The rock of an appropriate steep coast section has partly been blasted above and below water level. Into the produced cavity a concrete air chamber of a cross-section of 50 m² has been imbedded. The opening through which water penetrates into the air chamber is located between 3.5 and 7 m below the regular sea level. The oscillation range of the wave within the chamber (i.e. the difference between wave crest and wave trough) sums up to approximately 3.5 m. A closed steel cylinder has prolonged the upper section of the air chamber which has been designed as a concrete casing. At its top a Wells turbine coupled with a generator has been installed. To enhance power generation two moles have been installed in front of the plant. They intensify the breakers by resonance.

The turbines applied in conjunction with OWC breaker-powered generators can be operated according to two different models:

- Air flows through the turbine only in one direction. The OWC design has thus to ensure that a cyclical oscillating flow is converted into a cyclical pulsating flow. Nearly all early systems were based on such an approach. But in the meantime turbines maintaining the same rotation direction at alternating flow directions have replaced them.
- Air flows through the turbine in alternating directions. For this purpose a turbine design needs to be developed which is driven in one direction even if the direction of the air flow changes. This challenge can so far only be solved with low efficiency cross-flow turbines – of which nowadays only Wells turbines, named after their inventor – are used. This low efficiency could so far only be increased within certain limits and at high cost. Nevertheless, all such plants are equipped with a Wells turbine, since the valves, which must open and close within a period of about 10 s under sea conditions, seem to be technically reliable.

A.1.3 Further approaches

In the past, a series of additional system has been developed that are generally suitable for exploiting wave energy for electricity generation. However, up to now, they have not been successful in the marketplace. Nevertheless some of them are briefly introduced within the following considerations.

Cockerell's raft consists of articulated coupled floating bodies that are similar to a pontoon. Between these floating bodies joining piston pumps are installed to compress the working fluid (water or air). This pressurised working fluid is subsequently used to drive a turbine coupled with a generator for electricity generation (Fig. A.4). To date, plants of this type have only been designed or realised as miniature versions /A-6/. From the current viewpoint it is very unlikely that such a design will ever gain the importance of large-scale plants.

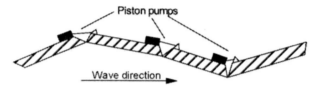

Fig. A.4 Operating principle of Cockerell's raft (according to /A-5/, /A-6/)

The Salter wave energy converter is characterised by numerous rotating blade-type floating bodies arranged next to each other around a horizontal axis (Fig. A.5). By means of a support structure the whole system is anchored on the seabed so that the blades operate under semi-submerged conditions. With each advancing wave they are moved to an upright position. Once the wave has passed underneath they return to their original position. The blades thus continuously nod with the rhythm of the wave period. This motion is transferred to a working fluid, which is compressed to drive a turbine and a generator /A-6/. It is very unlikely that this system will even gain far-reaching significance or will be applied on large scale since the TAPCHAN and the OWC systems are much more promising.

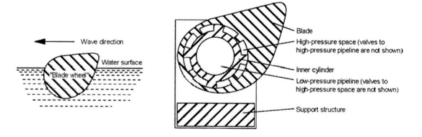

Fig. A.5 Operating principle of a Salter wave energy converter (left: operating principle; right: basic design; according to /A-5/, /A-6/)

A.2 Energy from tides

In association with the rotation of the earth, the gravitation forces of the moon and the sun (i.e. movement and gravitation of planets; see Chapter 2) periodically alter the water level of the ocean. In the open sea tide waves are characterised only by height differences of a little above 1 m. Yet, the mainland has a braking effect on the tide wave and generates backwaters at the shoreline, so that maximum water level changes of 10 m and more are possible. In certain coastal areas, such as bays and river estuaries, the tidal range may rise up to 20 m due to resonances or funnelling.

With regard to tides there are two different methods for power generation: utilisation of the potential energy of backwaters by a tidal power stations or marine current exploitation. Both methods will be briefly described within the following explanations.

A.2.1 Tidal power plants

Tidal energy, i.e. the tidal range, can be exploited according to different methodological approaches.

The easiest method consists of a one-basin-system that is only used in one direction (Fig. A.6, top). A bay is separated from the open sea by means of a dam, but remains connected to the sea by a floodgate and a turbine. The control system of the turbine and the floodgate ensures that water can only enter the bay via the floodgate and can only exit the bay through the turbine. This turbine is connected with an electrical generator providing the electrical energy. A fundamental disadvantage of such types of tidal power plants is that energy can only be generated during comparatively short time periods (Fig. A.6, top). Its advantage is the simple turbine design.

To avoid this disadvantage a cross-flow turbine can be applied for power generation (Fig. A.6, centre). The floodgate, which is also needed for this system design, accelerates the inflow and outflow of water during periods when there is almost no water level difference between the basin and the sea (i.e. high tide and low tide). This design ensures power generation over much longer time periods (Fig. A.6, centre). This is only possible with a higher technical expenditure. But in general the higher possible energy provision overcompensates the higher costs.

Tidal power stations can also be designed as a two-basin-system (Fig. A.6, bottom). Between the two basins another turbine is integrated. This turbine is either built into a dam or into a connecting canal located between the two basins connected to the sea also via turbines. Water inflow and outflow is controlled as water enters one basin at high tide and exits through the other basin at low tide, after having passed through the corresponding turbines. Energy production is thus even more regular and can be performed without any interruptions. However, the power plant design is more demanding and the two basins require more space.

In most cases projects for tidal power plants equipped with such storage reservoirs that hold back the water until the water level outside of the reservoir has reached a lower level are often not put into practice due to high costs and significant environmental effects of this space-consuming design. Currently, there are only very few tidal power plants in operation worldwide. The power station located at the estuary of the river Rance, near St. Malo in France has been operated since 1966. The Canadian 20 MW prototype plant, located at Fundy Bay, is in successful operation since late 1984. Moreover, there is one plant in Koslogubsk in Russia and two plants in China. However, only the first one can be regarded as prototype of large-scale tidal power plant, as for an average tidal range of approximately 8.5 m, 240 MW have been installed.

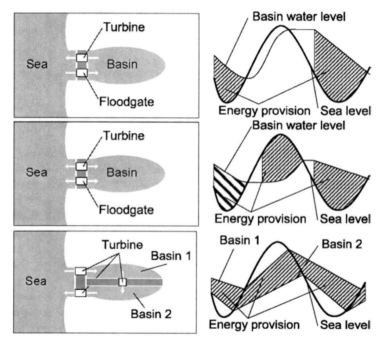

Fig. A.6 Operating principles of tidal power stations (top: one-basin tidal power station using one flow direction; centre: one-basin tidal power station using opposite flow directions; bottom: two-basin tidal power station; according to /A-2/)

On the whole, the technical potential of tidal power stations worldwide is too low to largely contribute to global energy provision. However, local conditions may be favourable.

A.2.2 Harnessing high and low tide streams

Since harnessing the energy of the sea by utilising a bay to be separated by a dam, is connected with enormous consequences for the natural environment, lately, exploitation methods based on the high and low tide stream, and thus on water motion caused by low and high tides, have been investigated. However, a major disadvantage of such systems is the comparatively low energy density of the ocean currents.

There are only little pressure differences in such currents with the result of relatively slow current speeds and big current cross-sections. Therefore turbines suitable for low and high tide streams must be developed, such as the large Savonius or Darrieus rotors, which perform satisfactorily under the described conditions. Fig. A.7 illustrates the example of a corresponding project study in comparison to an offshore wind power station (see Chapter 7.2).

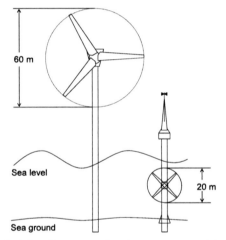

Fig. A.7 Operating principle of a rotor using the ocean current in comparison to an offshore wind power station (according to /A-7/)

As for any other stream, the high and low tide stream power P_{Wa} of sea water flowing through a given cross-section (in this case the rotor cross-section) is calculated according to the Equation (A.1) used for wind power stations, namely by the water density ρ_{Wa}, the current cross-section S_{Rot}, and the water current speed v_{Wa} in third power.

$$P_{Wa} = \frac{1}{2}\rho_{Wa} S_{Rot} v_{Wa}^3 \qquad\qquad (A.1)$$

According to Equation (A.1) it is sensible to exploit the low and high tide stream if sufficiently high current speeds are available. For water streams with a flow rate of e.g. 0.1 m/s the energy density results in 0.5 W/m². If the flow rate is increased by factor 4 to 2 m/s, energy density grows by factor 8,000 to 4,000 W/m².

Up to now, projects of pure high and low tide stream exploitation have not been put into practice due to still unsolved technical problems of the planned large-scale turbines. But first prototypes are in operation.

A.3 Further possibilities

Besides the discussed options, there exist further possibilities to exploit ocean energy for the purpose of power generation.

A.3.1 Thermal gradients

The major portion of the solar insulation is stored as heat in the atmosphere and in solid or liquid components of the earth's surface. About 20 % of the total radiated energy from the sun is converted into heat solely in tropical oceans. From a technical viewpoint it is possible to utilise this heat. Yet, due to the large water surfaces located within the equator belt, the theoretical potential of this option is relatively high.

Fig. A.8 shows the typical water temperature variations of equatorial oceans in relation to water depth. According to this illustration, the temperature varies roughly between 22 and 28 °C (within the course of a year) within the water layer close to the surface. The temperature of the deeper layers remains more or less constant over the entire year and is relatively low when compared to the water temperature at the surface.

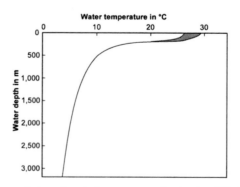

Fig. A.8 Temperature variations in relation to ocean depth, simplified illustration (according to /A-2/)

In principle, this type of thermal energy could be used for power generation by means of open or closed Rankine processes (ORC processes). This cycle is based on the temperature difference between the warm surface water of a maximum of 22 to approximately 28 °C and the cold deep waters of approximately 4 to 7 °C (Fig. A.9). The efficiency of such a cycle depends on the useful temperature difference. Due to the small available temperature differences of a maximum of about 20 K, these power plants can only achieve very low efficiencies that vary between 1 and a maximum of 3 %. In addition, for this electricity generation method enormous volumes of water need to be circulated. Furthermore, the water needs to be transferred from deep sea layers to the ocean surface and vice-versa. The design of these plants is thus very expensive.

To keep energy transfer to the customer to a minimum, ocean thermal energy conversion plants are mainly built near the coastline. In parallel, an easy access to cold deep water is required to run such a cycle.

As the energy yield is relatively low when compared to the large mass or water flows to be transferred, this technology, generally referred to as OTEC (Ocean Thermal Energy Conversion) has, to date, not been applied due to economic aspects. Since there still remain unsolved technical problems with regard to the successful operation of such ocean thermal energy conversion plants, their application in the sunny parts of our planet cannot be expected within the years to come.

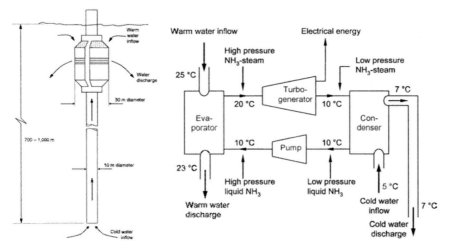

Fig. A.9 Example of an OTEC plant (left: layout; right: power plant cycle; according to /A-1/)

A variety of this technology suitable for geographical areas, where both cold ocean currents and geothermal energy are available, has only lately been introduced. This method benefits from the temperature differences between warm areas near volcanic structures and cold ocean water. However, it is unlikely that this method will be put into practice in the near future due to considerable technical problems.

A.3.2 Ocean currents

It is also possible to exploit ocean currents resulting from different temperatures in different areas. Such an utilisation is particularly sensible at straits where high current speeds are available.

For instance, the narrowest section of the Florida Current has a width of 80 km. At this point water throughput varies between 20 and 30 Mio. m³/s at an average speed of 0.9 m/s. Based on such an ocean current of an average width of 50 km, an average depth of 120 m and a speed of about 2 m/s within the current's core section, an electric power of 2,000 MW can be provided using appropriate converters /A-6/. However, the slow speed of ocean currents inevitably leads to low energy densities, which even within the current's core is below 2.2 kW/m². The corresponding conversion plants must thus be of considerable size. The principle of this renewable energy exploitation method corresponds to that of using the high and low tidal stream.

Since there are only little pressure differences but relatively high current speeds to be exploited, appropriate turbines need to be developed (similar to the rotors of wind power stations (Fig. A.7)), which perform satisfactorily under the above-mentioned conditions. The operating principle of most of the turbines designed for the application in ocean current energy conversion plants is similar to gigantic wind power stations, where wind is replaced by the ocean current. However, turbine blades only rotate twice or three times per minute, because of the low current speed. Due to the low energy density all proposals mentioned so far require turbines of a diameter of 150 m and more, which have never been built so far.

Contrary to the above concepts, already some time ago an "energy converter suitable for slow water currents" equipped with "sea parachutes" made of very stable material and arranged numerously on a circulating rope has been proposed /A-8/. The rope with the parachutes runs over a pulley wheel anchored on the seabed and coupled to a generator. The parachutes, whose cap diameters may amount to 100 m, are inflated and driven by the ocean current. When exceeding the apex of the rope, e.g. after 18 km referring to a design study, the parachutes shut down due to the lack of impact pressure and are pulled in this position against the current to the pulley wheel. The parachutes need to be locked in closed position by a corresponding device, otherwise they would open during retraction, due the current jam, inside the parachute and prevent the function of the overall system. After having passed the pulley wheel, the parachutes are again inflated within the ocean current and the described process starts over again. To date, there is still no pilot plant and it is questionable whether this concept will ever gain importance in the marketplace due to fundamental technical weak points.

To harness the comparatively high ocean current speed of the Strait of Messina (strait between Southern Italy and Sicily), between Ganzirri and Punta Pezzo, the proposal illustrated in Fig. A.10 has been elaborated. Altogether 100 turbines of the shown type need to be installed approximately 100 m below sea level, where the current reaches its maximum speed. However, also for this method it is questionable whether the technical challenges can be met.

Due to economic and environmental reasons, it is doubtful whether ocean current energy conversion plants will ever be commissioned for operation on a large scale. On the one hand, available project studies have shown very high investment costs and considerable technical challenges. On the other hand, a high-volume ocean current energy conversion plant may have a considerable impact on the environment, for instance, the respective ocean current could be diverted; for example the European climate may change fundamentally by a diversion of the Gulf Stream. From the current viewpoint this type of renewable energy is unlikely to be ever applied for energy provision on a large scale.

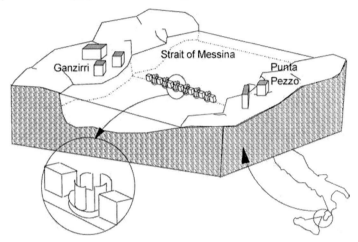

Fig. A.10 Conceptual study of an ocean current energy conversion plant designed for the Strait of Messina (according to /A-1/)

A.3.3 Salinity gradients

The global water cycle leads to the generation of large quantities of freshwater (Chapter 2). This freshwater is again mingled with salt water, as rivers flow into the sea, and therefore closes the global water cycle. Basically the separation of freshwater from salt water requires energy which is stored in the demixed water volumes. There are proposals to exploit the energy released at river estuaries where freshwater and salt water mix. Theoretically, chemical energy can be converted into potential energy (via the effect of osmotic pressure), which subsequently can be transferred into electricity by adapted hydropower stations. Despite the simple operating principle, this method is not technically feasible, since the required semi-permeable membranes are not available yet.

A.3.4 Water evaporation

Within the global water cycle water evaporates at sea level and rises to great heights, to condense to drops and precipitate (Chapter 2). This natural process can be imitated by energy generation systems by means of gigantic towers (so-called Mega-Power-Towers /A-9/) installed in the ocean. Other fluids, such as ammonia, replace the water. Such fluids evaporate at lower temperatures than water and need less thermal energy to maintain such a cycle. Within such an energy generation system during the evaporation e.g. ammonia absorbs thermal energy from the ocean and evaporates. Ammonia vapours rise inside the tower and condense at its top, due to the cold atmosphere in such heights. Liquid ammonia precipitates onto the floor and drives appropriate turbines in order to provide electricity.

The simplest design consists of a tower of a height of approximately 5 km, a basic diameter of 200 m and an ascending tube of a diameter of 50 m. This great elevation is necessary to achieve the sufficient temperature difference between bottom and top in order to start the cycle. At the bottom and top of the tower huge containers collect the precipitating or evaporated liquid.

In order to minimise the tower weight, it is advisable to use a plastic core covered with aluminium on both sides as construction material. Although such a construction would weigh up to 400,000 t it would only be submerged into the ocean by 1.5 m, since enormous hydrogen-filled tanks, similar to gas-filled balloons, would lift up the tower and keep it in upright position. By means of ropes of a thickness of 30 cm and a length of up to 8 km, the construction would be "secured" at three points.

Another variant consists of a tower, which is 7.5 km high. For this dimension, the diameters would amount to approximately 2,500 m at the bottom, to 750 m within the tube and to about 1,200 m at the plate-shaped condenser at the top.

Both designs are suitable to withstand heavy wind. Studies of both models have revealed divergences of 344 m, or respectively 57 m, at the top of the tower. This is due to the enormous weight of the plate-shaped condenser that stabilises the tube located below (see /A-9/).

Up to now, only conceptual studies have been conducted with regard to this method, and it is unlikely that this option will largely contribute to power provision in the near future.

Annex B: Energetic Use of Biomass

The term "biomass" describes organic (i.e. carbon containing) material. Biomass thus comprises
- natural phytoplankton and zooplankton (plants and animals),
- the resulting residues, by-products, and waste (e.g. animal excreta),
- dead (not yet fossil) phytoplankton and zooplankton (e.g. straw) and
- in a broader sense, all substances resulting from technical conversion and/or material utilisation of organic material (e.g. black liquor, paper and cellulose, abattoir waste, organic household garbage, vegetable oil, alcohol).

Delimiting biomass from fossil energy carriers starts with peat, namely the fossil secondary product of decomposition. According to this delimitation of terms, peat is not considered as biomass; in some countries (e.g. Sweden, Finland), however, peat is referred to as biomass.

Biomass is further divided into primary and secondary products /B-1/.
- Primary products arise from direct photosynthetic exploitation of solar energy and include the entire phytoplankton, e.g. agricultural and forestry products from energy plant cultivation (such as fast-growing trees, energy grass) or vegetable residues, by-products, and waste from agriculture, forestry, and the further processing industry (including straw, and residual wood from forests and industry).
- Secondary products, in contrast, only indirectly receive their energy from the sun; they are created by decomposition or conversion of organic matter in higher organisms (e.g. animals). They include, for instance, the entire zooplankton, its excrements (e.g. manure, solid waste), and sewage sludge.

B.1 Structure of a typical supply chain

An energy supply chain based on biomass encompasses all processes from cultivation of energy plants or provision of residues, by-products, or organic waste up to final energy supply (such as district heat or electrical energy). It thus covers the life-cycle of organic matters from production, i.e. primary energy, up to the provision of the corresponding useful energy (Fig. B.1).

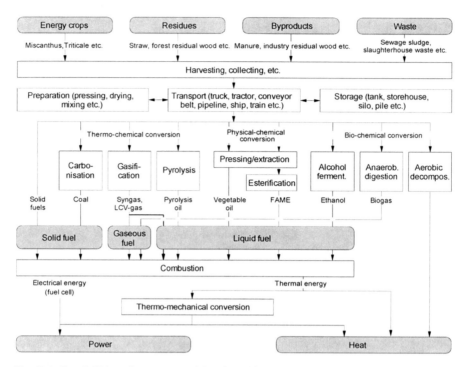

Fig. B.1 Possibilities of energy provision from biomass (grey shaded fields: energy carriers, not grey shaded fields: conversion processes; simplified presentation without considering light as useful energy; FAME fatty acid methyl ester; reactions that occur in fuel cells are regarded as "cold" combustion) (according to /B-1/)

The aim of such a biomass supply chain is to satisfy a possibly varying demand of final or useful energy and to provide the necessary conversion plants with the required quantity and quality of the respective organic material /B-1/.

Each supply chain consists of the life cycle sections biomass production respectively provision, conversion, utilisation and disposal. Generally, each section can be sub-divided into numerous individual processes. For instance, biomass production, among other things, requires seed bed preparation, application of fertilisers, and care. Since the different processes within an overall life cycle do not all occur in the same place the corresponding distances must be covered by appropriate transportation (e.g. trucks, tractors, pipelines).

In the end, a certain supply chain is determined by the framework conditions which in turn depend on biomass production (supply side) on the one hand and final energy provision (demand side) on the other hand. Further deciding factors are economic and technical (and administrative) framework conditions that have a significant effect with regard to putting a supply chain into practice. For instance, the choice of a certain conversion method is made, among other factors, by the final energy carrier to be provided (e.g. thermal energy, electrical energy), or the corresponding useful energy (e.g. heat, power) and – far and foremost – by the legal environmental regulations. Furthermore, a supply chain can be determined by the prescribed disposal of substances generated within the course of supply and/or utilisation (such as fermented manure remaining after biogas production, ashes remaining after the combustion of the solid biofuels). The different alternatives of disposing of residues, by-products, and waste and/or the applied conversion technology also require different biomass properties (such as size and shape, water content), which usually must be defined and ensured prior to conversion. It may thus be advantageous to first produce an appropriate secondary energy carrier (e.g. wooden pellets, wood chips, straw bales). Furthermore, biomass type (such as wood or herbaceous biomass), quality (e.g. water content, composition) and variations of energy demand and biomass supply with regard to seasonal differences are of significance. The above properties may result in different storage requirements; for instance, drying of biomass may be necessary to ensure storage stability. Moreover, the selected combination needs to be economically feasible under the given framework conditions, approvable, and socially acceptable /B-1/.

B.2 Conversion into final or useful energy

Within a supply chain available biomass can be processed and converted into the desired useful energy according to a great variety of methods and options /B-1/.

The easiest way is to burn lignocelluloses biomass directly within a furnace after mechanical preparation (such as chipping or pressing). However, for numerous other promising applications (such as fuel provision for car or truck engines or highly efficient power generation within a gas turbine) it is recommended or even required to first convert biomass into a liquid or gaseous secondary energy carrier. The actual conversion into final or useful energy is thus performed after one or several of the following biofuel properties have been specifically enhanced: energy density, handling, storage and transportation properties, environmental performance and energetic exploitation, potential to substitute fossil energy carriers, disposability of residues, by-products, and waste.

The available processes of converting organic matter into solid, liquid, or gaseous secondary energy carriers, performed prior to transformation into the desired final or useful energy, are generally divided into thermo-chemical, physical-chemical, and bio-chemical processes (Fig. B.1).

B.2.1 Thermo-chemical conversion

By thermo-chemical conversion processes (such as gasification, pyrolysis, and carbonisation) solid biofuels are transformed into solid, liquid, and/or gaseous secondary energy carriers primarily using heat /B-1/.

Gasification. Within a thermo-chemical gasification solid biofuels are preferably converted into a gaseous energy carrier. For this purpose an oxygen-containing gasification agent (such as air) is added under-stoichiometrically to convert e.g. the carbon of the biofuels into carbon monoxide, and thus into gaseous energy carriers. Simultaneously, the process heat required to run this process is provided by partial combustion of parts of the used solid biofuels. The produced low-caloric fuel gas is suitable for heat provision by means of burners as well as for power respectively a combined heat and power generation by gas engines or tur- bines respectively fuel cells. The produced gas may alternatively also be further converted into liquid or gaseous secondary energy carriers (e.g. methanol, Fischer-Tropsch-Diesel fuel, Bio-SNG) suitable for the use within the transporta- tion sector.

Pyrolysis. For pyrolysis solid biofuels are treated exclusively by the use of ther- mal energy with the goal to maximise the share of liquid products. Such pyrolysis processes are based on the pyrolytic decomposition of biomass at high tempera- tures in the absence of oxygen. Within this thermo-chemical conversion process the organic matter is transformed into gaseous products (e.g. carbon monoxide, carbon dioxide), liquid products (e.g. bio-oil), and solid products (e.g. charcoal). Assuming that the required technology is available, the resulting liquid secondary energy carriers could serve as a fuel for suitable furnaces, or respectively as fuel for power generation or combined heat and power provision in CHP-plants as well as for transportation purposes.

Carbonisation. Carbonisation refers to the thermo-chemical conversion of solid biofuels aiming at a maximum output of solid products (charcoal). Also for this process organic matter is thermally decomposed. The required process heat is of- ten provided by partial combustion of the used raw material (i.e. by thermal de- composition of released gaseous and liquid decomposition products). Carbonisation does thus not differ essentially from gasification or pyrolysis; how- ever, the conditions of this thermo-chemical conversion method are set to ensure a maximum output of solid reaction products. Subsequently, the generated carbon- ised biomass or charcoal serves primarily for heat provision by means of corre- sponding plants. Alternatively, it can also be used for other purposes besides en- ergy generation (e.g. as activated charcoal).

B.2.2 Physical-chemical conversion

Physical-chemical conversion includes all possibilities of a provision of energy carrier on the basis of oil seeds. For these processes biomass containing vegetable oil or fat serves as starting material (e.g. rape seeds, sunflower seeds, coconuts). First, the liquid oil phase is separated from the solid phase. Separation can, for in- stance, be performed by mechanical pressing (e.g. separating rape oil from solid rape residues, the so called rape cake). Alternatively, a separation of the oil is also possible based on extraction, using a solvent. The products remaining from this physical-chemical process are a mixture of the vegetable oil and the solvent as well as a mixture between the oil seed residues and the solvent. Once the solvent, which is used again within the process, has been removed the vegetable oil and the oil meal remains. Both processes are often combined. First, a mechanical pressing and then an extraction is performed. The obtained vegetable oil can be used as a fuel for engines and CHP-plants and is applied either in its pure form or after chemical conversion (i.e. transesterification) to Fatty Acid Methyl Ester (FAME)) /B-1/.

B.2.3 Bio-chemical conversion

Bio-chemical processes use micro-organisms or bacteria, and thus biological processes, to convert biomass into secondary energy carriers or useful energy /B-1/.

Alcohol fermentation. After appropriate preparation biomass containing sugar, starch, or cellulose can be decomposed into ethanol based on an alcoholic fermentation in an aqueous medium. The alcohol is removed from this slurry by distillation. Due the fact that ethanol and water form an azeotropic mixture pure alcohol is obtained by dehydration, subsequently using an entrainer. The resulting pure ethanol serves as a fuel in engines or CHP-plants for the provision of final or useful energy. In some countries ethanol is blended with low concentrations of conventional gasoline; such mixtures (like the so called E5-fuel) can be applied as a fuel for all vehicles equipped with an Otto engine.

Anaerobic digestion. By means of bacteria organic substances under anaerobic conditions (i.e. conversion in the absence of air) can be digested. One product of such an anaerobic digestion is a vapour-saturated gas mixture (biogas) consisting roughly of about 60 % of methane (CH_4) and 40 % of carbon dioxide (CO_2). This decomposition process occurs naturally e.g. at the bottom of lakes and is technically performed in biogas plants, sewage sludge plants, and land fills. The produced gas serves after gas cleaning as an energy carrier in gas burners or engines. After appropriate processing and compression, the gas mixture can in principle also be fed into natural gas pipelines and can provide the same duty as natural gas. Additionally, the use for transportation purpose is also possible (as it is true for natural gas).

Aerobic fermentation. For aerobic fermentation also biological processes are applied to decompose biomass in the presence of oxygen. The main oxidation product of this process is carbon dioxide (compost formation). Within this processes heat is released, which can, for instance, be tapped by heat pumps and be provided as low-temperature heat for covering a given heat demand. However, these possibilities have not yet been put into practice.

Annex C: Energy Units

Decimal prefixes

Atto	a	10^{-18}
Femto	f	10^{-15}
Piko	p	10^{-12}
Nano	n	10^{-9}
Mikro	μ	10^{-6}
Milli	m	10^{-3}
Zenti	c	10^{-2}
Dezi	d	10^{-1}
Deka	da	10^{1}
Hekto	h	10^{2}
Kilo	k	10^{3}
Mega	M	10^{6}
Giga	G	10^{9}
Tera	T	10^{12}
Peta	P	10^{15}
Exa	E	10^{18}

Conversion factors

	kJ	kWh	kg SKE	kg OE	m^3 gas
1 Kilo joule (kJ)		0.000278	0.000034	0.000024	0.000032
1 Kilo watt hour (kWh)	3,600		0.123	0.086	0.113
1 kg hard coal unit (SKE)	29,308	8.14		0.7	0.923
1 kg oil equivalent (OE)	41,868	11.63	1.486		1.319
1 m^3 natural gas	31,736	8.816	1.083	0.758	

The conversion factors refer to the net calorific value.

References

/1-1/   Kaltschmitt, M.: Renewable Energies; Lessons; Institute for Environmental Tech-
         nology and Energy Economics, Hamburg University of Technology, Summer Term
         2006 and Winter Term 2006/07
/1-2/   Hulpke, H. u. a. (Hrsg.): Römpp Umwelt Lexikon; Georg Thieme, Stuttgart, New
         York, Germany, USA, 2000, 2. Auflage
/1-3/   BP (Hrsg.): BP Statistical Review of World Energy 2005; BP, London, UK, June
         2006 (www.bp.com)
/1-4/   Kaltschmitt, M.; Hartmann, H. (Hrsg.): Energie aus Biomasse; Springer, Berlin, Hei-
         delberg, Germany, 2001
/1-5/   Wöhe, G.: Einführung in die Allgemeine Betriebswirtschaftslehre; Franz Vahlen,
         München, Germany, 2005, 22. Auflage

/2-1/   Aschwanden, M.: Physics of the Solar Corona; Springer, Berlin, Heidelberg, New
         York, Germany, 2004
/2-2/   Flemming, G.: Einführung in die angewandte Meteorologie; Akademie, Berlin,
         Germany, 1991
/2-3/   Liljequist, G. H.; Cehak, K.: Allgemeine Meteorologie; Springer, Berlin, Heidelberg,
         Germany, 1984
/2-4/   Kaltschmitt, M.; Huenges, E.; Wolff, H. (Hrsg.): Energie aus Erdwärme; Deutscher
         Verlag für Grundstoffindustrie, Stuttgart, Germany, 1999
/2-5/   Malberg, H.: Meteorologie und Klimatologie; Springer, Berlin, Heidelberg, New
         York, Germany, 2006, 4. Auflage
/2-6/   Häckel, H.: Meteorologie; UTB, Stuttgart, Germany, 2005, 5. Auflage
/2-7/   Kaltschmitt, M.: Renewable Energies; Lessons, Institute for Environmental Tech-
         nology and Energy Economics, Hamburg University of Technology, Summer Term
         2006 and Winter Term 2006/07
/2-8/   Duffie; J. A.; Beckman, W. A.: Solar Engineering of Thermal Processes; John Wiley
         & Sons, New York, Brisbane, USA, 1991, 2nd edition
/2-9/   Liu, B. Y. H.; Jordan, R. C.: The Interrelationship and Characteristic Distribution of
         Direct, Diffuse and Total Solar Radiation; Solar Energy 4 (1960), 3, S. 1 - 19
/2-10/   Taylor, P. A.; Teunissen, H. W.: The Askervein Project: Overview and Background
         Data; Boundary-Layer Meteorology 39(1987), S. 15 – 39
/2-11/   Intergovernmental Panel on Climate Change, IPCC Data Distribution Centre,
         http://ipcc-ddc.cru.uea.ac.uk/

/2-12/   DWD (Hrsg.): Deutsches Meteorologisches Jahrbuch; Deutscher Wetterdienst, Offenbach a. M., Germany, verschiedene Jahrgänge

/2-13/   Neubarth, J.; Kaltschmitt, M. (Hrsg.): Regenerative Energien in Österreich – Systemtechnik, Potenziale, Wirtschaftlichkeit, Umweltaspekte; Springer, Wien, Austria, 2000

/2-14/   Streicher, W.: Sonnenenergienutzung; Vorlesungsskriptum; Institut für Wärmetechnik, Technische Universität Graz, Austria, 2005

/2-15/   Christoffer, J.; Ulbricht-Eissing, M.: Die bodennahen Windverhältnisse in der Bundesrepublik Deutschland; Berichte des Deutschen Wetterdienstes Nr. 147; Selbstverlag des Deutschen Wetterdienstes, Offenbach a. M., Germany, 1989

/2-16/   Möller, F.: Einführung in die Meteorologie; Band 2; Bibliographisches Institut, Mannheim, Germany, 1973

/2-17/   Hellmann, G.: Über die Bewegung der Luft in den untersten Schichten der Atmosphäre; Meteorologische Zeitschrift 32(1915), 1

/2-18/   Hsu, S. A.: Coastal Meteorology; Academic Press, London, UK, 1988

/2-19/   Etling, D.: Theoretische Meteorologie; Vieweg & Sohn, Braunschweig/Wiesbaden, Germany, 1996

/2-20/   Tangermann-Dlugi, G.: Numerische Simulationen atmosphärischer Grenzschichtströmungen über langgestreckten mesoskaligen Hügelketten bei neutraler thermischer Schichtung; Wissenschaftliche Berichte des Meteorologischen Institutes der Universität Karlsruhe Nr. 2, Karlsruhe, Germany, 1982

/2-21/   DWD (Hrsg.): Karte der Windgeschwindigkeitsverteilung in der Bundesrepublik Deutschland; Deutscher Wetterdienst, Offenbach a. M., Germany, 2001

/2-22/   Troen, I.; Petersen, E. L.: European Wind Atlas; Risø National Laboratory, Roskilde, Denmark, 1989

/2-23/   Vischer, D.; Huber, H.: Wasserbau; Springer, Berlin, Heidelberg, Germany, 2002, 6. Auflage

/2-24/   DWD (Hrsg.): Leitfaden für die Ausbildung im deutschen Wetterdienst; Nr. 1: Allgemeine Meteorologie; Deutscher Wetterdienst, Offenbach a. M., Germany, 1987

/2-25/   Bundesanstalt für Gewässerkunde (Hrsg.): Deutsches Gewässerkundliches Jahrbuch; Bundesanstalt für Gewässerkunde, Bonn, Germany, verschiedene Jahrgänge

/2-26/   Cralle, H. T.; Vietor, D. M.: Solar Energy and Biomass; in: Kitani, O.; Hall, C. W. (Hrsg.): Biomass Handbook; Gordon and Breach Saina Publishers, New York, USA, 1989

/2-27/   Lerch, G.: Pflanzenökologie; Akademie, Berlin, Germany, 1991

/2-28/   Kaltschmitt, M.; Hartmann, H. (Hrsg.): Energie aus Biomasse; Springer, Berlin, Heidelberg, Germany, 2001

/2-29/   Knauer, N.: Grundlagen der Futterproduktion auf Weidegrünland; Schriftenreihe der Landwirtschaftlichen Fakultät der Universität Kiel, Germany, Heft 47, 1970

/2-30/   Lieth, H.: Phenology and Seasonality Modelling; Ecol Studies 8, Heidelberg, Germany, 1974

/2-31/   Larcher, W.: Ökophysiologie der Pflanzen; UTB, Stuttgart, Germany, 2001, 6. Auflage

/2-32/   Ludlow, M. M.; Wilson, G. L.: Photosynthesis of Tropical Pasture Plants, II; Illmi-
nance, Carbon Dioxide Concentration, Leaf Temperature and Leaf Air Pressure Difference; Australien Journal of Biological Science 24 (1971), S. 449 - 470

/2-33/   Brehm, D. R. u. a.: Ergebnisse von Temperaturmessungen im oberflächennahen Erdreich; Zeitschrift für angewandte Geowissenschaften 15 (1989), 8, S. 61 - 72

/2-34/   Haenel, R. (Hrsg.): Atlas of Subsurface Temperatures in the European Community;
Th. Schäfer, Hannover, Germany, 1980

/2-35/   Hurtig, E. u. a. (Hrsg.): Geothermal Atlas of Europe; Geographisch-Kartographische
Anstalt, Gotha, Germany, 1992

/2-36/   Haenel, R.; Staroste, E. (Hrsg.): Atlas of Geothermal Resources in the European
Community, Austria and Switzerland; Th. Schäfer, Hannover, Germany, 1988

/2-37/   Strasburger, E.: Lehrbuch der Botanik; Gustav Fischer, Stuttgart, New York, 1983,
32. Auflage

/2-38/   BP (Hrsg.): BP Statistical Review of World Energy 2005; BP, London, UK, June
2006 (www.bp.com)

/3-1/   Hahne, E.; Drück, H.; Fischer, S.; Müller-Steinhagen, H.: Solartechnik (Teil 2), Lessons; Institut für Thermodynamik und Wärmetechnik (ITW), Universität Stuttgart,
Germany, 2003

/3-2/   EN 410: Glas im Bauwesen – Bestimmung der lichttechnischen und strahlungsphy-
sikalischen Kenngrößen von Verglasungen; Beuth-Verlag, Berlin, Germany, 2004

/3-3/   EN 13790: Wärmetechnisches Verhalten von Gebäuden, Berechnung des Heizener-
giebedarfs, Wohngebäude; Beuth-Verlag, Berlin, Germany, 2004

/3-4/   Platzer, W.: Eigenschaften von transparenten Wärmedämmmaterialien; Tagungsband: Transparente Wärmedämmung, Arbeitsgemeinschaft Erneuerbare Energie,
Gleisdorf, Austria, 1995

/3-5/   Treberspurg, M.: Neues Bauen mit der Sonne; Springer, Wien, Austria, 1999, 2.
Auflage

/3-6/   Feist, W.: Das Niedrigenergiehaus - Neuer Standard für energiebewusstes Bauen;
C.F. Müller, Heidelberg, Germany, 2006, 6. Auflage

/3-7/   Heimrath, R.: Dynamische Simulation 'der Betonkernkühlung mit Hilfe eines Erd-
wärmetauschers; Report; Institut für Wärmetechnik; Technische Universität Graz,
2000

/3-8/   Kerschberger, A.: Transparente Wärmedämmung zur Gebäudeheizung: Systemaus-
bildung, Wirtschaftlichkeit, Perspektiven; Institut für Bauökonomie, Universität
Stuttgart, Bauök-Papiere Nr. 56, Stuttgart, Germany, 1994

/3-9/   Streicher, W.: Informatik in der Energie- und Umwelttechnik; Lessons; Institut für
Wärmetechnik, Technische Universität Graz, Austria, 2004

/3-10/   Streicher, W.: Sonnenenergienutzung; Lessons; Institut für Wärmetechnik, Technische Universität Graz, Austria, 2005

/3-11/   Heimrath, R.: Sensitivitätsanalyse einer Wärmeversorgung mit Erdreich-
Direktverdampfungs-Wärmepumpen; Diplomarbeit, Institut für Wärmetechnik,
Technische Universität Graz, Austria, 1998

/4-1/   Duffie; J. A.; Beckman, W. A.: Solar Engineering of Thermal Processes; John Wiley & Sons, New York, Brisbane, Australia, 1991

/4-2/   Ladener, H.; Späte, F.: Solaranlagen – Handbuch der thermischen Solarenergienutzung; Ökobuch, Staufen, Germany, 2003, 8. Auflage

/4-3/   Themeßl, A.; Weiß, W.: Solaranlagen Selbstbau – Planung und Bau von Solaranlagen – Ein Leitfaden; Ökobuch, Staufen, Germany, 2004, 4. Auflage

/4-4/   Fink, C.; Riva, R.; Pertl, M.: Umsetzungsstandard von solarunterstützten Wärmenetzen im Geschosswohnbau – Technik, Messergebnisse, Wirtschaftlichkeit; Otti Technologie Kolleg (Hrsg.): 16. Symposium Thermische Solarenergie, Kloster Banz, Staffelstein, 2006, S. 136– 141

/4-5/   VDI-Gesellschaft Verfahrenstechnik und Chemieingenieurwesen (Hrsg.): VDI Wärmeatlas; Springer, Berlin, Heidelberg, Germany, 2006, 10. Auflage

/4-6/   Streicher, W.: Sonnenenergienutzung; Skriptum, Institut für Wärmetechnik, Technische Universität Graz, Austria, 2005

/4-7/   Mittelbach, W.: Sorptionsspeicher – Neue Perspektiven für Solare Raumheizung; Tagung "Gleisdorf Solar 2000", Gleisdorf, Austria, 2000

/4-8/   Holter, C.; Streicher, W.: New Solar Control without External Sensors; ISES World Solar Conference, Gothenburg, Sweden, 2003

/4-9/   Streicher, W.: Minimizing the Risk of Water Hammer and Other Problems at the Beginning of Stagnation of Solar Thermal Plants – a Theoretical Approach; Solar Energy, Volume 69, Number 1-6, 2001

/4-10/   Clairent (Hrsg.): Stoffwerteprogramm von Antifrogenen; Clairent GmbH, Frankfurt, Germany, 1998

/4-11/   Peuser, F. A.: Zur Planung von Solarkollektoranlagen und zur Dimensionierung der Systemkomponenten; Wärmetechnik 31(1986), 7, 10, 12 und 32(1987), 2

/4-12/   Arbeitsgemeinschaft Erneuerbare Energie (Hrsg.): Heizen mit der Sonne; Gleisdorf, Austria, 1997

/4-13/   Frei, U.; Vogelsanger, P.: Solar thermal systems for domestic hot water and space heating; SPF Institut für Solartechnik, Prüfung, Forschung, Rapperswil, Switzerland, 1998

/4-14/   VDI (Hrsg.): VDI-Richtlinie 2067: Wirtschaftlichkeitsberechnungen von Wärmeverbrauchsanlagen, Blatt 4: Brauchwassererwärmung; VDI, Düsseldorf, Germany, 1974

/4-15/   Hertle, H.: Passivhaus schlägt Solarenergie? Evaluation des Solar- und Energiepreises Pforzheim/Enzkreis; Otti Technologie Kolleg (Hrsg.): 16. Symposium Thermische Solarenergie, Kloster Banz, Staffelstein, Germany, 2006, S. 256 – 261

/4-16/   Lange, F.; Keilholz, C.: Anlagenpreise contra Qualität? Wertschöpfung und Kostensenkung im Zeitspiegel; Otti Technologie Kolleg (Hrsg.): 16. Symposium Thermische Solarenergie, Kloster Banz, Staffelstein, Germany, 2006, S. 522 – 527

/4-17/ Streicher, W.; Oberleitner, W.: Betriebsergebnisse der größten Solaranlage Österreichs, Solarunterstütztes Biomasse-Nahwärmenetz Eibiswald; Otti Technologie Kolleg (Hrsg.): 9. Symposium Thermische Solarenergie, Kloster Banz, Staffelstein, Germany, 1999

/4-18/ Fink, C.; Heimrath, R.; Riva, R.: Solarunterstützte Wärmenetze; Teil Thermische Solaranlagen für Mehrfamilienhäuser; Report, Institut für Wärmetechnik, TU Graz, Austria, 2002 (www.hausderzukunft.at)

/4-19/ Thür, A.: Sonnige Herbergen, Markteinführung von Solaranlagen in Beherbergungsbetrieben; Erneuerbare Energie, 1/97, 1997

/4-20/ Themessl, A.; Kogler, R.; Reiter, H.: Erneuerbare Energie für die Stadt Villach; Arbeitsgemeinschaft Erneuerbare Energie, Gleisdorf, Austria, 1995

/4-21/ Vakuum-Beschichtungen von Solarabsorbern – Dünne Schichten, die es in sich haben; Sonnenenergie (2000), 6, S. 20 – 23

/4-22/ Markenzeichen: Dunkelblau; Sonne, Wind und Wärme 24(2000), 1, S. 18

/5-1/ Rabl, A.: Active Solar Collectors and Their Applications; Oxford University Press, UK, 1985

/5-2/ Baehr, H. D.: Thermodynamik – Grundlagen und technische Anwendungen; Springer, Berlin; Heidelberg, Germany, 2006, 13. Auflage

/5-3/ Weinrebe, G.: Technische, ökologische und ökonomische Analyse von solarthermischen Turmkraftwerken, IER Forschungsbericht 68, Institut für Energiewirtschaft und Rationelle Energieanwendung, Universität Stuttgart, Germany, 2000

/5-4/ Becker, M.; Klimas, P. (Hrsg.): Second Generation Central Receiver Technologies – A Status Report; C. F. Müller, Karlsruhe, Germany, 1993

/5-5/ Winter, C.-J. u. a.: Solar Power Plants; Springer, Berlin, Heidelberg, Germany, 1991

/5-6/ Pacheco, J. E.: Results of Molten Salt Panel and Component Experiments for Solar Central Receivers (SAND 94-2525); Sandia National Laboratories, Albuquerque, New Mexico, USA, 1995

/5-7/ Sánchez, M. et al: Receptor Avanzado de Sales (RAS) – Setup, Test Campaign and Operational Experiences Final Report of a 0.5 MWth Molten Salt Receiver at the Plataforma Solar de Almería, Ref. PSA-TR 02/97; CIEMAT-PSA, Tabernas, Spain

/5-8/ Becker, M.; Böhmer, M. (Hrsg.): Solar Thermal Concentrating Technologies; C.F. Müller, Heidelberg, Germany, 1997

/5-9/ Buck, R.; Bräuning, T; Denk, T; Pfänder, M.; Schwarbözl, P.; Tellez, F.: Solar-Hybrid Gas Turbine-based Power Tower System (REFOS); Journal of Solar Energy Engineering 124(2002), 3

/5-10/ Fesharaki, M.: Industrielle Anwendungen inverser Gasturbinenprozesse – Biomasse – Sonnenenergie und Industrielle Abgase; Dissertation, Institut für Thermische Turbomaschinen, Technische Universität Graz, Austria, 1997

/5-11/ Tyner, C. u. a.: Solar Power Tower Development: Recent Experiences; 8th International Symposium on Solar Thermal Concentrating Technologies, DLR, Köln, Germany, 1996, Proceedings

/5-12/   Steinmüller (Hrsg.): PHOEBUS Solar Power Tower; Steinmüller, Gummersbach, Germany, 1995

/5-13/   Haeger, M. u. a.: Operational Experiences with the Experimental Set-Up of a 2,5 MW$_{th}$ Volumetric Air Receiver (TSA) at the Plataforma Solar de Almeria; PSA Internal Report; CIEMAT, Almeria, Spain, 1994

/5-14/   Romero M. et al.: Design and Implementation Plan of a 10 MW Solar Tower Power Plant Based on Volumetric-Air Technology in Seville (Spain); in: Pacheco, J. E.; Thornbloom, M. D. (eds): Proceedings of Solar 2000: Solar Powers Life, Share the Energy; ASME, New York, USA, 2000

/5-15/   www.solarpaces.org (Mai 2006)

/5-16/   Schiel, W. et al: Collector Development for parabolic trough power plants at Schlaich Bergermann und Partner; 13th International Symposium on Concentrating Solar Power and Chemical Energy Technologies, June 2006, Seville, Spain

/5-17/   www.solarheatpower.com (Mai 2006)

/5-18/   www.schott.com (Mai 2006)

/5-19/   Cohen, G. u. a.: Recent Improvements and Performance Experience at the Kramer Junction SEGS Plants; ASME Solar Energy Conference, San Antonio, Texas, USA, 1996; Proceedings, S. 479 – 485

/5-20/   Keck, T. et al: Eurodish - Continuous operation, system improvement and reference units; 13th International Symposium on Concentrating Solar Power and Chemical Energy Technologies, June 2006, Seville, Spain

/5-21/   Stine, W. B.; Diver, R. E.: A Compendium of Solar Dish/Stirling Technology (SAND 93-7027); Sandia National Laboratories, Albuquerque, New Mexico, USA, 1994

/5-22/   Laing, D.; Goebel, O.: Natrium Heat Pipe Receiver der 2. Generation für ein 9 kW$_{el}$ Dish/Stirling System; 9. Internationales Sonnenforum, Stuttgart, Germany, 1994, Tagungsband

/5-23/   Walker, G.: Stirling Engines; Clarendon, Oxford, UK, 1980

/5-24/   Werdich, M.; Kübler, K.: Stirling-Maschinen: Grundlagen, Technik, Anwendung; Ökobuch, Stauffen, Germany, 2003, 9. Auflage

/5-25/   Mancini, T. et al: Dish-Stirling Systems: An Overview of Development and Status; Journal of Solar Energy Engineering, Vol. 125, May 2003

/5-26/   Günther, H.: In hundert Jahren – Die künftige Energieversorgung der Welt; Franckh'sche Verlagshandlung, Stuttgart, Germany, 1931

/5-27/   Schlaich, J.; Schiel, W.; Friedrich, K.; Schwarz, G.; Wehowsky, P.; Meinecke, W.; Kiera, M.: Aufwindkraftwerk, Übertragbarkeit der Ergebnisse von Manzanares auf größere Anlagen; Report (BMFT-Förderkennzeichen 0324249D), Schlaich, Bergermann und Partner, Stuttgart, Germany, 1990

/5-28/   Schlaich, J.; Bergermann, R.; Schiel, W.; Weinrebe, G.: The Solar Updraft Tower - An Affordable and Inexhaustible Global Source of Energy; Bauwerk-Verlag, Berlin, Germany, 2004

/5-29/ Gannon, A.J.; van Backström, T.W.: Solar Tower Cycle Analysis with System Loss and Solar Collector Performance; in: Pacheco, J. E.; Thornbloom, M. D. (eds): Proceedings of Solar 2000: Solar Powers Life, Share the Energy; ASME, New York, USA, 2000

/5-30/ Dos Santos Bernardes, M.A.; Voß, A.; Weinrebe, G.: Thermal and technical analysis of solar chimneys; Solar Energy, Vol. 75 (2003), 6, S. 511 – 524, Elsevier, New York, USA

/5-31/ Haaf, W.; Friedrich, K.; Mayr, G.; Schlaich, J.: Solar Chimneys, Part I: Principle and Construction of the Pilot Plant in Manzanares; Solar Energy 2(1983), S. 3 – 20

/5-32/ Haaf, W.; Lautenschlager, H.; Friedrich, K.: Aufwindkraftwerk Manzanares über zwei Jahre in Betrieb; Sonnenenergie 1(1985), S. 11 – 17

/5-33/ Weinrebe, G.: Solar Chimney Simulation; Proceedings of the IEA SolarPACES Task III Simulation of Solar Thermal Power Systems Workshop, Cologne, Germany, September 2000

/5-34/ Kumar, A.; Kishore, V.V.N.: Construction and operational experience of a 6,000 m² solar pond at Kutch, India; Solar Energy 65(4), S. 237 – 249 (1999)

/5-35/ www.ormat.com (September 2006)

/5-36/ www.rmit.edu.au/news (August 2001)

/5-37/ Xu, H. (Ed.): Salinity Gradient Solar Ponds, Practical Manual Part I: Solar Pond Design and Construction, El Paso Solar Pond Project, 1993

/5-38/ Schiel, W.; Keck, T.: Dish/Stirling-Anlagen zur solaren Stromerzeugung; BWK 53(2001), 3, S. 60

/5-39/ Budgetary Cost Estimate for a 10 kW Dish/Stirling System, Schlaich Bergermann und Partner, Stuttgart, 2006

/5-40/ Goebel, O.: Direct Solar Steam Generation in Parabolic Troughs (DISS) - Update on Project Status and Future Planning; PowerGen ′99, June 1999, Frankfurt, Germany

/5-41/ Tabor, H.: Solar Ponds; The Scientific Research Foundation, Jerusalem, Israel, 1981

/6-1/ Meissner, D. (Hrsg.): Solarzellen - Physikalische Grundlagen und Anwendungen in der Photovoltaik; Vieweg, Braunschweig/Wiesbaden, Germany, 1993

/6-2/ Schmid, J. (Hrsg.): Photovoltaik - Strom aus der Sonne; C. F. Müller, Karlsruhe, Germany, 1993, 3. Auflage

/6-3/ Shockley, W.: Electrons and Holes in Semiconductors; D. Van Nostrand, Princeton, New York, USA, 1950

/6-4/ Sze, S. M.: Physics of Semiconductor Devices; J. Wiley & Sons, New York, USA, 1981

/6-5/ Fonash, S. J.: Solar Cell Device Physics; Academic Press, New York, USA, 1981

/6-6/ Coutts, T. J.; Meakin, J. D. (Hrsg.): Current Topics in Photovoltaics; Academic Press, London, UK, 1985

/6-7/ Goetzberger, A.; Voß, B.; Knobloch, J.: Sonnenenergie: Photovoltaik; Teubner, Stuttgart, Germany 1997

/6-8/ Green, M. A.: High Efficiency Silicon Solar Cells; Trans Tech Publications, Aedermannsdorf, Switzerland, 1987

/6-9/   Luque, A.; Hegedus, S. (eds.): Handbook of Photovoltaic Sciences and Engineering;
          J. Wiley & Sons, New York, USA, 2003

/6-10/   Köthe, H. K.: Stromversorgung mit Solarzellen; Franzis, München, Germany, 1991,
          2. Auflage

/6-11/   Kaltschmitt, M.: Renewable Energy; Lessons; Institute for Environmental Technol-
          ogy and Energy Economics, Hamburg University of Technology, Summer Term
          2006

/6-12/   Henry, C. J.: J. Appl. Phys. 51, 4494 (1980)

/6-13/   Raicu, A.; Heidler, K.; Kleiß, G.; Bücher, K.: Realistic reporting conditions for site-
          independent energy rating of PV devices, 11th EC Photovoltaic Solar Energy Con-
          ference, Montreux, Canada, 1992

/6-14/   Archer, M. D.; Hill, R.: Clean Electricity from Photovoltaics; Imperial College
          Press, London, UK, 2001

/6-15/   Green, M. A.: Silicon Solar Cells: Advanced Principles and Practice; Bridge
          Printery, Sydney, Australia, 1995

/6-16/   Green, M. A.; Emery, K.; King, D. L.; Igari, S.; Warta, W.: Solar Cell Efficiency
          Tables (Version 19), Progr. in Photovoltaic Research & Applications 10(2002), 55

/6-17/   Shaped Crystal Growth 1986; J. Crystal Growth 82(1987)

/6-18/   Chao, C.; Bell, R. O.: Effect of Solar Cell Processing on the Quality of EFG Nona-
          gon Growth; 19th IEEE Photovoltaic Spec. Conference, New Orleans, USA, 1987;
          Proceedings

/6-19/   Taguchi, M.; Kawamoto, K.; Tsuge, S.; Baba, T.; Sataka, H.; Morizane, M.; Uchi-
          hasi, K.; Nakamura, N.; Kyiama, S.; Oota, O.: HITTM Cells – High Efficiency Crys-
          talline Si Cells with novel structure, Progr. in Photovoltaic Research & Applications
          8(2000), 503

/6-20/   Stäbler, D. L.; Wronski, C.: Optically induced conductivity changes in discharge-
          produced amorphous silicon; J. Appl. Phys. 51(1980), 3262

/6-21/   Rau, U.; Schock, H. W.: Cu(In,Ga)Se$_2$ Solar Cells; in: Archer, M. D.; Hill, R.: Clean
          Electricity from Photovoltaics; Imperial College Press, London, UK, 2001

/6-22/   Bonnet, D.: Cadmium Telluride Solar Cells; in: Archer, M. D.; Hill, R.: Clean Elec-
          tricity from Photovoltaics; Imperial College Press, London, UK, 2001

/6-23/   Keppner, H.; Meier, J.; Torres, P.; Fischer, D.; Shah, A.: Microcrystalline Silicon
          and Micromorph Tandem Solar Cells; Applied Physics A (Materials Science Proc-
          essing) A69, 169(1999)

/6-24/   Catchpole, K. R.; McCann, M. J.; Weber, K. J.; Blakers, A. W.: A Review of Thin-
          Film Crystalline Silicon for Solar Cell Applications. II. Foreign Substrates; Solar
          Energy Materials and Solar Cells 68(2001), 173

/6-25/   Brendel, R.: Review of Layer Transfer Processes for Crystalline Thin-Film Silicon
          Solar Cells; J. Apppl. Phys. 40(2001), 4431

/6-26/   Werner, J. H.; Dassow, R.; Rinke, T. J.; Köhler, J. R.; Bergmann, R. B.: From Poly-
          crystalline to Single Crystalline Silicon on Glass; Thin Solid Films 383(2001), 95 –
          100

/6-27/   Luque, A.: Solar Cells and Optics for Photovoltaic Concentration; Adam Hilger, Bristol and Philadelphia, USA, 1989

/6-28/   Hinsch, A.; Kroon, J. M.; Kern, R.; Uhlendorf, I.; Holzbock, J.; Meyer, A.; Ferber, J.: Long-term Stability of Dye-sensitised Solar Cells; Progr. in Photovoltaic Research & Applications 9(2001), 425

/6-29/   Roth, W. (Hrsg.): Netzgekoppelte Photovoltaik-Anlagen; Fraunhofer-Institut für Solare Energiesysteme, Freiburg, Germany, Juni 2001

/6-30/   Roth, W. (Hrsg.): Dezentrale Stromversorgung mit Photovoltaik; Seminarhandbuch Fraunhofer-Institut für Solare Energiesysteme, Freiburg, Germany, 2002

/6-31/   Sauer, D.U.; Kaiser, R.: Der Einfluss baulicher und meteorologischer Bedingungen auf die Temperatur des Solargenerators – Analyse und Simulation; 9. Symposium Photovoltaische Solarenergie, Staffelstein, Germany, 1994, S. 485 – 491

/6-32/   Götzberger, A.; Stahl, W.: Global Estimation of Available Solar Radiation And Costs of Energy for Tracking And Non-Tracking PV-Systems; Proceedings, 18th IEEE Photovoltaic Specialists Conference, Las Vegas, Nevada, USA, October 1985

/6-33/   Jossen, A. (Hrsg.): Wiederaufladbare Batterien – Schwerpunkt: stationäre Systeme; Seminarband OTTI-Technologiekolleg, Ulm, Germany, 2004

/6-34/   Wagner, R.; Sauer, D.U.: Charge strategies for valve-regulated lead/acid batteries in solar power applications, J. Power Sources 95(2001), S. 141 – 152

/6-35/   Hartmann, H.; Kaltschmitt, M. (Hrsg.): Biomasse als erneuerbarer Energieträger – Eine technische, ökologische und ökonomische Analyse im Kontext der übrigen erneuerbaren Energien; Schriftenreihe „Nachwachsende Rohstoffe", Band 3, Landwirtschaftsverlag, Münster-Hiltrup, Germany, 2002, vollständige Neubearbeitung

/6-36/   NN: Kostendaten zu Photovoltaiksystemen bzw. Systemkomponenten; Photon Special 2005, Solar Verlag, Aachen, 2005

/6-37/   Fachinformationszentrum Karlsruhe (Hrsg.): Photovoltaikanlagen – Untersuchungen zur Umweltverträglichkeit; BINE Projekt Info Nr. 6, September 1998

/6-38/   Bernreuter, J.: Strom von der grünen Wiese; Photon 6 (2001), 2, S. 28 – 32

/6-39/   Diefenbach G.: Photovoltaikanlagen als Naturschutzzonen - Landnutzung in Harmonie mit der Natur bei zentralen Photovoltaikanlagen am Beispiel der 340 kW Anlage in Kobern-Gondorf; Energiewirtschaftliche Tagesfragen 44(1994), S. 41 – 64

/6-40/   Degner, A. u. a.: Elektromagnetische Verträglichkeit und Sicherheitsdesign für photovoltaische Systeme - Das europäische Verbundprojekt ESDEPS; Ostbayrisches Technologie-Kolleg (OTTI), Regensburg (Hrsg.): 14. Symposium Photovoltaische Solarenergie. Staffelstein, Germany, 1999; S. 425 – 429

/6-41/   Moskowitz, P. D.; Fthenakis, V. M.: Toxic Material Release from PV-Modules during Fires; Brookhaven National Laboratory, Upton, New York, 1993

/6-42/   Möller, J.; Heinemann, D.; Wolters, D.: Integrierte Betrachtung der Umweltauswirkungen von Photovoltaik-Technologien; Ostbayrisches Technologie-Kolleg (OTTI), Regensburg (Hrsg.): 13. Symposium Photovoltaische Solarenergie. Staffelstein, Germany, 1998. S. 549 – 553

/6-43/   Moskowitz, P. D.; Fthenakis, V. M.: Toxic Materials Released from PV-Modules during Fires; Brookhaven National Laboratory, Upton, New York, USA, 1993

528   References

/6-44/   Kleemann, M.; Meliß, M.: Regenerative Energiequellen; Springer, Berlin, Heidelberg, 1993, 2. Auflage

/7-1/   Kaltschmitt, M.: Renewable Energies; Lessons; Institute for Environmental Technology and Energy Economics, Hamburg University of Technology, Summer Term 2006

/7-2/   Betz, A.: Das Maximum der theoretischen Ausnutzung des Windes durch Windmotoren; Zeitschrift für das gesamte Turbinenwesen, 20. 9. 1920

/7-3/   Hau, E.: Windkraftanalagen; Springer, Berlin, Heidelberg, Germany, 2002

/7-4/   Gasch, R.; Twele, J.: Windkraftanlagen; Teubner, Stuttgart, Germany, 2005, 4. Auflage

/7-5/   Molly, J. P.: Windenergie - Theorie, Anwendung, Messung; C. F. Müller, Heidelberg, Germany, 1997, 3. Auflage

/7-6/   Heier, S.: Windkraftanlagen; Teubner, Stuttgart, Germany, 2005, 4. Auflage

/7-7/   Landesumweltamt Brandenburg (Hrsg.): Geräuschemissionen und Geräuschimmissionen im Umfeld von Windkraftanlagen; Fachbeiträge des Landesumweltamtes, Potsdam, Germany, 1997

/7-8/   Osten, T.; Pahlke, T.: Schattenwurf von Windenergieanlagen: Wird die Geräuschabstrahlung der MW-Anlagen in den Schatten gestellt?; DEWI Magazin 7(1998), 13, S. 6 – 12

/7-9/   Krohn,   S.:   Offshore   Wind   Energy: Full   Speed   Ahead; www.windpower.org/aricles/offshore.htm

/7-10/   Hafner, E.: Kleine Windkraftanlagen haben Zukunft; Sonne, Wind und Wärme 10/2002

/7-11/   Crome, H.: Kriterienkatalog für kleine Windkraftanlagen, Erneuerbare Energien 8/2002

/7-12/   Michalak, J.: Schattenwurf des Rotors einer Windkraftanlage; Windenergie aktuell 5(1995), 2, S.17 – 23

/7-13/   Imrie, S. J.: The Environmental Implications of Renewable Energy Technology - Full Report. DG Research, The STOA Programme; Luxembourg, Selbstverlag, 1992

/7-14/   Galler, C.: Auswirkung der Windenergienutzung auf Landschaftsbilder einer Mittelgebirgsregion – Optimierung der Standortplanung aus landschaftsästhetischer Sicht; Schriftenreihe des Institutes für Landschaftspflege und Naturschutz am Fachbereich für Landschaftsarchitektur und Umweltentwicklung der Universität Hannover, Arbeitsmaterialien 43, Hannover, Germany, August 2000

/7-15/   Fachinformationszentrum Karlsruhe (Hrsg.): Windenergie und Naturschutz; BINE-Projektinfo Nr. 2, Karlsruhe, Germany, März 1996

/7-16/   Loske, K.-H.: Einfluss von Windkraftanlagen auf das Verhalten der Vögel im Binnenland; in: BWE (Hrsg.): Vogelschutz und Windenergie – Konflikte, Lösungsmöglichkeiten und Visionen; BWE, Osnabrück, Germany, 1999

/7-17/   Vögel brüten auch in der Nähe von Windparks; Neue Energie 10(2001), 4, S. 25

/7-18/   Menzel, C.: Mehr Hasen gezählt – Wildtiere lassen sich durch Windturbinen nicht stören; Neue Energie 10(2001), 4, S. 24

/7-19/ Bundesamt für Naturschutz (BfN): Empfehlungen des Bundesamtes für Naturschutz zu naturschutzverträglichen Windkraftanlagen; BfN, Bonn-Bad Godesberg, Germany, 2000

/7-20/ Umweltbundesamt (UBA): Hintergrundpapier zum Forschungsprojekt: Untersuchungen zur Vermeidung und Verminderung von Belastungen der Meeresumwelt durch Offshore-Windenergieparks im küstenfernen Bereich der Nord- und Ostsee (FKZ 200 97 106); UBA, Berlin, Germany, 2001

/7-21/ Institut für Tourismus- und Bäderforschung in Nordeuropa (Hrsg.): Touristische Effekte von On- und Offshore-Windkraftanlagen in Schleswig-Holstein; Institut für Tourismus- und Bäderforschung in Nordeuropa, Kiel, Germany, September 2000

/7-22/ Kehrbaum R.; Kleemann, M.; Erp van, F.: Windenergieanlagen – Nutzung, Akzeptanz und Entsorgung; Schriften des Forschungszentrums Jülich, Reihe Umwelt/Enviroment, Band 10, Jülich, Germany, 1998

/7-23/ Beitz, W.; Küttner, K.H. (Hrsg.): Dubbel – Taschenbuch für den Maschinenbau; Springer, Berlin, Heidelberg, Germany, 1981, 14. Auflage

/7-24/ Bundesverband Windenergie e.V. (Hrsg.): Windenergie 2005, Marktübersicht; BWE Service GmbH, Osnabrück, April 2005

/7-25/ Pohl, J.; Faul F., Mausfeld, R.: Belästigung durch periodischen Schattenwurf von Windenergieanlagen, Untersuchung im Auftrag des Landes Schleswig-Holstein; Institut für Psychologie, Christian-Albrechts-Universität Kiel, Germany, Juli 1999

/8-1/ Schröder, W. u. a.: Grundlagen des Wasserbaus; Werner, Düsseldorf, Germany, 1999, 4. Auflage

/8-2/ Giesecke, J.: Wasserbau; Skriptum zur Vorlesung, Institut für Wasserbau, Universität Stuttgart, Germany, 2000

/8-3/ Laufen, R.: Kraftwerke; Grundlagen, Wärmekraftwerke, Wasserkraftwerke; Sprin¬ger, Berlin, Heidelberg, Germany, 1984

/8-4/ Rotarius, T. (Hrsg.): Wasserkraft nutzen - Ratgeber für Technik und Praxis; Rotarius, Cölbe, Germany, 1991

/8-5/ Kaltschmitt, M.: Renewable Energy; Lessons; Institute for Environmental Technology and Energy Economics, Hamburg University of Technology, Summer Term 2006

/8-6/ Giesecke, J.; Mosony, E.: Wasserkraftanlagen – Planung, Bau und Betrieb; Springer, Berlin, Heidelberg, Germany, 2005, 4. Auflage

/8-7/ Bundesamt für Konjunkturfragen (Hrsg.): Einführung in Bau und Betrieb von Kleinstwasserkraftanlagen, Bern, Switzerland, 1993

/8-8/ Vischer, D.; Huber, A.: Wasserbau; Springer, Berlin, Heidelberg, Germany, 2002, 6. Auflage

/8-9/ Wasserwirtschaftsverband Baden-Württemberg (Hrsg.): Leitfaden für den Bau von Kleinwasserkraftwerken; Frankh Kosmos, Stuttgart, Germany, 1994, 2. Auflage

/8-10/ Voith (Hrsg.): Francis-Schachtturbinen in standardisierten Baugrößen; Werksschrift 2519, Voith, Heidenheim, Germany, 1985

/8-11/   Voith (Hrsg.): Peltonturbinen in standardisierten Baugrößen; Werksschrift 2517, Voith, Heidenheim, Germany, 1985

/8-12/   BMWi (Hrsg.): Bericht über den Stand der Markteinführung und der Kostenent-wicklung von Anlagen zur Erzeugung von Strom (Erfahrungsbericht zum EEG); Bundesministerium für Wirtschaft und Technologie (BMWi), Berlin, Germany, 2002

/8-13/   Zaugg, C.; Leutewiler, H.: Kleinwasserkraftwerke und Gewässerökologie; Bundesamt für Energiewirtschaft, Bern, Switzerland, 1998, 2. Auflage

/8-14/   Bunge, T. et al.: Wasserkraft als erneuerbare Energiequelle – rechtliche und ökologische Aspekte; UBA Texte 01/01, Berlin, Germany, 2001

/8-15/   Tönsmann, F.: Umweltverträglichkeitsuntersuchungen bei der Modernisierung von Wasserkraftwerken; In: Wasserbau und Wasserwirtschaft Nr. 75: Betrieb, Unterhalt und Modernisierung von Wasserbauten; Lehrstuhl für Wasserbau, Technische Universität München, Germany, 1992

/8-16/   Bayerisches Landesamt für Wasserwirtschaft (Hrsg.): Grundzüge der Gewässerpflege; Fließgewässer; Schriftenreihe Heft 21, Bayerisches Landesamt für Wasserwirtschaft, München, Germany, 1987

/8-17/   Strobl, T. u. a.: Ein Beitrag zur Festlegung des Restabflusses bei Ausleitungskraftwerken; Wasserwirtschaft 80(1990), 1, S. 33 – 39

/8-18/   Jorde, K.: Ökologisch begründete dynamische Mindestwasserregelungen in Ausleitungsstrecken; Mitteilungen des Instituts für Wasserbau, Universität Stuttgart, Heft 90, Stuttgart, Germany, 1997

/9-1/   Halozan, H.; Holzapfel K.: Heizen mit Wärmepumpen, TÜV Rheinland, Köln, Germany, 1987

/9-2/   Kruse, H.; Heidelck, R.: Heizen mit Wärmepumpen; Verlag TÜV Rheinland, Köln, Germany, 2002, 3. erweiterte und völlig überarbeitete Auflage

/9-3/   Cube, H. L.; Steimle, F.: Wärmepumpen – Grundlagen und Praxis; VDI-Verlag, Düsseldorf, Germany, 1984

/9-4/   Sanner, B.: Erdgekoppelte Wärmepumpen, Geschichte, Systeme, Auslegung, Installation; Fachinformationszentrum Karlsruhe, Karlsruhe, Germany, 1992

/9-5/   Gerbert H.: Vergleich verschiedener Erdkollektor-Systeme; Symposium Erdgekoppelte Wärmepumpen, Fachinformationszentrum Karlsruhe, Germany, 1991

/9-6/   Sanner, B.; Rybach, L.; Eugster, W.J.: Erdwärmesonden Burgdorf – ein Programm und viele Missverständnisse; Geothermie CH 1/97, S. 4 – 6

/9-7/   Eugster, W.J.; Rybach, L.: Langzeitverhalten von Erdwärmesonden – Messungen und Modellrechnungen am Beispiel einer Anlage in Elgg (ZH), Schweiz; IZW-Bericht 2/97, Karlsruhe, Germany, 1997, S. 65 – 69

/9-8/   Gerbert, H.: Vergleich verschiedener Erdkollektor-Systeme; IZW-Bericht 3/91, FIZ, Karlsruhe, Germany, 1991, S. 75 – 86

/9-9/   Messner, O. H .C.; De Winter, F.: Umweltschutzgerechte Wärmepumpenkollektoren hohen Wirkungsgrades; IZW-Bericht 1/94, FIZ, Karlsruhe, Germany, 1994

/9-10/ Bukau, F.: Wärmepumpentechnik, Wärmequellen – Wärmepumpen – Verbraucher – Grundlagen und Berechnungen; Oldenbourg, München, Germany, 1983

/9-11/ Hellström, G.: PC-Modelle zur Erdsondenauslegung; IZW Bericht 3/91, FIZ, Karlsruhe, Germany, 1991

/9-12/ Hackensellner, T.; Dünnwald, G.: Wärmepumpen; Teil VIII der Reihe Regenerative Energien, VDI, Düsseldorf, Germany, 1996

/9-13/ Gilli, P. V.; Streicher, W.; Halozan, H.; Breembroeck, G.: Environmental Benefits of Heat Pumping Technologies; IEA Heat Pump Centre, Analysis Report HPC – AR6, March 1999

/9-14/ ASUE (Hrsg.): Gas-Wärmepumpen; Broschüre der Arbeitsgemeinschaft für sparsamen und umweltfreundlichen Energieverbrauch e. V., Hamburg, Germany, 1996

/9-15/ Streicher, W. u. a.: Benutzerfreundliche Heizungssysteme für Niedrigenergie- und Passivhäuser, Endbericht zum Projekt in der Forschungsausschreibung „Haus der Zukunft"; Institut für Wärmetechnik, TU Graz, Austria, 2004

/9-16/ Katzenbach, R.; Knoblich, K.; Mands, E.; Rückert, A.; Sanner, B.: Energiepfähle – Verbindung von Geotechnik und Geothermie; IZW-Bericht 2/97, FIZ, Karlsruhe, Germany, 1997

/9-17/ Adnot, J.: Energy Efficiency of Room Air-Conditioners, Study for the Directorate-General for Transport and Energy, Commission of the European Union, Contract DGXVII4.1031/D/97.026, 1999

/9-18/ Neubarth, J.; Kaltschmitt, M. (Hrsg.): Regenerative Energien in Österreich – Systemtechnik, Potenziale, Wirtschaftlichkeit, Umweltaspekte; Springer, Wien, Austria, 2000

/9-19/ Kaltschmitt, M.; Huenges, E.; Wolff, H. (Hrsg.): Energie aus Erdwärme; Deutscher Verlag für Grundstoffindustrie, Stuttgart, Germany, 1999

/9-20/ Europäisches Komitee für Normung (Hrsg.): Kälteanlagen und Wärmepumpen - Sicherheitstechnische und umweltrelevante Anforderungen, EN 378-1, Beuth Verlag, Berlin, Germany, Juni 2000

/9-21/ SIA (Hrsg.): Grundlagen zur Nutzung der untiefen Erdwärme für Heizsysteme. Serie "Planung, Energie und Gebäude"; SIA-Dokumentation D0136, Zürich, Switzerland, 1996

/9-22/ Österreichischer Wasser- und Abfallwirtschaftsverband (Hrsg.): Anlagen zur Gewinnung von Erdwärme (AGE); ÖWAV-Regelblatt 207, Wien, Austria, 1993

/9-23/ Ingerle, K.; Becker, W.: Ausbreitung von Wärmepumpen-Kältemitteln im Erdreich und Grundwasser; Studie im Auftrag der Elektrizitätswerke Österreichs, Wien, Austria, 1995

/9-24/ Kraus, W. E.: Sicherheit von CO2-Kälteanlagen, Kohlendioxid – Besonderheiten und Einsatzchancen als Kältemittel; Statusbericht des Deutschen Kälte- und Klimatechnischen Vereins, Nr. 20, Stuttgart, Germany, November 1998

/9-25/ Fachinformationszentrum Karlsruhe (Hrsg.): CO_2 als Kältemittel für Wärmepumpe und Kältemaschine; BINE-Projektinfo 10/00, Karlsruhe, Germany, 2000

/10-1/   Bußmann, W. u. a. (Hrsg.): Geothermie – Wärme aus der Erde; C. F. Müller, Karlsruhe, Germany, 1991

/10-2/   Kaltschmitt, M.; Huenges, E.; Wolff, H. (Hrsg.): Energie aus Erdwärme; Deutscher Verlag für Grundstoffindustrie, Stuttgart, Germany, 1999

/10-3/   Rummel, F.; Kappelmeyer, O.: Erdwärme – Energieträger der Zukunft? Fakten, Forschung, Zukunft; C.F. Müller, Karlsruhe, Germany, 1993

/10-4/   Schulz, R. u. a. (Hrsg.): Geothermische Energie – Forschung und Anwendung in Deutschland; C. F. Müller, Karlsruhe, Germany, 1992

/10-5/   GFZ (Hrsg.): Evaluierung geowissenschaftlicher und wirtschaftlicher Bedingungen für die Nutzung hydrogeothermaler Ressourcen; Geothermie Report 99-2, Abschlussprojekt zum BMBF-Projekt BEO 0326969, Potsdam, Germany, 1999

/10-6/   Edwards, L.M.; Chilingar, G.V.; Rieke, H.H.; Fertl, W.H. (eds.): Handbook of Geothermal Energy; Gulf Publishing Company, Houston, London, Paris, Tokyo, USA, 1982

/10-7/   Seibt, P. u. a.: Untersuchungen zur Verbesserung des Injektivitätsindex in klastischen Sedimenten; Studie im Auftrag des BMBF; GTN Geothermie Neubrandenburg GmbH, Neubrandenburg, Germany, 1997

/10-8/   Poppei, J.: Entwicklung wissenschaftlicher Methoden zur Speicherbewertung und Abbauüberwachung; Bericht zum BMFT Vorhaben 0326912A, Bonn, Germany, 1994 (unveröffentlicht)

/10-9/   Brandt, W.; Kabus, F.: Planung, Errichtung und Betrieb von Anlagen zur Nutzung geothermischer Energie – Beispiele aus Norddeutschland; VDI Bericht 1236, VDI-Verlag, Düsseldorf, Germany, 1996

/10-10/   Poppei, J.: Tiefe Erdwärmesonden; Geothermische Energie - Nutzung, Erfahrung, Perspektive; Geothermische Fachtagung, Schwerin, Germany, Oktober 1994

/10-11/   Dickson, M.H.; Fanelli, M.: What is Geothermal Energy?; Instituto di Geoscienze e Georisorse, CNR, Pisa, Italy, 2004, http://www.earlham.edu/~parkero/Seminar/Geothermal_20Energy.pdf

/10-12/   Witt, J.; Kaltschmitt, M.: Weltweite Nutzung regenerativer Energien; BWK 56(2004), 12, S. 43 – 50

/10-13/   Jung, R. et al: Abschätzung des technischen Potenzials der geothermischen Stromerzeugung und der geothermischen Kraft-Wärmekopplung (KWK) in Deutschland; Bericht für das Büro für Technikfolgenabschätzung beim Deutschen Bundestag; BGR/GGA, Archiv-Nr. 122 458, Hannover, Germany, 2002

/10-14/   Ura, K.; Saitou, S.: Geothermal Binary Power Generation System; World Geothermal Congress, Kyushu-Tohoku, Japan, 2000

/10-15/   Bresee, J. C.: Geothermal Energy in Europe - The Soultz Hot Dry Rock Project; Gordon and Breach Science Publishers, Philadelphia, USA, 1992

/10-16/   Rummel, F.; Kappelmeyer, O. (Hrsg.): Erdwärme; C. F. Müller, Karlsruhe, Germany, 1993

/10-17/   Haenel, R. u. a.: Geothermisches Energiepotential; Pilotstudie zur Abschätzung der geothermischen Energievorräte an ausgewählten Beispielen in der Bundesrepublik Deutschland; Niedersächsisches Landesamt für Bodenforschung; Hannover, Germany, 1988

/10-18/   Schulz, R. u. a.: Geothermie Nordwestdeutschland – Endbericht; Niedersächsisches Landesamt für Bodenforschung, Hannover, Germany, 1995

/10-19/   Haenel, R; Staroste, E. (Hrsg.): Atlas of Geothermal Resources in the European Community, Austria and Switzerland; Th. Schäfer, Hannover, Germany, 1988

/10-20/   Kalfayan, S.: Production Enhancement with Acid Stimulation; PennWell Corporation, Tulsa Oklahoma, USA, 2000

/10-21/   Howard, G.C.; Fast, C.R.: Hydraulic-Fracturing; SPE Monograph, Houston, Texas, USA, 1970

/10-22/   Murphy, H.; Brown, D.; Jung, R.; Matsunaga, I.; Parker, R.: Hydraulics and Well Testing of Engineered Geothermal Reservoirs; Geothermics, Special Issue Hot Dry Rock/Hot Wet Rock Academic Review, Vol. 28, no. 4/5 Aug./Oct. 1999, S. 491 – 506

/10-23/   Iglesias, E.; Blackwell, D.; Hunt, T.; Lund, J.; Tamanyu, S. (eds.): Proceedings of the World Geothermal Congress 2000, Kyushu – Tohoku, Japan, 2000

/10-24/   Rafferty, K.: Geothermal Power Generation – A Primer on Low-Temperature, Small-Scale Applications; Geo-Heat Center, Oregon Institute of Technology, Klamath Falls, OR, USA, 2000

/A-1/   Boyle, G.: Renewable Energy; Oxford University Press, Oxford, UK, 1996

/A-2/   Graw, K.-U.: Nutzung der Tideenergie – Eine kurze Einführung; Geotechnik – Wasserbau – Wasserwirtschaft: Materialien No. 2, Professur Grundbau und Wasserbau, Universität Leipzig, Germany, 2001

/A-3/   Graw, K.-U.: Wellenenergie – Eine hydromechanische Analyse; Bericht Nr. 8 des Lehr- und Forschungsgebietes Wasserbau und Wasserwirtschaft, Bergische Universität - GH Wuppertal, Germany, 1995

/A-4/   www.uni-leipzig,de/welle/index.html (September 2002)

/A-5/   Laughton, M. A.: Renewable Energy Sources; Elsevier Applied Science, London, UK, 1990

/A-6/   Kleemann, M.; Meliß, M.: Regenerative Energiequellen; Springer, Berlin, Heidelberg, Germany, 1993, 2. Auflage

/A-7/   Hoppe-Kilpper, M.: Persönliche Mitteilung; Institut für Solare Energieversorgungssysteme (ISET), Kassel, Germany, 2002

/A-8/   Hoffmann, W.: Energie aus Sonne, Wind und Meer; Harri Deutsch, Thun und Frankfurt/Main, Germany, 1990

/A-9/   Ziegler, T.: Wolkenkratzer für die Nordsee; Die Welt, 27. 01. 1996

/B-1/   Kaltschmitt, M.; Hartmann, H. (Hrsg.): Energie aus Biomasse; Springer, Berlin, Heidelberg, 2001

Index

Breinigsville, PA USA
10 February 2010
232290BV00005B/37/P